Ecologically Sound Integrated Regional Energy Planning

ECOLOGICALLY SOUND INTEGRATED REGIONAL ENERGY PLANNING

T.V. RAMACHANDRA

Nova Science Publishers, Inc.
New York

Senior Editors: Susan Boriotti and Donna Dennis
Editorial Coordinator: Tatiana Shohov
Office Manager: Annette Hellinger
Graphics: Wanda Serrano
Editorial Production: Vladimir Klestov, Matthew Kozlowski, Tom Moceri, Anthony T. Slovak and Maya Columbus
Circulation: Ave Maria Gonzalez, Vera Popovic, Luis Aviles, Raymond Davis, Melissa Diaz, Marlene Nunez, Magdalene Nunez, and Jeannie Pappas
Communications and Acquisitions: Serge P. Shohov
Marketing: Cathy DeGregory

Library of Congress Cataloging-in-Publication Data
Ecologically sound integrated regional energy planning / T.V. Ramachandra.
p.cm.
Includes bibliographical references and index.
ISBN 1-56072-876-0.
1. Power Resources – India – North Kanara. 2. Energy policy – India – North Kanara. I. Title.

TJ163.25.I4 R35 2000
333.79'0954'87-dc21 00-049008

400 Oser Ave, Suite 1600
Hauppauge, New York 11788-3619
Tele. 631-231-7269 Fax 631-231-8175
e-mail: Novascience@earthlink.net
Web Site: http://www.novapublishers.com

Printed in the United States of America

To my parents and my teachers who empowered me at an early age with their trust, faith and hopes;

To Bharath and Suma

and

To my affectionate wife, Nagarathna, whose positive, steady and generous support is a critical and an essential ingredient in all my endeavours.

CONTENTS

List of Tables and Figures		**ix**
Preface		**xv**
Acknowledgements		**xix**
Chapter 1	Introduction	**1**
Chapter 2	Potential Assessment of Wind Energy in Uttara Kannada District	**25**
Chapter 3	Bioenergy Potential and Prospects	**55**
Chapter 4	Hydro Energy Potential and Prospects	**97**
Chapter 5	An Ecologically Friendly Optimal Design of Hydroelectric Projects	**127**
Chapter 6	Solar Energy Potential and Prospects	**187**
Chapter 7	Energy Consumption Patterns in the Domestic Sector	**237**
Chapter 8	Energetics in Paddy Cultivation	**309**
Chapter 9	Energy Utilisation in Industrial Sector	**331**
Chapter 10	Integrated Regional Energy Planning	**367**
Summary		**413**
Index		**419**

LIST OF TABLES AND FIGURES

TABLES

1.1 Percentage Population Increase (Talukwise) in Uttara Kannada District (1951 to 1991) 16
1.2 Geographical Area, Area Under Forest, Population Density, Livestock Density - Talukwise - Uttara Kannada District 16
2.1 List of Stations in Uttara Kannada District 29
2.2.1 Mean Wind Speed (kmph) at Sirsi (at 610 m Elevation) 29
2.2.2 Mean Wind Speed (kmph) at Sirsi (at 622 m Elevation) 30
2.3 Mean Wind Speed (kmph) at Sirsi II (Elevation 622 m, location Huliappa Gudda) and Kumta (Elevation 8 m) 30
2.4.1 Monthly Mean Wind Speed, Sd, Max, Min (computed for data of Meteorological Observatories located at Karwar, Honnavar and Shirali based on data for 1990-1994) 36-37
2.4.2 Mean Wind Speed at Honnavar, Elev. 26 MSL, based on data for 1939-1989 37
2.4.3 Mean Wind Speed at Karwar based on data for 1952-1989 38
2.5.1 Frequency Distribution (period of occurrence) and Cumulative Frequency Distribution at Sirsi 39
2.5.2 Frequency Distribution and Cumulative Frequency Distribution at Kumta 40
2.6 Weibull Parameters (c and k) for Sirsi and Kumta 41
2.7 Energy Pattern Factor (EPF) for Sirsi and Kumta 41
2.8 Mean Power Density for Kumta and Sirsi 45
2.9 Monthly Mean Power Density at Karwar, Honnavar and Shirali 46
2.10.1 Monthly Mean Power at Karwar, Kumta, Honnavar and Shirali 47-48
2.10.2 Monthly Mean Energy at Karwar, Kumta, Honnavar and Shirali 49-50
2.10.3 Rated Power and Energy that could be generated at Karwar, Kumta, Honnavar and Shirali 52
2.10.4 Cost per kWh and Production Cost per MW 53
3.1.1 Population (1991), Total Area and Area under Forests in Uttara Kannada 57
3.1.2 Details of Forest Area According to Legal Status as in 1980-81 in the District 61
3.1.3 Areas Lost in the District for Various Purposes 62
3.2 Litter Productivity in Tropical Moist Forests of Uttara Kannada District 65
3.3.1 Standing Biomass in Various Types of Vegetation 66
3.3.2 Estimates of Net Annual above Ground Productivity (t/ha) 67
3.4 Biomass Productivities in Various Types of Vegetation 67
3.5.1 Population (1991) and Area Under Different Types of Vegetation in Uttara Kannada (Talukwise) 68
3.5.2 Population (1991) and Area Under Different Types of Vegetation in Uttara Kannada (Talukwise) 68
3.6.1 Area (in hectares) under Different Crops, and Average Production and Yield in Uttara Kannada District 69-70
3.6.2 Average (Avg), Standard Deviation (Sd), Maximum and Minimum Production (in tonnes) during 1984-85 to 1993-94 70-73
3.6.3 Average (Avg) and Standard Deviation (Sd) - Production: Energy from Agricultural Residues for Fuel and Fodder in lakh kWh 74-76
3.6.4 Area (in hectares), Production (in tonnes) of different Horticultural Crops and Net Energy Available (in lakh kWh). 78-79
3.6.5 Fuel Biomass from Coconut Palm per Year 79
3.7.1 Talukwise Livestock Population and Estimated Dung and hence Gas Yield per Year 81-82
3.7.2 Talukwise Litter Production in tonnes 83
3.7.3 Bio-Residues Available from Various Crops 83-84
3.7.4 Fodder Requirement in Uttara Kannada 84
3.8 Biogas Value Compared with Other Energy Sources 86
3.9 Area Required for a 1000 MW Thermal Station 88
4.1 Development Projects in Uttara Kannada District 99
4.2.1 Annual Rainfall (mm) – Talukwise 105
4.2.2 Talukwise Computation of COV (Coefficient of Variation) 105
4.3.1 Catchment Area, Stream Slope and Catchment Slope computed for various Streams of Bedthi River 109-110
4.3.2 Catchment Area, Stream Slope and Catchment Slope computed for various Streams of Aghnashini River 111-112
4.4.1 Estimated Power in kW (monthwise) in Boosangeri Stream and Muregar Jog (based on Field Measurements) 113
4.4.2 Estimated Power in kW (monthwise) in Boosangeri Stream and Muregar Jog by the Rational Method - with the Knowledge of Catchment Area and Precipitation (100 Years) 115

4.5.1 Estimated Water Discharge and Power (MW) in Ungauged Streams of Bedthi Catchment Area 115-116
4.5.2 Estimated Water Discharge and Energy ('000 Units) in Ungauged Streams of Bedthi Catchment Area 117-118
4.6.1 Estimated Water Discharge and Power (MW) in Ungauged Streams of Aghnashini Catchment Area 119
4.6.2 Estimated Water Discharge and Energy ('000 Units) in Ungauged Streams of Aghnashini Catchment Area 120
5.1 Water Discharge (million cubic meters) at Magod (Bedthi river) 141
5.2 Water Discharge (million cubic meters) at Shivganga–Sonda (Bedthi River) 141
5.3 Water Discharge (million cubic meters) at Unchalli (Aghnashini river) 142
5.4 Water Discharge (million cubic meters) at Benne (Aghanashini river) 142
5.5.1 Submergence Area (sq. kms) for Various Heights of the Dam (in metres) at Magod (Bedthi River) 144
5.5.2 Probable Relationship Between Submergence Area vs. Height, Volume vs. Height-Magod 146
5.6.1 Submergence Area for Various Heights of the Dam at Shivganga (Bedthi River) 145
5.6.2 Probable Relationship Between Submergence Area vs. Height, Volume vs. Height-Shivganga 146
5.7.1 Submergence Area for Various Heights of the Dam at Unchalli (Aghnashini River) 145
5.7.2 Probable Relationship Between Submergence Area vs. Height, Volume vs. Height-Unchalli 146
5.8.1 Submergence Area for Various Heights of the Dam at Benne (Aghnashini River) 145
5.8.2 Probable Relationship Between Submergence Area vs. Height, Volume vs. Height-Benne 146
5.9 Power (MW) and Energy (mkWh) Available at Major Streams of Bedthi River Basin 149
5.10 Power (MW) and Energy (mkWh) Available at Major Streams of Aghnashini River Basin 150
5.11.1 Landuse Pattern in Submergence Area for Various Heights of Dam 152
5.11.2 Thermal Power (Mega Watts) Computation Based on Biomass Productivity in Each Category of Land Use 152
5.11.3 Energy from Biomass (in million Units) from Various Categories of Land Use 154
5.12.1 Hydro, Thermal Energy and Hence Net Energy Available from Submerged Area 154
5.12.2 Net Energy Computation for Various Heights of Dam 156
5.12.3 Net Energy Computation Considering Conversion Efficiency of Biomass as 35% 156
5.13.1 Net Energy Gain for Various Design and for Biomass Productivity of 6.5 t/ha/yr 159
5.13.2 Case 2 – Considering the Area of 95.03 Sq.Kms. divided into Forests/Fuel Wood and Water Storage 159
5.13.3 Net Energy: Thermal Efficiency into Consideration (If Wood is Converted to Electricity) 159
5.13.4 Case 3 – Enet = Eh'+Efw+0.35(Ebio-Efw), if Ebio > Efw and Eh'+Ebio if Efw > Ebio 160
5.14.1 Case 1 – Net Energy Gain for Various Designs and Biomass Productivity of 13.5 T/Ha/Yr from Forests 160
5.14.2 Case 2 – Considering Area of 95.03 sq.km divided into Forests/Fuel Wood (Fw) and Water Storage 160
5.14.3 Net Energy considering Thermal Efficiency (if all Wood is Converted to Electricity) 161
5.14.4 Case 3 – Enet = Eh'+Efw+0.35(Ebio-Efw), If Ebio > Efw and Eh'+Ebio if Efw > Ebio 161
5.15.1 Case 1 – Net Energy Gain for Various Designs and Biomass Productivity of 20.0 T/Ha/Yr from Forests 161
5.15.2 Case 2 – Considering Area of 95.03 sq.kms. divided into Forests/Fuel Wood (fw) and Water Storage 162
5.15.3 Case 3: Net Energy considering Thermal Efficiency (if all Wood is used for Electricity Generation) 162
5.15.4 Case 4 - Enet = Eh'+Efw+0.35(Ebio-Efw), if Ebio > Efw and Eh'+Ebio if Efw > Ebio 162
5.16.1 Case 1 – Net Energy Gain for Various Designs and Biomass Productivity of 27.5 T/Ha/Yr from Forests 163
5.16.2 Case 2 – Considering Area of 95.03 sq. kms. divided into Forests/Fuel Wood (Fw) and Water Storage 163
5.16.3 Net Energy Considering Thermal Efficiency (if all Wood is Converted to Electricity) 163
5.16.4 Case 3 – Enet = Eh'+Efw+0.35(Ebio-Efw), if Ebio > Efw and Eh'+Ebio if Efw > Ebio 164
5.17 Environment Benefits from One Hectare of Forest (During 50 Year Life Span) 167
5.18 Civil, Electrical (El), Environmental (Env) and Rehabilitation costs (million rupees) for Various Heights of Dam 168
5.19.1 Case 1 - Cost per kWh of Electricity (net) considering Biomass Productivity of 6.5 t/ha/yr 169
5.19.2 Case 2 – Cost per kWh of Electricity (net) considering Biomass Productivity of 13.5 t/ha/yr 170
5.20.1 Sectorwise Electricity (EI) Consumption ('000 units) and Average Tariff in Uttara Kannada District 172
5.20.2 Net Electrical Energy Generated for Different Submergence Areas (in million kWh) at Magod Site of Bedthi River (Ebio based on 6.5 t/ha/yr) 172
5.20.3 Anticipated Sectorwise Consumption of Electricity (million kWh) based on Present Consumption Patterns 174
5.20.4 Sectorwise Revenue (computed as per KEB Tariff) in million Rupees 174
5.21.1 Benefit to Cost ratio for Various Scenarios considering value of Forest Land as Rs.37,065/ha. 174
5.21.2 Benefit to Cost ratio for Various Scenarios considering value of Forest Land as Rs.1,11,200/ha. 174
5.21.3 Benefit to Cost ratio for Various Scenarios considering value of Forest Land as Rs.1,11,200/ha. + biomass, MFP and Timber as Rs. 12,000/year/hect 174
5.21.4 Benefit to Cost ratio considering Standing Biomass + Value of Land as Rs.2.25 million/ha 175
5.21.5 Benefit to Cost ratio considering Standing Biomass + Value of Land as Rs.4.55 million/ha 175
5.21.6 Benefit to Cost ratio considering Standing Biomass + Value of Land as Rs.2.25 million/ha. and Environmental Value as Rs.14.3 million/hect. 175
6.1.1 Monthwise Mean Hourly Global Solar Radiation (kWh/sq.m) at Goa 190
6.1.2 Monthwise Max. Value (highest value recorded during the hour) of Global Solar Radiation at Goa 190
6.1.3 Monthwise Standard Deviation of the Hourly Values of Global Radiation at Goa 191

6.1.4 Monthwise Mean Solar Elevation (Degree) at Goa 191
6.1.5 Monthwise Mean Values of Direct Solar Radiation (kWh/sq.m) at Goa 192
6.1.6 Monthwise Relative Airmass (The Total Mass of the Atmosphere Traversed by a Beam of Solar Radiation) at Goa 192
6.1.7 Monthwise Highest Value (recorded during the hour) of Diffuse Solar Radiation at Goa 193
6.1.8 Monthwise Mean Value of Diffuse Solar Radiation at Goa in kWh/sq.m 193
6.1.9 Monthwise Standard Deviation of Hourly Diffuse Radiation Values at Goa 194
6.2.1 Monthwise Maximum Values of Global Radiation at Mangalore 194
6.2.2 Monthwise Mean Hourly Values of Global Radiation at Mangalore 195
6.2.3 Monthwise Standard Deviation Values of Global Radiation at Mangalore 195
6.2.4 Monthwise Elevation of Direct Solar Radiation (Degree) at Mangalore 196
6.3.1 Global Solar Radiation in kWh/sq.m at Goa and Mangalore 199
6.3.2 Sunshine Duration at Goa and Mangalore 199
6.3.3 Mean Duration of Sunshine per Hour at Goa 200
6.3.4 Mean Duration of Sunshine per Hour at Mangalore 201
6.3.5 Diffuse Solar Radiation (kWh/sq.m) at Goa and Mangalore 202
6.3.6 Monthwise D/G Values at Goa and Mangalore 202
6.4.1 List of Regression Constants 'a' and 'b' Computed for (G'/ETR) and (n/N') 205
6.4.2 Comparison of Computed and Observed Values of Global Solar Radiation 205
6.4.3 List of Regression Constants 'c' and 'd' Computed for (D'/G') and (G'/ETR) 207
6.4.4 Comparison of Computed and Observed Values of Diffuse Solar Radiation 207
6.4.5 Comparison of Computed and Observed Values of Clear Sky Noon Global Radiation 208
6.4.6 Comparison of Computed and Observed Values of Clear Sky Noon Diffuse Radiation 208
6.5.1 Mean Values of Solar Radiation and Associated Parameters (Computed) for Goa 209
6.5.2 Mean Values of Solar Radiation and Associated Parameters (Computed) for Mangalore 209
6.5.3 Global Radiation Tilt Factors at Goa 210
6.5.4 Global Radiation Tilt Factors at Mangalore 210
6.6.1 Skewness and Kurtosis (monthwise) for Goa and Mangalore 213
6.6.2 Median Values of Hourly and Daily Global Solar Radiation in kWh/sq.m. at Goa 214
6.6.3 Median Values of Hourly and Daily Global Solar Radiation in kWh/sq.m. at Mangalore 214
6.6.4 Percentage Frequency and Cumulative Frequency Distribution of Daily Global Solar Radiation (at Goa) 215
6.6.5 Percentage Frequency and Cumulative Frequency Distribution of Daily Global Solar Radiation (at Mangalore) 216
6.7.1 Regression Constants, Correlation Coefficient Computed for Various Relationships (Goa) 221
6.7.2 Regression Constants, Correlation Coefficient Computed for Various Relationships (Mangalore) 222
6.8.1 Actual and Estimated Values of GR and % Error Computed for Goa and Mangalore (kWh/sq.m/day) 223
6.8.2 Estimated Values of GR at Karwar, Honnavar and Shirali using equations 11 and 13 223
6.9.1 Barren/Wasteland Available in Coastal Taluks of Uttara Kannada District 226
6.9.2 Solar Radiation Potential (kWh/month) by using 5% of Current Barren Land for Harnessing Solar Energy in Coastal Taluks 228
6.9.3 Solar Radiation Potential (kWh/month) by using 1% of Current Barren Land for Harnessing Solar Energy in Coastal Taluks 229
6.10.1 Solar Devices and Possible Applications 230
6.10.2 Cost per kWh at Various Efficiencies (Solar thermal systems) 231
6.11.1 Per Capita per Day Fuel Consumption 231
6.11.2 Seasonwise Fuel wood Consumption (in tonnes) for Water Heating 232
6.11.3 Fuel wood Consumption (in tonnes) per Season for Cooking 232
7.1 Mandalwise Population and its Percentage Increase during Four Decades 250
7.2 Mandalwise Number of Persons/hectare, Livestock/hectare and Percentage Forest Cover 251
7.3.1 Mandalwise and Communitywise Landholding Particulars (in hect) for Kumta Taluk 252
7.3.2 Mandalwise and Communitywise Landholding Particulars (in Percentage) for Kumta Taluk 253
7.4.1 Mandalwise and Categorywise Land Distribution (Frequency Distribution), Kumta Taluk 254
7.4.2 Mandalwise and Categorywise Land Distribution (Percentage), Kumta Taluk 255
7.5.1 Seasonwise and Regionwise Fuel wood Requirement for Cooking in Kumta Taluk 257
7.5.2 Seasonwise and Regionwise Water Heating Fuel Wood Requirement in Kumta Taluk 258
7.5.3 Kerosene, Biogas and LPG Requirement for Cooking in Domestic Sector 258
7.5.4 Kerosene and Electricity Requirement for Lighting in Domestic Sector 259
7.6.1 Villagewise and Seasonwise Cooking Fuel wood Consumption 260-61
7.6.2 Villagewise and Seasonwise Water Heating Fuel wood Consumption 262-63
7.6.3 Villagewise Kerosene Biogas and LPG Consumption for Cooking, and Kerosene and Electricity for Lighting 264-65
7.7.1 Communitywise and Seasonwise Cooking Fuel wood Consumption 266
7.7.2 Communitywise and Seasonwise Water Heating Fuel wood Consumption 267
7.7.3 Communitywise Kerosene, Biogas and LPG Consumption for Cooking, and Kerosene and Electricity for Lighting 268
7.8.1 Fuel wood Consumption for Cooking (kgs/person/day) in Various Landholding Categories 269
7.8.2 Fuel wood Consumption for Water Heating (kgs/person/day) in Various Landholding Categories 270
7.8.3 Kerosene, Biogas and LPG usage for Cooking, and Kerosene and Electricity for Lighting in Various Landholding Categories 270
7.9.1 Cooking Fuel wood Consumption in Various Household Income Categories 271
7.9.2 Fuel wood Consumption for Water Heating in Various Household Income Categories 271
7.9.3 Kerosene, Biogas and LPG Consumption for Cooking in Various Household Income Categories 271
7.9.4 Kerosene and Electricity for Lighting in Various Household Income Categories 272

7.10.1 Cooking Fuel wood Consumption in Various Categories of Number of Adult Equivalent/Household 272

7.10.2 Water Heating Fuel wood Consumption in Various Categories of Number of Adult Equivalent/Household 272

7.11.1 Cooking Fuel wood Consumption based on Type of Devices 273

7.11.2 Fuel wood Consumption for Water Heating based on Type of Devices 274

7.12.1 Energy Consumption for Cooking and Water Heating based on Household Income in Kumta Town 275

7.12.2 Energy Consumption for Lighting based on Household Income in Kumta Town 276

7.12.3 Fuel Consumption for Cooking and Water Heating based on Number of Persons/ Household 276

7.12.4 Communitywise Fuel Consumption for Cooking and Water Heating 276

7.12.5 Communitywise Monthly per Capita Electricity Consumption 277

7.13.1 Energy Consumption for Cooking and Water Heating in Mundgod 277

7.13.2 Energy Consumption for Space Heating and Lighting in Mundgod (based on Income) 277

7.14.1 Energy Consumption for Cooking and Water Heating in Mundgod (based on Number of Persons/Household) 278

7.14.2 Energy Consumption for Space Heating and Lighting Purposes in Mundgod Taluk 278

7.15.0 Regionwise Fuel Requirement for Cooking and Water Heating (Bbased on Average Values of Per Capita Fuel Consumption) 279

7.16.1 Talukwise Fuel wood Consumption for Cooking and Water Heating 281

7.16.2 Talukwise Commercial Energy Consumption for Cooking and Lighting 282

7.16.3 Total Requirement of Fuel wood in Uttara Kannada District 282

7.17 Thermal Efficiency and Quantity of Wood Required (in kgs) for Heating 50 Litres of Water in Water Heating Devices 289

7.18 Fuel wood Consumption using TC's and IC's (Communitywise) 292

7.19 Fuel Consumption (in kgs/person/day) Based on Number of Persons per Household 293

7.20 Fuel Consumption (kgs/person/day) Based on Type of Fuel wood 295

8.1 Energy Requirements (GJ/Hectare) for Various Operations in Paddy Cultivation - Coastal Zone 315

8.2 Energy Requirements (GJ/Hectare) for Various Operations in Paddy Cultivation - Hilly Zone 316

8.3 Energy Requirements (GJ/Hectare) for Various Operations in Paddy Cultivation - Interior Zone 317

8.4 Regionwise Variations in Different Energy Inputs and Outputs (GJ) for Paddy Cultivation 318

8.5 Energy Consumption Pattern in Paddy Cultivation Among Various Categories of Farmers (Landholding) 319

8.6 Energy Consumption Pattern in Various Categories of Landholding within a Region 323-24

8.7 Regression Analyses of Yield and Various Energy Input Parameters 324

8.8 Regionwise Analyses of Factors Affecting the Yield in Paddy Cultivation 325

8.9 Regression Analyses (Landholdingwise) of Factors Affecting the Yield 326-27

9.1 Energy in Karnataka during 1990-91 (in millions of tones of oil equivalent – mtoe) 333

9.2 Growth of Small-Scale Industries in Karnataka 335

9.3 Number of Units, Investment and Manpower Details in Small-Scale Industries (Districtwise - Cumulative) 336-38

9.4 Percentage Increase in Industrial Units in Karnataka (Districtwise) 339-41

9.5.1 Districtwise Data for the years 1982-83 and 1991-92 342

9.5.2 Districtwise Data for the year 1991-92 343

9.6.1 Dispersal of Units 344

9.6.2 Dispersal of Units by Nature of Activities 345

9.7 Sectorwise Details of Number of Units, Manpower and Production 346

9.8 Regression Analysis: Energy & Production 357

9.9 Regression Analysis: SEC & Production 357

9.10 Distribution of Industries in Different PPCU Ranges 358

9.11 Energy Auditing in the Cashew Processing Firms 359

9.12 Computation of SEC / EI and PPCU 361

10.1.1 Sectorwise Energy Consumption (in million kWh) in Uttara Kannada 388

10.1.2 Sourcewise Composition of Energy (in million kWh) in Uttara Kannada 390

10.1.3 Sourcewise and Typewise Composition of Energy in Uttara Kannada (mkWh) 390

10.1.4 Sectorwise and Sourcewise Consumption of Energy in Uttara Kannada (mkWh) 390

10.1.5 Energy Demand (during 1994) in Uttara Kannada District (mkWh) 390

10.2.1 Solar Energy Potential in Uttara Kannada District 391

10.2.2 Wind Energy Potential in Uttara Kannada District and Anticipated Electricity that could be Generated by using 5% of Wasteland 391

10.3.1 Monthwise Hydropower Potential in Bedthi and Aghnashini Basins 392

10.3.2 Talukwise Bioresource Demand and Availability taking biomass productivity as 3.6 t/ha/yr 392

10.3.3 Talukwise Bioresource Demand and Availability taking biomass productivity as 13.5 t/ha/yr 393

10.3.4 Talukwise Estimated Dung and hence Gas Yield per Year 393

10.4.1 Assumed Efficiency Values for Various Resource - Task Combinations 398

10.4.2 K_{ijk} Values for Various Resource - Task Combinations 398

10.4.3 Assumed Plant Load Factor and Diversity Factor 398

10.4.4 Cost per Unit of Energy Used in the Design 398

10.5.1 Optimal Allocation of Resources (mkWh) for Various Tasks – Minimisation of Costs - Year 1994 399

10.5.2 Optimal Allocation of Resources for Various Tasks - Maximisation of Efficiency 399

10.5.3 Optimal Allocation of Resources (mkWh) for Various Tasks – Maximisation of Efficiency and Minimisation of Costs - Year 1994 399

10.5.4 Saving by Optimal Allocation of Energy Sources for Various Tasks 400

10.6.1 Scenario I - Optimal Allocation of Resources for Various Tasks – Maximisation of Efficiency – Year 2000 401

10.6.2 Scenario I - Optimal Allocation of Resources for Various Tasks – Maximisation of Efficiency – Year 2005 401

10.6.3 Scenario I - Optimal Allocation of Resources for Various Tasks – Maximisation of Efficiency – Year 2010 401

10.6.4 Uttara Kannada Scenario I - Base Case Demand Results for all Fuels by Sector (millon kWh) 402
10.7.1 Scenario II - Optimal Allocation of Resources for 2000 (million kWh) 402
10.7.2 Scenario II - Optimal Allocation of Resources for 2005 (million kWh) 402
10.7.3 Scenario II - Optimal Allocation of Resources for 2010 (million kWh) 402
10.7.4 Scenario II - Demand Results for all Fuels by Sector (million kWh) 403
10.8.1 Scenario III - Optimal Allocation of Resources for 2000 (million kWh) 403
10.8.2 Scenario III - Optimal Allocation of Resources for 2005 (million kWh) 403
10.8.3 Scenario III - Optimal Allocation of Resources for 2010 (million kWh) 403
10.8.4 Scenario III - Transformation Case: Demand Results for all Fuels by Sector (million kWh) 404
10.9.1 Scenario IV - Optimal Allocation of Resources for 2000 (million kWh) 404
10.9.2 Scenario IV - Optimal Allocation of Resources for 2005 (million kWh) 404
10.9.3 Scenario IV - Optimal Allocation of Resources for 2010 (million kWh) 405
10.9.4 Scenario IV - State Energy Norms: Demand Results for all Fuels by Sector (million kWh) 405
10.10.1 Scenario V - Optimal Allocation of Resources for 2000 (million kWh) 405
10.10.2 Scenario V - Optimal Allocation of Resources for 2005 (million kWh) 406
10.10.3 Scenario V - Optimal Allocation of Resources for 2010 (million kWh) 406
10.10.4 Scenario V - State Energy Norms+ Efficiency: Base Case Demand Results for all Fuels by Sector (million kWh) 406

FIGURES

1.1.1 Geographical Location of Uttara Kannada District 15
1.1.2 Population and Livestock Density - Uttara Kannada District 17
1.1.3 Talukwise Population Share – Uttara Kannada District 17
1.1.4 Population Increase (%) – (1951-1991) Talukwise 18
1.2.0 Forest and Total area in a Taluk - Uttara Kannada District 18
1.3.0 Talukwise Percentage of Forest - Uttara Kannada District 19
1.4.0 Taluk's Forest Cover Share – Uttara Kannada District 19
2.1 Mean Daily Wind Speed at Sirsi - Elevation 610m 31
2.2.1 Mean Hourly Wind Speed (year) at Sirsi - Elevation 610m 32
2.2.2 Mean Daily Wind Speed - Karwar 32
2.2.3 Mean Daily Wind Speed - Honnavar-Elev 26m 33
2.3 Hourly Wind Speed at Sirsi – Elevation 610m 33
2.4 Monthly Wind Speed at Karwar 34
2.5 Monthly Wind Speed at Honnavar 34
2.6 Monthly Wind Speed at Shirali, Uttara Kannada District 35
2.7 Monthly Mean Wind Speed - Karwar, Kumta, Honnavar and Shirali 35
2.8.1 Frequency Distribution at Sirsi (period of occurrence) 42
2.8.2 Frequency Distribution at Kumta (period of occurrence) 42
2.9.1 Cumulative Frequency Distribution (CDF) for Sirsi (January-June) 43
2.9.2 CDF for Sirsi (July-December) 43
2.9.3 CDF for Kumta (January-June) 44
2.9.4 CDF for Kumta (July-December) 44
2.10 Mean Wind Power Density - Karwar, Kumta, Shirali and Honnavar 46
3.1 Population (Talukwise) – Geographical and Forest Area (hectares) 58
3.2 Availability to Demand (Demand at 0.7 t/person/year) Ratio - Woody Biomass (Talukwise) 58
3.3 Availability to Demand (Demand at 1.1 t/person/year) Ratio - Woody Biomass (Talukwise). 59
3.4 % Population Requirement (met with Biogas Option) 59
4.1 Mean Annual Rainfall (Talukwise)-Uttara Kannada District 106
4.2 Power in Muregar and Boosangeri (Field Data and Estimates) 112
4.3 Power in Muregar and Boosangeri (Based on Rational Method) 114
5.1.1 Height versus Submergence Area, Dam at Magod (Bedthi River) 147
5.1.2 Height versus Storage Volume, Dam at Magod (Bedthi River) 147
5.2 Area vs. Energy Generated 148
5.3.1 Total Submergence Area (for various dam heights) 153
5.3.2 Land-use of Submerged Area (for various dam heights) 153
5.4.1 Hydro Energy Generated and Bioenergy Lost – mkWh (for various dam heights) 155
5.4.2 Net Energy available – mkWh for various dam heights (Eh – Ebio) 155
5.5.1 Fuel Wood Requirement and Ebio Available – mkWh 165
5.5.2 Net Energy from 95.03 sq.kms Hydro + Thermal (Ebio @6.5 t/ha/yr) 165
5.5.3 Net Energy from 95.03 sq.kms Hydro (Eh) + Thermal (different Ebio) 169
5.5.4 Net Energy from 95.03 sq.kms Hydro (0.7*Eh) + Thermal (different Ebio) 166
5.6.1 Cost/unit of Electricity – En=Eh'+ 0.35*Ebio 171
5.6.2 Cost/unit of Electricity - En=Eh + Ebio 171
5.7.1 Benefit Cost Ratio (for various environmental costs) 173
5.7.2 Annual Costs & Eh generated (for various environmental costs) 173
6.1.1 Mean Hourly Global Solar Radiation at Goa 197
6.1.2 Mean Hourly Global Solar Radiation at Mangalore 197
6.2.1 Duration of Sunshine per Hour at Goa 203
6.2.2 Duration of Sunshine per Hour at Mangalore 203

6.3.1 Computed and Observed Values of Global Solar Radiation 206
6.3.2 I(H), Direct Solar Radiation - D: diffuse, G:Global Solar Radiation 211
6.3.3 Extra Terrestrial Radiation, Actual Sunshine and Maximum Possible 212
6.3.4 Daily Global Radiation Tilt Factors at Goa and Mangalore 212
6.4 Median Values of Hourly, Daily Global Solar Radiation 217
6.5.1 Cumulative Percentage Frequency Distribution (December-May) at Goa 217
6.5.2 Cumulative Percentage Frequency Distribution (June-November) at Goa 218
6.6.1 Cumulative Percentage Frequency Distribution (December-May) at Mangalore 218
6.6.2 Cumulative Percentage Frequency Distribution (June-November) at Mangalore 219
6.7.1 Global Radiation (Estimated and Actual Values) at Goa 224
6.7.2 Global Radiation (Estimated and Actual Values) at Mangalore 224
6.8.1 Global Radiation at Goa, Karwar, Honnavar, Shirali and Mangalore (as per equation.11) 225
6.8.2 Global Radiation at Goa, Karwar, Honnavar, Shirali and Mangalore (as per equation.13) 225
7.1 Percentage Population Increase - Mandalwise and Regionwise (1951-1991) 249
7.2 Human and Livestock Density Regionwise 249
7.3 Fuel wood Consumption (for Cooking) Regionwise 280
7.4 Fuel wood Consumption (for Water Heating) Regionwise 280
8.1 Area, Production, Yield - Paddy (Temporal Changes in Uttara Kannada) 313
8.2 Talukwise Cropped Area (%) 313
8.3 Energetics in Paddy Cultivation - Zonewise, Kumta Taluk 319
8.4 Energetics in Paddy Cultivation - Landholdingwise, Kumta Taluk 320
8.5 Energy Input (GJ/hect) for Paddy Cultivation 320
8.6 Energy Output (GJ/hect) of Paddy Cultivation 321
8.7 Output/Input Ratios for Various Categories of Landholding 321
8.8 Paddy Yield (0.1 tonne/hect) – Zonewise 322
9.1 SEC and PPCU 359
9.2 SEC and EI Variation for Different PPCU 360
10.1.1 Regional Energy Planning Model Structure (Overall) 372
10.1.2 Detailed Structure of Integrated Regional Energy Planning 372
10.1.3 Methodology for Assessment of Regional Bioresource Status 374
10.1.4 Design of Hydroelectric Station with Optimal Storage Option: Detailed Operation Module 375
10.2 Integration of Sources and Devices to Perform Various Tasks 379
10.3.1 Sectorwise Energy Consumption - Uttara Kannada District 388
10.3.2 Energy Consumption Sourcewise – Uttara Kannada District 389
10.3.3 Sourcewise Consumption of Energy (Source - Types) 389
10.3.4 Sourcewise, Sectorwise (C: Commercial, NC: Noncommercial) 389
10.4.1 Hydroelectric Energy (Streams + River) Bedthi and Aghnashini Basins 395
10.4.2 Total Energy Estimated in Bedthi and Aghnashini River Basins 396
10.4.3 Net Energy Available in Bedthi Basin - from an Area of 95.8 sq.kms 396
10.5.1 Biomass Availability and Demand - Talukwise, Uttara Kannada District 397
10.5.2 Biomass Availability and Demand - Talukwise, Uttara Kannada District 397

PLATES

3.1.0 Vegetation Types in Uttara Kannada District- NVDI (NRSA Imagery) 94
3.2.0 Vegetation Types in Uttara Kannada District (Based on Forest Map of South India) 95
4.1.0 Rivers and Streams in Uttara Kannada District 122
4.2.0 Distribution of Tanks in Uttara Kannada District 123
4.3.0 Catchment Area and Streams at Muregar Jog 124
4.4.0 Catchment Area and Streams at Bhushangeri Jog (Boosangeri) 125
5.1.0 Tributaries and Streams in Bedthi River Basin 178
5.2.0 Tributaries and Streams of Aganashini River Basin 179
5.3.0 Contours of Bedthi River (Downstream of Magod Falls) 180
5.4.0 Contours of Bedthi River (Upstream of Magod Falls) 181
5.5.0 Various Contours (at 440, 460, 480 and 500 m) at Magod (Proposed Dam Site) 182
5.6.0 Various Contours (at 340, 360, 380, 400, 420, 440, 460 and 480m) at Shivaganga 183
5.7.0 Catchment Area, Tributaries and Streams in Aghnashini River Basin 184
5.8.0 Contours of Aghnashini River 185

PREFACE

Energy has been known to man in its most magnificent incarnations - sunlight, wind, water and fire. Its subtler form has been no less influential, the muscular energy that is in man himself, which has been sustained by his daily intake of food, water and air. Influence of energy over man has been sublimal and savage, heralding prosperity as well as destruction. Since the beginning of his existence, it has played an overwhelming role in his day to day affairs. The glorious sunshine that awakens us every morning or the gentle rippling of the streams, the breeze in the mellow of the evening or the tempestuous winds during a storm, the frothing waves, fiery fountains (volcanoes), hot and bubbling springs, deafening thunder and lightning reminds of its great power and capacity to influence lives and the landscape. Early man learnt to use the energy hidden in water and wood for his sustenance. He burnt wood to cook food, keep himself warm and light the surrounding, and used the power of the wind to sail on high seas. He domesticated animals and used them for transportation and agriculture. He innovated and developed wind mills and used them to draw water. Energy played a vital role in shaping civilisations and cultures. There was no stopping the industrial man. He burrowed into the bosom of the earth for oil and coal, and exploited energy in all its available forms. He developed techniques and technology to limit the power of energy according to his needs. Modern man has refined it to match his lifestyle and luxury. From the tiny microchips to mammoth rockets, energy has transformed every facet of our life. From animal-drawn carts to fuel powered concordes, wind mills to wind turbines, sailing boats to luxury liners, man has come a long way in allowing energy to shape his life. The only difference between then and now is the fact that modern man has began to lay his hands on the energy reserves of the future generations.

Energy plays a vital role in our society. It effects and affects all that we do. Its role in economic development of a region needs no emphasis; exercising greater influence because of its qualitative effects. At the basic level, all our activities involve the use of energy, whether it comes from the warmth of the sun, the stored energetic potential of flowing water, fossil fuels extracted from the earth, or from the calories contained in the food we eat. It is useful to think about energy enabled services, such as lighting, cooking, heating and cooling, or transportation. Ultimately the availability of such "services" is what counts for satisfying basic socio-economic needs, and providing the basis for all economic production and human activities.

Harnessing bioenergy is in effect a way of exploiting nature's solar collectors-living plants that use photosynthesis to turn the energy of sun light into carbohydrate molecules. Animals gain some of these energy when they feed on vegetable matter. Throughout history, human society have relied on bioenergy when they burn wood in order to cook food or for space heating. Since the emergence of fossil fuels as a major energy in the late nineteenth century, bioenergy has been largely ignored by energy planners, because much of its use falls outside the commercial markets that government keep tabs on. Although bioenergy (biomass) is a renewable source, much of it is currently used in ways that are neither renewable or sustainable. In many regions, firewood is in increasingly short supply as growing populations convert forests to agricultural lands. The increasing demand for firewood by the rural populations is only one reason, although an important one, for the continuous deforestation. Other factors are overgrazing, expansion of arable land, industrial wood felling and growing urbanisation. The resulting ecological effects are heavy soil erosion, desertification and periodic flooding. The fuel shortages have forced women and children to spend much of their time in collecting fuel wood, while in many areas, crop residues and animal dung - which otherwise are valuable bio fertilisers are burned in cookstoves.

The production process is non existent without energy; the lack or shortage of which has serious impact on our economy. It is the central concept in most branches of natural science, social science and engineering as it plays a crucial role in diverse processes such as chemical reactions, cloud formation, functioning of cells, production of goods, transportation, etc. Fossil fuels such as coal, oil and natural gas are the basis of industrial society but they are disappearing at an increasing pace. Present fossil fuel potential is unable to meet the growing demand of our society. There is a need to look for viable alternatives to meet the scarcity. In this regard rational decision making is necessary to eliminate wasteful use of resources. The development of renewable sources of energy will increase the diversity of energy sources in a region and thereby increase the security of meeting the energy service needs. The regional planning machinery could play a leading role in the sensitive development of renewable resources to the benefit of the environment and the local community. However, at present, the vast majority of the region's planning machinery (at state and district level) have little or no knowledge of the renewable energy potential that exists within their boundaries.

Energy utilisation in Karnataka State, India, differs radically from that of other, industrialised countries. Non commercial sources of energy such as firewood and agricultural residues constitute the main source of energy in rural areas. There are enormous variations in the amount of energy used within India in different states and regions. There exist regional disparities both in terms of energy supply and demand. Linkages between various energy sub sectors such as coal, oil, electricity and non commercial energy sources are non existent. Energy planning in our country is not an integrated activity. The plans for electricity, oil, coal, and firewood are separate exercises. Secondly, the planning activity considers only the demand and projects it over a period of years. The efficiency in utilisation has not been investigated. The primary goal of planners is for supply expansions based on the assumption of linkage between energy consumption and the economic growth of a region. With this approach, energy becomes an end in itself and the focus shifts on meeting increased energy consumption through energy supply expansion alone. This supply and demand based planning for each individual energy form has resulted in problems like inefficiency, loss, more conversions and other environmental problems. This conflict between energy demand and environmental quality goals can be solved by an integrated approach to the problem of energy planning with emphasis on minimising the consumption of non renewable sources of energy such as coal, oil, etc., maximising the efficiency of end use devices in all sectors and organising an approach towards optimal use of renewable sources.

This path would certainly help in boosting economic growth of a region and equally promote structural changes that will be conducive to overall development. Opportunities for energy conservation exists at all levels of the energy system, including the supply side (production and conversion from energy sources) and the demand side (final consumption of these carriers by end use technologies). Thus, adoption of energy conservation and renewable energy sources helps in maintaining the sustainability of a region.

The term sustainable development was coined by the UN Bruntland Committee to describe a development which satisfies the needs of the present without compromising the ability of future generations to meet their own needs. The sound environmental policy focuses on the environmental aspects and is a part of a wider concern aimed at well being and living standards. Sustainable development demands clean and pollution free environment and enshrines good quality natural resources for both present and for the future. Energy, biodiversity and physical space are the critical resources for present and future human demands. The international community has espoused the objective of sustainable development in various agreements which include - Agenda 21 (Rio), convention on biological diversity and the climatic conference at Kyoto recently.

The continuing demographic and unplanned economic growth are the major threats to sound environmental policy. The environmental gains achieved by more environmentally efficient production are offset by increased pollution due to growths in the volume of production and consumption. Socio-cultural, administrative (integrated planning approaches) and technological breakthroughs are needed if we are to achieve the objectives of sustainable development. Sound environmental policy of a region should focus on:

- continuously improving the efficiency with which the environment is used by the desired social and technological breakthroughs,
- judicious use of science and technology. Science and technology plays a pivotal role in bringing sustainable development closer and in providing insights in to the technological and social breakthroughs,
- focussing on quality of overall environment,
- integrated, customised and flexible approaches,

- internalising environmental costs in prices. This encourages citizens and industry to use energy and raw materials efficiently and stimulates sustainable production and consumption, proper mechanism to improve enforcement of policies, thereby ensuring that policy is fully implemented and environmental crime reduced,
- emphasis on renewable resources,
- energy efficiency measures, which are the cheapest way of protecting the environment, since they are themselves cost-effective. Energy efficiency by nature is a decentralised activity since it concerns multitude of factors: rural and urban, domestic, industrial and transportation sectors, local authorities, public and private services,
- integrated energy planning - to execute the conceptual scheme to seek the optimal balance between energy efficiency action programmes and supply action programmes (through renewable sources of energy), by comparing their impacts on the economy and the environment, which influence the development pattern and its sustainability.

Global energy supply policies are generally elaborated at the national level. However, depending on the administrative and political structure of the country, the administrative regions can have considerable degree of autonomy in their energy choices. In our country, district is the most appropriate and quite pertinent for elaborate regional energy management plan. The regional energy plan has following stages - consumption analyses and determinants of demand, assessment of renewable energy resources available, exploration of future energy demand, action plan and follow up action and evaluation.

Integrated Regional Energy Planning is a planning process where utilities evaluate available demand and supply side options to provide energy services (which includes purchased power) and determine an optimal energy service strategy, given economic and environmental factors. The essential concept of integrated regional energy planning is the equal treatment, or integration, of renewable energy and conservation based energy strategies. Ecologically Sound Integrated Energy Planning basically demands the following:

i. A procedural, informational and regulatory background.
ii. Identification of resources, economic costs, ecological implications and sensitivity analyses, which need to be carried out on both demand and resource sides.
iii. Optimal allocation has to be identified and the decision taken regarding implementation of what has been decided on demand side management and capacity building.

In this regard, the proposed ecologically sound integrated energy planning at a district level is an endeavour to maintain the balance between energy and man, and influence the environment to retain its sustainability.

T. V. Ramachandra

ACKNOWLEDGMENTS

I am grateful to Prof. D.K. Subramanian and Prof. Niranjan V. Joshi for their guidance, encouragement, constant support and generosity during my research career. Their immense experience and intellect enriched this research immensely. I express my gratitude to my family for their understanding and patience during the course of completing this book. I thank Prof. K.M. Chetty, for reviewing my manuscripts, for helpful advice and suggestions.

I thank the Indian Institute of Science (I.I.Sc) for encouraging the continued pursuit of my energy research and for providing the necessary logistic support and infrastructure. Internet and E mail facility at I.I.Sc has truly revolutionised the process of information assimilation and dissemination. I thank SERC officials for understanding my needs and for providing me an account in Hamsadvani to access FTP and Gopher Facility as early as 1993. Through E Mail, I have received a great deal of help and advice from many experts and organisations over the past five years. I thank all those who have been involved in this research in many ways. I thank IIEC (International Institute of Energy Conservation), Energy and Environment Division (Lawrence Berkeley National Laboratory), TERI (Tata Energy Research Institute), New and Renewable Energy Enquiries Bureau, Water Resources Centre, EPA (Environmental Protection Agency), IEA (International Energy Agency), EPRI (Electric Power Research Institute), World Bank, UNEP Collaborating Centre on Energy and Environment, Gate/GTZ, IT (Intermediate Technology), ORNL (Oak Ridge National Laboratory), ESCAP, United Nations, CADDET, CREST, FAO, OECD, UNIDO, UNDP, WHO, WMO, East West Centre, The Stockholm Environment Institute, IRN (International River Network), IAHS (International Association for Hydrological Sciences), Colorado State University, Solar Energy Society of India, World Watch Institute, World Resources Institute, and World Energy Council for providing me requested information and publications.

I thank my colleagues at Centre for Ecological Sciences for many rewarding discussions and providing a stimulating research environment. Prof. R. Sukumar's timely generosity is appreciated for providing the AECC computer for data keying and analyses in the initial stages. My sincere gratitude to Dr. Subhash Chandran, Dr. Sri Niwas Singh, Mr. D.M. Bhat, Mr. P.R. Bhat, Mr. C.M. Shastri, Mr. R.B. Harikantra and Mr. Srikanth M. Gunaga for their valuable help and suggestions. I thank Mr. Rosario M. Furtado, Mr. Deepak M. Shetti, Mr. Raghavendra Rao and Mr. Vishnu Mukri for their help in field data collection in some very trying conditions. I thank Mr. Janardanan N. Pillai for his cooperation while sharing office space and for valuable tips in programming and computer related work. I thank Mrs. Maria Joseph for helping me in my daily correspondence, proof-reading of data and other timely assistance. I thank all my project colleagues for their kind assistance and support during the field work. My gratitude to the GIS team members - Ms. G. Savitha, Ms. Samapika Padhy and Mr. Srinivas Prasad, for providing GIS outputs and flow charts.

For helping me in developing DSS through VB, I thank Mr. Rajiv Kumar Jha and Mr. B.K. Vijay Prasad.

Mr. Joshua David helped me in the final stage of the manuscript preparation with precision and dedication, which involved proof-reading and editing of the drafts. This entailed much persuasion to keep Delay David from delaying further. I am grateful to my computer which sustained 12-13 hours of continuous usage everyday during my research and writing of the book. I thank Mr. Nair for keeping my computer in a healthy condition by servicing it promptly.

And for the continued financial support for this research provided by the Ministry of Environment and Forests and Ministry of Science and Technology, Government of India and the Indian Institute of Science, I sincerely thank them all. Last but not the least, I wish to express my deepest gratitude to all those whom I might have inadvertently failed to acknowledge.

Chapter 1

INTRODUCTION

1.0 INTRODUCTION

Energy plays a vital role, affecting all the activities that take place in the society. Its role in economic development of a region needs no emphasis; because of its qualitative effects, it exercises greater influence. It is the key element in the production process, and the lack or shortage of it has a serious impact on the economy. It is a central concept in most branches of natural science, social science and engineering, since it is essential to make things happen. It plays a crucial role in diverse processes like chemical reactions, cloud formation, functioning of cells, production of goods, transportation, etc. Fossil fuels, such as, coal, oil and natural gas lay the foundation for an industrial society, but they are disappearing at an alarming rate. Present fossil fuel potential is unable to meet the growing demands of the society. There is a need to look for viable alternatives to meet this scarcity. In this regard, rational decision making is necessary to eliminate wasteful use of resources.

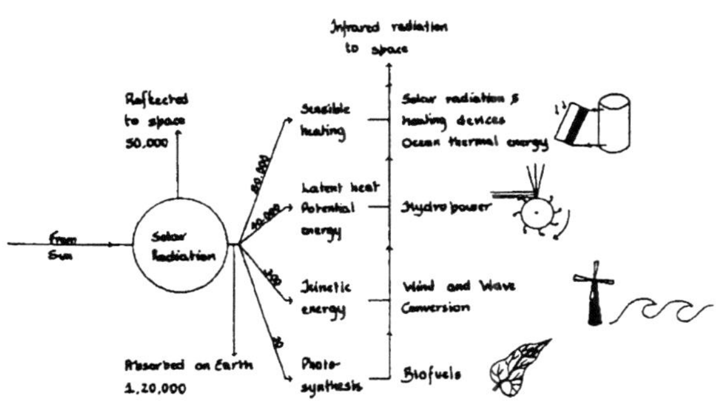

1.1.0 Energy Use versus Time

The history of man's emergence from the ecological niche, appropriate to medium size omnivorous mammals, to his present position as the earth's dominant species is attributed to his increasing skill in harnessing and manipulating energy. Each stage of his evolution has been marked by an extension of his ability to control the natural flows and accumulated energy resources. Initially, he used only food energy, as did all other species of animals. Primitive man's use of energy was of the order of 2000 kcal/day, mainly as food energy. Hunting man used some fuel wood for heating and cooking, his use of energy being about 3000 kcal/day. Early agricultural man used animate sources of energy for agriculture and transportation, with his consumption roughly being about three times that of hunting man. Gradually, man discovered new ways of using energy supplies to his own advantage by controlling fire, growing crops and domesticating animals. Cultural Development and increase in human population have led to the use of other energy sources and increase in human population. Thus, the influence of energy availability is seen in the form and structure of the human society itself.

In a subsistence agricultural economy, majority of the people spend most of their time working to obtain the food necessary for their own survival. Agriculture is essentially a process by which the stored energy of photosynthesis is directed towards man. It begins with the selection and cultivation of plants. Irrigation, manure and fertiliser have enabled him to maximise food production. With these, he has opened wider the gates of photosynthesis, through which solar energy becomes his food. The domestication of animals widen the control of natural energy flows. With this domestication, grasslands have become energy gatherers. Grass otherwise cannot be used by man as fuel or food. With domestication, the cattle is allowed to graze, and the subsequent milk and meat produced is used by man. Muscle energy is used for ploughing, to lift water and for transportation. The use of human beings as slaves was a feature of the early civilisation. They were given energy only in the form of food for the work carried out. Both the slave and domestication of animals were precursors of the mechanical engine: they were means of converting the energy of food or fuel into work.

Village and urban settlements were common by 2000 BC and the pace of change gradually increased. The growing ability to harness energy and therefore reliable food supplies gave impetus to civilisation. Gradually, around 100 BC, waterwheel (in the West, China and India) and windmills (in the Arab countries) acquired the place to provide energy in the medieval economy. This hastened the pace of social and technical development. Tapping of fossil fuels like coal, oil, etc., led to an endless catalogue of achievements and conquests of the physical world. This led to the industrial revolution. In the nineteenth century, a host of new inventions took place in response to the growing availability of energy, for harnessing more energy sources. This resulted in advancement of science and technology and emergence of many disciplines like civil, structural, mechanical and electrical engineering. But, a declining trend in available stock of fossil fuels like coal, oil and natural gas in the later half of the nineteenth century has necessitated the search for viable alternatives. In looking at the earth's energy resources, it is necessary to go beyond the mere question of their absolute magnitude or their theoretical potential, if the technology to harness them could be devised. The practical availability of energy resources is limited by social, geographical, political, economic and technical constraints. Abundant coal resources in Russia or oil resources in the Middle East and constraints associated in using these resources by other countries have clearly shown that proving the existence of resources does not always promise accessibility to everyone.

1.1.1 Technical Aspects of Energy

Energy is the capacity to do work - to exert a force 'F' through a distance 'd'; to accelerate a mass 'm' from rest to some velocity 'v'; to move an electric charge 'q' across a voltage difference 'V'; to lift a mass 'm' to height 'h'. In each case, a specific amount of energy - Fd, $(1/2)mv^2$, qV or mgh - is required. Energy is available in many different forms:

Potential energy -	objects, because of their position, capable of doing work on other objects are said to possess potential energy;
Kinetic energy -	the energy of moving body, proportional to its mass and the square of its speed;
Heat energy -	the kinetic energy of random thermal motion of matter on an atomic scale;
Mechanical energy -	the kinetic energy of organised motion of bulk matter;
Electrical energy -	the energy due to the forces that charged bodies exert on one another;
Chemical energy -	the energy of chemical bonds, due to electrical forces acting at the atomic level in accord with quantum mechanics;
Nuclear energy -	the energy of nuclear binding due to nuclear forces;
Gravitational energy -	the energy due to the force of gravitational attraction - such as the potential energy that a body of water has because of its height above sea level;
Light energy -	the energy of visible, ultraviolet, infrared, electromagnetic radiation, etc.

Energy undergoes transformations from one form to another. These transformations are governed by the laws of thermodynamics, which state:

(i) Energy is a conserved quantity - while it can move from one place to another, it is neither created nor destroyed. That is, the total amount of energy does not change.

(ii) In energy transformations, energy tends to pass from concentrated forms to dilute forms. It changes in such a way that the amount of work that can be obtained from it decreases.

The first law of thermodynamics is based on laws of motion of Isaac Newton, whose work was instrumental in providing theoretical view of the universe, in which energy is the central concept linking the whole range of observed phenomena. Einstein later, through his theory of relativity was able to transcend Newtonian concept of the physical universe, widening the perception of the nature of physical reality. Einstein's special theory of relativity equated mass and energy and expressed it as $E = mc^2$, that is, energy is equal to the square of velocity of light. Joule, electronvolt, kilocalories, kilowatt-hour and kilogram mass are the units commonly used in the different disciplines of electricity, atomic physics, dietetics and engineering to express energy. All these units that can be expressed in terms of each other, can also be reduced to units of mass. For example, 1 kilowatt-hour is 3.6 Giga joules, 859.8 kilo calories, 2.247×10^{25} electronvolt or 4.007×10^{-11} kilogram mass. This highlights the unifying power of the concept of energy.

Accounting of energy, using the first law, allows one to translate energy from various sources into thermal equivalent units such as kcal, while energy measured using units (such as MTOE: million tonnes of oil equivalent, MTCE: million tonnes of coal equivalent) based on the second law of thermodynamics takes into account efficiency of conversion, as well as, differing efficiencies of combustion.

Usefulness and transformability of energy are explained in terms of "entropy" by Clausis. Energy at a high temperature is obviously more useful, or transformable, than energy at low temperature. This leads to the definition of 'energy efficiency'. Efficiency is defined as the amount of useful energy produced by a system as a proportion of the total energy input. In assessing the resources as alternatives, limitations imposed by the second law need to be clearly appreciated.

1.1.2 Classification of Energy Sources

All the sources of energy, currently available for harnessing, can be linked to two fundamental forces in nature - gravitational and nuclear. Nuclear fusion is the source of solar energy - the driving force for much of the energy consumed on earth today.

Renewability or non-renewability of a solar driven process is distinguished based on the energy storage or cycling time involved. Renewable resources have a cycling time less than 100 years, while for non-renewable resources, it is greater than a million years. The depletable resources are fossil fuels, which are non-renewable since the rate of their utilisation far exceeds the rate at which they are formed. Examples of renewable resources are hydro energy, solar energy, wind, biomass, and energy from wastes (such as biogas, agrowastes, etc.).

The renewable solar energy is subdivided into direct and indirect types. Sunlight used directly can produce electricity, heat or derive a chemical reaction. It is used indirectly when it drives other processes, biological-chemical or climatic-mechanical, which in turn are used as sources of energy.

The energy sources can be classified in a number of ways based on the nature of their transaction, as commercial and noncommercial sources of energy. All energy resources, particularly the commercial ones, are natural. Coal, oil and nuclear sources constitute commercial sources, while firewood, biomass and animal dung constitute non- commercial sources. Also, the energy sources are classified based on animate and inanimate characteristics.

Energy sources could also be classified as exhaustible/depletable or non-depletable/renewable resources. The distinguishing feature of an exhaustible resource is that, it gets exhausted when used as an input of a production process, and at the same time, its undisturbed role of growth is nil. That is, the temporal services provided by a given stock of an exhaustible resource are finite.

Further, based on conventionality in deriving energy, energy sources could be classified as conventional (coal, oil, hydro, nuclear, etc.) and non-conventional (solar, wind, tidal, geothermal, biogas, etc.) sources.

They are also classified as primary or secondary types - coal, firewood, etc., being primary sources and electricity, a secondary source. Energy in its primary form can be of different kinds. The main types are Chemical (fossil fuels - coal, oil, natural gas, peat; biomass - wood, agricultural residues, etc.), Potential (water at a certain height), Kinetic (wind, waves), Radiation (sun), Heat (geothermal reservoirs, ocean thermal reservoirs) and Nuclear (uranium). The primary form of energy must generally be converted into secondary or final forms of energy before it can be used. For instance, the potential energy of a waterfall (primary energy) is converted into electricity (secondary energy), which is transmitted and transformed to supply (final) energy to a factory, where it is converted into mechanical energy (useful energy) for productive operations.

Important types of secondary energy are electricity and mechanical energy. But chemical energy is also important as a secondary energy, for instance, in the form of refined oil products. Final energy is the energy that reaches the consumer. This can be electricity at a suitable voltage, or chemical energy in kerosene or batteries. The consumer, finally, uses certain equipment to convert the final energy he buys, into useful energy for one of his end use activities, e.g., irrigation, transport, cooking, etc.

Most of the energy sources are substitutable to each other due to the fact that some form of energy can be converted to other - such as coal to electricity, use of photo electricity to drive a chemical reaction, wind energy to pump and store water that could be used to produce electricity when required, or solid biomass to produce liquid or gaseous fuels of higher calorific value. All forms are ultimately converted into heat. This gives rise to the inter-fuel substitution process with which an economy can substitute its abundantly available resources to the scarcely endowed one.

1.1.3 Components of Energy System

Energy is a complex process as it is possible to convert it into different forms, transport it, store it in some forms and use it in various end use modes in innumerable kinds of places. Thus, an energy system essentially consists of four components. They are generation, transmission/transport, distribution and consumption/utilisation.

Generation involves conversion of energy sources into acceptable forms that can be easily transported. Energy can be drawn from water falls, sun, wind and from chemical energy stored in coal, oil and biomass resources. The major factors which influence the generation components of an energy system are the technologies of generation and their scales, efficiency and costs. Examples of these are electricity generated from thermal, hydro or nuclear stations, coal extraction and oil refinement. The size and site of generating facilities depend on a number of factors like availability of raw materials, land, water and other infrastructural facilities.

Transmission/transport and distribution are needed to supply the useful forms of energy to various places of consumption. Electricity from centralised power stations is more conveniently transmitted over long distances by transmission and distribution networks. Sources like coal, oil, LPG (liquified petroleum gas) and wood can be transported to different places either for direct use or for further conversion into other useful forms, e.g., transportation of coal or oil to run a thermal power station located near a major load centre. Consumption/utilisation relates to that component of an energy system where energy is utilised in several end use devices required to carry out specific tasks. For example, wood or LPG is used for cooking stoves, while electricity is used for lighting.

Many energy sources can be used either in direct or indirect mode to perform a task. Electricity, kerosene and firewood can be used in the direct mode for cooking. In indirect mode, energy from resources need to be converted in many stages to the form required by an end use device. An example is conversion of coal to heat (in a boiler to raise steam), heat to mechanical energy (in a turbine), mechanical to electrical energy (in a generator), electricity to mechanical energy (in a motor), and mechanical to potential form (in a pump) to lift water from a well. There may be multiple paths from a source to an end use device, such as, biogas can be used for lighting either directly through a mantle lamp or converted to electricity and used to light a bulb.

In the case of a decentralised system, energy resources are transported directly to places of use, and used in a suitable end use device. Some examples of decentralised systems are - use of firewood for cooking and water heating in the domestic sector, small industries like brick kilns, tobacco curing, sericulture, etc. However, before the advent of technologies for large-scale exploitation in a centralised system, decentralised systems were the only mode of energy used. Because of convenience, easy maintenance, diversity of usage and reduced costs of operation, energy production became centralised. Efficiencies, costs, and conversion possibilities of end use devices are some of the factors that influence the final stage of energy system.

1.2.0 REVIEW OF RENEWABLE SOURCES OF ENERGY

The main types of useful energy are heat and mechanical energy. Heat though predominant in household activities like cooking and water heating, is also needed for productive activities like drying and industrial applications. Mechanical power is predominant in the productive sectors, e.g., industry, agriculture and transport.

Efficient use of energy is achieved when unnecessary energy conversions are avoided, as each conversion has limited efficiency and, therefore, implies a certain loss of energy as wasted heat. For instance, if secondary energy can immediately serve as final energy, or even as useful energy, substantial losses can be avoided. Examples are wind machines in irrigation or hydro turbines powering a shaft. This principle favors decentralised energy generation and is particularly relevant with new and renewable energy sources. Very high efficiencies can be achieved with cogeneration, where heat as a by-product of power is not wasted, but put to good use on the spot. This can be applicable to biomass gasification. The counterbalance originates from economics of scale which generally favor large-scale energy conversion.

The foregoing sections make clear that renewable energy technologies cannot be limited to the technologies per se. It is therefore important as to what form of energy is provided, and how the technology can serve a particular end use activity. Sometimes, a combination of technologies would give an optimum energy mix, to secure a continuous and reliable energy supply. In the following section, some renewable energy technologies, such as, wind energy, hydro energy, solar energy, geothermal energy and draught animal power are reviewed.

1.2.1 Wind Energy

Part of the kinetic energy of wind can be captured by a rotor and converted into mechanical or potential energy. The major drawback of a wind power system stems from the nature of the source - wind has a low power density, and wind speed and direction are highly variable. Therefore, wind power systems are best for applications which can tolerate varying power input, such as, pumping of water from wells, the traditional windmill application in many parts of the world. The power that can be achieved by a wind system is proportional to the cube of the wind speed and the square of the windmill diameter. That is,

$$P = n\, C\, d^2\, v^3$$

where,
P = theoretical wind power (kW)
d = windmill diameter (m)
v = wind speed (m/sec)
C = constant (5.10^4 kg/m^3)
n = windmill efficiency (20-40 per cent)

There are three principle dimensions to wind energy technology, namely:

(a) Type of machine - horizontal axis, from sail and multibladed to fast-running rotor types, and vertical axis from Savonius to fast-running Darrieus and variations,

(b) Applications - water pumping, desalination, heating, cooling, autonomous power generation and power generation with grid connection, and

(c) Size of machine - small (less than 10 kW), medium (10-100 kW), medium-large (100-1000 kW) and large (more than 1 MW).

(i) *Wind Energy Driven Waterpumping Systems (Windpumps)*

A small windmill with direct mechanical drive to a water pump is an old technology, and commercial machines are still in extensive use in many parts of the world. In most cases, water is pumped whenever the wind blows and is stored in a tank for use as needed. Careful matching of the windmill to pump characteristics is important, and the windmill, pump and storage must be treated as a system.

The conventional multibladed windmill is a type that has been widely used for this purpose in the past, but has not found widespread application in developing countries because of economic and financial reasons. In the past few years, some new all-metal multibladed windmill designs have been developed by

Intermediate Technology Development Group (ITDG) in the United Kingdom and Consultancy Services Wind Energy Developing Countries (CWD) and Working Group on Development Technology (WOT) in the Netherlands. Compared to the traditional multibladed windmills, they are considerably lighter in weight, more simple to construct and have slightly higher over-all efficiency. These windmills, after adapting the design to locally available materials, can be manufactured in sparsely equipped workshops and are cheaper than commercially available windmills by a factor of 2 to 4. Many such windmills are operating in pilot projects in India, Kenya, Pakistan, Peru, Tunisia and Sri Lanka. Experience in these projects are showing that extensive field testing is essential, and is resulting in several design modifications to ensure the reliability of the machines.

Besides the design of all-metal windmills, improved types of sail and/or wooden windmills have been developed in Colombia, India and The Netherlands, to try and meet the demands of the subsistence farmer. These mills, which can be fabricated in village workshops, are cheaper than commercial mills, even allowing for replacement of sails every two years. Drawbacks of this type of windpumps are relatively low efficiency and higher time requirement for operation and maintenance.

(ii) *Wind Energy Driven Electricity Generation (Wind Generators)*

Autonomous wind power systems in 10-1000 kW range, with diesel back-up and possibly hydroelectric storage, are already an economic alternative to conventional diesel generation for isolated communities in good wind regimes. With existing technology, the wind turbine can be used to give significant fuel savings, although it cannot replace the diesel. Systems are of low solidity (high speed) type and comprise horizontal, as well as, vertical axis windmills.

1.2.2 Hydro Energy

Potential energy of water at a certain height can be converted into mechanical energy by using a turbine. The falling water hits or exerts a pressure on the blades of the turbine. The power available is proportional to the head and the rate of discharge of the water. That is,

$$P = \eta\, C\, H\, Q$$

where
P = power available (kW)
η = hydropower system efficiency (70 per cent)
C = constant ($\approx 9.81 \times 10^3$ kg/m^{2}s^2)
H = head of the water (m)
Q = flow of the water (m^3/s)

The flow through the turbine is not independent of the water head. For a given turbine size, the flow will actually be faster if the head is larger. As a result, high heads are favored for hydropower generation. In practice, water heads are used from less than one meter to several hundred meters. With low heads, obviously, a large flow must be available to get a substantial amount of power.

The capacity of hydropower plants can vary between a few kW to about 1,000 MW. The appropriate scale of a hydropower unit depends on local geographic and climatological conditions, and the characteristics of the demand for power. The scale and local conditions also, to a large extent, determine the type of construction and machinery required, and accordingly the investments. Hydropower generation is highly site-specific. This implies that proper attention must be given to site selection and local surveys. For big hydro schemes, it would require elaborate feasibility studies that have to be done by specialised consultants. For small hydro units simple survey methods can be adequate.

Another classification of hydro plants is based on different systems of water usage. That is,

i. River power plants, where the head is created by weirs or dams,
ii. Diversion schemes that basically utilize naturally available heads,
iii. Run-of-river plants with little or no control of discharge, and
iv. Storage power plants with high dam and large reservoir for flow regulation.

The construction of a hydropower unit essentially consists of a water intake, a penstock that leads the water to the turbine, a control system, the turbine itself, and except micro units, a generator to convert the mechanical energy into electrical energy. The hydropower system also includes civil works like dams, weirs, canals, etc.

Various types of turbines are available. They can be of the reaction type, like the Francis turbine and the Kaplan turbines, where the turbine is completely submersed in the water, or of the impulse type, like the Pelton turbine, where a jet of water hits the turbine blades. The traditional water wheel can also be considered as an impulse type system. The cross flow turbine that is used for mini and micro units have the characteristics of both the types. Special types are the Segner turbines that can be used in micro units, and the hydraulic ram. The latter does not produce mechanical power, but can be used to pump relatively small quantities of water to relatively higher heights.

Turbines for big hydro units are tailor made, i.e., specifically adapted to the prevailing head, flow and local conditions. To a lesser extent, the same is true for smaller hydro units. Such adaptations determine the efficiency that can be achieved. Hydro technology is a mature technology and local manufacturing is possible in many countries. Turbines have a long life time (30 years or more) with little maintenance required and their operation is relatively simple.

(i) *Hydroelectricity:*

The utilisation of hydropower is mainly for electricity generation. The generator is linked to a system for the transmission and distribution of electricity. Big hydropower units are generally linked to the national or central electric grid. The electricity generation costs per kWh can be low, e.g., for units bigger than 10,000 kW, one third the costs of fossil fuel powered systems. Small hydro units often serve a local electric grid. The problem here is to achieve an acceptable capacity utilisation. Specific demands (e.g., for lighting) together with low off-peak demands, result in low average load factors for the unit. Load factors as low as 15 to 20 per cent are not uncommon, which result in a very unattractive price per kWh. This is sometimes aggravated by high maintenance costs and breakdowns. The position of mini and micro units that produce electricity is different, as will be discussed in the next section. Furthermore, hydropower schemes often serve irrigation, fisheries and flood control systems as well.

(ii) *Shaft Power and Multipurpose Hydro Units:*

It is increasingly being accepted that mini and micro hydro units do not have to be built on the same standards as big hydro systems. For instance, relatively quick survey methods for site selection can be used, protection measures be less strict, control mechanisms simplified, and lower efficiency accepted. This greatly reduces the investment costs and still leaves a useful option for small-scale power supply. The power of micro hydro plants is best used when the larger part is consumed on the spot as mechanical shaft power. Applications can be small-scale agro-processing industries like grinding, rice hulling and oil pressing, or sawing and water lifting. In addition, cheap dynamos can be connected for supplying electricity in the evenings. Civil works and operation of the system can be provided by semi-skilled local labour. Cross flow turbines are particularly promising for these units, because of their versatility and simple construction.

1.2.3 Solar Energy

The radiation energy of the sun can be captured and converted into heat and/or electricity. A distinction must be made between beam and diffuse radiation. The former reaches the receiving surface in a straight line from the sun, while the latter is reflected in the atmosphere by clouds, dust, etc. This leads to different systems of solar energy conversion.

(i) *Solar Energy Technologies:*

(a) Passive Thermal Systems:

The simplest method for the collection of solar energy is to design windows of buildings for systematic collection of available solar energy in the cold season. Solar collectors that make no use of ancillary equipment like fans and pumps are known as passive systems. Passive collection is the simplest and usually the most cost-effective way of gathering solar energy (beam and diffuse), and many of these techniques have been used for centuries. They became neglected by architects and builders during the period of low-energy costs. Technically speaking, passive heating techniques are well understood. They have the advantage of using traditional materials and building techniques, and can be used in a wide range of climates. Since people live in the "collector", the approach is chiefly of interest for heating buildings in temperate climates and upland and mountainous areas in lower latitudes.

(b) Flat Plate Collectors:

By far most of the stationary solar collectors are flat-plate type. Simple flat-plate collectors have been able to achieve temperatures of 40° - 80°C with conversion efficiencies of between 40 and 80 per cent. To achieve higher temperatures, the absorber is often covered with a special selective coating. For temperatures in excess of about 100°C, special designs of stationary collectors, such as the evacuated tube collectors, have been designed and tested. Advantages of the flat-plate collector are its simplicity and the absorption of beam, as well as, diffuse radiation. Disadvantages are the low maximum temperatures that can be reached (40-120°C) and the rather low efficiencies at "elevated" temperatures.

Constructed in a modular form (units 1-4 m^2), the flat-plate collectors can be coupled to make collector arrays of various sizes, depending on the application. However, there is a maximum size for collector arrays, owing to their thermal losses. It is therefore possible to design a heating facility for a block of houses, but it would be necessary to install several units for a whole agglomeration.

(c) Tracking Collectors:

It is very difficult to obtain fluid temperatures in excess of 150° C using even the most sophisticated flat-plate collectors, because their surface thermal losses become appreciable. A possible and practical means of achieving higher temperatures is by using concentrating type of solar collectors. This necessarily implies that practically all of the diffuse component of solar radiation is lost to the system. Therefore, attempts must be made to utilize as much of the direct solar beams as possible. Hence, the concentrator is normally equipped with a tracking device which, in effect, ensures that it "follows the sun" continuously. The absorber in this case is located close to the geometric focus of the concentrator, to ensure that it intercepts most of the incident direct radiation. There are, in general, two types of concentrators: the linear-focussing concentrators and the point-focussing ones. The former is generally equipped with a single-axis tracking system, and the latter with a two-axis system. In the former case, the absorber is a tube on which the solar image shifts as a function of the sun's position, while in the latter, the absorber covers the area around the focal point. In practice, linear concentrations are generally limited to a factor of 30 to 40, while point concentrators range from 500 to 2,000. For the two classes of concentrators, the efficiencies of direct solar beam conversion is usually in the order of 60-70 per cent. Point-focusing systems are far more sophisticated and complex than linear-focusing systems, and can only be justified when high temperatures (300°C to 1,000°C) are to be reached.

(d) Photovoltaic Generators (Solar Cells):

Solar cells, usually in the form of thin films or wafers, are semiconductor devices that convert 3-25 per cent of the incident solar energy into DC electricity, with efficiencies depending on illumination-spectrum intensity, solar-cell design, materials and temperature. A solar cell behaves much like a low-voltage (about 0.5 volts) battery, whose charge is continuously replenished at a rate proportional to incident illumination. Connection of such cells into series-parallel configurations allows the design of solar panels with high currents and voltages as high as several kilovolts. The extraordinary simplicity of a solar-voltaic system would make it a highly desirable energy system, both in developing areas and in industrialized nations. The attractions of photovoltaic arrays include the absence of moving parts, very slow degradation of properly sealed cells, possibility for modular systems at sizes from a few watts to megawatts, and extreme simplicity of use.

Until now, however, high costs of development and fabrication of solar arrays have discouraged widespread application. There is evidence that, with appropriate technological development and mass production techniques, the cost of solar arrays can be lowered to the point where a complete system - solar conversion, power conditioning and transmission/distribution - can compete on a lifecycle cost basis with other large-scale energy-system alternatives, and perhaps be even useful in small-scale applications in remote rural areas.

(ii) *Applications of Solar Energy*

(a) Solar Cookers

Three types of solar cookers have been developed: solar heat boxes, which must be oriented manually and are composed of an insulated box with double glazing; parabolic disk-type reflectors that concentrate the direct beams on a focal zone where the pot is placed; and a system in which solar collection is separated from the cooking stove and heat is transported by a working fluid. The latter makes it possible to cook inside the house and to install a heat storage system for cooking when the sun is not shifting. It is well known that a number of promotional demonstrations of solar cookers failed. Such factors as high unit costs, lack of storage facilities for supplying heat to cook the food during evening hours, unreliable units which could not withstand rural conditions and inadequate social acceptability was responsible for the failure. In the last two years, however, it seems that the insulated box cooker in certain parts of India (especially Gujarat state) has been introduced with a reasonable amount of success. The collector surface and, therefore, costs of solar cookers not only depend on average overall solar radiation, types of fuel and cooking habits, but also on the thermal efficiency of the cooker. These thermal efficiencies are again related to design, materials and state of maintenance of the equipment. Devices and corresponding efficiencies are as follows.

Flat plate solar cooker:	10-20 per cent (of the total with heat transport solar radiation)
Parabolic disk reflector:	60-70 per cent (of direct cooker beam radiation)
Insulated box cooker:	35-45 per cent (of direct beam radiation)

(b) Solar Dryers

Of all the direct uses of solar energy, sun drying of crops is perhaps the most ancient and widespread. Traditionally, drying is done by spreading the products on the ground in the open and exposing them to the sun. However, at the same time, they are exposed to bad weather and insects. In order to reduce the risk resulting from such hazards, solar dryers have been designed, built and used extensively. The earlier units were based on the "greenhouse effect". They reduced the drying time, enhanced the quality of the products, and were found particularly useful for products harvested in the wet seasons. Solar dryers have also been used for timber and wood drying. Generally, the principle of the solar dryer is to heat air in a suitable solar collector and circulate it by natural or forced means through the product to be dried. The solar collector or air heater, as it is normally called, may be external to or form an integral part of the dryer. Solar dryers are particularly suited to small homesteads and villages in developing countries, where they can play an important role in crop preservation. As an added incentive, the technology involved in the construction and operation of a typical dryer is simple enough to be universally applied. One attraction of solar dryers is that several designs can be built by using mostly locally available materials. Further research is required, however, with regard to materials particularly, to ensure that the units last for more than ten years. Perhaps, the main inhibitions to the use of solar dryers have to do with lack of information and the acceptability of dried products. The latter factor is of great importance when the products are cash crops.

(c) Solar Power

Electricity can be produced in two different ways: photovoltaic or thermodynamic. Today, only decentralized production of solar electricity can be envisaged as an economic possibility. For very small amounts of power (to 5kW), photocells are definitely most appropriate. Beyond that level, photovoltaic and thermodynamic conversion may be competitive. While it is possible for the situation to change, it must be noted that at present, solar-generated electricity is often too costly. However, photocells are an ideal power source for remote areas and for equipments that require only little power: school televisions (20-30 W), Hertzian relays (1-100 W), air radio-navigation systems (6-400 W), railroad signals (to 500 W), and telephones in rural areas (0.5-60 W).

The production of high levels of electricity (1 MW and more) may be justified only when the electricity produced is integrated into one utility grid. The difficulty of storing energy for night use may impede the isolated use of solar plants. Fortunately, solar electricity works very well as a supplement to the traditional electricity sources: hydropower, fuel fired and nuclear plants. The thermodynamic conversion systems are classic: alternative engines (Rankine); steam turbines (water and organic); gas turbines. All these generation systems have been field tested, and the only problem that arises is economic in nature. However, the economics of solar power production may be improved by using cogeneration techniques, providing both thermal energy and mechanical or electrical power, and using the normally wasted heat properly. A good deal of research has been carried out to establish the capability of solar electricity (photovoltaic or thermal) to compete in the future with conventional electricity sources. In the short and medium term, solar electricity will be widespread in isolated applications only.

(d) Solar Pumps

Recent technological progress makes solar energy for water lifting, in the 100-800 W range, a realistic option under certain conditions. The solar pump system consists of a photovoltaic array, a motor and a pump. Particularly, the pump efficiencies have been improved up to 65 to 70 per cent, and the overall costs have been reduced. Typical applications are irrigation of landholdings up to two hectares with 60 m^3 of water per day through static heads of 2 to 10 m, or the supply of some 20 m^3 of drinking water per day through static heads of 10 to 30 m for a village of up to 1,500 people. However, solar pumps are still expensive, which implies that this technology is attractive only when alternatives (diesel, wind, etc.) are lacking. Investment costs are still such that pay back periods may exceed 5 years. It is expected that in a few years time the costs will fall further so as to make solar pumps more competitive among prevailing alternatives.

1.2.4 Draught Animal Power (DAP)

The draught power of an animal depends on the species, breed, size, body weight and the intensity of feeding. The energy output is also influenced by climate and terrain, training and management factors, nature of work and finally, but decisively, the implement (agriculture tools) efficiency itself. Majority of draught animals (DAs) produce between 0.3 to 0.6 kW on a sustained basis. Assuming a conservative average of 0.4 kW per DA, the power output for an estimated 300 million DAs in the world may be in the order of 120 million kW. Replacement of a majority of these animals by tractors appears unfeasible in the foreseeable future in most low-income countries.

The DAP system can fulfill its potential role only if conscious attempts are made to upgrade it. The following procedures would certainly enhance DAP's contribution to the rising demand for energy in farming:

(a) increase the number of days of utilization of DAs,
(b) better food and health care for animals should mean increased output,
(c) better implements and carts enable better utilization of DA energy potentials,
(d) better recovery and use of by-products during the animal working life and after death, will improve the economics of the DAP and
(e) breeding schemes can improve on motive power and efficiency of the DAP system.

(i) *Draught Animal Pumping*

The gross energy expenditure of a working bullock can be calculated as the energy consumed daily as food by the animal, minus the energy in its dung. This method gives a daily gross energy input for a bullock as about 97,200 kJ. If this energy is supplied to the animal as cereal, it will need approximately 6.7 kg/day. Useful work production of a bullock is estimated at 7,920 kJ/day, which puts the efficiency of an animal at around 8 per cent.

(ii) *Draught Animal Transport*

Freight in most regions of the developing countries is often small in weight, if not in volume, and the produce in the rural areas rarely exceed 750 kg (an average discerned from a survey of Indian cart operations). Animal drawn vehicles are economical for loads ranging from half to two tonnes over distances up to 40 km in the course of a single day. Trucks require at least five tonne loads over distances of 100 km or more for 250 to 300 days of the year, if operations are to break even. In contrast, animals work only for 50 to 100 days in conjunction with carts. In the carrying of sugar-cane, cotton, tobacco, wood for fuel, hay, bran for dry feed, fertiliser, and manure in the rural areas, animal drawn vehicles enjoy a distinct advantage over trucks. They are thus extensively used for intra-village and inter-village transport. The animals are used for ploughing, tillage, carting and a variety of farm operations so that the farmer is able to spread the cost of maintenance over a number of operations. Cart designs are highly commodity-specific. Large-volume and low-density loads call for a different kind of platform structure than concentrated commodities like firewood, coal, brick, etc. In order to improve on the efficiency of animal drawn vehicles for rural transportation, following measures should be taken:

- improvement of vehicle design (pneumatic tyres, bearings, axles, etc.),
- improvement of harnessing of animals (in present designs the animal not only hauls the combined weight of cart and freight, but also carries an appreciable weight of its own body),
- improvement of animal breeding (this measure eventually will lead to increased and more efficient motive power).

The proper audit of existing energy sources and the pattern of energy end use by different activity sectors of regional economy is essential for framing the future plan for energy strategy for alternative resources.

1.3.0 ENERGY PLANNING

Currently, energy planning in India is not an integrated activity. Since there are many energy sources and end uses, many organisations and agencies deal with different aspects of energy. The plans for electricity, coal, oil and fuel wood are done by respective organisations mainly based on the projection of energy demand. The primary goal of this approach is to go in for energy supply expansions on the assumption that there is a correlation between energy use and gross domestic product. With this approach energy becomes an end in itself, and the focus shifts on meeting increased energy consumption through energy supply expansion alone. This supply and demand based planning approach for each individual energy form has resulted in problems like more losses, more conversions and low efficiencies. This is evident from the disappearance of forests, village wood lots, road side trees, construction of giant hydroelectric dams, fossil fuel based power plants and controversial nuclear plants. This conflict between the energy demand and environmental quality goals can be solved by having an integrated approach to the problem of energy planning, with a view to minimise consumption of non-renewable resources of energy and maximise efficiency of energy usage and harnessing of renewable sources of energy in an ecologically sound way. Another aspect that has to be considered in the planning process is that of matching energy resources and end uses. Because of convenience, current usage of high quality energy, such as, electricity for low quality activities like bath water heating is to be discouraged. Hence, strategies for integrated energy planning should include (a) improvements in efficiencies of end use devices and/or conversion equipments, (b) optimising energy source - end use matching, (c) organised approach towards optimal use of renewable resources, (d) proper exploitation of biomass energy resources, and (e) discourage use of depletable resources (by penalising).

Ecologically sound development of the region is possible when energy needs are integrated with environmental concerns at the local and global levels, for which an integrated planning framework would be necessary. The central theme of the integrated energy plan would be to prepare and implement the area based decentralised energy plans for meeting energy needs for subsistence, development of alternate energy sources (renewable, non-renewable) at the least cost to the economy and environment, and linking the micro level plans with national economic planning and development programs, including those for energy, agriculture and rural development sectors.

In this study, a district is taken as the unit of energy planning since (i) district is the basic administrative unit for implementing all developmental programs and (ii) district level planning will facilitate integration between national planning exercises and planning at lower levels.

1.4.0 OBJECTIVE

The main objective of this study is to develop an "Ecologically Sound Integrated Energy Plan" at district level to arrive at an optimal mix of energy sources that can meet the energy demand in a region by keeping in view the development priorities of the region.

1.4.1 Justification for Selection of Uttara Kannada District

Uttara Kannada district in Western Ghats region is chosen for carrying out this study. Western Ghats provides a variety of ecological niches and resources, but is under threat of deterioration due to short term perspectives and narrow sectoral approaches in planning process through indiscriminate pursuit of developmental activities. In Uttara Kannada district, various developmental projects involving intensification of resource use have led to large-scale soil erosion, siltation, salination, lowering of ground water table, gradual depletion of fertility of agricultural fields, extensive deforestation and overgrazing. These problems are compounded with increasing human population in this region. This necessitates prevention and control of soil erosion and integrated watershed management through promotion of conservation activities of local communities and the application of environmentally sound technologies. The procurement of energy is also responsible in varying degrees for much of the ongoing deforestation, loss of vegetation and top soil. While energy availability is a determining factor in agricultural productivity, the use of agriculture residues for energy purposes leads to soil impoverishment. The current inefficient use of energy in various sectors is also responsible for the detrimental effects on the environment. Therefore, sound policy and management decision should involve three components - energy, economics and environment.

In this context, micro level planning exercise was carried out in mid 1980's by Centre for Ecological Sciences, I.I.Sc, India successfully with the involvement of local people and nongovernmental organisations (NGO's) from a cluster of villages at Sirsi and Kumta taluks of Uttara Kannada district. This involved (1) survey and inventorying of the status of resources including land, water, crops, livestock, etc., (2) assessment of problems associated with degradation in the light of socio-economic structure, (3) identification of causes leading to degradation and (4) identification of economically viable and socially acceptable technologies, such as, dissemination of fuel efficient stoves, biogas, etc. This has resulted in providing important information such as - strategies for successful dissemination of environmentally sound devices, factors that play dominant role in success of devices, etc. With this background and experience, an attempt is made to develop integrated energy plan for Uttara Kannada district, taking into account present state of resources.

1.4.2 Energy Plan

The energy plan consists of the following components:

(i) Sectorwise energy demand: This involves the survey of the present energy consumption in different sectors for various end uses covering the type, magnitude and composition of fuel, trends, seasonal constraints and preferences in consumption; and estimation of energy demand based on the sample survey data.

(ii) Assessment of the supply situation: This involves analysis of present energy supply system; assessment of woody and other bioresources; assessment of renewable sources potential, such as, solar, wind and hydro; and study of supply system of commercial energy.

(iii) Development of an energy plan for the district: Based on the estimated supply and demand, an energy plan to meet the energy demand would be worked out in accordance with the development priorities. Techno-economic analyses of various energy technologies would be carried out to find out the technical and economic viability of the system. An energy plan at district level would be proposed based on the Decision Support System approach. Analyses of the importance of community participation in energy conservation, planning and identification of measures that will enhance the level of participation.

(iv) Implementation and management: With the knowledge of administrative structure at district level and agencies implementing the energy development program, a suitable institutional structure would be suggested for implementing and managing the energy plan.

These components were studied over a period of 24 months - collection of secondary data (6 months) and primary data (18 months).

(a) *Collection of Secondary Data*

Involves collection of data from various government agencies pertaining to demographic, geographic, social, economic and ecological parameters - which have direct or indirect influence on energy supply and consumption in a region. This enables to construct the profile of a region including a broad picture of the energy system. This also helps to find talukwise land use pattern, climatological parameters, etc. In this connection, extensive data collection from various organisations at national, State, district, taluk, village and block level has to be undertaken. Nature of data and source are:

Villagewise demographic, geographic, occupational and infrastructural information (Directorate of Census Operation, Bangalore, Tahsildar's Office at each taluk),
Topographical information (Survey of India Toposheets 1:50000),
Talukwise demographic, geographic, main crops, etc. (The statistical outline of Uttara Kannada, District Statistical officer, Karwar),
Climatological parameters (India Meteorological Department at Pune and Bangalore),
Taluk level agriculture, horticulture information - area under each crop, yearwise production (Agriculture and Horticulture Department in each taluk),
Livestock information (Veterinary Department at each taluk),
Electricity consumption - sectorwise (Electricity Boards at each taluk),
Supply of kerosene (Food Department, Tahsildar's office),
Forest vegetation - type, fuel wood, minor forest produce, timber yield details (Forest Departments - Sirsi, Honnavar and Bangalore),
Renewable energy activities (Directorate of Economics and Statistics, Bangalore),
Landholding details (respective village Accountant's Office in Kumta taluk and Sirsi taluk).

This detailed data collection and compilation enables a detailed district profile. Taluk and village level information helps in selection of taluks and villages for conducting primary survey for demand estimate and assessment of energy supply situation. Householdwise landholding information collected from village accountant's office helps in selection of households for energy demand survey.

(b) *Literature* Survey

Extensive literature survey was carried out at libraries of various institutions and internet facilities using file transfer protocol (FTP), Gopher, Lynx (text browser), etc. Libraries having wealth of information in this area are in: Centre for Ecological Sciences, Karnataka State Council for Science and Technology, Centre for Renewable Energy Development Agency, Application of Science and Technology to Rural Areas, Department of Management Studies, Main Library (J.R.D. Tata Memorial Library) at the Indian Institute of Science, Indian Institute of Management (Bangalore), Institute for Social and Economic Change, Karnataka Forest Department, University of Agricultural Sciences, Jawaharlal Nehru Centre for Advanced Scientific Research (Bangalore), Tata Energy Research Institute (TERI) at Delhi, Directorate of Economics and Statistics Library, National Science Information Centre, INSDOC (Bangalore) and Raman Research Institute (Bangalore).

(c) *Primary Survey*

The primary survey consists of a series of surveys in various sectors to examine different components of the energy system

- survey for demand estimation in household, agriculture, industry and commercial sectors,
- survey for assessment of bioresource availability for energy purposes,
- survey for assessing commercial fuel supply,

- hydrological data collection at regular intervals in sample streams to assess hydro potential in Bedthi and Aghnashini river catchments,
- wind data collection at regular intervals in Sirsi and Kumta taluks.

Methods adopted and details about each of these surveys are explained in subsequent chapters.

In data collection from the domestic sector (in sampled villages), stratified random sampling methods are adopted based on community and landholding of a household. Data collection is designed so as to account for regional and seasonal variation in availability and demand of energy, fuel purchased and fuel gathered free of cost from nearby forests and grazing lands. Emphasis is placed on consistent measurement of consumption of different types of traditional fuels, estimation of moisture content, accounting for different types of fuels (twigs and agriculture residues) separately, classification of fuel consumption based on their end uses, efficiency and type of different end use devices.

(d) *Techno-Economic Analysis*

Technical and economic details of various renewable technologies (suitable to Indian condition), and costs are to be compiled from various industries, organisations and nodal agencies such as M/s Jyothi Industries, M/s Bharath Heavy Electricals Limited, M/s Tata BP Solar India Limited, M/s National Hydro Energy Centre (Engineering and Construction section), Delhi Energy Development Agency, Tamilnadu Energy Development Agency (TEDA), Tamilnadu Electricity Board (TNEB), Gujarath Energy Development Agency (GEDA), Karnataka Electricity Board (KEB), Tata Energy Research Institute (TERI), Centre for Renewable Energy Development Agency (CREDA), Karnataka State Council for Science and Technology (KSCST), Karnataka Power Corporation (KPC), Indian Renewable Energy Development Agency (IREDA) and Department of Non conventional Energy Sources (DNES). These information enable a detailed techno-economic analysis for various energy technologies for different end uses. Detailed techno-economic analyses carried out for alternatives are discussed in the respective chapters.

(e) *Energy Plan*

Towards the goal of implementing analytical methods for integrated energy planning, computerised Decision Support System (DSS) would be used. This provides useful assistance in the analyses of available information, optimal allocation of resources for various end uses, the projection of future scenarios and the evaluation of alternative scenarios.

1.5.0 STUDY AREA: UTTARA KANNADA DISTRICT

The district is located in the mid-western part of the state. It lies between 74° 9' to 75° 10' east longitude and 13° 55' to 15° 31' north latitude and extends over an area of 10,291 sq.kms. (Figure 1.1.1) which is 5.37 per cent of the total area of the state and ranks 10th among the districts. It extends to about 328 kms north-south and about 160 kms east-west. Most parts of the district are hilly and thickly forested. The coast stretches long in a near straight line to the south-east except at the shallow Karwar and Belekeri bays. The district, a long and narrow strip of territory, is surrounded by Belgaum district and Goa territory in the north, Dharwad district in the east, Dakshina Kannada and Shimoga districts in the south and the Arabian sea in the west. There is an extent of only 10 km of plain land in the south and it is here that the southern boundary of this district and the northern boundary of the Dakshina Kannada district meet.

1.5.1 Area and Population

The area of the district was 1,029,171 hectares and the population 12,43,526 in 1991 according to a census report. Supa is the biggest taluk with an area of 1,89,514.23 hectares, while Bhatkal is the smallest with an area of 35,731.96 hectares. For administrative purposes, the district has been divided into 11 taluks. There are 13 towns and 1,283 inhabited villages and 55 uninhabited villages in the district. Table 1.1 lists talukwise geographical area, population of the district since 1951 up to 1991 (10 years interval), and percentage increase in population at 10 years interval and 40 years interval. Talukwise forest cover, livestock population, human density (number of persons/hectare) and livestock density (livestock/hect) computed are listed in Table 1.2. It is seen that coastal taluks such as Bhatkal (3.61 person/hect), Kumta (2.29), Karwar (1.95) and Honnavar (1.88) have dense population compared to other taluks such as Mundgod (1.17), Sirsi (1.15) and Siddapur (1.10). Thickly forested taluks like Yellapur (0.63) and Supa (0.53) have low population densities. Coastal taluks have livestock density of 0.78 (Bhatkal), 0.57 (Honnavar), 0.45 (Karwar) and 0.40 (Kumta). This is depicted in Figure 1.1.2. Among 11 taluks in terms of human population, Sirsi has a share of 12.3% (1,52,935 persons) followed by Honnavar (11.7%), Karwar (11.5%) and Kumta (10.8%). Mundgod with 77,939 persons constitutes least population. Figure 1.1.3 illustrates these facts. Percentage increase in population in last four decades is presented in Figure 1.1.4.

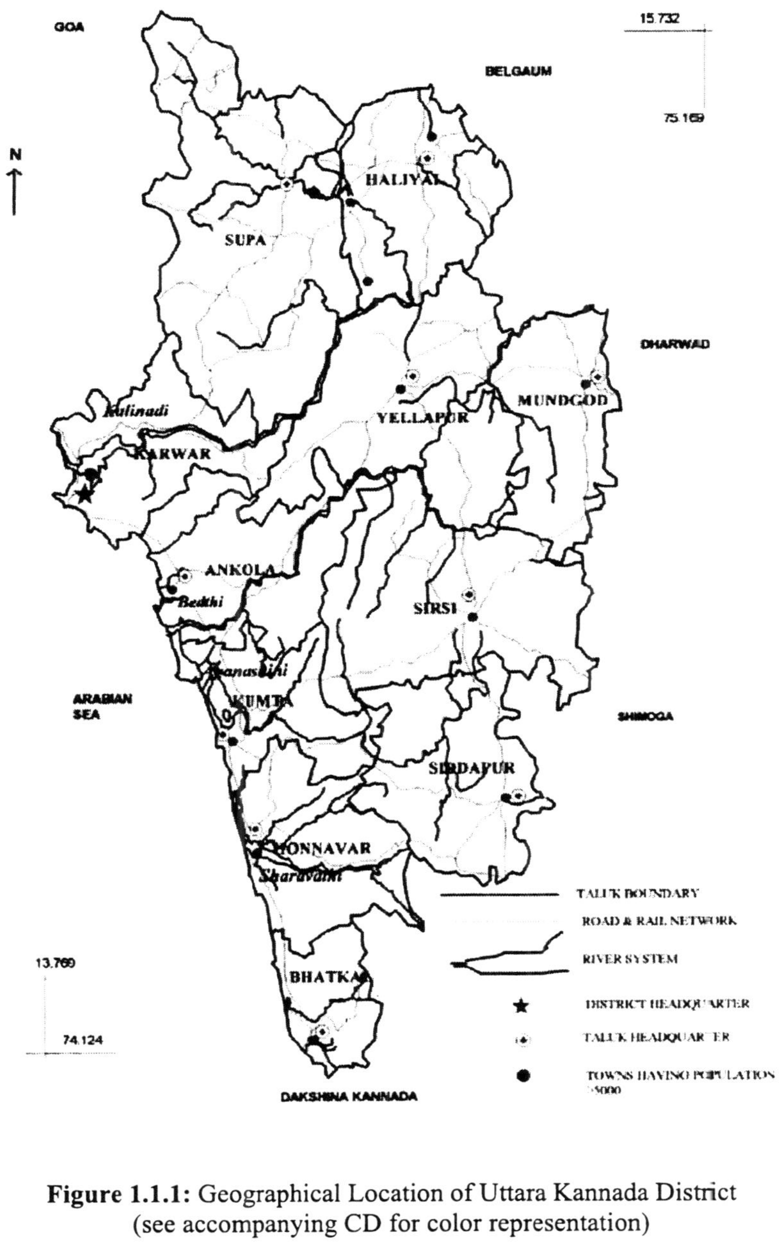

Figure 1.1.1: Geographical Location of Uttara Kannada District (see accompanying CD for color representation)

During last two decades, due to minor irrigation schemes, area under rice has increased, and there is a significant increase in human population.

Extent of forest cover in Yellapur is 87.18%, Supa 84.75% while Kumta has the lowest (67%). Talukwise extent of forest cover is illustrated in Figure 1.2. and Figure 1.3 depicts percentage forest cover in each taluk. Total area under forests is about 800044 hectares, which constitutes about 77.74% of the total geographic area. Supa with 20.1% and Yellapur taluk with 14.2% of total district forest area constitute good forest cover. Bhatkal with 3.2% and Kumta with 4.9% of total forest area have least share of forests in the district. Talukwise share of area under forests is depicted in Figure 1.4.

Table 1.1: Percentage Population Increase (Talukwise) in Uttara Kannada District (1951 to 1991)

	Taluk	Area	Total Population					% Population Increase				
		(in hectares)	1951	1961	1971	1981	1991	1951-1961	1961-1971	1971-1981	1981-1991	1951-1991
1	Ankola	91870.00	41848	57020	67871	81057	91310	36.26	19.03	19.43	12.65	118.19
2	Bhatkal	35731.96	38513	58501	71656	96896	129017	51.90	22.49	35.22	33.15	235.00
3	Haliyal	86489.33	25314	62666	85107	90792	94363	147.55	35.81	6.68	3.93	272.77
4	Honnavar	77546.13	64776	89928	106537	127988	145842	38.83	18.47	20.13	13.95	125.15
5	Karwar	73210.00	51510	91641	104477	125635	142845	77.91	14.01	20.25	13.70	177.32
6	Kumta	58486.81	52577	87556	100743	117325	134144	66.53	15.06	16.46	14.34	155.14
7	Mundgod	66890.00	12335	26814	44464	58745	77939	117.38	65.82	32.12	32.67	531.85
8	Siddapur	85930.00	35702	61561	73802	84525	94202	72.43	19.88	14.53	11.45	163.86
9	Sirsi	133342.62	42170	88579	114230	135107	152935	110.05	28.96	18.28	13.20	262.66
10	Supa	189514.23	17991	34385	35762	52215	99519	91.12	4.00	46.01	90.59	453.16
11	Yellapur	130110.00	15419	30898	44456	59989	81410	100.39	43.88	34.94	35.71	427.98
	District	1029121.08	398155	689549	849105	1030274	1243526	73.19	23.14	21.34	20.70	212.32

Table 1.2: Geographical area, Area under Forest, Population Density, Livestock Density - Talukwise-Uttara Kannada District

	Taluk	Area	Population	% incr	Forest	Forest		Pop/Hect.
		(in hectares)	1991	1951-1991	Area	Area (%)	LS/Hect.	
1	Ankola	91870.00	91310	118.19	75432.4	82.11	0.38	0.99
2	Bhatkal	35731.96	129017	235.00	25390.5	71.06	0.78	3.61
3	Haliyal	86489.33	94363	272.77	59099.8	68.33	0.64	1.09
4	Honnavar	77546.13	145842	125.15	57416.7	74.04	0.57	1.88
5	Karwar	73210.00	142845	177.32	52808.7	72.13	0.40	1.95
6	Kumta	58486.81	134144	155.14	39187.6	67.00	0.45	2.29
7	Mundgod	66890.00	77939	531.85	51167.4	76.49	0.87	1.17
8	Siddapur	85930.00	94202	163.86	62274.6	72.47	1.28	1.10
9	Sirsi	133342.62	152935	262.66	103227	77.41	0.64	1.15
10	Supa	189514.23	99519	453.16	160609	84.75	0.21	0.53
11	Yellapur	130110.00	81410	427.98	113430	87.18	0.40	0.63
	District	1029121.08	1243526	212.32	800043.7	77.74	0.55	1.21

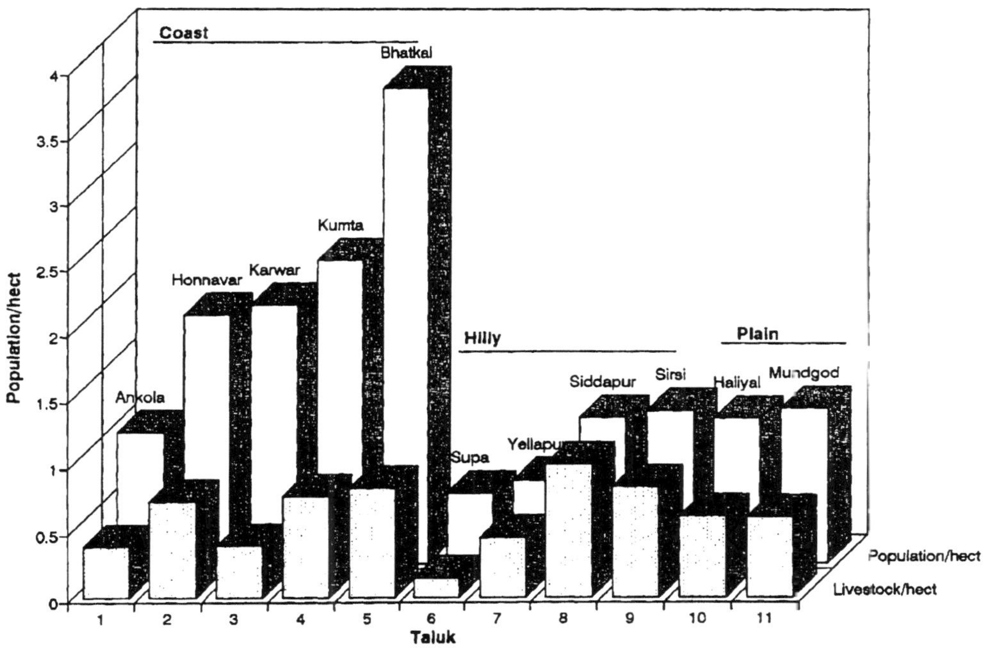

Figure 1.1.2: Population and Livestock Density, Uttara Kannada District

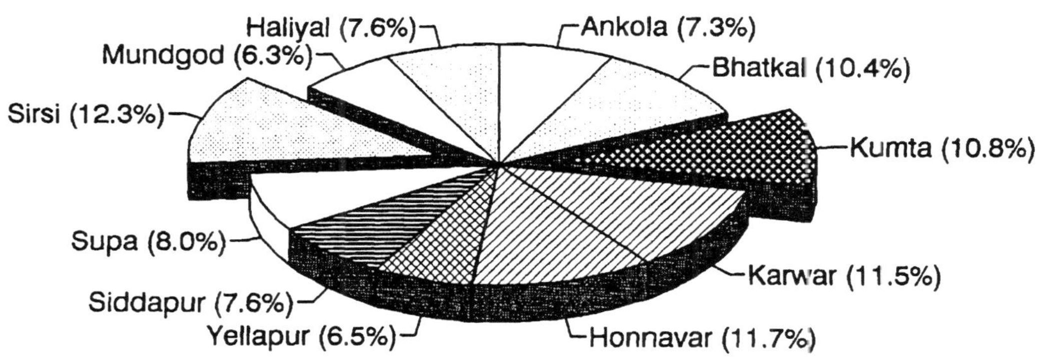

Figure 1.1.3: Talukwise Population Share – Uttara Kannada District

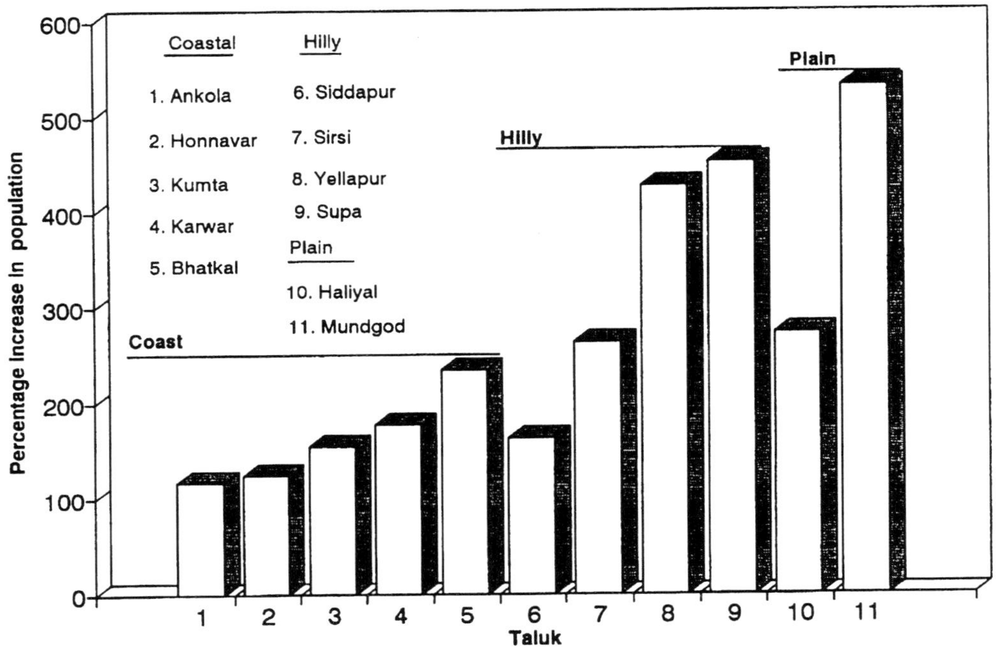

Figure 1.1.4: Population Increase (%) – (1951-1991) Talukwise

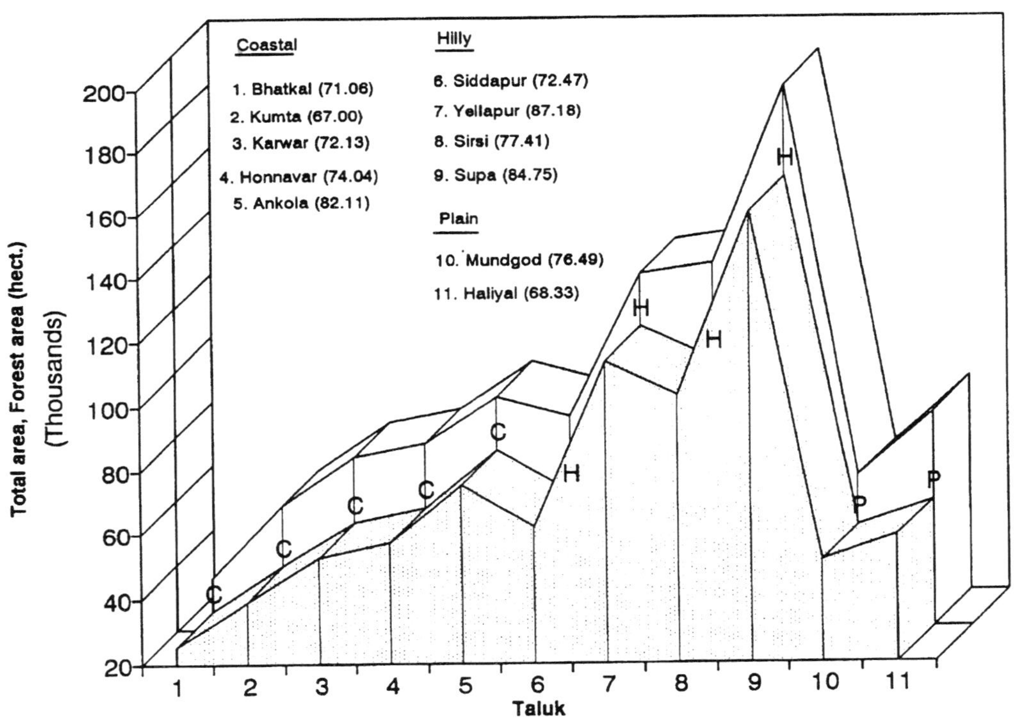

Figure 1.2: Forest and Total area in a Taluk - Uttara Kannada District

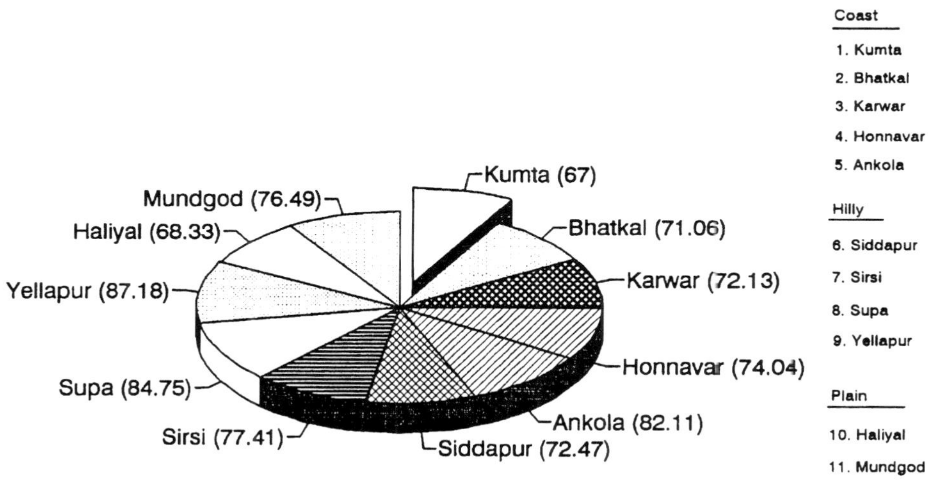

Figure 1.3: Talukwise Percentage of Forest - Uttara Kannada District

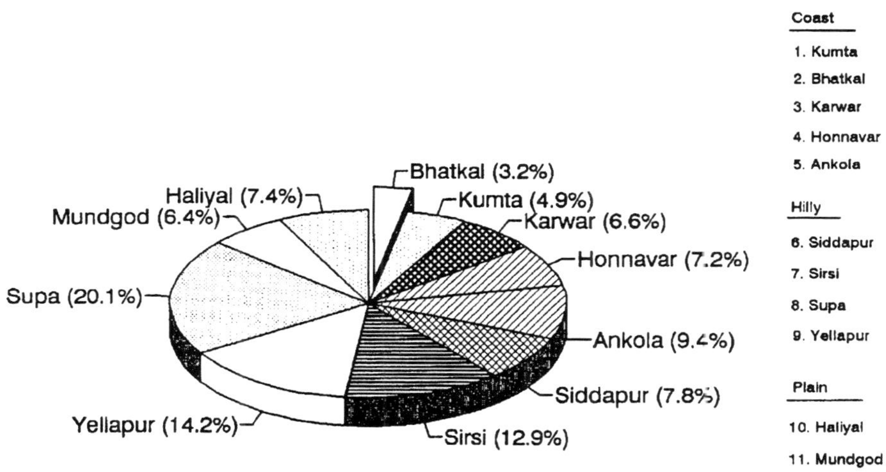

Figure 1.4: Taluk's Forest Cover Share – Uttara Kannada District

1.5.2 Physiography

The district is hilly and thickly wooded in most parts. Its major part is essentially highland, the lowland being restricted to pockets along the courses of rivers. A seemingly broken and irregular Sahyadri range of central hills, with an average height of 700 meters, divides the district into two parts, the uplands or the regions above the Ghat with an area of nearly 7,770 sq.kms (600 to 700 meters above sea level) and the low lands covering about 3,370 sq. kms. Except the shallow Karwar and Belekeri bays in the north, the 144 km of the Uttara Kannada coastal stretch is nearly a long straight line to the south east. The district has three main and distinctive regions - the coastlands, the Sahyadrian interior and the eastern margin where the tableland begins. Relief and climate have introduced these regional differences. The coast lands are the best developed areas with a high degree of economic development and a high density of population. It is in this region that the taluks Karwar, Ankola, Kumta, Honnavar and Bhatkal are situated. The Sahyadrian region is mostly forested, and the roads crossing the Ghats sustain human activity, though the valleys have special significance as belts of spice and areca gardens for which the district has been famous since antiquity. The eastern margin is an undulating land, partly under forest and partly cleared up for agriculture. It is a transitional zone between the forests proper and the cultivated uplands of the Dharwad district. Parts of Haliyal, Yellapur and Mundgod taluks are plain country, comparable with the western half of Dharwad district.

The Sahyadrian Tract: Beyond the low level plateaus of the coastal regions, lies the Malenadu (hill country) of North Karnataka. This is physically an extension of the Sahyadrian main range from Maharashtra in the north, and continues to the south into the districts of Shimoga and Hassan. From the Chandgad border in the north to the border of Siddapur taluk in the south, it has a length of about 110 km. As in the Maharashtra region, the Sahyadris rise in a series of steps from the coastal lowlands, but the scarp face is not so bold as in the districts of Kolhapur and Ratnagiri. This geological composition gives a greater variety and a definite 'break' which accounts for a more favourable rainfall in the Dharwad district. The 'rainshadow' area is not so sharp and immediate in northern Karnataka as it is in Maharashtra. Another interesting feature is the eastward shift of the main watershed: the Kali and the Sharavati rivers drain a large portion of the plateau area and through their captured courses divert these waters to the Arabian Sea. Faulted topography is typical of the western Sahyadris. This has given rise to waterfalls and rapids of which the more noteworthy are the Jog - already harnessed for electricity, and Unchalli (Lushington), Magod, Burde Jog and Lalguli, these and other waterfalls are the potential sources of power. The central portion of the Sahyadris mainly consists of the crest line ridges and upper reaches of rivers developing more mature valley forms before they plunge into deeper chasms and gorges to join the coastal levels. The eastern margin has a rounded topography and broader valley features which mark the transition between the main Sahyadrian landscape and the drier plateau of the upghat regions. This three fold division of the Sahyadrian region persists in its features of vegetation and is well reflected in its regional economy. In the western spurs with low-level laterite plateaus, the red soils favour stunted vegetation, but in the higher terraces, vegetation improves in both density and quality to an astonishing degree. Here, one comes across the best strands of teak trees, and stately patches of tall evergreen trees in hollows wherever there is copious under-ground water supply to supplement the seasonal downpour. In the central belt, there is a marked contrast between the hill tops and the backbone of ridges that support a thinner vegetation cover and the river valleys where luxuriant deciduous vegetation is to be found. Evergreen strands are a frequent feature, but the river banks almost invariably have bamboo thickets. In the eastern part of the Sahyadrian region, increasing dryness is the key feature of the landscape. It is a landscape with poorer types of forest having a parkland scenery: these are, in the forester's terminology, the 'teak pole' forests. Valleys open out towards the plateau margin of Uttara Kannada district. Increasing cultivation and receding forest cover is a recurring pattern in the landscape.

Hills: The district is almost covered with hills that may be arranged into three groups, viz., the flat-topped blocks of laterite from 60 to 90 meters high which roughen the coast belt; the westerly spurs from the central hills from 300 to 600 meters high stretch rugged and woody to the coast; and the main range and eastern spurs of the central hills. The west face of the Sahyadris in Uttara Kannada unlike the Sahyadris in Konkan does not rise in a single scarp, but is approached by numerous spurs and lower ridges. It is not much lower as it averages about 600 meters and in some places reaches 900 meters, but it is no longer the even wall-like crest of trap, unbroken by a river channel. In the district, granite has taken the place of trap, and through the rugged granite cliffs, large inland rivers force their way to the sea.

Locally, the hills in the district are considered a break between two main ranges, the Sahyadris to the north that end at the Kalinadi near Karwar and the Malabar Hills or Malaya Parvata that stretches south from the Sharavati river. Of the eleven peaks in the Sahyadris of the district, varying in height from 458 to 915 meters, the Gudehalli and Shirvegudda are in Karwar taluk, the Bhedasgaon in Mundgod, the Menshigudda in Sirsi, the Hukali, Rakshasa and Mavinagudda in Siddapur, and the Motigudda, Kaltigudda, Darshanigudda and Nishanigudda are respectively in Ankola, Kumta, Supa and Yellapur taluks. The highest of these hills, the Darshanigudda in Supa, about 915 meters above the sea, rises near the meeting of the boundaries of Goa and Belgaum with Uttara Kannada. The Gudehalli in Karwar, 549 meters above the sea and the Kaltigudda in Kumta, about 763 meters high, are health resorts; and the Nishanigudda a small peak in Yellapur, 458 meters above the sea and Bhedasgaon in Mundgod about 765 metres above the sea are Trigonometrical Survey Stations.

Rivers: The larger rivers in the district, unlike the Konkan rivers, drain a large area of the uplands, east of the Sahyadri scarp. There are four leading rivers, Kalinadi in the north, Bedthi or Gangavali about 32 km south, Aghnashini or Tadri rising far to the south but falling into the sea only about six miles south of Gangavali, and Sharavati or Gerusoppa river, about 24 km south of the Tadri, where it becomes a tidal creek on reaching the foot hills. Each of these rivers takes a second name from the chief town on its banks. Thus the Kalinadi (Karihóle) is locally known also as the Sadashivagad river, Bedthi as Gangavali, Aghnashini as Tadri, and Sharavati as Gerusoppa. In the hills, the channels of all the rivers are broad and rocky, showing the force of their monsoon torrents. At the foot of the hills, they are broad back-waters, the mouths stopped by bars of sand, which during heavy rains block the passage of the flood waters till they overflow the lowlands along their banks.

Soils: The various soils of the district are derived from laterites, granites and gneisses shales and traps. Patches of coastal alluvium are found in parts of the coastal region. Two clear-cut soil zones can be differentiated based on elevation, namely, the coastal taluks and the upghats. Soils in Kumta, Honnavar, Bhatkal and Siddapur are derived from laterites while some soils derived from granites also occur in Ankola and Karwar. Red loamy soils derived from mixed laterite and traps are found in parts of Supa, Haliyal and Mundgod taluks. In other parts of Mundgod, Haliyal and Supa taluks, patches of black soil are derived from trap rocks. There are also Gazni lands which are saline sandy soils along the coast. The types of soils may be derived into five groups.

1.5.3 Cropping Pattern

The total cropped area of the district is estimated at 12 per cent of the total area as against 55 per cent of the State. The percentage of area sown more than once is about 16 per cent of the net area sown as compared to 58 per cent of the State. The most important aspect of the district's cropping pattern is the predominance of food grains, particularly paddy. Its share in the cropped area is slightly less than 72 per cent. Uttara Kannada is the fourth largest paddy growing district in the State, the first three in order being Dakshina Kannada, Shimoga and Dharwad. The district accounts for about 8 per cent of the total area under paddy in the State. There is some diversion of land from this crop to high value crops like sugarcane and groundnut. Next to paddy, plantation crops like arecanut, cashew, coconut, pepper and cardamom occupy an important place. The area brought under all these crops has shown an upward trend during recent years. The cultivation of arecanut is confined to the taluks of Sirsi, Siddapur, Honnavar, Kumta and Yellapur. In the coastal belt, coconut is the important cash crop. However, the area under coconut and arecanut remained almost constant. With a steady increase in the price of groundnut, it has emerged as an important crop. Area under the groundnut crop has substantially increased by 68 per cent, while under sugarcane it has shown only a marginal increase of 3 per cent. Of the total foodgrains production, rice alone accounts for about 91 per cent. Though the district is self-sufficient in respect of paddy, it is deficit in pulses, ragi and coconut.

Manure: The manures that are in common use in the district are farmyard manure, green manure, compost and fertilisers. Farmyard manure is conserved by sectional filling method or by dumping it in one place till it is carried to the fields. Green manure is added by bringing green leaves from soppinabettas and from forests. Green leaves are procured from these areas in the evening and are spread over in the cattle shed and lifted to compost pits after 3-4 days. In addition, raising of green manure crops like sunhemp, kolinji and glyricidia is also taken up.

Paddy: The staple produce of the district is rice which on some lands is grown as a later/cold weather and an early/rain crop. Paddy is grown all over the district, the earliest grown area being near Karwar; in the rest of the low land coast, harvest is done a little later, then comes the upland crop, and last of all the eastern areas. The paddy lands in the coast are divided into gazni, honda, aremakki and makki lands. Gazni lands are along the coast, yielding one crop in the year. Honda lands are the good rice plots in the lower portions or valleys that are being watered by small streams and yielding every year two crops of rice or one of rice and another of pulse. Khar lands are the low fields along the rivers and salt water inlets flooded during the height of the rains so that paddy cannot be sown till the water recedes. Makki lands are higher lands entirely depending on monsoon rains and apt to loose the crop if the later rains fail. Above the Sahyadris, most rice plots are located in the valleys on the eastern flank of the Sahyadris. From there onwards, the rice lands stretch east a little beyond the boundary of the low woodlands as far as the heavy rain reaches, which supply many small reservoirs with enough water to last till January or February. All the rice fields are in the form of terraces surrounded by small banks to pond the water when the fields are flooded. The terraces vary from 1/4 hectare to a patch of 1/20th of a hectare according to the steepness of the ground. Coconut palms are sometimes grown along the bunds of the fields, their thick matted roots forming a valuable support to the embankments.

There are three types of paddy cultivation in the district, viz., drilling in the eastern parts of Sirsi, Siddapur and in Haliyal and Mundgod taluks, broadcasting in the coastal belt where the soil is sandy and also in salt land paddy areas in the taluks below the ghats, and transplanting in the rest of the tract. In few places, direct broadcasting with germinated seeds is done. Soon after the harvest of paddy, the fields are ploughed and harrowed, and planks are passed for levelling. When pre-monsoon showers are received, preparatory operations are completed including application of organic manures like Farm Yard Manure (FYM) and compost. In a few places, green manuring is practised and also dry and green leaves are incorporated into the soil. In Haliyal and Mundgod drill sowing is done in dry condition. Sowing commences during the second fortnight of May and continues up to June. Transplanting commences from June and continues up to the middle of August. In upghat taluks of Haliyal, Mundgod, parts of Sirsi and Yellapur, "hodata" operation is done. In this operation, when the crop is 2 to 2 1/2 months old, under wet conditions, wooden planks are passed over the crop. This helps to injure the crop (stem region) and stimulates tillering in paddy seedlings. In upghat, threshing is done by passing bullocks over the paddy straw, but in coastal taluks only manual labour is used for threshing. Power tillers and tractors have also come into use in the district. Whenever ample irrigation facilities are available, second summer crop is being raised. Soon after the harvest of kharif paddy, the lands are prepared and transplanting is done in the main fields. Short duration and mid-late varieties are planted.

Groundnut: Groundnut (*Arachis hypogea*) is grown on a small-scale in the transitional belt of the district in the kharif season. But in recent years, the area under it is on the increase during kharif season. It is also grown on uplands and makki lands where water does not stagnate. It has been introduced as a second crop on paddy fallow in the coastal taluks.

1.5.4 Type of Forests

Evergreen, semi-evergreen, moist deciduous, scrub and thorny vegetation constitute the different types of forest in Uttara Kannada district. Area under forests as per official records is about 77.74% of the total geographical area of the district.

The evergreen forests are found in places where the rainfall is more than 2250 mm, and form a narrow strip along the Western Ghats. It is also found in places where rainfall is between 1500 to 2500 mm. Moist deciduous forests are in the region where rainfall is about 1000 mm and more. Deciduous forests are found in areas where rainfall is below 900 mm. The evergreen and semi-evergreen patch consists of soft wood species (used in plywood industries, packing industries, etc.). Timber species such as rose wood, matti (*Terminalia crenulata*), honne (*Pterocarpus marsupium*), nandi (*Lagerstroemia microcarpa*), etc., are found in the deciduous patch.

1.5.5 Climate

The climate is determined largely by topography, geography (with respect to the sea) and the monsoon winds. Western Ghats acts as a divide between the west coast and the adjoining hilly regions with heavy rainfall, and the areas of dry low rainfall to the east. The elevations vary from sea level to about 1800 meters, and therefore the climate is characterised by high humidities in the coastal strip and the Western Ghats region throughout the year. The drier climate prevails in the area east of the ghats.

The season of high temperature occurs during April - May (summer), overcast skies and heavy rainfall in the coastal strip during June to September (South-West monsoon), retreating monsoon during October and November (post monsoon) and dry spell with clear sky and low humidity and temperature during December to February period. The mean annual rainfall is highest in the coastal taluks. Bhatkal among coastal taluks receives maximum rainfall of 3131.59 to 3941.55 mm annually. Hilly taluks namely Siddapur, Sirsi, Yellapur and Supa follow the coastal taluks with mean annual rainfall in the range 2470.11 to 2996.53 mm. Haliyal and Mundgod receive minimum rainfall of the order 1174.39 to 1234.58 mm. On an average, there are 103 rainy days (that is days with rainfall greater than 2.5 mm) in a year in the district. The number of rainy days varies from 87 days (Mundgod) to 114 days (Bhatkal).

1.6.0 Power Generation in Uttara Kannada District

Among the four major rivers, hydro potential of Kali and Sharavati has been tapped already for power generation. Nagajari Power house and Supa dam (of Kali river) has 810 MW and 100 MW hydro plants respectively. Work has been completed on the 120 MW hydro plant (at Kodasalli) and 150 MW (at Kadra) of Kali river. Now there is a proposal to tap hydropower of the Bedthi and Aghnashini rivers. Apart from these, a Nuclear power plant is being set up at Kaiga, near Karwar. Two units of 235 MW each are expected to be in operation by mid 2000.

1.7.0 Industries in Uttara Kannada District

Industrial sector accounts for 83.85% of the total electricity consumption in the district. There are seven large and medium scale industries. The prominent ones are West Coast Paper Mills Ltd., Indian Plywood Manufacturing Co. Ltd., Dandeli Ferro Alloys Pvt. Ltd., Jai Hind Saw Mill (Dandeli), Ballarpur Industries Ltd (Binaga, Karwar), Karnataka State Vaneers Ltd (Kavanchur), and Green Dehydration and Pelletisation (Siddapur). Apart from these, there are 1575 small-scale industries, 145 handloom industries, 9200 cottage industries and 275 agro based industries, 245 forest and plantation based industries and 82 mineral based industries in the district. Major mineral resources are manganese ore (at Haliyal), iron ore (at Supa), lime stone (at Yellapur), silica and moulding sand (at Karwar, Kumta, Bhatkal, Honnavar and Supa), lime shell (at Karwar, Kumta, Honnavar) and quartz (at Haliyal and Honnavar).

1.8.0 Energy Consumption

The domestic sector is the major energy consuming sector and is mainly dependent on fuel wood and other bioresources. Cooking, water heating and space heating (during heavy rainfall season) are the end use operations dependent on fuel wood. In the hilly regions, the fuel wood requirement is met mainly from forests and betta lands. In bioresource scarce coastal taluks, the forest department has made arrangement to supply fuel wood through the fuel wood depot's, and agricultural and horticultural residues. Quantity of fuel wood provided is 15 kgs/person/month (against ration card) at the rate of Rs. 25.50 per quintal.

The energy problem in this region is mainly related to non-availability of bioresources in some coastal taluks. Basic needs emphasise integrated planning, hence a balanced attention to all energy sources, communities and localities. This necessitates a need to decrease the demand for energy through efficiency improvement of end use devices in household, agriculture, and industrial sectors, and augmenting production of bioresources through energy plantations, agro forestry, etc. Tapping local renewable resources, such as, solar, wind and small hydel plants helps in decentralised energy generation, which in turn helps in setting up employment generating industries in the region (such as agro processing, etc.). In this regard, large-scale integrated studies of alternative approaches and inventive schemes can be valuable tools in enhancing energy conservation and use of renewable energy technologies.

This book presents a conceptual design for energy system which could meet demand of all sectors in the region. The proposed design in principle, supplies enough energy for the sustainable development of a region. Plan proposed based on Decision Support System approach (DSS) is flexible, adaptable and ecologically sound. Design of hydroelectric systems of sites at Magod (Bedthi river) and Unchalli sites (Aghnashini river) are done within social and ecological restraints of the region. Apart from these, feasible sites have been identified for setting up mini, micro and small hydel plants in Bedthi and Aghnashini river catchments. Minimum submergence of land ensures that the fuel wood requirement of the region is met, while generating electricity.

1.9.0 Organisation of the Book

Chapter 1 discusses the theme of Energy, components of energy system and the broad outline of a regional energy plan.

Chapters 2 to 6 discuss the energy resources availability in the district and detail the techno-economic analyses of wind energy, bioenergy, hydel energy and solar energy systems.

Chapters 7 to 9 assess sectorwise energy demand of various resources in the district.

Chapter 10 presents a conceptual design for the energy plan for Uttara Kannada district which could meet demand of all sectors in the region. The proposed design in principle could supply sufficient energy for the sustainable development of the region. Linear programming formulation is carried out for optimal allocation of resources for various tasks based on minimisation of costs and maximisation of efficiency. Integrated Regional Energy Plan proposed for Uttara Kannada district, Karnataka State based on Decision Support System approach (DSS) is flexible, adaptable and ecologically sound.

Chapter 2

POTENTIAL ASSESSMENT OF WIND ENERGY IN UTTARA KANNADA DISTRICT

2.0 SUMMARY

Availability of wind energy and its characteristics has been studied at Kumta and Sirsi based on primary data collected at these sites for 24 months. Wind regime at Karwar (for 1952-89), Honnavar (1939-89) and Shirali (1974-89) has also been analysed based on data collected from the respective meteorological observatories of the India Meteorological Department (IMD). Wind energy conversion systems would be most effective in these taluks during May to August. The monthly frequency distributions of wind speed have been analysed for Kumta and Sirsi, where hourly wind speed recording is available. It is shown that two-parameter Weibull distribution is a good representation of the probability density function for the wind speed. Energy Pattern Factor (EPF) and Power Densities are computed for sites at Kumta and Sirsi. With the knowledge of EPF and mean wind speed, mean power density is computed for Karwar, Honnavar and Shirali. The analyses show that coastal taluks Karwar and Kumta have good wind potential, which if exploited would help local industries, and coconut and areca plantations. Pre-monsoon availability of wind energy would help in irrigating these orchards, making wind energy a desirable alternative.

Keywords: Wind Energy, Energy Pattern Factor, Wind Farms, Weibull Distribution.

2.1.0 INTRODUCTION

Windmills have been used for centuries to grind grain and pump water in rural areas. Winds are caused by rotation of the earth and heating of the atmosphere by the sun. The total annual kinetic energy of air movement in the atmosphere is estimated to be about 3 x 10^{15} kWh or about 0.2% of the solar energy reaching the earth. The maximum technically usable potential is theoretically estimated to be 30 trillion kWh per year, or about 35% of current world total energy consumption (Wilbur, 1985). The power in the wind blowing at 25.6 km/h is about 200 W/m^2 of the area swept by the windmill. Approximately 35% of this power can be captured by the windmill and converted to electricity. However, it is important to note that the power output from the windmill varies with the cube of the wind speed. Consequently, only windy locations on mountains and coasts are suitable for the economic generation of electricity by wind power. Harnessing of wind energy could play a significant role in the energy mix of a region. Wind energy is renewable and environmentally benign. It has the advantage of being harnessed locally for applications in rural and remote areas. Pumping of water for agriculture and plantations is probably the most important application that contributes to rural development through multiple cropping. Wind driven electric generators could be utilised as an independent power source, and for purposes of augmenting the electricity supply from grids. In densely populated coastal taluks Karwar and Kumta, decentralised production of electricity would help local industries, especially seasonal agro-processing industries like cashew, etc.

The extent to which wind can be exploited as a source of energy depends on the probability density of occurrence of different speeds. To optimise the design of a wind energy device, data on speed range over which the device must operate to maximise energy extractions are required, which requires the knowledge of frequency distribution of the wind speed. Data on mean monthly and annual wind speeds for a long time (30 - 50 years) are available at meteorological observatories and the data on frequency distribution is available from sites at Sirsi and Kumta, where primary data has been collected for the last 24 months. The wind measurements were made using cup anemometer with mechanical counter fixed to a guyed mast at a height of 5 m at Sirsi and Kumta.

2.1.1 Wind Energy Resource Parameters

Wind cannot be transported and, therefore, wind turbines must be located where the wind resource is present. The energy content of the wind, being related to the cube of the wind speed, varies significantly with only small changes in wind speed. This fact demands the importance of having accurate wind speed data when the wind energy resource is being evaluated. Since wind speed is a continuously varying parameter, it is customary to average the wind speeds during each hour and use the hourly mean wind speed as the basic parameter in calculations of wind power.

2.1.2 Mean Wind Speed and Energy Resource

The annual wind speed at a location is useful as an initial indicator of the value of the wind resource. The relationships between annual mean wind speed and potential value of the wind energy resource are listed below.

Annual mean wind speed @ 10 m Height	Indicated value of wind resource
< 4.5 m/s	Poor
4.5 - 5.4 m/s	Marginal
5.4 - 6.7 m/s	Good to Very Good
> 6.7 m/s	Exceptional

In locations where data are not available, a qualitative indication of a high annual mean wind speed can be inferred from geographical location, topographical features, wind-induced soil erosion, and deformation of vegetation. However, accurate determination of the mean annual wind speed requires anemometer data for at least 12 months. Earlier studies have revealed that the mean wind speed for a given year varies from year to year but within $\pm 10\%$ of the long-term mean. A $\pm 10\%$ variation in mean wind speed would show a variation in energy content of $\pm 30\%$ or more.

The power P due to the kinetic energy of wind is proportional to (1/2)*[Mass x $(\text{Velocity})^2$], that is, the mass of air passing through an area A at velocity V. The mass of air passing through an area A per unit time is ρAV and the total power available from wind is

$$P = [(1/2)*(\rho AV)*V2] = [(1/2)*(\rho AV^3)] \quad (1)$$

where ρ is the mass density, ρAV air mass flow rate, and V the airstream velocity.

2.1.3 Adjustment of Heights of Anemometer

Anemometers at different meteorological stations are set at different levels. The measurements and wind speed recorded at each station, prior to any analysis, had to be adjusted to the same height. The standard height according to the World Meteorological Organization is 10 meters above the ground level (WMO, 1964). This height is adopted in the analyses. The horizontal component of the wind velocity can vary a great deal with height under the influence of frictional and impact forces on the ground. The most common model for the variation of horizontal velocity with height (Lysen, 1983), is given by the logarithmic wind profile equation as

$$(V1/V2) = (H1/H2)^{\alpha} \quad (2)$$

where V1 is wind speed at height H1 of 10m above ground level, V2 wind speed at height H2 above ground level, and α the roughness factor, determined by substituting the wind speed data obtained with anemometer height in various wind directions, found to be 0.30.

2.1.4 Wind Speed Frequency Distribution

Because the energy in the wind is proportional to the cube of wind speed, wind speeds above the annual mean will have a disproportionate amount of wind energy. As a result, two locations with identical mean wind speeds that have differing wind speed distributions could have quite different levels of wind energy resources. Furthermore, the frequency of occurrence of various wind speeds, including the extremes, must be known to determine how well the operating range of a particular wind turbine under consideration matches the wind speed distribution. The wind speed frequency distribution at a given location is either tabulated from wind speed data measured as a function of time or approximated by a probability distribution function based on measured data or assumed wind resources characteristics. In recent years, much effort has been made to construct an adequate statistical model for describing the wind frequency distribution. Earlier studies have shown that the Weibull distribution gives a good fit to the experimental wind speed data (Lysen, 1983; Justus, 1975). The probability density function F(v) for wind data is given by

$$F(v) = (k/c)*(V/c)^{k-1}*\text{Exp}[-(V/c)^{k}] \quad (3)$$

The cumulative distribution function (CDF) is

$$M(v) = 1- \text{Exp}[-(V/c)^{k}] \quad (4)$$

where k is a dimensionless shape factor, c scale factor with units of speed, and F(v) the probability or percentage of occurrence per unit speed of the wind speed V. The first and second moments of probability density function (PDF) in terms of c and k are

$$\bar{x}_v = c\,\Gamma(1+1/k) \quad (5)$$

$$\sigma_v^2 = c^2\,[\Gamma(1+2/k) - \Gamma^2(1+1/k)] \quad (6)$$

where $\bar{x}_v$ is the mean wind speed, σ_v^2 variance and Γ the gamma function.

There are several methods for determining the Weibull distribution parameters, such as (a) method of moment, (b) method of using energy pattern factor, (c) method of maximum likelihood and (d) method of least squares fit of the cumulative probabilities (Stevens and Smulders, 1980). The method of least squares fit of the cumulative probabilities is employed in the analyses of Sirsi and Kumta data, where hourly wind speed is available.

The CDF of wind speed is $M(v) = 1\text{-Exp}[-(V/c)^{k}]$. By taking natural logarithm to linearise the CDF equation,

$$\ln\{-\ln[1-M(v)]\} = K \ln V - K \ln c, \quad (7)$$

which is of the form y = ax + b. By plotting different values of ln{-ln[1-M(v)]} vs ln V, a straight line is fitted to the points. The slope of the line is k, with the intercept on the ln{-ln[1-M(v)]} axis being [-k ln(c)], giving c.

2.1.5 Energy Pattern Factor (EPF)

The power density per unit area is given by,

$$P = K_{Em} * [1/2(\rho * Vm^3)] \text{ for the month and} \quad (8)$$

$$P = K_{Ea} * [1/2(\rho * Va^3)] \text{ for the year} \quad (9)$$

where ρ is the density, and K_{Em} and K_{Ea} the energy pattern factor [EPF] for month and year respectively (Mani and Mooley, 1983).

$$K_{Em} = \frac{\text{Total amount of power available in the wind}}{\text{Power calculated by cubing mean wind speed}}$$

$$= \frac{\text{Mean Power Density for the month}}{\text{Mean Power Density at monthly mean speed}}$$

$$= \frac{1/2 * \rho \sum V_i^3/Nm}{1/2 * \rho * Vm^3} = \frac{\sum V_i^3/Nm}{Vm^3} \quad (10)$$

where V_i is the hourly wind speed during the month, Nm the number of hourly wind speed values during the month and Vm the monthly mean wind speed = $\Sigma V_i/Nm$,
while the annual energy pattern factor is

$$K_{Ea} = \frac{\sum V_i^3/Na}{Va^3}$$

The Mean Power Density for the month = $K_{Em} * 1/2 * \rho * Vm^3$, and

for the year $= K_{Ea} * 1/2 * \rho * Va^3$

2.2.0 OBJECTIVE

The primary goal is to assess the potential of wind energy that can be exploited to meet the regional energy demand in a decentralised way. Out of 11 taluks in Uttara Kannada district, sites were chosen to monitor wind velocity at Sirsi and Kumta. Availability of infrastructure facilities like a research station with computers, etc., has also played a dominant role in site selection.

2.3.0 METHODS

The work carried out in Uttara Kannada district was based on:

(a) Data from meteorological observatories at Karwar (for 1952-1989), Honnavar (1939-1989) and Shirali (1974-1989) were obtained from India Meteorological Department (IMD), Government of India, Pune, and daily wind data for the period 1990-1994 for these observatories from IMD, Bangalore.

Cup counter anemometers with hemispherical cups measuring 7.62 cms in diameter were used in India Meteorological Department observatories until 1973. During 1973-1979, these anemometers were replaced with 3 cup anemometers with 127 mm diameter conical cups, in conformity with international practice. The wind instruments are at 10 m above ground over open terrain in all these observatories.

The primary data was obtained by installing cup counter anemometers with mechanical counter fixed on 5 meter tall guyed masts at Sirsi and Kumta. The anemometer readings were noted down every three hours during the day, at the synoptic hour observation from 05.30 to 20.30 hours. Sites for fixing masts were selected in flat open terrain where the safety of the instruments was assured. Table 2.1 gives the location details of these stations.

Table 2.1: List of Stations in Uttara Kannada District

Location	Latitude N	Longitude E	Elevation
Karwar	14° 47'	74° 08'	4
Honnavar	14° 17'	74° 27'	26
Shirali	14° 05'	74° 32'	45
Kumta	14° 26'	74° 25'	8
Sirsi	14° 62'	74° 85'	610

To get synoptic hour wind speed, readings were taken twice at 3 minute intervals, that is, after 3 and 6 minutes, in order to check the correctness of the readings recorded. At Sirsi initially, the anemometers were installed in an open play ground (Arts and Science College Campus, Sirsi) at two locations, where readings were taken for 36 months (Table 2.2.1 and Table 2.2.2), and on a hillock (Huliappa Gudda) at elevation 622 m (Table 2.3), where readings were recorded for 24 months.

Table 2.2.1: Mean Wind Speed (kmph) at Sirsi (at 610 m elevation)

Hours	0530	0830	1130	1430	1730	2030	w	w1	w2
Jan	3.6	8.1	12.2	11.4	7.7	4.5	7.7	11.8	5.3
Feb	2.6	7.6	10.4	10.2	9.8	4.3	7.3	11.3	4.8
Mar	2.1	5.4	8.8	11.6	11.8	6.3	7.3	10.6	5.3
Apr	3.5	7.0	9.5	12.1	13.2	9.4	8.6	11.0	7.2
May	3.8	9.6	11.2	13.4	12.8	9.6	10.0	13.1	8.2
June	5.8	11.7	13.5	14.3	11.9	7.9	10.7	14.2	8.6
July	6.3	9.6	13.5	14.3	12.8	8.7	11.9	15.2	10.
Aug	6.4	9.5	13.5	14.9	12.4	8.7	11.0	13.9	9.1
Sept	2.3	6.7	10.0	12.1	10.4	6.2	8.0	11.2	6.1
Oct	1.9	7.3	9.9	9.6	7.2	4.5	6.3	9.4	4.5
Nov	4.9	10.4	13.1	11.2	5.8	4.7	8.4	12.1	6.1
Dec	5.7	11.4	12.6	10.2	6.4	5.2	9.4	12.7	7.4
Annual	4.1	8.2	11.5	12.1	10.2	6.6	8.9	12.2	6.9

Table 2.2.2: Mean Wind Speed in (kmph) at Sirsi (at 622 m elevation)

Hours	0530	0830	1130	1430	1730	2030	w	W1	w2
Jan	5.3	8.0	13.0	11.0	7.0	4.9	7.9	11.6	5.6
Feb	3.2	7.7	12.2	10.2	10.4	5.3	7.6	11.8	5.0
Mar	3.0	5.8	8.0	11.5	12.4	6.2	7.2	10.1	5.5
Apr	4.3	7.9	10.2	11.4	13.2	9.3	8.7	11.1	7.3
May	5.1	9.3	12.0	13.3	12.6	9.6	10.5	13.3	8.8
June	6.6	11.0	12.4	14.1	11.8	8.2	10.4	13.6	8.7
July	6.6	10.5	13.2	16.4	13.7	10.8	11.8	14.7	9.9
Aug	7.0	11.0	15.6	16.8	13.7	9.0	11.9	15.4	9.8
Sept	2.0	6.0	8.8	10.1	9.88	5.7	7.2	10.4	5.3
Oct	1.2	6.0	8.6	9.6	9.1	5.1	5.5	8.3	3.9
Nov	6.5	13.0	15.6	13.9	5.6	5.9	9.8	14.1	7.2
Dec	6.2	11.4	12.5	8.6	7.14	5.2	9.3	11.8	7.6
Annual	4.8	9.0	11.8	12.2	10.4	7.1	9.0	12.2	7.1

Table 2.3: Mean Wind Speed (kmph) at Sirsi II (elevation 622 m, location Huliappa Gudda) and Kumta (elevation 8 m)

	Sirsi	Sirsi	Kumta	Kumta
Month	Avg.	Sd.	Avg.	Sd.
January	6.92	3.25	5.95	3.12
February	6.88	2.68	7.76	4.16
March	7.20	3.74	9.09	1.64
April	8.38	5.57	9.42	2.55
May	9.09	5.48	9.87	3.15
June	11.19	7.59	11.83	2.79
July	18.17	7.61	13.03	2.38
August	14.19	7.15	11.54	5.45
September	11.14	7.74	6.71	4.49
October	8.39	6.66	6.59	3.62
November	7.72	6.50	6.29	2.29
December	8.42	6.88	7.73	3.27

2.4.0 WIND DATA PRESENTATION, DISCUSSION AND ANALYSES

The annual and monthly means of wind speed (w) for 24 hours ending at 08.30 IST, the mean wind speed (w1) for the period 08.30 hours to 17.30 hours and the mean wind speed (w2) for the period 17.30 hours to 08.30 hours, as well as, the wind speeds at the six synoptic hours of observation are listed in Table 2.2.1 and Table 2.2.2 for Sirsi. Figure 2.1 is the pictorial representation of monthwise mean daily wind speed in kmph. It is noticed that at Sirsi wind speed is high from May to August. It ranges from 13.1 kmph during May and reaches a peak of 15.2 kmph in July. While at site II in Sirsi, it ranges from 10.5 in May and reaches a peak in July and August at 15.4 kmph. Figure 2.2.1 gives a diurnal variation of mean wind speed in a day. It is seen that the day is windy between 08.30 and 17.30 hours, the maximum occurring around 14.30 hours. Figures 2.2.2 and 2.2.3 give pictorially the variation in mean daily wind speed (monthwise) for Honnavar and Karwar. Figure 2.3 gives monthwise hourly wind speed which substantiates the earlier result. The May to August period is windy during the year and the variation is most pronounced during this period. The diurnal variation is least during the monsoon season for all sites studied.

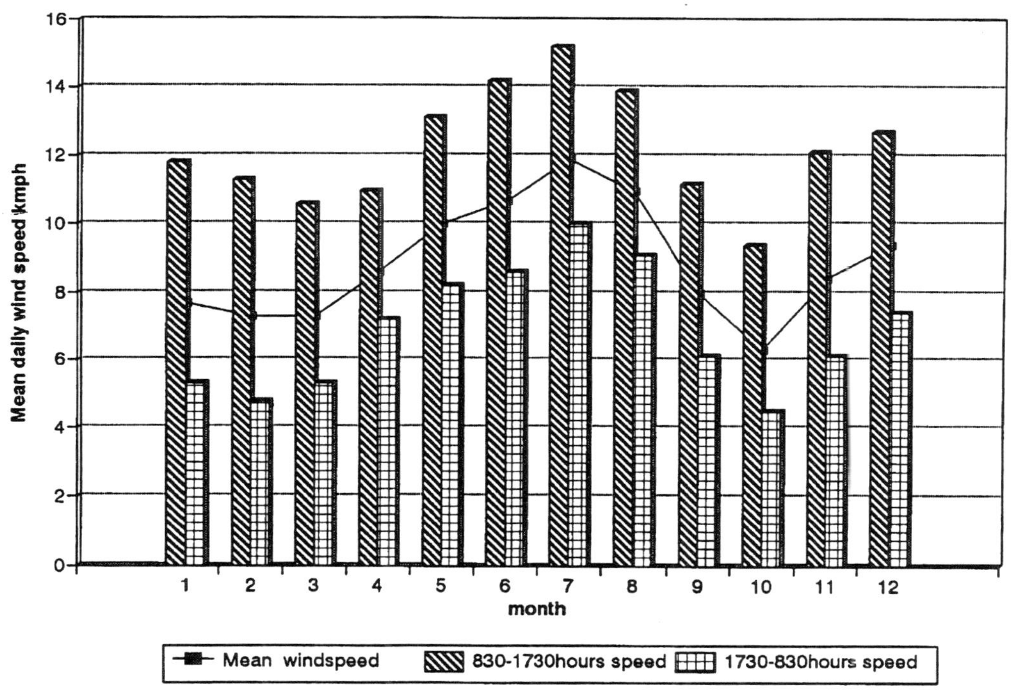

Figure 2.1: Mean Daily Wind Speed at Sirsi - Elevation 610m

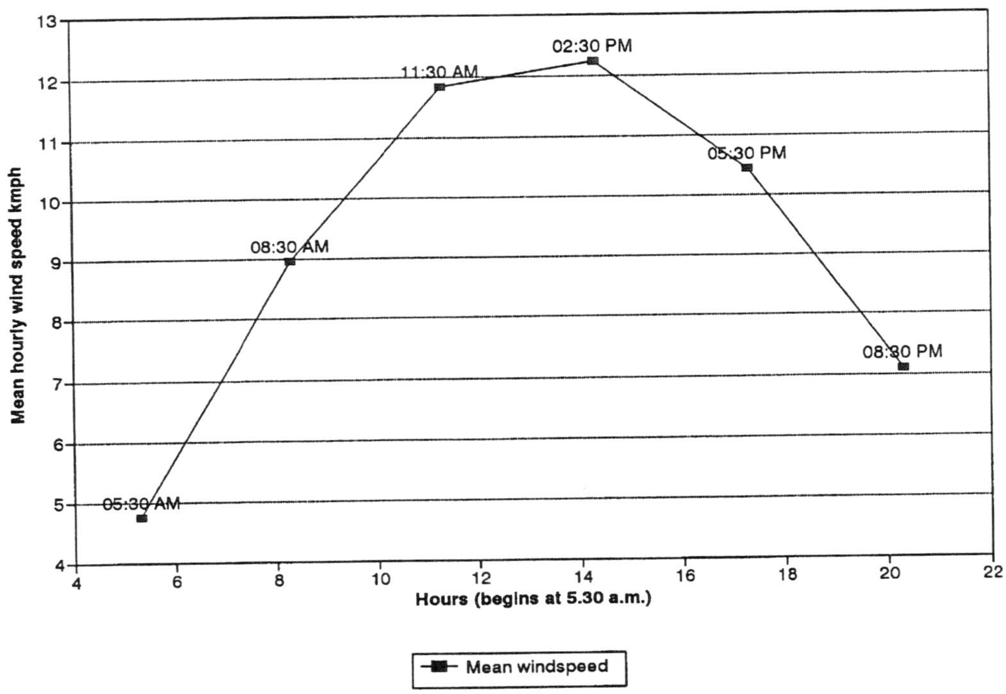

Figure 2.2.1: Mean Hourly Wind Speed (year) at Sirsi-Elevation 610m

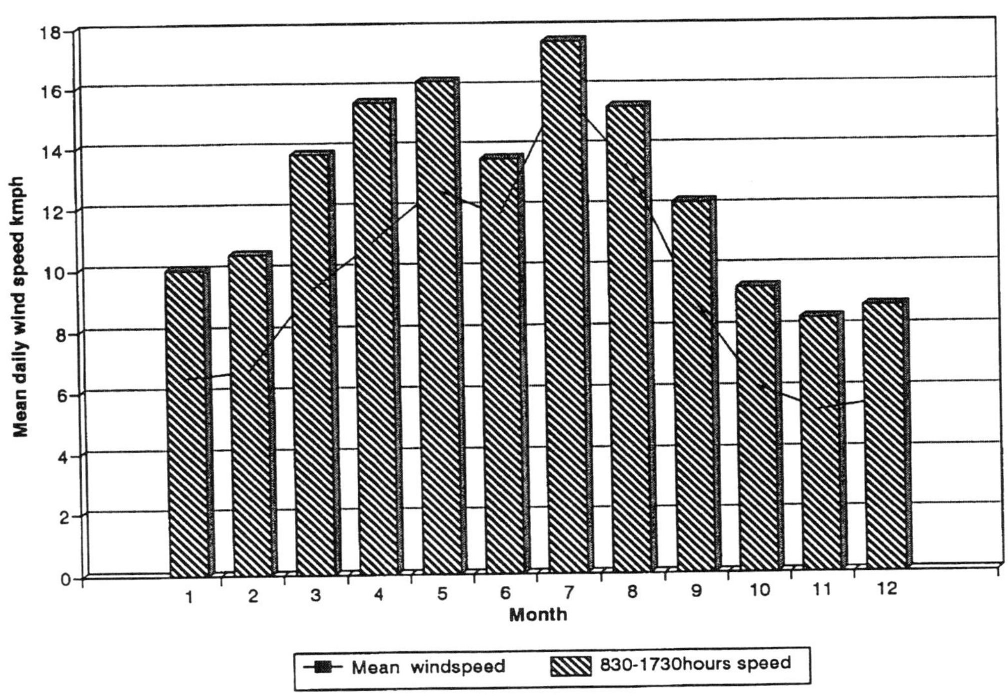

Figure 2.2.2: Mean Daily Wind Speed – Karwar

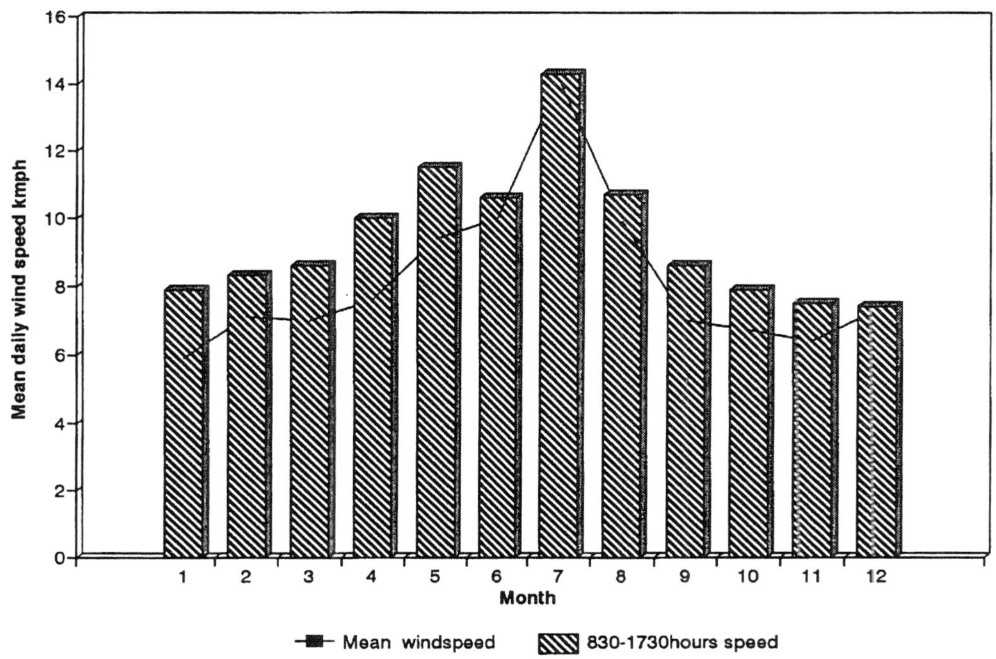

Figure 2.2.3: Mean Daily Wind Speed – Honnovar-Elev 26m

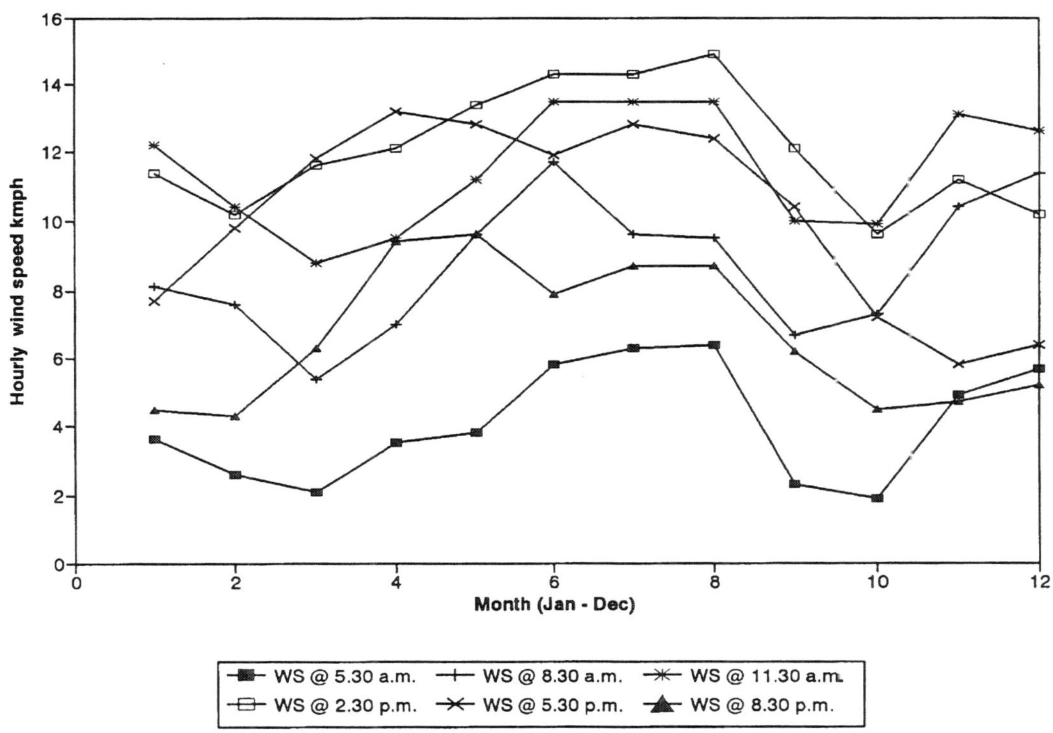

Fig 2.3: Hourly Wind Speed at Sirsi – Elevation 610m

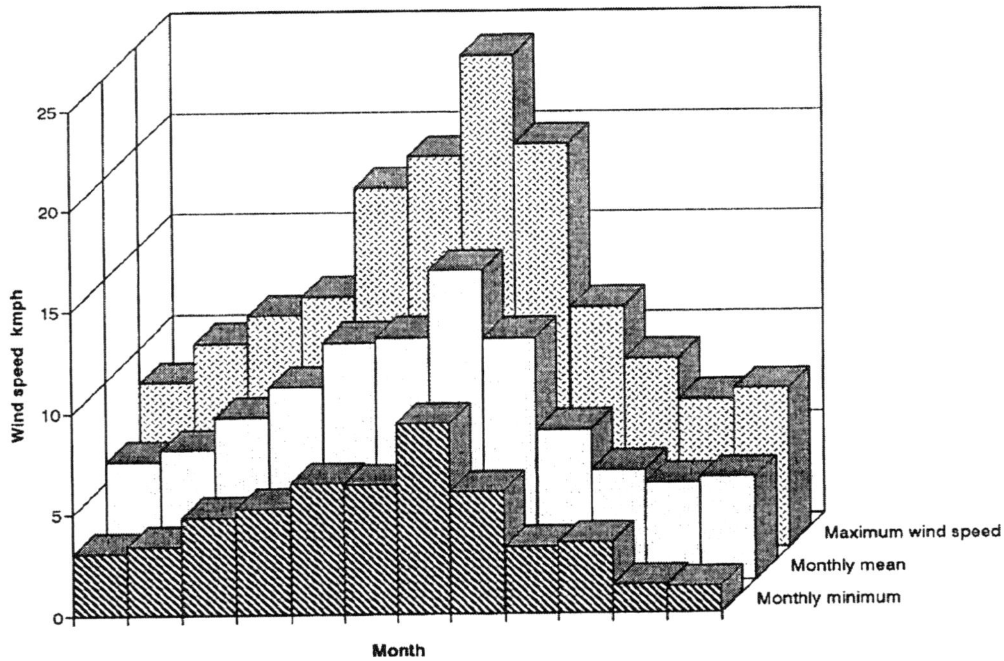

Figure 2.4: Monthly Wind Speed at Karwar

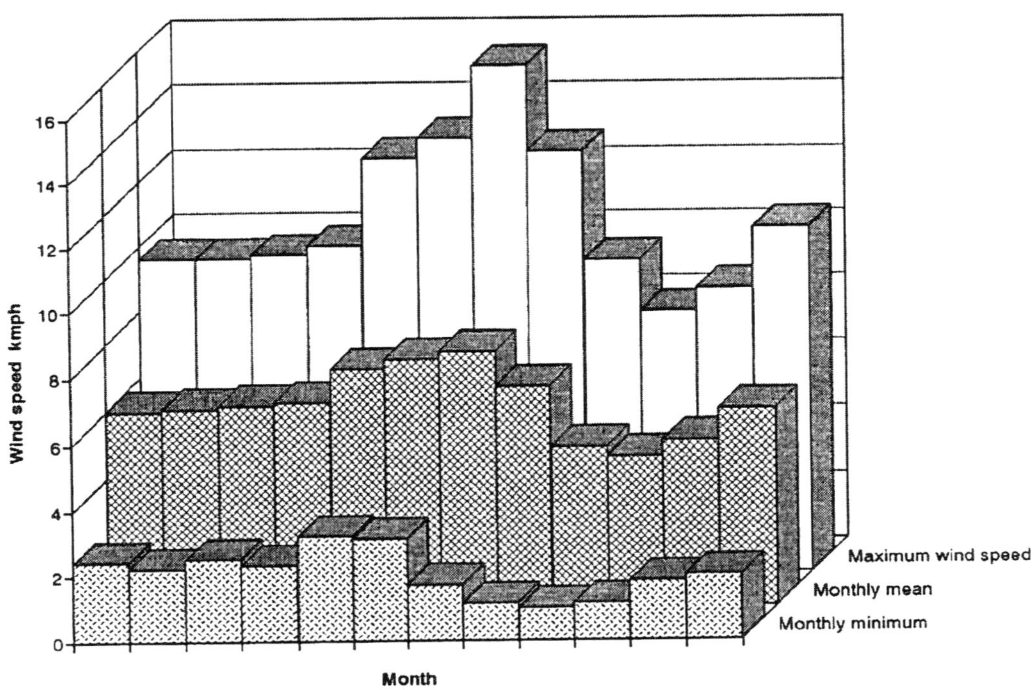

Figure 2.5: Monthly Wind Speed at Honnavar

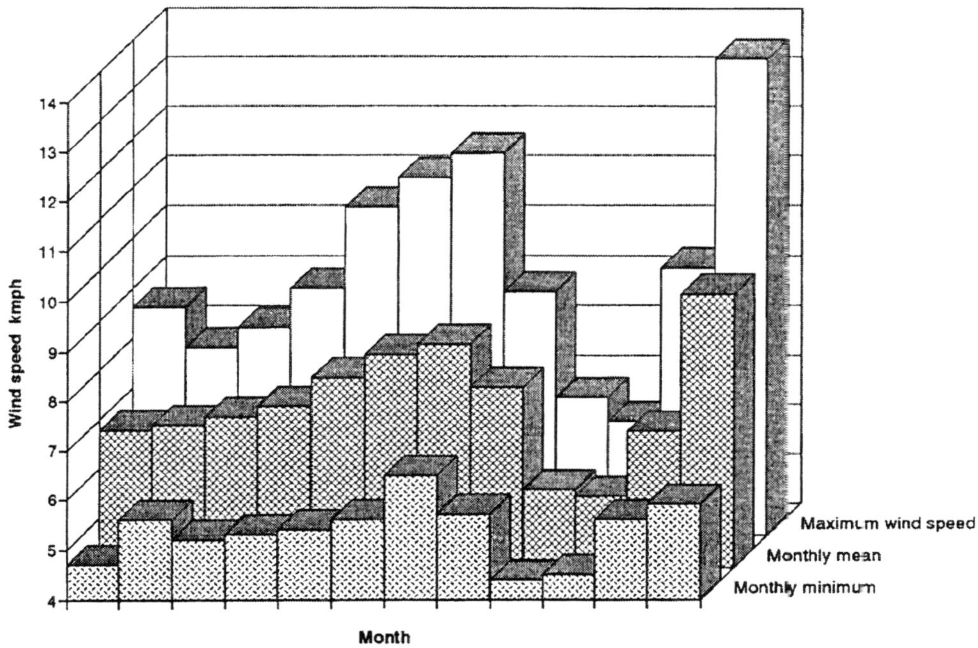

Figure 2.6: Monthly Wind Speed at Shirali, Uttara Kannada District

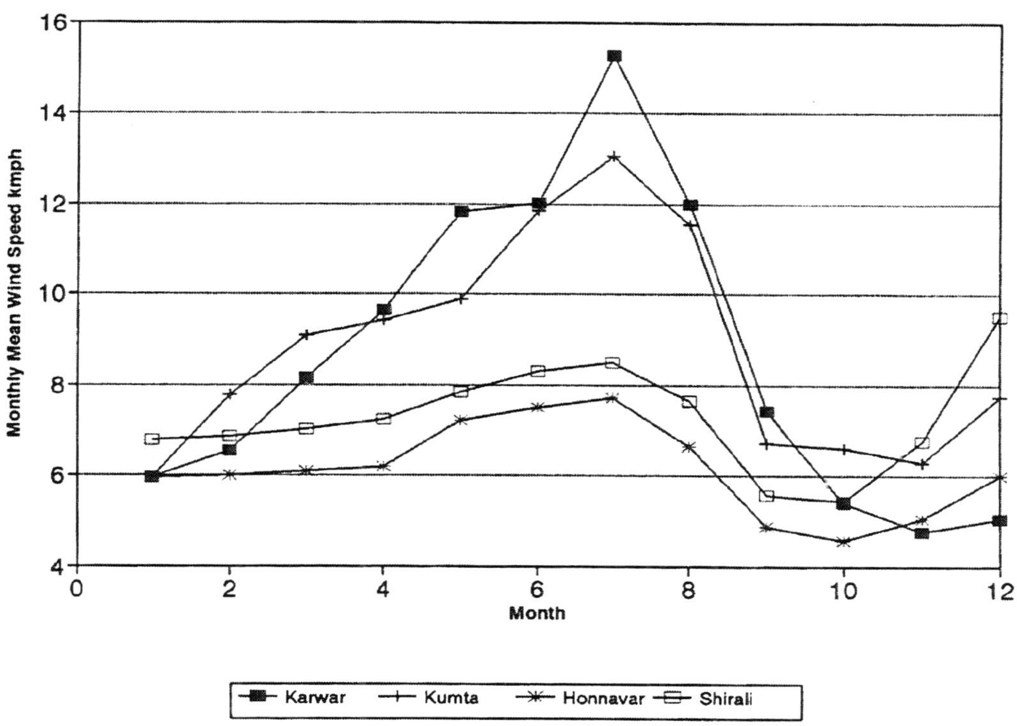

Figure 2.7: Monthly Mean Wind Speed – Karwar, Kumta, Honnavar and Shirali

Table 2.3 gives the monthly mean wind speed and standard deviation at sites located in Sirsi (622 m) and Kumta (8 m). At site II at Sirsi, the wind speed varies from a minimum of 6.88 ± 2.68 kmph (for February) to a maximum of 18.17 ± 7.61 kmph (for July). Table 2.4.1 gives the mean wind speed, standard deviation, maximum and minimum wind speed, calculated based on daily wind data at meteorological observatories at Karwar, Honnavar and Shirali for the period 1990-1994. Wind speed is maximum during July in all these stations. It ranges from 15.2 kmph for Karwar, 8.5 for Shirali and 7.7 kmph for Honnavar. The same is shown by bar graphs in Figures 2.4, 2.5 and 2.6 respectively.

Table 2.4.1: Monthly Mean Wind Speed, Sd, Max, Min
[computed for data of Meteorological Observations located at Karwar, Honnavar and Shirali based on data for 1990 – 1994]

Place: Karwar

	Avg	Sd	Max	Min
Jan	5.96	1.43	8.30	3.10
Feb	6.55	2.03	10.20	3.40
Mar	8.15	1.99	11.50	4.80
Apr	9.65	2.17	12.40	5.20
May	11.82	2.73	17.80	6.50
June	12.01	3.09	19.40	6.40
July	15.27	4.03	24.40	9.40
August	11.98	3.95	20.00	6.00
Sept	7.44	2.53	11.80	3.30
Oct	5.41	1.62	9.30	3.50
Nov	4.75	1.56	7.30	1.40
Dec	5.04	1.63	7.80	1.30

Place: Honnavar

	Avg	Sd	Max	Min
Jan	5.95	1.93	9.60	2.40
Feb	6.00	2.04	9.60	2.20
Mar	6.10	1.98	9.70	2.50
Apr	6.20	2.03	10.00	2.30
May	7.21	2.39	12.70	3.20
June	7.50	2.64	13.30	3.10
July	7.72	3.30	15.50	1.70
August	6.66	2.81	12.90	1.10
Sept	4.87	2.05	9.50	1.00
Oct	4.55	2.02	7.90	1.10
Nov	5.04	1.79	8.60	1.80
Dec	6.00	2.23	10.50	2.00

Table 2.4.1: Monthly Mean Wind Speed, Sd, Max, Min [computed for data of Meteorological Observations located at Karwar Honnavar and Shirali based on data for 1990 – 1994] (Cont'd)

Place: Shirali

	Avg	Sd	Max	Min
Jan	6.78	1.10	8.60	4.70
Feb	6.87	0.72	7.80	5.60
Mar	7.03	0.83	8.20	5.20
Apr	7.25	1.06	9.00	5.30
May	7.84	1.59	10.60	5.40
June	8.30	1.66	11.20	5.60
July	8.50	1.78	11.70	6.50
August	7.64	0.90	8.90	5.70
Sept	5.56	0.67	6.80	4.40
Oct	5.42	0.50	6.30	4.50
Nov	6.76	1.07	9.40	5.60
Dec	9.51	2.35	13.60	5.90

Table 2.4.2 gives the monthly mean wind speed for 08.30-17.30 hours and 17.30-08.30 hours at Honnavar based on data for 1939-1989 and Table 2.4.3 gives details for Karwar based on data for 1952-1989. Variation in monthly wind speed is quite evident in the graph shown in Figure 2.7. Among the coastal taluks, Karwar seems to have better wind regime, followed by Kumta and Shirali.

Table 2.4.2: Mean Wind Speed at Honnavar, Elev. 26 MSL, based on data for 1939-1989

			Mean Wind Speed (kmph)	
Hours	8:30 a.m.	5:30 p.m	Bet 0830 - 1730 Hours	Bet 1730 - 2030 Hours
Jan	4.8	6.3	5.9	7.9
Feb	4.5	6.7	7.1	8.3
March	3.8	7.7	7	8.6
April	3.2	7.8	7.6	10
May	4.2	9.2	9.4	11.5
June	4.5	7.2	10	10.6
July	6.8	9.1	14.1	14.3
August	4.9	6.7	9.9	10.7
Sept	3	6	7	8.6
Oct	3.2	5.7	6.7	7.9
Nov	4.2	4.8	6.4	7.5
Dec	5	5.2	7.3	7.4
Annual	4.34	6.87	8.20	9.44

Table 2.4.3: Mean Wind Speed at Karwar based on data for 1952 - 1989

			Mean Wind Speed	
	8:30 a.m.	5:30 p.m.	Bet. 0830 – 1730 Hours	Bet. 1730 - 2030 Hours
Jan	4.4	8.9	6.5	10
Feb	3.5	11.9	6.7	10.5
March	3.9	15.8	9.3	13.8
April	3.9	16.7	10.9	15.5
May	6.9	14.4	12.6	16.2
June	8.6	13.1	11.7	13.6
July	13.3	15.9	15.9	17.5
August	11.1	14.6	13.4	15.3
Sept	5.3	12.1	9	12.1
Oct	4.3	8.7	6.1	9.3
Nov	5.1	6	5.3	8.3
Dec	5	6.8	5.6	8.7
Annual	6.28	12.08	9.42	12.5

The frequency distribution of wind speed for each month at Sirsi and Kumta is given in Tables 2.5.1 and 2.5.2 respectively. The speed range over which a windmill should be designed to operate and the maximum wind its structure has to withstand depends on frequency distribution of the wind speed. Attempts have been made to fit simple distributions to the observed frequency distributions. Analysis is carried out for the sites at Sirsi and Kumta where primary data has been collected for 24 months. Tables 2.5.1 and 2.5.2 list the number of occurrences and cumulative frequency distributions of wind speeds in the range 0-4, 4-8, 8-12, 12-16, 16-20, 20-24 and >24 kmph. It is seen that the percentage occurrence of wind in the class interval 0-8 kmph is more during January-April (>50%) and September-December (44 to 60%), while 8-12 kmph wind is predominant during May-August (25 to 40%). This is illustrated in Figures 2.8.1 (Sirsi) and 2.8.2 (Kumta). The observed cumulative distribution for January-June and July-December at stations located in Sirsi and Kumta are shown in Figures 2.9.1, 2.9.2, 2.9.3 and 2.9.4 respectively. It is seen that the Weibull fit is good for both these stations. The goodness of fit is tested using the Chi square test. The parameters of the Weibull distribution obtained (c and k) are given in Table 2.6.

Table 2.5.1: Frequency Distribution (period of occurrence) and Cumulative Frequency Distribution at Sirsi

Month	Data Available	0-4 kmph	0-8 kmph	8-12 kmph	12-16 kmph	16-20 kmph	20-24 kmph	24 kmph
Jan	744	269	134	212	85	34	10	0
Feb	672	254	129	172	84	26	7	2
Mar	744	310	155	162	101	12	4	0
April	720	247	124	195	120	22	12	0
May	744	142	71	293	156	60	22	0
June	720	128	66	271	190	42	23	2
July	745	97	58	256	176	98	60	43
Aug	743	89	46	270	206	76	56	2
Sept	720	319	159	154	81	5	2	2
Oct	744	383	191	139	27	2	2	0
Nov	720	234	116	165	111	64	30	3
Dec	744	232	118	214	100	45	35	2

Cumulative frequency distribution, Sirsi

Month	0-4 kmph	0-8 kmph	8-12 kmph	12-16 kmph	16-20 kmph	20-24 Kmph	>24
Jan	0.36	0.54	0.83	0.94	0.94	0.99	1
Feb	0.38	0.57	0.83	0.95	0.95	0.99	1
Mar	0.42	0.63	0.84	0.98	0.98	0.99	1
April	0.34	0.52	0.79	0.95	0.95	0.98	1
May	0.19	0.29	0.68	0.89	0.89	0.97	1
June	0.18	0.27	0.65	0.91	0.91	0.97	1
July	0.13	0.21	0.55	0.79	0.79	0.92	1
Aug	0.12	0.18	0.55	0.82	0.82	0.92	1
Sept	0.44	0.66	0.88	0.99	0.99	1.00	1
Oct	0.51	0.77	0.96	0.99	0.99	1.00	1
Nov	0.33	0.49	0.72	0.87	0.87	0.96	1
Dec	0.31	0.47	0.76	0.89	0.89	0.95	1

Table 2.5.2: Frequency Distribution and Cumulative Frequency Distribution at Kumta

Month	Data Available	0-4 kmph	4-8 kmph	8-12 kmph	12-16 kmph	16-20 kmph	20-24 kmph	>24 kmph
Jan	752	129	257	224	94	32	16	
Feb	738	133	261	190	100	40	14	2
Mar	801	152	303	187	134	14	11	
April	752	131	259	195	127	20	20	2
May	748	65	126	304	176	40	37	2
June	799	69	120	288	216	66	40	8
July	760	24	57	305	232	82	60	34
Aug	760	52	80	290	202	74	62	12
Sept	777	131	259	202	142	22	21	2
Oct	706	121	237	190	124	18	16	3
Nov	734	123	240	188	119	44	20	3
Dec	769	116	228	224	121	36	44	2

Cumulative frequency distribution, Kumta

Month	0-4 kmph	4-8 kmph	8-12 kmph	12-16 kmph	16-20 kmph	20-24 kmph	>24 kmph
Jan	0.17	0.51	0.81	0.94	0.94	0.98	1
Feb	0.18	0.53	0.79	0.93	0.93	0.98	1
Mar	0.19	0.57	0.80	0.97	0.97	0.99	1
April	0.17	0.52	0.78	0.95	0.95	0.97	1
May	0.09	0.26	0.66	0.90	0.90	0.95	1
June	0.09	0.24	0.60	0.87	0.87	0.95	1
July	0.03	0.11	0.51	0.81	0.81	0.92	1
Aug	0.07	0.17	0.56	0.82	0.82	0.92	1
Sept	0.17	0.50	0.76	0.94	0.94	0.97	1
Oct	0.17	0.51	0.78	0.95	0.95	0.98	1
Nov	0.17	0.49	0.75	0.91	0.91	0.97	1
Dec	0.15	0.45	0.74	0.90	0.90	0.94	1

Table 2.6: Weibull Parameters (c and k) for Sirsi and Kumta

	Sirsi		Kumta	
Month	C	k	c	k
January	8.38	2.09	8.10	1.96
February	8.74	2.24	11.45	2.34
March	10.46	2.97	9.47	2.54
April	8.53	2.14	8.17	3.44
May	10.62	2.95	8.12	2.93
June	10.72	2.00	11.71	2.02
July	10.69	1.21	10.41	1.32
August	13.36	1.73	10.23	1.47
September	12.24	3.74	11.69	3.43
October	11.06	2.59	11.14	3.21
November	9.23	1.23	10.65	3.01
December	9.78	1.42	11.07	3.18

The monthly mean wind Energy Pattern Factor for the sites at Sirsi and Kumta is listed in Table 2.7. The concept of EPF is useful in calculating the available energy in the wind along with the knowledge of annual mean speed. It is also useful while choosing a location with limited wind data, because long-term data from neighbouring sites can be correlated with on-site short time measurements.

Table 2.7: Energy Pattern Factor (EPF) for Sirsi and Kumta

Month	EPF-Kumta	EPF-Sirsi
January	2.30	1.84
February	2.29	1.99
March	2.23	2.24
April	2.62	2.78
May	2.58	2.80
June	2.56	2.88
July	2.33	3.02
August	2.92	2.78
September	2.42	2.40
October	2.21	1.79
November	2.07	1.48
December	2.18	1.90

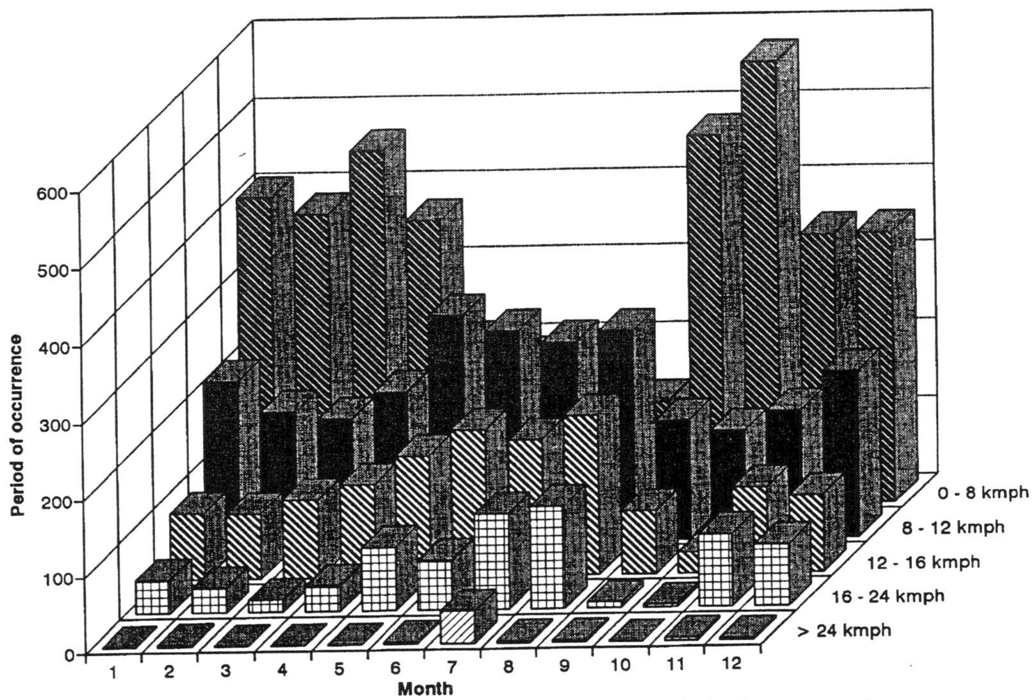

Figure 2.8.1: Frequency Distribution at Sirsi (period of occurrence)

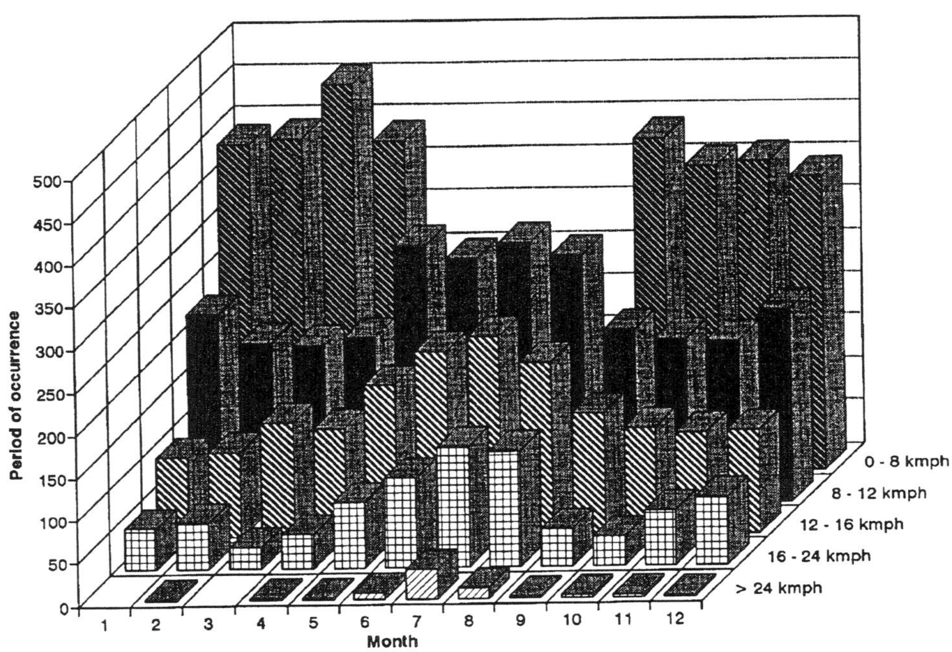

Figure 2.8.2: Frequency Distribution at Kumta (period of occurrence)

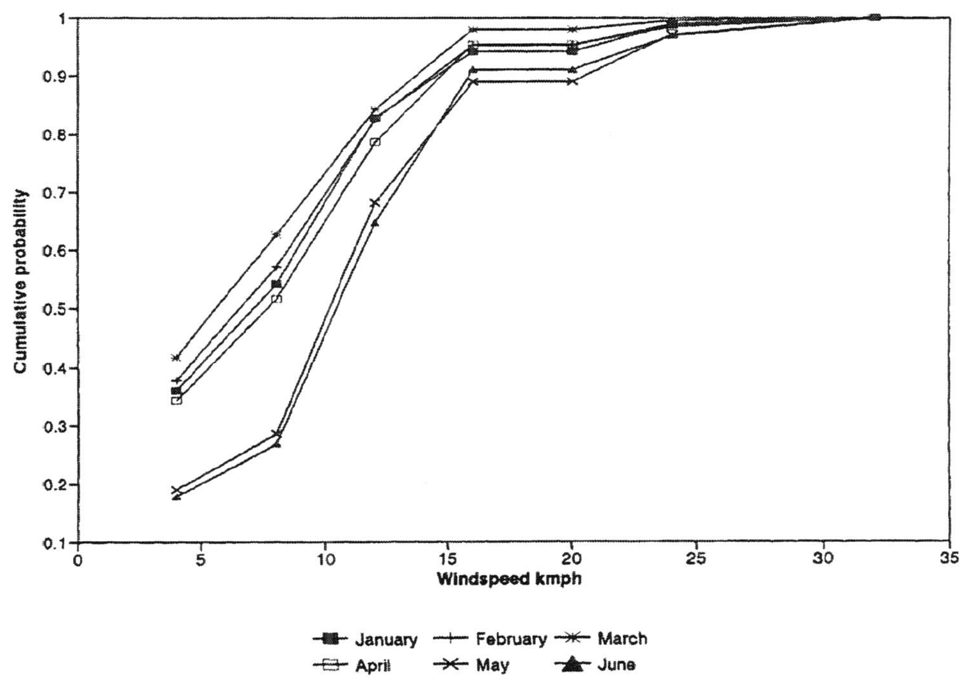

Figure 2.9.1: Cumulative Frequency Distribution (CDF) for Sirsi (January-June)

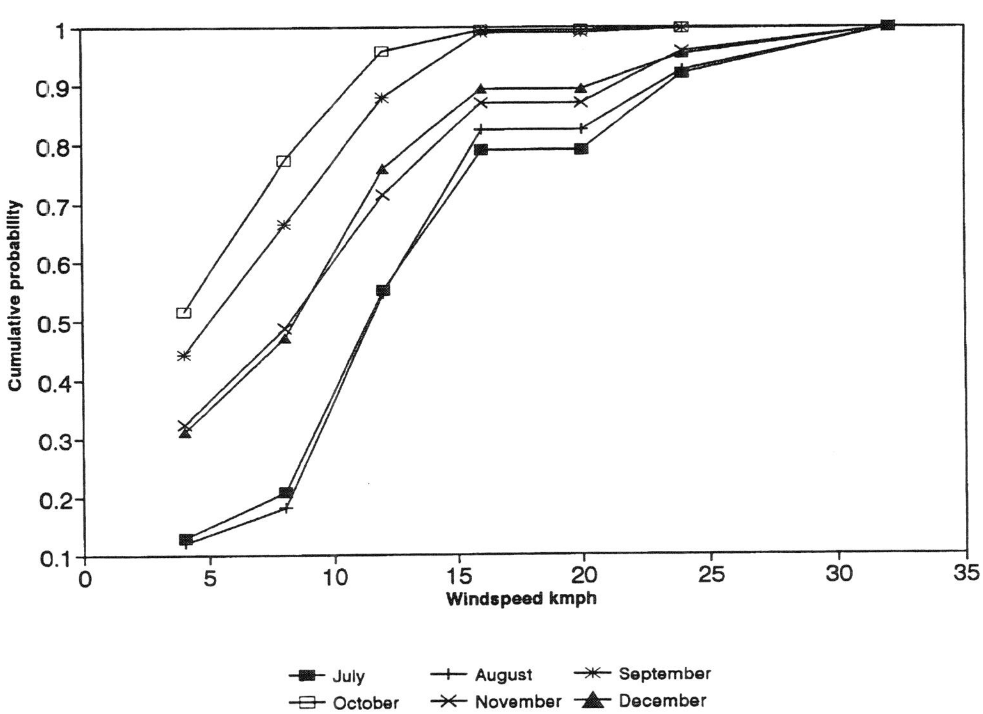

Figure 2.9.2: CDF for Sirsi (July-December)

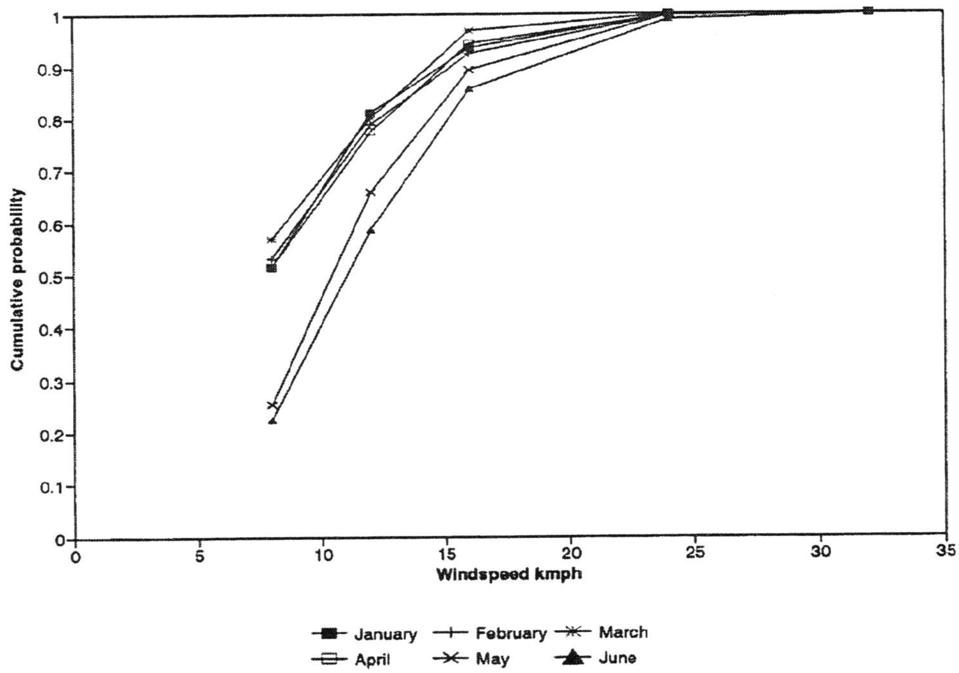

Figure 2.9.3: CDF for Kumta (January-June)

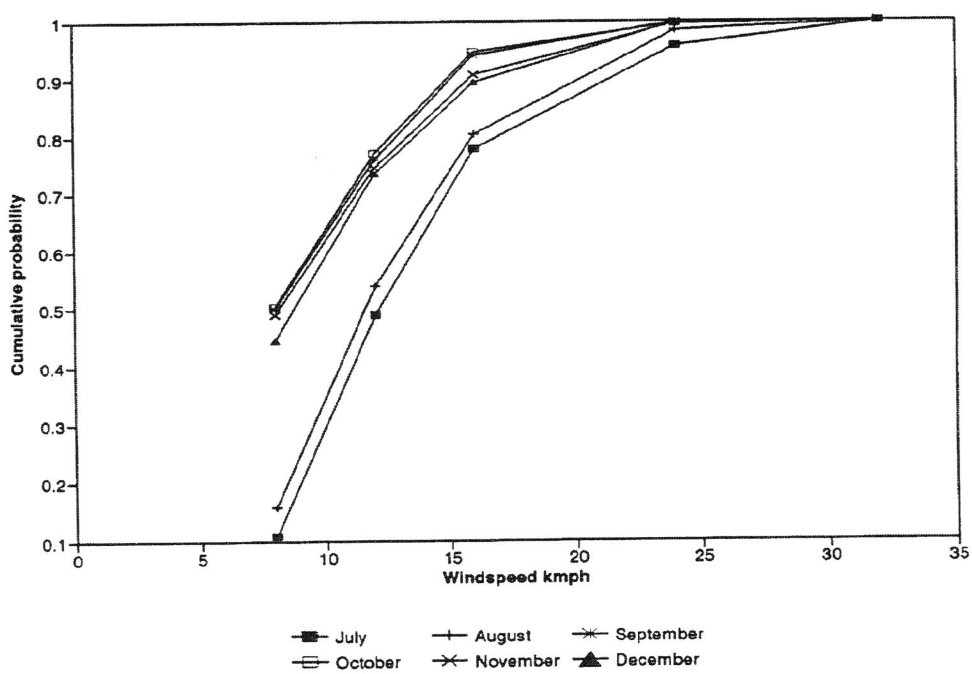

Figure 2.9.4: CDF for Kumta (July-December)

The monthly average hourly wind speed with monthly average power density (W/m^2) is given in Table 2.8. Monthly average hourly wind speed is maximum in the month of July for Kumta (13.03 kmph) and Sirsi (18.17 kmph). The mean annual wind speed at Kumta is 8.82 kmph and the corresponding power density at Kumta is 26.49 W/m^2. Similarly at Sirsi, annual mean wind speed is 9.81 kmph and corresponding power density 45.18 W/m^2.

Table 2.8: Mean Power Density for Kumta and Sirsi

	Kumta		Sirsi	
	V (kmph)	Pd (W/m2)	V (kmph)	Pd(W/m2)
Jan	5.95	6.25	6.93	7.87
Feb	7.77	13.79	6.89	8.37
March	9.10	21.59	7.20	10.77
April	9.42	28.26	8.38	21.06
May	9.88	31.99	9.10	27.12
June	11.84	54.79	11.20	51.97
July	13.03	66.36	18.17	233.06
August	11.54	57.77	14.20	102.29
Sept	6.71	9.41	11.15	42.73
Oct	6.60	8.15	8.40	13.63
Nov	6.30	6.65	7.72	8.77
Dec	7.74	12.99	8.42	14.6
Annual Mean	8.82	26.49	9.81	45.18
Annual Sd.	2.27	20.82	3.26	62.39
Maximum	13.03	66.36	18.17	233.06
Minimum	5.95	6.25	6.89	7.87

Hourly wind speed data are not available for coastal taluks Karwar, Honnavar and Shirali. However, primary hourly wind speed data are available for Kumta, which is also located in the coastal belt. Therefore, EPF is computed for Kumta with hourly wind speed data and is used to compute mean power density at Karwar, Honnavar and Shirali, listed in Table 2.9. The mean annual wind speed at Karwar is 8.67 kmph and the corresponding power density is 30.64 W/m^2. The variation in monthly mean power density for Karwar, Kumta, Honnavar and Shirali is illustrated in Figure 2.10.

With this background, potential assessment of wind energy has been carried cut along the coastal tract of Uttara Kannada district. For this purpose, the coastal tract along the seashore was divided into three zones. Zone-A consists of land space about 5 km wide along the coastal tract, parallel to the sea. Zone-B and Zone-C correspond to 7.5 km and 10 km width along the coastal tract. Villages in these zones were compiled from village maps. Corresponding wastelands available in these villages are according to a 1991 census. It is assumed that 1% of wastelands could be made available for harnessing wind energy. With this assumption, monthwise wind potential has been computed at Karwar, Kumta, Honnavar and Bhatkal taluks. Zone-D and Zone-E refer to 5% and 10% of the total wasteland available in 10 km width along the coastal tract. Last two columns in Tables 2.10.1 and 2.10.2 indicate wind potential if land space available is about 5% and 10% of total wasteland in these zones. It is found that about 0.75 to 2 MW could be generated in Karwar and Kumta taluks during seasons in this tract.

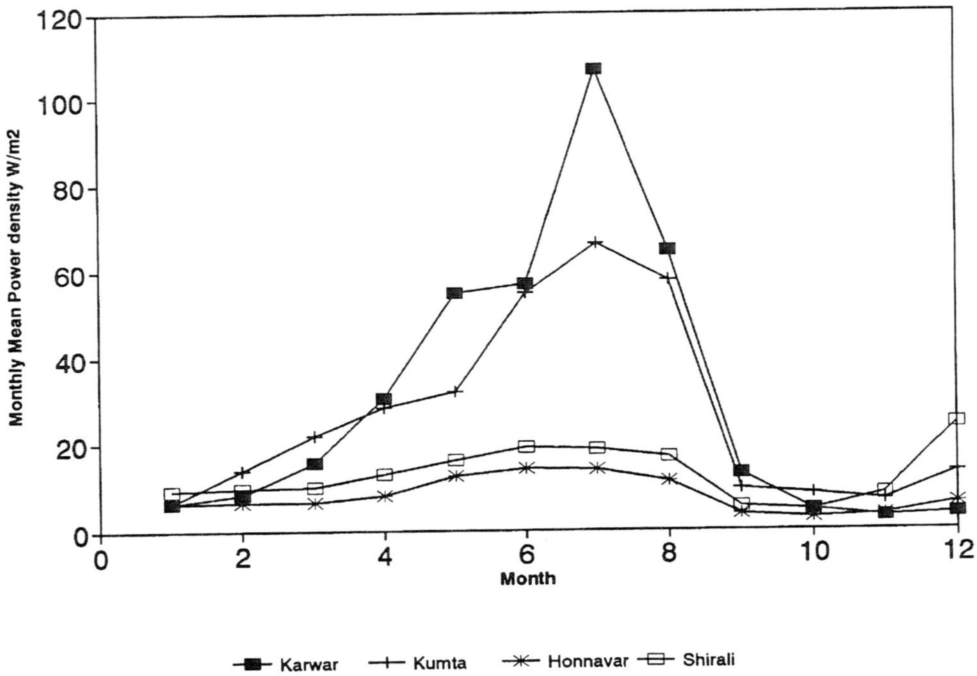

Figure 2.10: Mean Wind Power Density – Karwar, Jumta, Shirali and Honnavar

Table 2.9: Monthly Mean Power Density at Karwar, Honnavar and Shirali

	Kumta	Karwar		Honnavar		Shirali	
Month	EPF	V (kmph)	Pd (W/m2)	V (kmph)	Pd (W/m2)	V (kmph)	Pd (W/m2)
Jan	1.84	5.96	6.26	5.95	6.23	6.78	9.23
Feb	1.99	6.55	8.27	6.00	6.37	6.87	9.53
Mar	2.24	8.15	15.55	6.10	6.52	7.03	9.95
Apr	2.78	9.65	30.38	6.20	8.06	7.25	12.90
May	2.8	11.82	54.84	7.21	12.45	7.84	16.03
June	2.88	12.01	57.18	7.50	13.93	8.30	18.88
July	3.02	15.27	106.80	7.72	13.78	8.50	18.41
Aug	2.78	11.98	64.63	6.66	11.09	7.64	16.78
Sept	2.4	7.44	12.83	4.87	3.58	5.56	5.35
Oct	1.79	5.41	4.51	4.55	2.68	5.42	4.53
Nov	1.48	4.75	2.85	5.04	3.41	6.76	8.22
Dec	1.9	5.04	3.59	6.00	6.06	9.51	24.11
Average	2.33	8.67	30.64	6.15	7.85	7.29	12.83

Table 2.10.1: Monthly Mean Power at Karwar, Kumta, Honnavar, and Shirali

Karwar		Estimated power in kW			5%	10%
Month	Pd W/m2	Zone-A	Zone-B	Zone-C	Zone-D	Zone-E
Land sq.m.		1376.93	1786.93	2134.73	10673.65	21347.30
Jan	6.26	8.61	11.18	13.35	66.77	133.55
Feb	8.27	11.38	14.77	17.65	88.25	176.50
Mar	15.55	21.41	27.78	33.19	165.94	331.88
Apr	30.38	41.83	54.28	64.85	324.24	648.48
May	54.84	75.51	98.00	117.07	585.36	1170.72
June	57.18	78.73	102.18	122.07	610.33	1220.66
July	106.80	147.05	190.84	227.99	1139.93	2279.86
Aug	64.63	88.99	115.49	137.97	689.83	1379.67
Sept	12.83	17.66	22.92	27.39	136.93	273.86
Oct	4.51	6.20	8.05	9.62	48.09	96.17
Nov	2.85	3.92	5.09	6.08	30.41	60.83
Dec	3.59	4.94	6.41	7.66	38.31	76.62
Average	30.64	42.19	54.75	65.41	327.03	654.07

Kumta		Estimated power in kW			5%	10%
Month	Pd (W/m2)	Zone-A	Zone-B	Zone-C	Zone-D	Zone-E
Land		1493.31	1956.36	2893.80	14469.00	28938.00
Jan	6.25	9.33	12.23	18.09	90.43	180.86
Feb	13.79	20.59	26.98	39.91	199.53	399.06
Mar	21.59	32.24	42.24	62.48	312.39	624.77
Apr	28.26	42.20	55.29	81.78	408.89	817.79
May	31.99	47.77	62.58	92.57	462.86	925.73
June	54.79	81.82	107.19	158.55	792.76	1585.51
July	66.36	99.10	129.82	192.03	960.16	1920.33
Aug	57.77	86.27	113.02	167.17	835.87	1671.75
Sept	9.41	14.05	18.41	27.23	136.15	272.31
Oct	8.15	12.17	15.94	23.58	117.92	235.84
Nov	6.65	9.93	13.01	19.24	96.22	192.44
Dec	12.99	19.40	25.41	37.59	137.95	375.90
Average	26.49	39.57	51.84	76.69	333.43	766.86

Table 2.10.1: Monthly Mean Power at Karwar, Kumta, Honnavar, and Shirali (Cont'd)

Honnavar		Estimated power in kW			5%	10%
Month	Pd W/m2	Zone-A	Zone-B	Zone-C	Zone-D	Zone-E
Land		1178.82	1426.82	2122.87	10614.35	21228.70
Jan	6.23	7.34	8.89	13.23	66.13	132.27
Feb	6.37	7.51	9.09	13.53	67.63	135.26
Mar	6.52	7.69	9.30	13.84	69.20	138.40
Apr	8.06	9.51	11.51	17.12	85.60	171.20
May	12.45	14.67	17.76	26.43	132.13	264.26
June	13.93	16.42	19.88	29.58	147.88	295.76
July	13.78	16.24	19.66	29.25	146.24	292.48
Aug	11.09	13.07	15.82	23.53	117.67	235.34
Sept	3.58	4.23	5.11	7.61	38.05	76.10
Oct	2.68	3.16	3.82	5.69	28.45	56.90
Nov	3.41	4.03	4.87	7.25	36.25	72.49
Dec	6.06	7.14	8.64	12.86	64.28	128.55
Average	7.85	9.25	11.20	16.66	83.29	166.59

Shirali		Estimated power in kW			5%	10%
Month	Pd W/m2	Zone-A	Zone-B	Zone-C	Zone-D	Zone-E
Land		1186.04	1588.28	1748.58	8742.90	17485.80
Jan	9.23	10.94	14.65	16.13	80.66	161.32
Feb	9.53	11.31	15.14	16.67	83.36	166.73
Mar	9.95	11.80	15.81	17.40	87.01	174.02
Apr	12.90	15.30	20.49	22.55	112.77	225.54
May	16.03	19.02	25.47	28.04	140.19	280.38
June	18.88	22.40	29.99	33.02	165.09	330.19
July	18.41	21.83	29.24	32.19	160.95	321.91
Aug	16.78	19.90	26.64	29.33	146.66	293.33
Sept	5.35	6.34	8.50	9.35	46.77	93.53
Oct	4.53	5.37	7.19	7.91	39.56	79.13
Nov	8.22	9.75	13.06	14.38	71.90	143.79
Dec	24.11	28.60	38.30	42.16	210.81	421.63
Average	12.83	15.21	20.37	22.43	112.14	224.29

Table 2.10.2: Monthly Mean Energy at Karwar, Kumta, Honnavar, and Shirali

Karwar		Estimated Energy in kWh			5%	10%
Month	Pd W/m2	Zone-A	Zone-B	Zone-C	Zone-D	Zone-E
Land		1376.93	1786.93	2134.73	10673.65	21347.30
Jan	6.26	2243.06	2910.96	3477.53	17387.66	34775.32
Feb	8.27	2677.65	3474.96	4151.31	20756.57	41513.13
Mar	15.55	5574.28	7234.10	8642.12	43210.58	86421.16
Apr	30.38	10540.67	13679.30	16341.78	81708.88	163417.76
May	54.84	19663.61	25518.72	30485.57	152427.84	304855.68
June	57.18	19840.98	25748.91	30760.56	153802.82	307605.64
July	106.80	38292.92	49695.17	59367.61	296838.05	593676.11
Aug	64.63	23173.14	30073.26	35926.58	179632.92	359265.85
Sept	12.83	4451.38	5776.84	6901.21	34506.07	69012.14
Oct	4.51	1615.33	2096.32	2504.34	12521.70	25043.40
Nov	2.85	988.69	1283.08	1532.81	7664.07	15328.15
Dec	3.59	1286.90	1670.09	1995.15	9975.76	19951.51
Sum in kilo units (k kWh)		130.35	169.16	202.09	1010.43	2020.87

Kumta		Estimated Energy in kWh			5%	10%
Month	Pd (W/m2)	Zone-A	Zone-B	Zone-C	Zone-D	Zone-E
Land		1493.31	1956.36	2893.80	14469.00	28938.00
Jan	6.25	2430.36	3183.98	4709.66	23548.30	47096.60
Feb	13.79	4843.41	6345.27	9385.77	46928.87	93857.74
Mar	21.59	8395.44	10998.73	16269.05	81345.24	162690.48
Apr	28.26	10634.64	13932.26	20608.25	103041.27	206082.55
May	31.99	12439.56	16296.86	24105.92	120529.61	241059.21
June	54.79	20618.25	27011.62	39954.93	199774.64	399549.28
July	66.36	25804.61	33806.18	50005.28	250026.40	500052.81
Aug	57.77	22464.32	29430.13	43532.32	217661.62	435323.25
Sept	9.41	3541.12	4639.16	6862.13	34310.63	68621.26
Oct	8.15	3169.19	4151.90	6141.40	30706.98	61413.96
Nov	6.65	2502.49	3278.47	4849.43	24247.15	48494.30
Dec	12.99	5051.26	6617.58	9788.56	48942.78	97885.56
Sum in kilo units (k kWh)		121.89	159.69	236.21	1181.06	2362.13

Table 2.10.2: Monthly Mean Energy at Karwar, Kumta, Honnavar, and Shirali (Cont'd)

Honnavar		Estimated Energy in kWh			5%	10%
Month	Pd W/m2	Zone-A	Zone-B	Zone-C	Zone-D	Zone-E
Land		1178.82	1426.82	2122.87	10614.35	21228.70
Jan	6.23	1912.54	2314.90	3444.19	17220.94	34441.88
Feb	6.37	1766.61	2138.27	3181.39	15906.97	31813.93
Mar	6.52	2001.31	2422.35	3604.05	18020.27	36040.54
Apr	8.06	2395.67	2899.67	4314.23	21571.14	43142.28
May	12.45	3821.24	4625.15	6881.45	34407.26	68814.51
June	13.93	4138.76	5009.48	7453.26	37266.32	74532.64
July	13.78	4229.22	5118.97	7616.17	38080.85	76161.70
Aug	11.09	3403.05	4118.98	6128.35	30641.77	61283.53
Sept	3.58	1064.86	1288.89	1917.65	9588.27	19176.53
Oct	2.68	822.72	995.80	1481.59	7407.95	14815.91
Nov	3.41	1014.41	1227.83	1826.80	9134.00	18268.00
Dec	6.06	1858.84	2249.90	3347.48	16737.41	33474.83
Sum in kilo units (k kWh)		28.43	34.41	51.20	255.98	511.97

Shirali		Estimated Energy in kWh			5%	10%
Month	Pd W/m2	Zone-A	Zone-B	Zone-C	Zone-D	Zone-E
Land		1186.04	1588.28	1748.58	8742.90	17485.80
Jan	9.23	2849.38	3815.73	4200.84	21004.22	42008.43
Feb	9.53	2659.83	3561.90	3921.39	19606.95	39213.89
Mar	9.95	3073.61	4116.02	4531.43	22657.16	45314.33
Apr	12.90	3855.09	5162.52	5683.56	28417.80	56835.60
May	16.03	4952.16	6631.67	7300.98	36504.90	73009.81
June	18.88	5643.82	7557.89	8320.69	41603.43	83206.85
July	18.41	5685.69	7613.96	8382.42	41912.09	83824.17
Aug	16.78	5180.89	6937.97	7638.20	38190.99	76381.97
Sept	5.35	1598.76	2140.97	2357.05	11785.27	23570.54
Oct	4.53	1397.57	1871.55	2060.44	10302.21	20604.42
Nov	8.22	2457.84	3291.41	3623.60	18118.00	36236.00
Dec	24.11	7447.01	9972.63	10979.14	54895.69	109791.39
Sum in kilo units k kWh		46.80	62.67	69.00	345.00	690.00

2.5.0 Windfarm and Turbine Size

Because windfarms consist of modular units of single design, it is thought that the wind power technology offers significant economies of scale. This is very much true for turbine manufacturing and not for turbine deployment. The analysis is carried out with data and information provided by Tamilnadu Energy Developmental Agency (TEDA, India). This suggests that large windfarms are no more attractive compared to the small ones. Although construction and mobilisation costs may fall per unit for a larger windfarm, (i) total costs for turbines and civil works remain proportional to windfarm size, (ii) larger windfarm requires proportionally higher electric interconnection costs because of the need for more costly high voltage step-up facilities and (iii) array efficiency decreases with windfarm size. It is found at present that the wind turbines of 225 kW each are economically attractive. If 10% of wasteland (in the region of 10 km width along the coastal tract) is made available for power generation, the calculations show that along the coastal tract of Karwar and Kumta, windfarms of rated capacity of 1.35 megawatt (6 units of 225kW each) can be established.

The land requirements for wind units were calculated assuming that turbines were to be erected in rows with the distance between the rows at 7 times the rotor diameter and the distance between the turbines 5 times the rotor diameter. The rows are normally aligned perpendicular to the prevailing wind direction. For a turbine of about 200 kW, the rotor diameter would be about 30 meters. With these assumptions, it is noticed that for one megawatt installed capacity, land required is about 20 hectares. Of this area, 10% is directly used for foundations, roads and other facilities.

2.6.0 Estimation of Output

The monthly energy output at respective sites based on wind data is listed in Table 2.10.2. It is seen that approximately two million units could be generated at Karwar and Kumta by using 10% of the wasteland along the 10 km wide coastal tract.

Turbine efficiency of 40% and electric generator efficiency of 85% are used in the calculation. With this, an overall efficiency of 35% in wind-electric generator units could be achieved. A load factor of 0.3 is assumed in the computation of energy output.

Apart from well known day-night and seasonal variation of wind regime, it is to be noted that the wind is fairly reliable compared to hydel resources (rainfall). The coefficient of variation (COV) of mean wind speed is about 5.85 per cent compared to a higher COV of rainfall (of about 28%).

Table 2.10.3 lists rated capacity in each region and energy that could be generated using wind energy conversion system. When wind speeds are low, the wind machine will remain idle. As wind speeds rise to cut-in wind speed, the generating unit will start functioning. The rated output is computed by taking the average of the windy months in a year. It is the power output that the machine produces when it is operating at rated wind speed. Wind speeds above the rated output would not produce greater power output levels.

2.7.0 Cost Estimates

The cost estimates include (a) wind turbines including control systems, panels, environmental protection, capacitor banks, assembly and erection, (b) electrical works including transmission and distribution, and (c) civil works, fencing and land acquisition.

The amortised total annual cost (ATAC) for the wind generation system is given by

$$ATAC = \frac{C*r*(1+r)*t}{(1+r)^t - 1} + O\&M,$$

where C = total capital cost, t = expected life (assumed as 20 years), r = discount rate (10% is assumed) and O&M = annual operation and maintenance charge.

Table 2.10.3: Rated Power and Energy that could be generated at Karwar, Kumta, Honnavar and Shirali

If LF = 0.45 Energy in kilo units '000 kWh

		Zone-A	Zone-B	Zone-C	Zone-D	Zone-E
Karwar	63	119.68	155.32	185.55	927.77	1855.53
Kumta	48	98.55	129.11	190.98	954.90	1909.81
Honnavar	12	19.29	23.35	34.74	173.71	347.42
Shirali	20	27.16	36.38	40.05	200.24	400.48

If LF = 0.3

		Avg Power in kW		1%	5%	10%
		Zone-A	Zone-B	Zone-C	Zone-D	Zone-E
Karwar	63	86.42	112.16	133.99	669.94	1339.88
Kumta	48	71.43	93.58	138.42	692.11	1384.22
Honnavar	12	13.98	16.92	25.18	125.91	251.81
Shirali	16.60	19.69	26.37	29.03	145.13	290.27
		Energy in kilo units '000 kWh				
		Zone-A	Zone-B	Zone-C	Zone-D	Zone-E
Karwar	63	79.49	103.16	123.24	616.21	1232.42
Kumta	48	65.70	86.08	127.32	636.60	1273.21
Honnavar	12	12.86	15.57	23.16	115.81	231.62
Shirali	17	18.11	24.25	26.70	133.49	266.99

Cost per unit is obtained by dividing this (ATAC) by annual generated energy. Cost computed based on the data of Kayathar windfarm (TNEB, 1994) in Tamilnadu works out to be Rs.1.30/unit (kWh). The details of electrical, civil and the annual maintenance costs are listed in Table 2.10.4.

2.8.0 CONCLUSIONS

Human pressure on forests to meet the daily energy requirement in the form of fuel wood and fodder for domestic purposes and rural industries is quite evident from the barren hill tops in the coastal belt of Uttara Kannada district. Therefore, harnessing of a renewable source like wind at feasible sites would help in the eco-development of the region. Analyses of wind data clearly indicate that the coastal area, especially Karwar and Kumta taluks, has good wind potential. The availability of wind resources is quantified and characterised in this chapter. Local need for energy in a decentralised way for industrial needs, and for irrigation purposes (in the pre-monsoon period for agriculture and plantations like areca, coconut, etc.) makes the exploitation of wind energy feasible and desirable.

Table 2.10.4: Cost per kWh and Production Cost per MW

Turbines (million Rs.)	15
Electrical works	1.31
Civil works	1.75
Land acquisition	0.75
Total (million Rs.)	18.81
Levelized capital	
(20 years life time, Discount rate 10%)	2.21
Annual maintenance	0.33
(2% capital costs)	
Annual Wages	0.07
Total per year	2.61
Total per MW (million Rs.)	2.61
Total Energy mWh	2014.00
Cost /unit (kWh)	1.30

2.9.0 ACKNOWLEDGEMENTS

The Centre for Renewable Energy Development Agency (CREDA), Bangalore supported this research partly by providing equipment. Mr. Pyannavar, Technical Officer, CREDA helped in exploratory survey of sites in Kumta and Sirsi. Mr. Rosario, M. Furtado, Mr. Deepak Shetti and Mr. P.R. Bhat from CES Field Research Station at Sirsi and Mr. Vishnu Mukri at Kumta extended their support in collecting field data regularly and monitoring sites at Sirsi and Kumta. Historical monthly surface data collected in meteorological observatories were provided by the India Meteorological Department, Government of India, Pune. Daily surface data for the recent years was provided by the India Meteorological Department, Bangalore office. I thank Dr. Tulsidas, Director, IMD, Bangalore and Dr. Mokashi and Dr C.D. Padmanabhan at IMD, Pune for their kind cooperation in providing the data.

2.10.0 REFERENCES

1. Wilbur, L.C., 1985. *Handbook of Energy Systems Engineering - Production and Utilisation*, John Wiley & Sons. Inc., New York, U.S.A.
2. WMO, 1964. *Guide to Meteorological Instrument and Observing Practices,* 4th Edition, World Meteorological Organisation, No. 8, TP. 3, Geneva, Switzerland.
3. Lysen, H., 1983. *Introduction to Wind Energy*, 2nd Edition, Consultancy Services, Wind Energy, Developing Countries (CWD) 82-1, The Netherlands.
4. Op cit, Ref 3.
5. Justus, C.G., 1975. *Wind Data Collection and Assessment*, Proc. Initial Wind Energy Data Assessment Study, Ashville, NC, NSF-RA-75-020, Department of Energy, USA.
6. Stevens, M.J.M. and Smulders, P.T., 1980. *The Estimation of the Parameter of the Weibull Wind Speed Distribution for Wind Energy Utilization Purposes*, Wind Energy Group, Department of Physics, University of Technology, Eindhoven, The Netherlands.

7. Mani, A. and Mooley, D.A., 1983. *Wind Energy Data for India*, Allied Publishers Private Limited, New Delhi.
8. TNEB, 1994. *Tamilnadu Electricity Board Statistics at a Glance*, Tamilnadu, India.

Chapter 3

BIOENERGY POTENTIAL AND PROSPECTS

3.0 SUMMARY

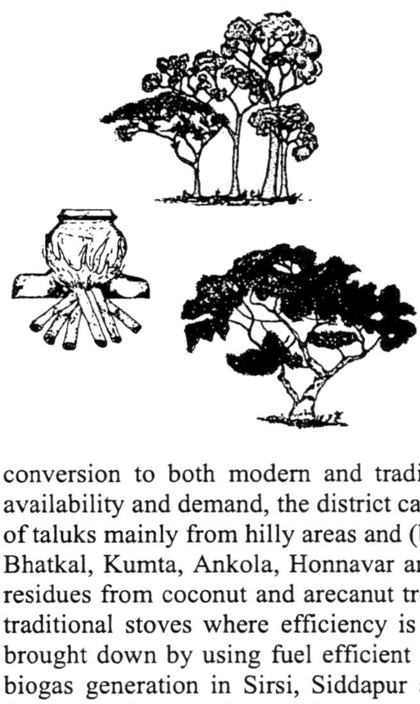

Bioenergy is the energy released when organic carbon material reacts with oxygen. This organic material derived from plants and animals is also referred to as biomass. Biomass is a flexible feedstock capable of becoming solid, liquid or gaseous fuel by chemical and biological processes. These intermediate biofuels (like methane, ethanol, charcoal, etc.) can substitute fossil fuels. Wood and charcoal are important as household fuels and for small-scale industries like brick making, cashew processing, etc. The scarcity of biofuels has far-reaching implications on the environment. Hence, expansion of bioenergy systems could influence the betterment of both socio-economic and environmental conditions of the region.

This section examines the present role of biomass in the region's (Uttara Kannada district) energy supply, and calculates the potential for future biomass provision and scope for its conversion to both modern and traditional fuels. Based on detailed investigation of biomass resource availability and demand, the district can be categorised into two zones: (a) Biomass surplus zone consisting of taluks mainly from hilly areas and (b) Biomass scarce zone consisting of thickly populated coastal taluks Bhatkal, Kumta, Ankola, Honnavar and Karwar. Fuel wood is mainly used for cooking, and horticulture residues from coconut and arecanut trees for heating water. Most of the households in the region still use traditional stoves where efficiency is less than 10%. The present inefficient fuel consumption could be brought down by using fuel efficient stoves (saving of approx. 27%). Availability of animal residues for biogas generation in Sirsi, Siddapur and Yellapur taluks promises a viable alternative for cooking and lighting fuels, apart from their use as useful fertiliser. However, to support the present livestock population, fodder from agricultural residues is insufficient in these taluks. There is a need to supplement the fodder availability with fodder crops as successfully tried in Banavasi village by some progressive farmers.

Keywords: Bioenergy, Fuel wood Consumption, Integrated Energy Planning, Energy Efficient Devices, Per Capita Fuel Consumption, Techno-Economic Analyses, Biogas, Wood Gasification, Energy Plantation.

3.1.0 INTRODUCTION

Biomass refers to solid carbonaceous material derived from plants and animals. These include the residues of agriculture and forestry, animal wastes and wastes from food processing operations. A small amount of solar energy is used by plants in the process of photosynthesis and this trapped energy can be used in various ways. Wood and grass can be dried and then burnt to release heat. Plant material, particularly those rich in starches and sugars such as sugar-cane, wheat, etc., can be fermented to produce ethanol. Alternately, methanol can be produced by the distillation of biomass which contains considerable amount of cellulose such as wood and bagasse (residue from sugar-cane). Both these alcohols can be used to fuel vehicles and machinery, and can be mixed with petrol to make a petrol/alcohol blend. Although biomass energy is predominantly used in rural areas, it provides an important fuel source for the urban poor and many rural small and medium scale industries. In order to meet the growing demand for energy, it is imperative to focus on efficient production and use of biomass energy to meet both traditional (as a heat supplier) and modern fuel requirements (like electricity and liquid fuels). This production of biomass, in all its forms for fuel, food and fodder, demands environmentally sustainable land use and integrated planning approaches.

Detailed planning would be required from National to State, District, Taluk and Village levels. The inappropriate selection and site matching of species or management strategies can have adverse effects and lead to degradation and abandonment of land. However, the correct selection of plant species can allow the economic production of energy crops in areas previously capable of only low plant productivities. Simultaneously multiple benefits may accrue to the environment. Such selection strategies allows synergistic increases in food crop yield and decreases fertiliser applications, while providing local source of energy and employment (Hall, 1991; Hall and Rosillo-Calle, 1991; Hall, et al., 1992). In this section, the resource base in Uttara Kannada district under each sector, such as forests, agriculture, horticulture and animal residues, is explained talukwise. An attempt is also made to illustrate the present role of biomass energy, resource base for future development, and some promising conversion technologies and their uses. Biogas, wood based steam power generation, energy plantation and biomass gasification are the most promising bioenergy technologies.

3.1.1 Present Role of Bioenergy

Bioenergy is one of the primary sources of fuel in India. A recent study on energy utilisation in Karnataka considering all types of energy sources and sectorwise consumption reveals that, traditional fuels such as firewood (7.440 million tonnes of oil equivalent (mtoe) - 43.62%), agro residues (1.510 mtoe - 8.85%), and biogas and cowdung (0.250 mtoe - 1.47%) account for 53.20% of total energy consumption. In rural areas, the dependence on bioenergy to meet domestic requirements like cooking and water heating is as high as 80-85%. Fuel wood and agricultural residues are also widely used as fuel in rural industries like cashew and other agro processing industries, brick kilns, and in commercial sectors like hotels, etc. Detailed investigation of energy consumption in ninety villages of Kumta taluk reveals that annual per capita fuel wood consumption for domestic purposes (cooking, water heating, etc.) is 0.7-1.1 tonnes, and in rural industries, such as cashew processing (fuel wood consumption is 4.5-8.5 kgs per kg of cashew kernels), jaggery making (one tonne of fuel wood (logs) for processing 5 tonnes of sugar-cane), brick manufacture (400-800 kgs of fuel wood for making 1000 bricks), etc.

3.2.0 BIOENERGY

3.2.1 Resources from Forests

The District of Uttara Kannada is situated between lat. 13°55'N and 15°31'N and long. 74°9'E and 75°4'E covering an area of 10,291 sq.kms and is subdivided into 11 taluks that can be grouped according to their geographical characteristics as:

i) coastal region, which has hot and humid climate (rainfall varies between 2500-3556 mm), comprising Karwar, Ankola, Kumta, Honnavar and Bhatkal taluks,

ii) hilly region of the Western Ghats (500-1000 meters high), which is very humid to the south (rainfall varies from 3048-3556 mm), comprising the taluks of Sirsi, Siddapur, Supa and Yellapur, and

iii) region of transition, which is drier (rainfall varies between 1016-1524 mm), comprising the taluks of Mundgod and Haliyal.

This geographical diversity is responsible for four vegetation types found throughout the district as follows.

- Evergreen forest, found mainly in Sirsi, Siddapur and the hilly eastern region of Honnavar, Kumta, Ankola and Karwar.
- Semi deciduous forest on the slopes of Ankola, Kumta, Karwar, Honnavar, Siddapur and Sirsi.
- Deciduous forest in Haliyal, Supa and Mundgod.
- Moist deciduous forest in Ankola, Bhatkal, Yellapur, West Karwar and the coastal region of Kumta.

The district has about 80.58% of the geographical area under forests (8292.65 sq.kms). The per capita forest and land area is about 0.64 and 0.83 hectares respectively. Total geographical area in hectares, population as per 1991 census, area under forests, per capita land area, per capita forest area and ratio of forest area to total land area - talukwise, are listed in Table 3.1.1. It is seen that Bhatkal among coastal taluks is thickly populated (3.61), while Supa is thinly populated (0.53). Table 3.1.1 also provides information on talukwise vegetation cover. Yellapur taluk has 87% vegetation cover, while the minimum cover of 67% is in Kumta. These are shown pictorially in Figure 3.1. The forest area according to legal status in the district could be classified as (1) reserve forests (7,727.84 sq.kms, 93.18% of total forest area), (2) protected forests (542.77 sq.kms), (3) village forests (20.20 sq.kms) and (4) unclassified forests (1.84 sq.kms).

Table 3.1.1: Population (1991), Total Area and Area Under Forests in Uttara Kannada

TALUK	Total Area sq.kms	Total Area hect.	Population Total (1991)	Forest hect.	No of Persons per Geo. Area	Forest/ Total Area	Forest Area /Person	Total Area /Person
Supa	1895.1	189514.23	99519	160608.71	0.53	0.85	1.61	1.90
Yellapur	1301.1	130110	81410	113429.56	0.63	0.87	1.39	1.60
Ankola	918.7	91870	91310	75432.37	0.99	0.82	0.83	1.01
Haliyal	864.9	86489.33	94363	59099.78	1.09	0.68	0.63	0.92
Siddapur	859.3	85930	94202	62274.6	1.10	0.72	0.66	0.91
Sirsi	1333.4	133342.62	152935	103226.57	1.15	0.77	0.67	0.87
Mundgod	668.9	66890	77939	51167.36	1.17	0.76	0.66	0.86
Honnavar	775.5	77546.13	145842	57416.68	1.88	0.74	0.39	0.53
Karwar	732.1	73210	142845	52808.68	1.95	0.72	0.37	0.51
Kumta	584.9	58486.81	134144	39187.59	2.29	0.67	0.29	0.44
Bhatkal	357.3	35731.96	129017	25390.46	3.61	0.71	0.20	0.28
District	10291.2	1029121.08	1243526	800042.36	1.21	0.78	0.64	0.83

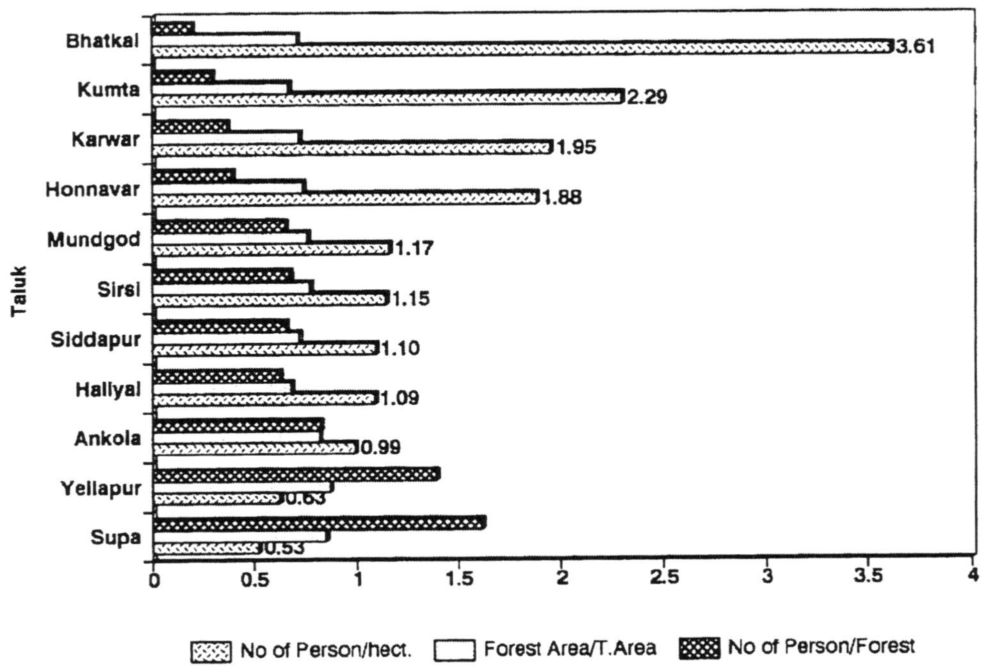

Figure 3.1: Population (Talukwise) – Geographical and Forest Area (hectares)

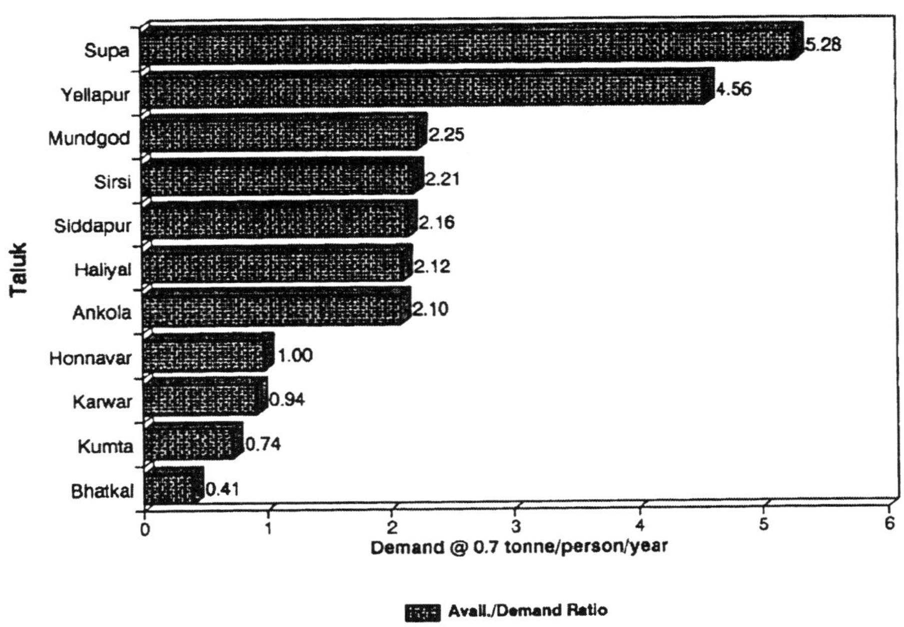

Figure 3.2: Availability to Demand (Demand at 0.7 t/person/year) Ratio – Wood Biomass (Talukwise)

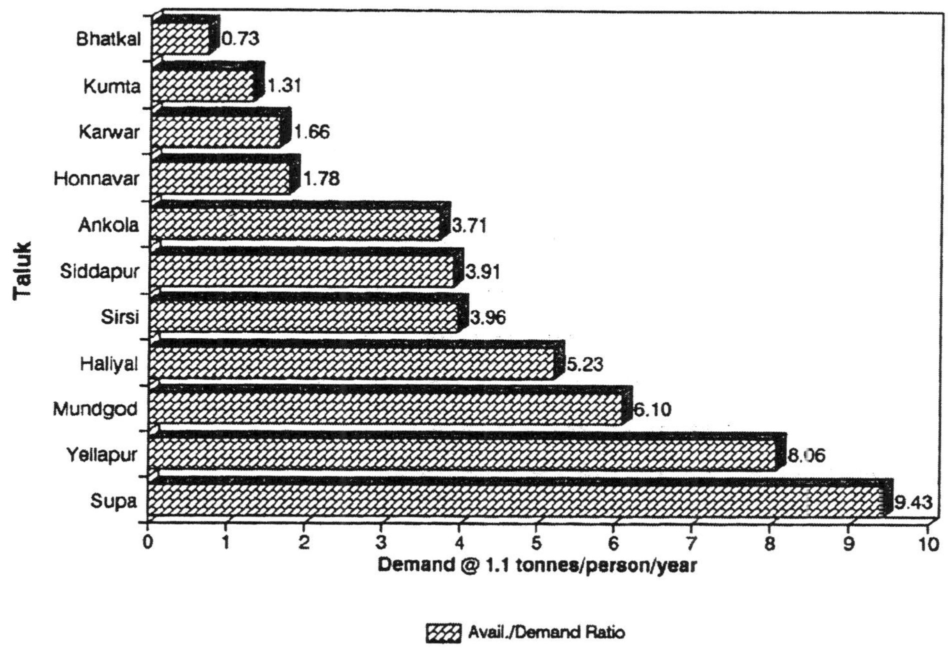

Figure 3.3: Availability to Demand (Demand at 1.1 t/person/year) Ratio – Woody Biomass (Talukwise)

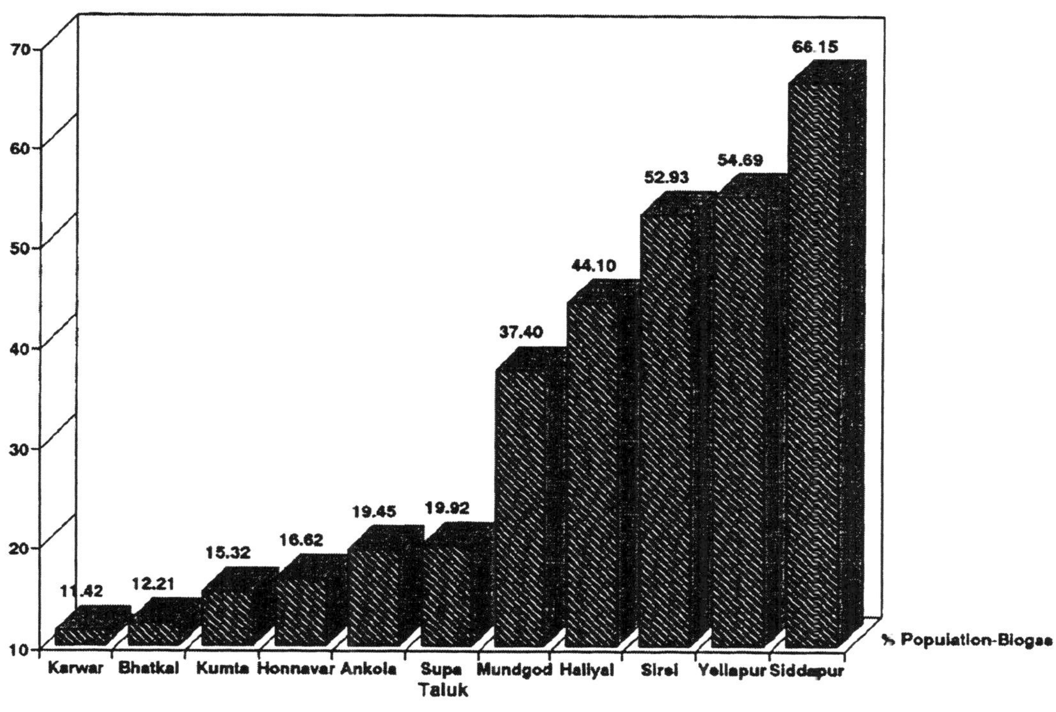

Figure 3.4: % Population Requirement – (met with Biogas Option)

A significant majority of people in Uttara Kannada meet the bulk of their resource needs from plant and animal matter, either gathered directly from non-cultivated land (from forests), water (fallen logs in a river during rainy season) or produced on crop lands and orchards mostly under rainfed conditions. The biomass needs in this region varies from a wide range of needs other than fuel, fodder and food. Eighty different plant species from natural vegetation are used for purposes ranging from food, condiments, masticatories, fuel for domestic cooking, bath water heating and agro processing, fodder for livestock, thatch and small timber for houses and cattlesheds, matting for protecting the houses during the rainy season, poles for fencing, leaf manure for fields and orchards, mulch for weed control, leaves for control of insect pests, wood for manufacture of carts and agricultural implements, fishing traps, brooms, mats, baskets, ropes, idols, garlands, decorative articles, etc. (Gadgil, 1990). This dependence on biomass is mainly due to necessity apart from a diverse tradition.

3.2.2 Working Plans

The concept of working plans was established in India (1890) during the colonial rule, to document the forest policies of the State, both scientific and rational forest management. The working plans prepared so far since 1890 focus mainly on ecological information of the area, floristic information, forest produce and its market potential, availability of labour, growth and yield statistics, technical proposals for future management, financial forecast, details of felling succession and cost of the plan. The choice of the criteria in working plans, although subjective, is intended to give an accurate view of the forest policy and the management system followed. The working plans reveal that, only some areas like the valleys of the Kalinadi, Gangavali, Aghnashini and Sharavati rivers have been covered because of easy access to the banks from the river and cheaper transport facilities. Also most of the wood depots were either located on the coasts at the river mouths or nearby places like Kodibag, Karwar, Honnavar, Bhatkal, etc., or above the Ghats, near the logging sites such as Kirwatti, Londa, Haliyal, Dandeli, etc. The second major area of interest is the drier tract of Haliyal and Mundgod which was easily accessible and contained rich deciduous species like Teak (*Tectona grandis*), *Nandi (Lagerstroemia microcarpa), Honne (Pterocarpus marsupium), Kindal (Terminalia paniculata), Matti (Terminalia tomentosa), Dindal* (*Anogeissus latifolia*) and Jungle wood. The limiting of the economically interesting species to a small proportion of the total forest would have helped in sustainable management if the foresters had contented themselves only with the felling of these species. But the silvicultural systems followed actively discouraged the growth of non economic species to the detriment of the floristic wealth of the tropical forest.

3.2.3 Vegetation Types in Uttara Kannada District

Flora of Uttara Kannada forests was studied by various persons. Buchanan (1801) gives a brief account of the plants encountered in the course of his journey through this district. Botany of Uttara Kannada forests of recent times is provided by Talbot (1909). A good deal of floristic information on coastal, interior and inland high forest is available from the forest working plans (Aitchison, 1910; Mavinkurve, 1955; Shanmukhappa, 1966 and 1977). Denudation of Kanara coast is well accounted by Dhareshwar (1941).

Subash Chandran M.D. (1993) after a detailed survey of various types of forests in Uttara Kannada, links the forest composition with management of the forest ecosystem and makes certain recommendations for more sustainable utilisation of forest resources and betterment of the degraded forest ecosystem. He also discusses the effect of slash and burn agriculture on forest composition practiced in the interior hills and subsequently banned by the close of the 19th century. He highlights the return of evergreen forests with the stoppage of slash and burn agriculture, considered one of the major causal factors for the lower incidence of deciduous trees like *Tectona grandis* in the evergreen forest belt. The forest map of South India (Pascal, et al., 1983), published by Institut Francais de Pondicherry, covering Belgaum-Dharwar-Panaji and Shimoga is used to identify the vegetation types in Uttara Kannada district. Secondary moist deciduous forest could be seen in the river valleys of Kali, Gangavali and Aghnashini. These are secondary forests as the evergreen patch in this region is clear felled as part of Working Plans of the government in early 19th century. Many such patches in yellow colour could be seen with the evergreen forest fragments.

From this analyses, it is evident that the percentage evergreen patch is about 51.1%, moist deciduous 23.2%, dry deciduous 8.6% and coastal 17.1%.

The forests are divided according to convenience of administration by the Forest Department as:

(1) Reserve forests maintained by the Forest Department,
(2) Forest plantations maintained by the Forest Department. Out of those areas notified under the Forest Act, the barren lands and hillocks denuded of tree growth are being afforested by raising plantations like cashew, etc.,
(3) Reserve forests under Section 4 of the Karnataka Forest Act,
(4) Revenue plantations maintained by the Revenue Department representing the afforestation works done in the unreserved lands for which a notification under Section 4 of the Act is issued,
(5) Unreserved lands maintained by the Revenue Department. Protection and management of tree growth in unreserved forests are vested with the Forest Department. Suitable areas of reserved forests are vested with the Forest Department. Suitable areas of unreserved lands in compact blocks have been surveyed and demarcated and the notification under Section 4 of the Act has been issued for constituting them as reserve forests and
(6) Several panchayats in the recent years have started maintaining certain areas of wastelands to develop forests under the scheme "Farm Forestry", which envisages the creation of fuel reserves to help people get their supply of fuel, small-sized timber to meet their agricultural demands and to provide green manure, fodder and grazing to village cattle. Table 3.1.2 gives details of forest area according to legal status as in 1980-81 in the district.

Table 3.1.2: Details of Forest Area According to Legal Status as in 1980-1981 in the District

Forest Types	Area (sq.km.)
Reserve Forests	7727.84
Protected Forests	542.77
Village Forests	20.20
Unclassified Forests	1.84
Total	8292.65

3.2.4 Loss of Forest Area

The increasing population has a direct bearing on increased demand for agricultural lands and this has always had an adverse effect on valuable forests. Unauthorised cultivation poses a serious problem. The high tension electricity lines riddling the forests, increased communications, settlement of expropriated ryots and refugees of lands released for irrigation and hydroelectric projects have a heavy impact on forests. People settling down in colonies in the interior depend on forests for their timber and firewood needs. Grazing also has deleterious effect on forests. The release of forest lands has now been discouraged and any such proposal has to be approved by both the Houses of Legislature. According to the recently enacted Central Forest Conservation Act, no forest land can be released for non-forestry purpose in the State without specific approval of the Government of India. Table 3.1.3 gives the purpose and the areas so far lost in the district.

Table 3.1.3: Details of Forest Area According to Legal Status as in 1980-1981 in the District

Purpose	Area Lost (in hect.)
Submersion	304
Released for rehabilitation	4202
Under power lines	626
Given for cultivation	50158
Mines	26713
Township	1097
Non-agricultural activities	7383
Kali power project	14176
Bedthi project	300
Colony and roads	273
Total	105232

Pascal (1986) classifies vegetation in this region as follows.

Group I. Evergreen or semi-evergreen climax and potentially related forests –

(a) Evergreen or semi-evergreen:

(1) Low elevation (0-850 m in the south and 0-650 m in the north).
Dipterocarpus indicus-Diospyros candolleana-Diospyros oocarpa type;
Honnavar and Kumta (scanty distribution).
Persea macrantha-Diospyros spp.-Holigarna type;
Honnavar and Kumta.
Diospyros spp.-Dysoxylum malabaricum-Persea macrantha type;
Eastern Sirsi.

(2) Medium elevation (650-1400 mm).
Memecylon umbellatum-Syzigium cumini-Actinodaphne angustifolia type;
Transition type: Supa and Joida.

(b) Secondary or degraded stages:

(1) Evergreen and semi-evergreen forests;
Disturbed low elevation: all taluks except Haliyal and Mundgod.
Disturbed transition: Supa and Joida.
Secondary low elevation: Bhatkal, Honnavar, Kumta, Sirsi and Siddapur.

(2) Secondary moist deciduous forests;
Dense forest: Joida, Yellapur and Sirsi.
Woodland to savanna woodland: all except Haliyal and Mundgod.

(3) Other degraded stages;
Thicket low elevation: all except Haliyal and Mundgod.
Thicket medium elevation: Supa and Joida.
Tree savanna to grass savanna low elevation: all except Haliyal and Mundgod.
Scattered shrubs low elevation: Bhatkal, Honnavar, Kumta, Ankola and Karwar.

Group II. Deciduous climax forests and degradation -
Occur in north-eastern parts of the district.

Group III. Plantations -
Acacia catechu, Casuarina equisetifolia, Eucalyptus spp., Tectona grandis, etc.

Group IV. Scrub type -
Common along the minor forest tract which runs as a belt for 8 to 16 kms along the coast.

3.2.5 Types of Vegetation in Various Taluks

Talukwise distribution of various vegetation (Campbell, 1883; Kamath, 1985) is discussed in the following section.

Ankola taluk: As one goes from east to west, forest type changes from laterite thorn to moist deciduous, laterite semi-evergreen in stream pockets and interior depressions. The western part, adjoining the coast, is also denuded due to unrestricted exercise of privileges and shifting cultivation in the past. The inland areas of moist deciduous and semi-evergreen vegetation are closed as fuel forest and high forest area, yielding firewood and timber respectively. Around Hattikeri, in laterite thorn forests, one can come across the *khair* (*Acacia catechu*) trees which yield valuable economic forest produce called *katha*. The Gangavali river valley area supports valuable teak forests having the most common undergrowth, bamboo.

Bhatkal taluk: The forest type changes from laterite thorn to laterite evergreen in this taluk. There is very little moist deciduous forests. The barren hills around Murdeshwar and Bhatkal are a testimony to the acts of unrestricted fellings. As one goes to the interior, the vegetation improves gradually. The evergreen forests around Kop village in the north-east of the taluk contain valuable timber for matches and plywood. The laterite thorn forests situated in the north-west and south-west of the taluk contain a large number of *khair (Acacia chindra)* trees. These are bigger in girth in the Balke forests in the south-west part of the border of Dakshina Kannada district.

Haliyal taluk: The eastern and north-western parts of the taluk comprise teak pole area tending to scrub type towards the border of Dharwad district. The western half comprises of high forests, yielding valuable timber, extracted and transported to Dandeli and Alnavar forest depots. There is sandalwood in drier parts of the area and is extracted annually on a sustained yield basis. There are patches of evergreen forest towards the western side in the lower portions of the valleys of rivers and perennial streams. Bamboo is considered one of the most valuable constituents of economic forest produce.

Honnavar taluk: The forest type changes from laterite to laterite semi-evergreen and evergreen. There is very little of moist deciduous type which can be seen only on top of small hills in the western part of the belt. The coastal strip of forests is all denuded and in many parts, the land has become unfit even to bear poor grass. As one advances to the interior, the forest growth improves gradually. These forests contain valuable timber trees like *poon (Calophyllum Sp.), matti (Terminalia enenulata), dindal (Anogeissus latifolia), bobbi (Hopea Sp.), honne (Pterocarpus marsupium), kindal (Terminalia paniculata), jamba (Xylia xylocarpa), nandi (Lagestroemia microcarpa), bharangi (Vitex altissima)* and others, suitable for matches and plywoods. The laterite semi-evergreen forests and evergreen forests in the north-east corner in Mahime and Jankadkal villages of the taluk, contain *thale (Pandanas Sp.)* palms. The belt of *Acacia catechu* also passes in this taluk, mostly confined to the south-west part. The evergreen forests of Gerusoppa contain varieties of canes which are exported.

Karwar taluk: As one goes from west to east the forest type gradually changes from laterite thorn to moist deciduous and laterite semi-evergreen to evergreen. The forest to the west of Honkane village has been depleted due to unrestricted exercise of privileges. The deciduous forests in the lower slopes tend to be high forests, yielding valuable timber of teak, *sissum (Dalbergia sissoo), honne (Pterocarpus marsupium), kindal (Terminalia paniculata)*, etc. *Jamba (Xylia xylocarpus)* is the predominant species of this tract. The upper slopes, lower valleys and banks of perennial streams contain patches of evergreen forests and large quantities of canes (*Calamum*) that are exported to various places. The upper slopes are not worked due to their inaccessibility. Reserved forests of the moist deciduous type in patches of laterite semi-evergreen in the interior, situated on the steep hills around Karwar, have been classified as "Karwar Town Five Miles Special Reserves".

Kumta taluk: The forest type changes from laterite thorn to moist deciduous, laterite semi-evergreen and evergreen as one advances from west to east. The timber bearing high forests are confined to the south-east part of the taluk at the foot of the Nilkund and Dodmane Ghats around Soppinahosahalli village. The laterite thorn forests around Mirjan, contain *khair (Acacia chindra)* trees which yield valuable catechu. Bamboo occurs in the Aghnashini valley around Soppinahosahalli.

Mundgod taluk: The forest type changes from scrub in the south-west near Sirsi taluk boundary and improves from east to west. The eastern half is comprised of teak pole area, while the western half comprises high forests. The deep valleys in the south-west and the perennial stream belts are covered with patches of semi-evergreen forests. The drier parts of the teak pole arch towards the border of Dharwad district and contain sandalwood.

Siddapur taluk: Due to scanty growth in the eastern side and a major part of the taluk being hilly, no part is organised except the area covered by sandalwood trees towards the north-east, east and south-east. This sandalwood belt also extends to Sirsi and Mundgod taluks. The eastern part is drier and the forest type improves to semi-evergreen from east to west towards the Ghats. There are many large patches of evergreen forests called *kans* in this taluk mostly confined to the west around Dodmane, Nilkund and Malemane Ghats, which contain valuable matchwood.

Sirsi taluk: The forests are mainly semi-evergreen and evergreen type. The evergreen forests are scattered all over the area. The sandalwood belt of Siddapur taluk runs into this taluk and is mostly confined to the south-eastern part bordering Siddapur taluk and Shimoga district.

Supa taluk: The greater part of the Supa tract is very hilly and precipitous. The forest area falls into two different types. The south-eastern part near Gund, parts of Nagjhari valley, Kalinadi and Kaneri slopes contain high forests, yielding mainly timber of valuable species. Gund has the finest teak plantations. Evergreen patches are also found in the valleys. The forests of the northern point near Castlerock yield only fuel and it merges into scrub forests, wherever the soil is very poor. Bamboo grows abundantly in this taluk.

Yellapur taluk: The northern parts of the taluk have valuable teak forests. Bamboo is also abundant here, confined to the catchment area of Gangavali. The bamboo belt also extends to Ankola taluk.

With the knowledge of vegetation in each taluk, the litter and above ground biomass availability in Uttara Kannada district were estimated.

3.3.0 Litter Production in Tropical Moist Forests of Western Ghats

Based on three year study in Attapadi evergreen forests, Pascal (1988) estimated the total mean production of litter as 8.5 t/ha/yr and that of leaves and foliar fragments alone as 6.4 t/ha/yr, which is nearly 74.75% of the total production.

A study on litter production in Uttara Kannada forests was carried out by Bhat, D.M. (1990) at Sonda reserve forest (evergreen species 44.77% and deciduous species 55.23%), Santgal reserve forest (evergreen 86.76%, deciduous 13.24%), Nagur reserve forest (evergreen 66.66%, deciduous 33.34%) and Bhairumbe leaf manure forest (evergreen 43.83%, deciduous 56.17%). The estimation of micro litterfall (leaf matter, reproductive parts, small wood $\leq$ 2 cm diameter), total ground litter (micro litter and dead herbs) and large wood litterfall (wood between 2-10 cms girth) based on two year study is given in Table 3.2. In all the sites studied, leaf litter is a major component which constitutes 65-92% of the total micro litter and 75-94% of the total ground litter. The micro litter estimated by others in similar type of vegetation is 8.8-12.0 t/ha/yr (Proctor, 1983), 6.00 t/ha/yr (Garg and Vyas, 1975), 3.44-4.20 t/ha/yr (Rai and Proctor, 1986 a & b) and 12.9-14.1 t/ha/yr (Songwe, et al., 1988).

Table 3.2: Litter Productivity in Tropical Moist Forests of Uttara Kannada District

Place	Micro litterfall	Ground litter	Wood litterfall
	<--------------------- Production (t/ha/yr)	-------------------	--->
Sonda	$9.783 \pm .20$	21.14 ± 4.71	0.41 ± 0.08
Bhairumbe	6.78 ± 1.62	10.37 ± 2.98	0.71 ± 0.02
Santgal	10.24 ± 0.22	19.16 ± 0.82	1.24 ± 0.36
Nagur	8.29 ± 0.95	15.77 ± 1.94	0.61 ± 0.24
Average	8.04 ± 0.88	15.44 ± 1.74	0.62 ± 0.16

3.4.0 Estimation of Standing Biomass and Biomass Productivity

To estimate standing biomass of any vegetation (tree), knowledge of the girth of a tree (at 130 cm height, also referred as GBH-Girth at breast height), height, branch diameter, etc., is needed. These parameters are also required to compute relationship between standing biomass and tree dimensions. The practices so far followed for this purpose are: (1) destructive sampling (some sample trees belonging to same species are cut and the woody components of the tree measured), (2) computing volume of a tree by actual measurement, that is, actual measurement of all woody components of the tree and its shape, and (3) referring all available literatures.

Standing biomass and biomass productivity from forests in all taluks of Uttara Kannada district are estimated by adopting the relationship and productivity data available in the literature. Other methods for this purpose through destructive (socially unacceptable) or non-destructive (time consuming) sampling, involves primary survey and long-term monitoring, which is unnecessary as researchers at the Centre for Ecological Sciences, Indian Institute of Science, have already carried out a long-term monitoring (since early 80s) of sample plots (1 hectare size) in this region and have many publications on these aspects. Also, other researchers like Rai. S.N. of Karnataka Forest Department have carried out research in this region based on data from evergreen forest sites that were clear felled (submergence due to hydroelectric projects).

Greenland and Kowal (1960) estimated the above ground biomass in Khade tropical forest at Ghana as 265.8 t/ha. Whitemore (1975) studies in tropical forests of Malaysia gives a standing biomass in the range 230-290 t/ha. Rai's (1984) study in Western Ghats at locations in Agumbhe, Bannadpare and Kagneri gives an average standing above ground biomass of 485.67 t/ha (by using the relationship $V=0.0790+0.4149 D^2H$, where V=volume of stem wood, D=diameter at breast height (at 130cm) and H=standing height). He also arrived at an annual basal area increment of 2% based on measurements of basal area of four preservation plots in Uttara Kannada district. Prasad, et al., (1987) estimate average standing biomass in the reserve forests at Nagur - Santgal (Kumta taluk) and Sonda - Bidralli (Sirsi taluk) as 243.25 t/ha, and average standing biomass estimates for minor forests at Bengle - Bhairumbe (Sirsi) and Chandavar - Mirzan (Kumta) as 129.92 t/ha. Lele, S.M., (1993), using regression relationship of $V=0.0250+0.5 D^2H$, estimates standing biomass in leaf manure forests (*Soppinabetta*) at Sirsi in the range 99-201 t/ha. The variation in the biomass estimated by Pascal according to bio volume (D^2H) obtained from 7 sampling plots of 1600 sq.m each in different vegetation is given in Table 3.3.1.

Table 3.3.1: Standing Biomass in Various Types of Vegetation

Vegetation Types	Biomass (t/ha)
Dense evergreen and semi-evergreen	518.40-833.22
Low evergreen	226.55
Secondary semi-evergreen	226.55
Dense deciduous	258.12
Savanna woodland	74.25

3.4.1 Net Annual above Ground Productivity (t/ha) of Tree Layer in Reserve Forests

The net primary productivity (NPP) for reserve forests at four locations in Western Ghats varies from 6.5 to 11.1 t/ha/yr (these values exclude losses due to dead trees). Estimations at sites located in Uttara Kannada district vary from a minimum of 3.95 to a maximum of 7.25 t/ha/yr. And in minor forests a variation of 1.50 to 3.5 t/ha/yr is seen. These are summarised in Table 3.3.2. The annual above ground productivities computed for various types of vegetation, considering wood litterfall and dead trees, are given in Table 3.4, which is used to compute the annual availability of woody biomass in Uttara Kannada district.

3.4.2 Bioenergy Availability from Forests and Demand in the District

Table 3.5.1 lists woody biomass annual availability in various taluks of Uttara Kannada district taking into account woody biomass productivity of 3.6 t/ha/yr (evergreen, semi-evergreen), 3.9 t/ha/yr (deciduous) and 0.9 t/ha/yr (coastal). While, Table 3.5.2 lists woody biomass availability based on higher values of productivity such as 13.5 t/ha/yr for deciduous, 6.5 t/ha/yr for evergreen and semi-evergreen and 1.5 t/ha/yr for coastal region of the district. Based on field data collected from villages in Kumta, Sirsi and Siddapur fuel demand in the region was found to be approximately 1.1 t/capita/yr (higher estimate) and 0.7 t/capita/yr (lower estimate). Talukwise fuel demand computed based on the survey result is listed in 10th column of Tables 3.5.1 and 3.5.2 respectively. Last column in these Tables (3.5.1 and 3.5.2) gives ratio of bioresource availability and fuel wood demand. This ratio if less than one means there is fuel wood scarcity in the taluk and if greater than one, indicates surplus bioresource / fuel wood in the taluk. Fuel wood scarcity is evident in thickly populated coastal taluks Karwar (0.94), Kumta (0.74) and Bhatkal (0.41) as listed in Table 3.5.1. Taking into account inaccessibility of dense forests and degradation of forests close to the villages, the available bioresource computed in Table 3.5.1, based on lower productivity values in each type of vegetation, seems to be more realistic than the values computed in Table 3.5.2. The ratio of talukwise fuel wood available from forests to demand is shown pictorially in Figures 3.2 and 3.3 (based on computations in Tables 3.5.1 and 3.5.2).

It appears that in many taluks household energy consumption does not in itself trigger the removal of tree cover and land degradation. In fact, where energy demand puts a heavy and directly observable damage on the environment, it is the overall framework of development rather than energy demand of the residential sector which is the prime cause of problem. In coastal taluks, the forest based industries, grazing pressure (due to non-availability of fodder), encroachment and conversion of forest land for other purposes have resulted in decrease of vegetation cover and increase in barren hill tops.

Table 3.3.2: Estimates of Net Annual above Ground Productivity (t/ha)

Forest type	Locality	NPP (t/ha/yr) of oven dry matter	Reference
Reserve	Bidralli	3.946	Narendra Prasad, et al., 1987
	Nagur	7.250	
	Santgal	5.296	
	Sonda	5.088	
Minor	Bengle	3.533	
	Bhairumbe	3.314	
	Chandavar	3.476	
	Mirzan	1.501	
Reserve	Agumbe	11.10	Rai, 1986
	Bannadpare	6.500	
	Kagneri	9.20	
	South Bhadra	9.500	
Reserve	Malenalli	6.2 (thick veget.)	Lele, 1993
		3.6 (medium type)	
		0.9 (low density)	
	Arasapura	6.2 (high density)	
		3.6 (medium type)	
		0.9 (low density)	
Shrub layer	Sonda	1046.88 kg/ha	Narendra Prasad
Reserve	Bidralli	969.75 kg/ha	
	Nagur	224.14 kg/ha	
Herb layer	Minor forest	8.016 ± 3.50 t/ha	
	Reserve	3.150 ± 3.69 t/ha	

Table 3.4: Biomass Productivities in Various Types of Vegetation

Vegetation Types	Biomass (t/ha/yr)
Dense Evergreen and Semi-Evergreen	13.41-27.0
Low Evergreen	3.60-6.50
Secondary Evergreen	3.60-6.50
Dense Deciduous Forest	3.90-13.50
Savanna Woodland	0.50-3.50
Coastal	0.90-1.50

3.5.0 BIOENERGY FROM AGRICULTURE

In this section, assessment of residues from major crops such as paddy, oil seeds, maize, etc., is carried out.

3.5.1 Agricultural Population in Uttara Kannada District

As per 1991 census, out of 4,28,663 workers, 1,41,345 were cultivators (32.98%), 82,283 agricultural labourers (19.19%), and 56,937 depend on livestock and forestry (13.28%). Thus, the population depending on agriculture and allied jobs amounts to 65.45%. The talukwise land utilisation for various agricultural crops, production and yield are given in Table 3.6.1. The net cropped area in the district is about 1084.28 sq.km., which amounts to 10.54% of the total area (compared to 55% of Karnataka State) as against 12% in 1980-81. The percentage area sown more than once is about 15% of the net area sown. Agriculture and plantations are the major sources of district revenue (nearly 50%). It is noticed that paddy yield is minimum in Mundgod (0.91 t/ha for improved variety, 1.62 for high yielding (HY) variety), while it is maximum for Sirsi (3.04 t/ha for HY variety, 1.42 for improved variety) and Siddapur (1.87 t/ha for HY variety, 1.62 t/ha for improved variety). Based on 1984-85 to 1993-94 talukwise crop data, average production (in tonnes), standard deviation, and maximum and minimum values for major crops computed are listed in Table 3.6.2.

Table 3.5.1: Population (1991) and Area under Different types of Vegetation in Uttara Kannada (Talukwise) [Computation of bioresource availability/demand ratio (considering demand as 1.1t/person)]

Taluk	Population Total (1991)	Forest hect.	Dry Deciduous	Moist Deciduous	Evergreen	Semi-evergreen	Coastal	Biomass Production	Demand @1.1 t/person	Avail./ Demand Ratio
Bhatkal	129017	25390.46			5078.1	8124.9	12187.4	58499.6	141918.7	0.41
Kumta	134144	39187.59		7837.5	11756.3	7837.5	11756.3	109333.4	147558.4	0.74
Karwar	142845	52808.68		10561.7	15842.6	10561.7	15842.6	147336.2	157129.5	0.94
Honnavar	145842	57416.68		2870.8	20095.8	17225.0	17225.0	160192.5	160426.2	1.00
Ankola	91310	75432.37		15086.5	22629.7	15086.5	22629.7	210456.3	100441	2.10
Haliyal	94363	59099.78	23639.9	23639.9	11820.0			219851.2	103799.3	2.12
Siddapur	94202	62274.6			40478.5	21796.1		224188.6	103622.2	2.16
Sirsi	152935	103226.57		15484.0	46452.0	41290.6		371615.7	168228.5	2.21
Mundgod	77939	51167.36	28142.0	23025.3				192645.1	85732.9	2.25
Yellapur	81410	113429.56		51043.3	62386.3			408346.4	89551	4.56
Supa	99519	160608.71		40152.2	120456.5			578191.4	109470.9	5.28
District	1243526	800042.36	51782.0	189701.3	356995.7	121922.4	79641.0	2680656.3	1367878.6	1.96

Table 3.5.2: Population (1991) and Area under Different types of Vegetation in Uttara Kannada (Talukwise) [Compulsion of bioresource availability/demand ratio (considering demand as 0.7t/person)]

Taluk	Population Total (1991)	Forest hect.	Dry Deciduous	Moist Deciduous	Evergreen	Semi-evergreen	Coastal	Biomass Production	Demand @0.7 t/person	Avail. /Demand Ratio
Supa	99519	160608.71		40152.2	120456.5			1031911.0	109470.9	9.43
Yellapur	81410	113429.56		51043.3	62386.3			721979.1	89551	8.06
Mundgod	77939	51167.36	28142.0	23025.3				522674.6	85732.9	6.10
Haliyal	94363	59099.78	23639.9	23639.9	11820.0			542536.0	103799.3	5.23
Sirsi	152935	103226.57		15484.0	46452.0	41290.6		666327.5	168228.5	3.96
Siddapur	94202	62274.6			40478.5	21796.1		404784.9	103622.2	3.91
Ankola	91310	75432.37		15086.5	22629.7	15086.5	22629.7	372635.9	100441	3.71
Honnavar	145842	57416.68		2870.8	20095.8	17225.0	17225.0	286222.1	160426.2	1.78
Karwar	142845	52808.68		10561.7	15842.6	10561.7	15842.6	260874.9	157129.5	1.66
Kumta	134144	39187.59		7837.5	11756.3	7837.5	11756.3	193586.7	147558.4	1.31
Bhatkal	129017	25390.46			5078.1	8124.9	12187.4	104100.9	141918.7	0.73
District	1243526	800042.36	51782.0	189701.3	356995.7	121922.4	79641.0	4562742.2	1367878.6	3.34

Table 3.6.1: Area (in hectares) under Different Crops and Average Production and Yield in Uttara Kannada District

(i) Area (in hectares) under different crops

	Paddy		Ragi	Pulses	Sugar-cane	Jowar	Maize	Oil seed	Total
	High Yield Variety	Improved Variety							
Ankola	4328	2274	5	291	61			1151	8110
Bhatkal	3468	1322	0	461	74			1048	6373
Haliyal	9666	7188	0	3358	630	125	28	345	21340
Honnavar	4184	1585	0	222	192	345	0	1649	8177
Karwar	4242	3079	12	77	16	0	22	143	7591
Kumta	4333	3005	218	183	127	0	0	931	8797
Mundgod	6680	6954	0	904	23	0	0	140	14701
Siddapur	5391	2350	10	142	215		0	33	8141
Sirsi	4725	6469	107	828			0	382	12511
Supa	3661	1745	0	45	260			20	5731
Yellapur	3367	3071		413	84			21	6956
Total	54045	39042	352	6924	1682	470	50	5863	108428

(ii) Average Production (Talukwise)

	Paddy		Ragi	Pulses	Sugar-cane	Jowar	Maize	Oil seed
	High Yield Variety	Improved Variety						
Ankola	11489.82	2889.91	8.57	103.13	4840.00			1800.56
Bhatkal	6739.34	1523.29	9.00	185.67	4756.04			2799.95
Haliyal	17194.45	7473.86	16.86	1126.15	32852.29	218.86	110.57	276.29
Honnavar	8202.86	1698.16	6.71	95.82	15704.00	431.00	0.36	1780.12
Karwar	10608.30	4232.36	35.70	54.24	1940.86	31.86	252.79	330.76
Kumta	9821.14	4522.90	82.74	93.93	7684.18	8.14	3.57	1341.70
Mundgod	10848.78	6351.72	2.71	538.08	2932.29	41.00	19.64	180.23
Siddapur	10066.04	3851.71	15.97	496.90	13170.86		18.36	279.50
Sirsi	14347.56	9181.39	210.61	380.30			4.07	676.54
Supa	6036.36	2438.72	6.73	49.78	11955.00			28.66
Yellapur	8014.98	5853.02		211.89	6189.71			52.56
Total	113369.65	50017.06	395.61	3335.89	102025.22	730.86	409.36	9546.86

Table 3.6.1: Area (in hectares) under Different Crops and Average Production and Yield in Uttara Kannada District (Cont'd)

(iii) Average Yield (Talukwise)

	Paddy		Ragi	Pulses	Sugar-cane	Jowar	Maize	Oil seed
	High Yield Variety	Improved Variety						
Ankola	2.65	1.27	1.71	0.35	79.34			1.56
Bhatkal	1.94	1.15		0.40	64.27			2.67
Haliyal	1.78	1.04		0.34	52.15	1.75	3.95	0.80
Honnavar	1.96	1.07		0.43	81.79	1.25		1.08
Karwar	2.50	1.37	2.98	0.70	121.30		11.49	2.31
Kumta	2.27	1.51	0.38	0.51	60.51			1.44
Mundgod	1.62	0.91		0.60	127.49			1.29
Siddapur	1.87	1.64	1.60	3.50	61.26			8.47
Sirsi	3.04	1.42	1.97	0.46				1.77
Supa	1.65	1.40		1.11	45.98			1.43
Yellapur	2.38	1.91		0.51	73.69			2.50
Total	23.66	14.69	8.63	8.91	767.78	3.00	15.44	25.33

Table 3.6.2: Average (avg), Standard Deviation (Sd), Maximum and Minimum Production (in tonnes) during 1984-85 to 1993-94

Crop: Paddy - High Yield Variety

	Avg	Sd	Max	Min
Ankola	11489.82	3159.50	18513.00	8179.00
Bhatkal	6739.34	785.40	7838.00	5303.00
Haliyal	17194.45	6315.85	23607.00	5676.00
Honnavar	8202.86	2248.29	11494.00	3838.00
Karwar	10608.30	3162.13	16276.00	7300.00
Kumta	9821.14	2408.96	15152.00	6693.00
Mundgod	10848.78	6030.95	20375.00	1275.00
Siddapur	10066.04	1573.61	13075.00	7964.00
Sirsi	14347.56	4337.03	22761.00	10535.94
Supa	6036.36	2131.31	8758.55	1788.00
Yellapur	8014.98	2243.38	10637.00	3650.00

Table 3.6.2: Average (avg), Standard Deviation (Sd), Maximum and Minimum Production (in tonnes) during 1984-85 to 1993-94 (Cont'd.)

Crop: Paddy - Improved Variety

	Avg	Sd	Max	Min
Ankola	2889.91	742.50	3874.00	1545.00
Bhatkal	1523.29	315.86	2010.00	1094.00
Haliyal	7473.86	3074.39	11511.00	3015.00
Honnavar	1698.16	650.97	2473.00	406.00
Karwar	4232.36	999.74	6027.00	2997.00
Kumta	4522.90	1225.35	6340.00	2883.00
Mundgod	6351.72	4154.18	13217.00	640.00
Siddapur	3851.71	702.39	5275.00	2900.00
Sirsi	9181.39	2652.26	12095.00	4633.00
Supa	2438.72	1401.81	5468.00	938.00
Yellapur	5853.02	2136.59	9755.00	2631.00

Crop: Jowar

	Avg	Sd	Max	Min
Haliyal	218.86	56.82	300.00	138.00
Mundgod	431.00	271.03	1080.00	204.00
Sirsi	31.86	61.58	181.00	0.00
Supa	8.14	10.96	30.00	0.00
Yellapur	41.00	57.25	168.00	0.00

Crop: Maize

	Avg	Sd	Max	Min
Haliyal	110.57	52.52	220.00	56.00
Honnavar	0.36	0.87	2.50	0.00
Mundgod	252.79	207.40	577.00	28.00
Siddapur	3.57	8.75	25.00	0.00
Sirsi	19.64	27.61	85.50	0.00
Supa	18.36	16.50	40.50	0.00
Yellapur	4.07	6.91	20.50	0.00

Table 3.6.2: Average (avg), Standard Deviation (Sd), Maximum and Minimum Production (in tonnes) during 1984-85 to 1993-94 (Cont'd.)

Crop: Oil seeds

	Avg	Sd	Max	Min
Ankola	1800.56	160.76	2142.00	1620.00
Bhatkal	2799.95	3532.12	11420.00	717.00
Haliyal	276.29	101.72	449.26	124.25
Honnavar	1780.12	363.05	2122.00	1027.00
Karwar	330.76	143.98	570.00	198.00
Kumta	1341.70	711.78	2739.92	130.00
Mundgod	180.23	95.99	318.00	56.00
Siddapur	279.50	197.14	616.00	60.47
Sirsi	676.54	175.50	951.00	397.00
Supa	28.66	16.31	47.00	3.00
Yellapur	52.56	27.48	111.00	29.00

Crop: Pulses

Ankola	103.13	22.22	141.00	73.00
Bhatkal	185.67	39.82	248.00	132.00
Haliyal	1126.15	214.76	1419.00	794.00
Honnavar	95.82	24.54	137.00	61.00
Karwar	54.24	34.28	109.00	25.00
Kumta	93.93	29.68	129.00	52.31
Mundgod	538.08	358.77	1365.00	247.00
Siddapur	496.90	573.99	1858.00	73.32
Sirsi	380.30	117.55	587.00	196.00
Supa	49.78	40.65	139.00	16.00
Yellapur	211.89	81.54	364.00	110.00

Crop: Ragi

	Avg	Sd	Max	Min
Ankola	8.57	6.19	21.00	0.00
Bhatkal	9.00	10.41	28.00	0.00
Haliyal	16.86	10.89	31.00	0.00
Honnavar	6.71	4.83	14.00	0.00
Kumta	35.70	19.82	69.00	0.00
Mundgod	82.74	80.62	259.00	0.00
Siddapur	2.71	4.49	12.00	0.00
Sirsi	15.97	10.90	32.00	0.00
Supa	210.61	152.01	424.00	0.00
Yellapur	6.73	7.61	24.00	0.00

Table 3.6.2: Average (avg), Standard Deviation (Sd), Maximum and Minimum Production (in tonnes) during 1984-85 to 1993-94 (Cont'd.)

Crop: Sugar-cane

Ankola	4840.00	1109.79	6717.00	3658.00
Bhatkal	4756.04	2130.00	6717.00	50.26
Haliyal	32852.29	10118.43	47520.00	14286.00
Honnavar	15704.00	8109.91	35250.00	10348.00
Karwar	1940.86	1100.44	3500.00	928.00
Kumta	7684.18	3616.40	12500.00	86.26
Mundgod	2932.29	3480.65	10604.00	0.00
Siddapur	13170.86	5861.12	22776.00	5250.00
Supa	11955.00	5771.85	19493.00	4425.00
Yellapur	6189.71	742.05	7308.00	4890.00

3.5.2 Agricultural Residues

(i) *Paddy Residues*: Paddy (*Oryza sativa*) is a major cereal crop (85.85%, 93,087 ha) in Uttara Kannada district. Rice husk and stalks are the major residues of paddy cultivation. The talukwise potential availability of husk and stalk is given in Table 3.6.3. The quality of husk depends on the type of rice mill. In the single huller, the husk is obtained in fine broken state and is always mixed with bran and broken rice. This husk-bran mix is used by farmers as cattle feed and by mill owners as boiler fuel. In the sheller cum huller or sheller and modern rice mills, the husk is free from bran and broken rice. The average higher calorific value of rice husk ranges from 2937.5 to 3461.31 kcals (the higher calorific value includes the latent heat of water vapour in the product of combustion-obtained with the help of bomb calorimeter). The lower calorific value ranges from 2637.29 to 3161.25 kcals. Proximate analysis of paddy husk shows that volatile matter ranges from 66.6-69.3%, fixed carbon 12.7-16.9% and ash 16.5-18.6%. Based on quantity of paddy production in each taluk, energy equivalent in kWh is computed for husk and stalk. Stalk is mainly used as fodder. The energy equivalent of paddy husk (both HY and improved variety) as indicated in Table 3.6.3 is about 1701.23 lakh kWh.

Table 3.6.3: Average (avg) and Standard Deviation (Sd) – Production Energy from Agricultural Residues for Fuel and fodder in Iakh kWh

	Production Average	Sd	Residues in Tonnes		Energy in lakh kWh that could be generated out of this waste	
Crop: Paddy - High Yield Variety			Husk	Stalk	Husk - Fuel	Stalk - Fodder
Ankola	11489.82	3159.50	3431.89	11486.32	119.64	400.41
Bhatkal	6739.34	785.40	2012.97	6737.29	70.17	234.86
Haliyal	17194.45	6315.85	5135.80	17189.21	179.03	599.22
Honnavar	8202.86	2248.29	2450.11	8200.36	85.41	285.86
Karwar	10608.30	3162.13	3168.59	10605.07	110.46	369.69
Kumta	9821.14	2408.96	2933.47	9818.15	102.26	342.26
Mundgod	10848.78	6030.95	3240.42	10845.48	112.96	378.07
Siddapur	10066.04	1573.61	3006.62	10062.97	104.81	350.80
Sirsi	14347.56	4337.03	4285.47	14343.19	149.39	500.00
Supa	6036.36	2131.31	1803.00	6034.52	62.85	210.36
Yellapur	8014.98	2243.38	2393.99	8012.53	83.45	279.32
Total	113369.65		33862.31	113335.09	1180.44	3950.86
Crop: Paddy - Improved Variety			Husk	Stalk	Husk - Fuel	Stalk - Fodder
Ankola	2889.91	742.50	863.19	2889.03	30.09	100.71
Bhatkal	1523.29	315.86	454.99	1522.83	15.86	53.09
Haliyal	7473.86	3074.39	2232.36	7471.58	77.82	260.46
Honnavar	1698.16	650.97	507.22	1697.64	17.68	59.18
Karwar	4232.36	999.74	1264.16	4231.07	44.07	147.50
Kumta	4522.90	1225.35	1350.94	4521.53	47.09	157.62
Mundgod	6351.72	4154.18	1897.19	6349.79	66.14	221.35
Siddapur	3851.71	702.39	1150.47	3850.54	40.11	134.23
Sirsi	9181.39	2652.26	2742.38	9178.59	95.60	319.97
Supa	2438.72	1401.81	728.42	2437.98	25.39	84.99
Yellapur	5853.02	2136.59	1748.24	5851.24	60.94	203.97
Total	50017.06		14939.57	50001.82	520.79	1743.06
Crop: Jowar				Stalk		Fodder
Haliyal	218.86	56.82		262.60		9.15
Mundgod	431.00	271.03		517.14		18.03
Sirsi	31.86	61.58		38.22		1.33
Supa	8.14	10.96		9.77		0.34
Yellapur	41.00	57.25		49.19		1.71
Total	730.86	457.64		876.92		30.57

Table 3.6.3: Average (avg) and Standard Deviation (Sd) – Production Energy from Agricultural Residues for Fuel and fodder in lakh kWh (Cont'd)

	Production Average	Sd	Residues in Tonnes		Energy in lakh kWh that could be generated out of this waste	
Crop: Maize			Husk	Stalk	Fuel	Fodder
Haliyal	110.57	52.52	110.57	221.14	4.50	8.99
Honnavar	0.36	0.87	0.36	0.71	0.01	0.03
Mundgod	252.79	207.40	252.79	505.57	10.28	20.56
Siddapur	3.57	8.75	3.57	7.14	0.15	0.29
Sirsi	19.64	27.61	19.64	39.29	0.80	1.60
Supa	18.36	16.50	18.36	36.71	0.75	1.49
Yellapur	4.07	6.91	4.07	8.14	0.17	0.33
Total	409.36		409.36	818.71	16.65	33.30
Crop: Oil seeds			Husk		Fuel	
Ankola	1800.56	160.76	539.78		21.95	
Bhatkal	2799.95	3532.12	839.37		34.14	
Haliyal	276.29	101.72	82.83		3.37	
Honnavar	1780.12	363.05	533.65		21.70	
Karwar	330.76	143.98	99.15		4.03	
Kumta	1341.70	711.78	402.22		16.36	
Mundgod	180.23	95.99	54.03		2.20	
Siddapur	279.50	197.14	83.79		3.41	
Sirsi	676.54	175.50	202.81		8.25	
Supa	28.66	16.31	8.59		0.35	
Yellapur	52.56	27.48	15.76		0.64	
Total	9546.86		2861.97		116.40	
Crop: Pulses	Average	Sd		Stalk		Fodder
Ankola	103.13	22.22		51.57		1.80
Bhatkal	185.67	39.82		92.84		3.24
Haliyal	1126.15	214.76		563.08		19.63
Honnavar	95.82	24.54		47.91		1.67
Karwar	54.24	34.28		27.12		0.95
Kumta	93.93	29.68		46.97		1.64
Mundgod	538.08	358.77		269.04		9.38
Siddapur	496.90	573.99		248.45		8.66
Sirsi	380.30	117.55		190.15		6.63
Supa	49.78	40.65		24.89		0.87
Yellapur	211.89	81.54		105.94		3.69
Total	3335.89			1667.94		58.14

Table 3.6.3: Average (avg) and Standard Deviation (Sd) –
Production Energy from Agricultural Residues for Fuel and fodder in Iakh kWh (Cont'd)

	Production Average	Sd	Residues in Tonnes	Energy in lakh kWh that could be generated out of this waste	
Crop: Ragi			Stalk		Fodder
Ankola	8.57	6.19	4.29		0.15
Bhatkal	9.00	10.41	4.50		0.16
Haliyal	16.86	10.89	8.43		0.29
Honnavar	6.71	4.83	3.36		0.12
Kumta	35.70	19.82	17.85		0.62
Mundgod	82.74	80.62	41.37		1.44
Siddapur	2.71	4.49	1.36		0.05
Sirsi	15.97	10.90	7.99		0.28
Supa	210.61	152.01	105.31		3.67
Yellapur	6.73	7.61	3.36		0.12
Total	395.61		197.81		6.90
Crop: Sugar-cane			Bagasse	Fuel	
Ankola	4840.00	1109.79	1462.83	59.49	
Bhatkal	4756.04	2130.00	1437.45	58.46	
Haliyal	32852.29	10118.43	9929.18	403.82	
Honnavar	15704.00	8109.91	4746.33	193.03	
Karwar	1940.86	1100.44	586.60	23.86	
Kumta	7684.18	3616.40	2322.44	94.45	
Mundgod	2932.29	3480.65	886.25	36.04	
Siddapur	13170.86	5861.12	3980.72	161.90	
Sirsi	8865.65	2341.35	2679.53	108.98	
Supa	11955.00	5771.85	3613.24	146.95	
Yellapur	6189.71	742.05	1870.76	76.08	
Total	110890.87		33515.33	1363.07	
Crop: Cotton			Stalk		Fodder
Haliyal	6376.75	2564.12	22318.63		778.03
Mundgod	4406.33	2300.12	15422.16		537.62
Siddapur	9.13	6.22	31.96		1.11
Sirsi	497.67	243.45	1741.85		60.72
Supa	124.57	43.56	436.00		15.20
Yellapur	591.37	231.12	2069.80		72.15
Total	12005.82	5388.59	42020.37		1464.83
Total Energy (lakh kWh) from Agricultural Residues in Uttara Kannada				3197.35	7287.66

(ii) *Bagasse*: Sugar-cane (*Saccharum officinarum*) is one of the important cash crops in Karnataka's economy. The area under sugar-cane in Uttara Kannada is about 1682 hectares with a production of 1,02,025 tonnes. Talukwise area, production and yield of sugar-cane is listed in Table 3.6.1, while Table 3.6.2 provides quantity of bagasse available in the district. Bagasse is the fibrous residue left after extraction of sugar-cane juice, used to make jaggery. The quantity of bagasse depends on the fibrous content, ranging from 30-32%. The bagasse produced is used as fuel along with logs of wood for making jaggery. The average gross calorific value of bagasse is about 3500 kcals. One tonne of bagasse generates about 2.5 tonnes of steam. Bagasse is used to generate methane, about 200 m^3 of which is generated out of a tonne of bagasse. In sugar industry, bagasse is used for generating steam in the boilers to drive the prime movers as well as for boiling and concentrating juice. The quantity of bagasse available in Uttara Kannada is about 33,515 tonnes and its energy equivalent is 1363 lakh kWh.

(iii) *Oil Seed*: Ground-nut (*Arachis hypogea*) in terms of weight is the most important oil seed crop, ranking next to cotton in economic value. Uttara Kannada district has about 5863 hectares (5.48%) under oil seed crop. Depending on variety, season and soil conditions, about 30% of ground-nut pod consists of shell. On average about 2861.97 tonnes of shell is available in the district. Average higher calorific value of shell is about 4532.15 kcal/kg and lower calorific value about 4248.58 kcal/kg. The energy equivalent of total available ground-nut husk is about 116.4 lakh kWh. Ground-nut shells are also reported to be good raw materials for the manufacture of activated carbons for bleaching purposes, and compare favourably well with other good quality carbons.

(iv) *Maize Residues*: Maize cobs constitute about 30% of maize grain (*Zea mays*). The cobs are used as fuel or as a supplementary feed for cattle. About 409.36 tonnes of maize husk and 818.71 tonnes of stalk (mainly used as fodder) are available. The energy equivalent of husk works out to be about 16.65 lakh kWh. While, residues of ragi (*Eleusine coracana*) also referred as finger millet (197.81 tonnes), pulses (1667.84 tonnes), cotton-*Gossypium hirsutum* (42,020.37 tonnes) and jowar-*Sorghum vulgare* (876.92 tonnes) are used mainly as fodder. Energy equivalents of these residues are listed in Table 3.6.3.

3.6.0 BIOENERGY FROM HORTICULTURE

Next to paddy, plantation crops such as areca (*Areca catechu*), cashew (*Anacardium occidentale*), coconut (*Cocos nucifera*), pepper (*Piper nigrum*) and cardamom (*Elletaria cardamomum*) occupy an important place. The area brought under these crops shows an upward trend during recent years. The cultivation of areca is confined to the taluks of Sirsi, Siddapur, Honnavar, Kumta and Yellapur. In the coastal belt, coconut is the important cash crop. Talukwise area and production of areca, coconut and cashew are listed in Table 3.6.4. Area under areca, coconut and cashew in Uttara Kannada are 8499, 5685 and 1749 hectares respectively.

3.6.1 Horticulture Residues

(i) *Coconut Residues*: The fuel biomass of coconut palm are leaves (12/tree/year), inflorescence (12-15/tree/year), shells (100/tree/year), husk (100/tree/year) and leaf sheath. Weight of these constituents per tree are given in Table 3.6.5.

Table 3.6.4: Area (in hectares) and Production (in tones) of different Horticultural Crops and Net Energy Available (in lakh kWh)

Crop: Areca

	Area hectares	% share	Production tonnes	Leaves tonnes	Inflore. tonnes	Nuts+Husk tonnes	Leaf Sheath tonnes	[husk (30%)+inflore. +leaf sheath (50%)+leaf (50%)] in lakh kWh
Ankola	235	2.77	293.75	1443.49	1443.49	875.61	951.75	101.00
Bhatkal	234	2.75	292.50	2874.69	2211.30	871.88	947.70	150.21
Haliyal		0.00						
Honnavar	922	10.85	1152.50	11326.77	8712.90	3435.37	3734.10	591.85
Karwar	23	0.27	28.75	282.56	217.35	85.70	93.15	14.76
Kumta	459	5.40	573.75	5638.82	4337.55	1710.23	1858.95	294.64
Mundgod	77	0.91	96.25	945.95	727.65	286.90	311.85	49.43
Siddapur	2400	28.24	5280.00	29484.00	22680.00	8942.40	9720.00	1540.61
Sirsi	2925	34.42	6435.00	35933.63	27641.25	10898.55	11846.25	1877.62
Supa	94	1.11	206.80	1154.79	888.30	350.24	380.70	60.34
Yellapur	1130	13.30	2486.00	13882.05	10678.50	4210.38	4576.50	725.37
Total	8499	100.00	16845.30	102966.73	79538.29	31667.27	34420.95	5405.83

Crop: Coconut

	Area hectares	% share	Nuts in '000s	Leaves tonnes	Inflore. tonnes	Husk tonnes	Shells Tonnes	Leaf Sheath tonnes	[Shell +30%husk+leaf sheath +inflore+leaves (40%)] in lakh kWh
Ankola	641	11.28	7692.00	3772.93	699.20	1153.80	3030.65	396.14	278.55
Bhatkal	642	11.29	7704.00	3778.81	700.29	1155.60	3035.38	396.76	278.98
Haliyal	14	0.25	168.00	82.40	15.27	25.20	66.19	8.65	6.08
Honnavar	1535	27.00	18420.00	9035.01	1674.38	2763.00	7257.48	948.63	667.04
Karwar	785	13.81	9420.00	4620.51	856.28	1413.00	3711.48	485.13	341.13
Kumta	1200	21.11	14400.00	7063.20	1308.96	2160.00	5673.60	741.60	521.47
Mundgod	53	0.93	636.00	311.96	57.81	95.40	250.58	32.75	23.03
Siddapur	45	0.79	540.00	264.87	49.09	81.00	212.76	27.81	19.55
Sirsi	373	6.56	4476.00	2195.48	406.87	671.40	1763.54	230.51	162.09
Supa	81	1.42	972.00	476.77	88.35	145.80	382.97	50.06	35.20
Yellapur	316	5.56	3792.00	1859.98	344.69	568.80	1494.05	195.29	137.32
Total	5685	100.00	68220.00	33461.91	6201.20	10233.00	26878.68	3513.33	2470.44

Table 3.6.4: Area (in hectares) and Production (in tones) of different Horticultural Crops And Net Energy Available (in lakh kWh) (Cont'd)

	Area hectares	% share	Production tonnes	Husk tonnes	Fuel wood tonnes	(husk+fuel wood) in lakh kWh
Ankola	304	17.38	212.80	319.20	1064.00	63.70
Bhatkal	188	10.75	131.60	197.40	658.00	39.39
Haliyal	10	0.57	7.00	10.50	35.00	2.10
Honnavar	340	19.44	238.00	357.00	1190.00	71.24
Karwar	140	8.00	98.00	147.00	490.00	29.33
Kumta	627	35.85	827.64	1241.46	2194.50	148.36
Mundgod	8	0.46	5.60	8.40	28.00	1.68
Siddapur	13	0.74	9.10	13.65	45.50	2.72
Sirsi	69	3.95	48.30	72.45	241.50	14.46
Supa	33	1.89	23.10	34.65	115.50	6.91
Yellapur	17	0.97	11.90	17.85	59.50	3.56
Total	1749	100.00	1613.04	2419.56	6121.50	383.45

Table 3.6.5: Fuel Biomass from Coconut Palm per Year

Components	Productivity/tree/yr	Biomass per tree (in kgs)
Leaves	12	48.50
Inflorescence	12-15	10.00
Shells	100	14.91
Husk	100	39.55
Leaf Sheath	12	6.55
Total		119.51

Coconut husk is generally used for making rope, coir products and for mulching in coconut plantations in Uttara Kannada district. Coir dust, which constitutes 70% of husk, decomposes very slowly in the soil as its pentosan-lignin ratio is less than 0.5. It absorbs water about eight times its weight and parts with it comparatively slowly (Thampan, 1981). The plaited leaves of coconut palm are used for thatching houses, fencing and for making baskets. Unplaited leaves are also used for fencing, mulching and for shading nurseries. Lower hard portion of leaves is used as fuel, which constitutes 40 - 50% of total weight of leaf.

Coconut shells are mainly used for fuel and to a lesser extent for the manufacture of various domestic utensils, fancy items, etc. The calorific value of shell is in the range of 4500-4800 kcals/kg. The commercial utilisation of coconut shell for the production of shell charcoal, activated carbon and shell flour is gaining importance now. The output of charcoal by ordinary processes is about 30% of the original shells. The average charcoal output has been found to be 35 kgs from 1000 whole shells. Talukwise energy equivalent of coconut residues is estimated assuming that 100% shells, 100% inflorescence, 30% husk and 40% leaves would be available as fuel, as listed in Table 3.6.4. It is seen that coastal taluks contribute a major share of 84.49% of the total. The energy equivalent of coconut residues for Uttara Kannada is about 2470.44 lakh kWh, with 667.04 lakh kWh for Honnavar followed by Kumta with 521.47 lakh kWh.

(ii) *Areca Residues*: The fuel biomass of Arecanut palm are leaves (6-7/tree/year), inflorescence (4-5/tree/year), husk and leaf sheath.

Areca husk: Husk is the outer cover of areca fruit. It constitutes about 60-80% of the total volume and weight (fresh weight basis) of the fruits (Bavappa, et al., 1982). Currently, this is being largely wasted in Uttara Kannada except for use as an inferior fuel and mulch. The husk fibers are predominantly composed of cellulose with varying proportions of hemicellulose (35-64.8%), lignin (13.0-26.0%), pectin and protopectin. These properties makes it a good raw material for manufacturing hard boards, paper boards, activated carbon, etc. Properly composted husk could be a good organic manure. It consists of 1.0-1.1% N_2, 0.4-0.5% P_2O_5 and 1.0-1.5% K_2O. It has been estimated (Bidappa, 1960) that one lakh tonne of composted husk provides about 1000 tonnes of N_2, 500 tonnes of P_2O_5 and 1000 tonnes of K_2O. It is also reported that husk is highly resistant to microbial degradation because of the presence of ligno cellulose.

Leaf sheath: Areca palm sheds about 6-7 leaf sheath per year. About 34,420.95 tonnes of leaf sheath is available in the district (talukwise availability is listed in Table 3.6.4). The sheath measures about 75-85 cms long, 35-40 cms wide at the center and 15-20 cms wide at the stalk end. Fresh fallen leaves contain about 55-60% moisture. The major constituents of leaf sheath are cellulose (43%), crude fibre (33%) and ash (5%). In Sirsi and Siddapur taluks leaf sheath is also used as cattle feed by some farmers. From organic manure point of view it consists of 0.7% N_2 , 0.3% P_2O_5 and 1.0% K_2O.

In coastal taluks like Honnavar, sheaths are used to manufacture hats for farm workers. In some places it is used to manufacture throw-away cups and plates, packing cases, leaf sheath plyboards, decorative veneer panels and picture mounts.

Leaf: Quantity of areca leaves available is about 1,02,966.7 tonnes. At present, it is used for thatching and also as mulch in areca gardens. The leaves are a good source of organic manure. Approximate chemical composition is 0.94% N_2 , 0.096% P_2O_5 and 1.0% K_2O. Other residues such as inflorescence are used mainly as fuel, about 79,538.39 tonnes of which is available annually. With the assumption that 30% husk, 50% leaf sheath and 100% inflorescence is available for fuel, the energy equivalent computed from areca residues works out to be 5405.83 lakh kWh.

(iii) *Cashew*: The cashew nut tree was introduced for the purpose of checking soil erosion on coastal lands and hill slopes. It does not grow satisfactorily at elevations higher than 300 meters (Jaiswal, et al., 1980). Cashew nut has gained commercial importance during the last two decades on account of the increased demand for its edible kernel and shell oil in the international markets.

Area under cashew plantation in Uttara Kannada is about 1749 hectares with annual production of 1613 tonnes of kernels. Cashew shell husk is an important residue with annual availability of 2419.56 tonnes. Considering net primary productivity of cashew wood from plantations as 3 t/ha/yr, total quantity of fuel wood available annually is about 6121.5 tonnes. The energy equivalent of cashew shell husk and fuel wood is 383.45 lakh kWh. Talukwise information of area under cashew production, quantity of husk and fuel wood available annually is given in Table 3.6.4.

3.7.0 BIOENERGY FROM ANIMAL RESIDUES

3.7.1 Livestock

Livestock is an important component of agro ecosystems. For instance, livestock provide the critical energy input to crop lands that is required for ploughing, threshing and other farm operations. Animal dung provides essential nutrients required for soil fertility and crop yields in the form of organic manure. Uttara Kannada farmers are a distinct example of this practice of mix of agriculture, animal care and silviculture which requires intensive use of crop lands, grazing lands and forest lands adjoining the village (which forms a part of the integrated village ecosystem). The quantity of animal residue available and the option of converting it to biogas for cooking purpose is discussed in the following section.

Data on livestock population collected from the taluk's Veterinary Department is tabulated in Table 3.7.1. There are about 4,63,729 cattle, 1,00,786 buffaloes, 18,853 goats and 4997 sheep in Uttara Kannada. Quantity of dung yield per cattle per day varies from place to place. It is seen that the cattle dung available per animal in coastal taluk is about 3-4.5 kgs/adult animal. In hilly taluks of Sirsi and Siddapur, the dung yield per animal ranges from 8-10 kgs, buffaloes 12-15 kgs, stall fed buffaloes 15-18 kgs and hybrid ones

15-18 kgs. By considering lower figures (such as 3 kgs per animal for coastal cattle, 8 kgs/animal for cattle in Sirsi and Siddapur, 12 kgs/animal for buffaloes), the total cattle and buffalo dung available is about 8,44,414.18 and 4,41,442.7 tonnes/year respectively. With the assumption of 0.036 m^3 of biogas yield per kg of cattle/buffalo dung (Khandelwal and Mahdi, 1986), total quantity of gas available (if all is used for biogas) is estimated to be about 46,290.85 thousand m^3. The estimated per capita requirement of gas is 0.34-0.43 m^3/day for domestic purposes. Which means that gas generated by animal dung is sufficient to meet the requirement of 30% of total population of Uttara Kannada district. Talukwise availability of gas and percentage of population's requirement computed is listed in Table 3.7.1 and the same is shown pictorially in Figure 3.4. From this analysis it is evident that in hilly taluks such as Sirsi, Siddapur and Yellapur, biogas if opted, can meet the requirement of more than 50% population of the respective taluks. It is seen that the daily dung yield in sheep is about 1.5 kgs, goat 1.5 kgs and pig about 2.5 kgs. Gas yield per kg of pig dung is about 0.062 m^3, and sheep and goat about 0.054 m^3. As indicated in Table 3.7.1, total quantity of gas available from sheep, goat and pig is about 8,97,340 m^3. Livestock sustain on fodder from grasslands, agricultural residues and forest lands. In view of this, the quantity of fodder available in the district from various categories of vegetation has been looked at.

Table 3.7.1: Talukwise Livestock Population and Estimated Dung and hence Gas yield per Year

Taluk	Cattle				Buffalo			
	Male	Female	Total	Dung Produced tonnes	Male	Female	Total	Dung Produced tonnes
Ankola	15943	11749	27692	30322.74	2951	4107	7058	30914.04
Bhatkal	10613	11838	22451	24583.85	2936	3846	6782	29705.16
Haliyal	21978	17118	39096	85620.24	3572	9635	13207	57846.66
Honnavar	19811	28863	48674	53298.03	2103	4807	6910	30265.8
Karwar	11220	8932	20152	22066.44	3654	4144	7798	34155.24
Kumta	18269	19066	37335	40881.83	2014	4831	6845	29981.1
Mundgod	18757	15122	33879	74195.01	1331	4673	6004	26297.52
Sirsi	44298	48670	92968	203599.92	3925	13305	17230	75467.4
Siddapur	36927	36646	73573	161124.87	2451	9807	12258	53690.04
Supa	12390	9178	21569	47235.45	2720	2100	4820	21111.6
Yellapur	21883	24457	46341	101485.80	4263	7611	11874	52008.12
District	232090	231639	463729	844414.18	31920	68866	100786	441442.68

Taluk	Total (cattle+buffalo)		Number of persons requirement	Total Population	Percentage of population that could use Biogas
	Dung tones	Biogas Thousand cubic meters			
Ankola	61236.78	2204.52	17764.09	91310	19.45
Bhatkal	54289.005	1954.40	15748.62	129017	12.21
Haliyal	143466.9	5164.81	41618.12	94353	44.10
Honnavar	83563.83	3008.30	24240.91	145842	16.62
Karwar	56221.68	2023.98	16309.27	142845	11.42
Kumta	70862.925	2551.07	20556.52	134144	15.32
Mundgod	100492.53	3617.73	29151.74	77939	37.40
Sirsi	279067.32	10046.42	80954.26	152935	52.93
Siddapur	214814.91	7733.34	62315.37	94202	66.15
Supa	68347.053	2460.49	19826.70	99519	19.92
Yellapur	153493.92	5525.78	44526.84	81410	54.69
District	1285856.9	46290.85	373012.45	1243526	30.00

Table 3.7.1: Talukwise Livestock Population and Estimated Dung and hence Gas yield per Year (Cont'd.)

Taluk	Sheep	Goats	Pigs	Waste Produced in tonnes/yr	Biogas Thousand cubic mtrs	Number of persons require./yr
Ankola	46	458	2	277.77	17.22	138.77
Bhatkal	196	701		491.11	30.45	245.36
Haliyal	850	2230	987	2586.94	160.39	1292.43
Honnavar	75	338	40	262.62	16.28	131.20
Karwar	357	303	286	622.33	38.58	310.91
Kumta	384	196	67	378.69	23.48	189.19
Mundgod	741	2779		1927.20	119.49	962.82
Sirsi	853	7245	55	4483.84	278.00	2240.11
Siddapur	415	3975		2403.53	149.02	1200.79
Supa	619	155	8	431.07	26.73	215.36
Yellapur	461	473	106	608.09	37.70	303.80
District	4997	18853	1551	14473.16	897.34	7230.75

3.7.2 Fodder Availability

Litter production in forests under various vegetation types is estimated talukwise and is listed in Table 3.7.2, while Table 3.7.3 is a compilation of agro residues such as stalk of paddy, cotton, maize, pulses, etc. It is found that about 2,08,918.67 tonnes of agro residues is available in the district.

3.7.3 Fodder Demand

The average weight of cattle and buffaloes are approximately about 200 and 260 kgs respectively. Taking the daily fodder requirement as about 2.5% of body weight, the total requirement computed for the district is about 10.85 lakh tonnes. As indicated in Table 3.7.4, total agro residues available is about 2.08 lakh tonnes. Balance of 8.77 lakh tonnes of fodder is being provided by forests and grasslands. The productivity of grasslands is about 5-15 t/ha/yr. In Sirsi, Siddapur and Yellapur yield is as high as 15-60 t/ha/yr. Grass is generally available during monsoon period. In Uttara Kannada 20,369 hectares is under permanent pasture and grass production is about 1.01 lakh tonnes. With the onset of dry months grass availability declines, and crop residues obtained from agriculture and leaf fodder from trees help tide over the scarcity. In some parts of Sirsi, like Banavasi, fodder crop such as sunhemp is grown during *kharif* season as second crop. A portion (about 25-30%) of the fodder crop is left behind to provide nutrients for *rabi* crop. This practice has contributed to higher yields of paddy and has also helped overcome the fodder scarcity. In recent years crop lands have intruded into grazing and forest lands. Reduction of grazing land area has resulted in overgrazing in the remaining grazing lands, turning them to wastelands. Animals now graze heavily in forest land for fodder. In areas under forests, about 50% has either closed canopy or is inaccessible. Which means, the remaining 50% of forests has to meet the fodder demand of 7.75 lakh tonnes. Whenever a forest is logged, the grazing animals suppress all regeneration and the logged area gradually turns into wasteland. The grazing pressure on forests has resulted in depletion of grass and natural regeneration, which has resulted in ecological imbalance. This necessitates integrated planning at village level to meet the growing demand for fuel and fodder. Fodder demand can be met by growing suitable fodder crop as has been successfully tried in Banavasi by some progressive farmers.

Table 3.7.2: Talukwise Litter Production in tonnes

Taluk	Forest hect	Dry Deciduous	Moist Deciduous	Evergreen	Semi Evergreen	Coastal	Litter Production Biomass production
	<---------area in hectares --------- under various vegetation --------->						Total tonnes
Ankola	75432.37		15086.5	22629.7	15086.5	22629.7	481107.7
Bhatkal	25390.46			5078.1	8124.9	12[illegible]87.4	120706.2
Haliyal	59099.78	23639.9	23639.9	11820.0			571258.5
Honnavar	57416.68		2870.8	20095.8	17225.0	17225.0	351935.5
Karwar	52808.68		10561.7	15842.6	10561.7	15842.6	336813.8
Kumta	39187.59		7837.5	11756.3	7837.5	11756.3	249938.4
Mundgod	51167.36	28142.0	23025.3				513362.1
Siddapur	62274.6			40478.5	21796.1		510807.4
Sirsi	103226.57		15484.0	46452.0	41290.6		868496.7
Supa	160608.71		40152.2	120456.5			1391273.0
Yellapur	113429.56		51043.3	62386.3			1016385.6
District	800042.36	51782.0	189701.3	356995.7	121922.4	79641.0	2680656.3

Table 3.7.3: Bio-Residues available from Various Crops

Taluk	Crop: Paddy				Total Residues (in tonnes)			
	High Yield Variety		Improved Variety		Paddy		Paddy + others	
	Husk	Stalk	Husk	Stalk	Husk	Stalk	Husk	Stalk
	2	3	4	5	6=2+4	7=3+5	8=6+16	9=7+17
Ankola	3431.89	11486.32	863.19	2889.03	4295.07	14375.35	4295.07	14431.20
Bhatkal	2012.97	6737.29	454.99	1522.83	2467.96	8260.12	2467.96	8357.45
Haliyal	5135.80	17189.21	2232.36	7471.58	7368.16	24660.79	7478.73	48034.66
Honnavar	2450.11	8200.36	507.22	1697.64	2957.33	9898.00	2957.69	9949.98
Karwar	3168.59	10605.07	1264.16	4231.07	4432.75	14836.14	4432.75	14863.26
Kumta	2933.47	9818.15	1350.94	4521.53	4284.42	14339.68	4284.42	14404.49
Mundgod	3240.42	10845.48	1897.19	6349.79	5137.61	17195.26	5390.40	33950.54
Siddapur	3006.62	10062.97	1150.47	3850.54	4157.09	13913.5[illegible]	4160.66	14202.42
Sirsi	4285.47	14343.19	2742.38	9178.59	7027.85	23521.78	7047.49	25539.27
Supa	1803.00	6034.52	728.42	2437.98	2531.42	8472.5[illegible]	2549.78	9085.18
Yellapur	2393.99	8012.53	1748.24	5851.24	4142.22	13863.77	4146.30	16100.21
Total	33862.31	113335.09	14939.57	50001.82	48801.88	163336.9[illegible]	49211.24	208918.67

Table 3.7.3: Bio-Residues available from Various Crops (Cont'd.)

Taluk	Ragi	Jowar	Maize		Pulses	Cotton	Total (others)	
	Stalk	Stalk	Husk	Stalk	Stalk	Stalk	Husk	Stalk
	10	11	12	13	14	15	16	17
Ankola	4.29				51.57		0.00	55.85
Bhatkal	4.50				92.84		0.00	97.34
Haliyal	8.43	262.60	110.57	221.14	563.08	22318.63	110.57	23373.87
Honnavar	3.36		0.36	0.71	47.91		0.36	51.98
Karwar					27.12		0.00	27.12
Kumta	17.85				46.97		0.00	64.82
Mundgod	41.37	517.14	252.79	505.57	269.04	15422.16	252.79	16755.27
Siddapur	1.36		3.57	7.14	248.45	31.96	3.57	288.91
Sirsi	7.99	38.22	19.64	39.29	190.15	1741.85	19.64	2017.49
Supa	105.31	9.77	18.36	36.71	24.89	436.00	18.36	612.68
Yellapur	3.36	49.19	4.07	8.14	105.94	2069.80	4.07	2236.44
Total	197.81	876.92	409.36	818.71	1667.94	42020.37	409.36	45581.76

Table 3.7.4: Fodder Requirement in Uttara Kannada

Taluk	Total		Total (Cattle + Buffalo) Body Weight in tonnes	Fodder Demand (@2.5% of Body Weight) Tonnes
	Cattle	Buffalo		
Ankola	27692	7058	7373.48	67283.01
Bhatkal	22451	6782	6253.52	57063.37
Haliyal	39096	13207	11253.02	102683.81
Honnavar	48674	6910	11531.4	105224.03
Karwar	20152	7798	6057.88	55278.16
Kumta	37335	6845	9246.7	84376.14
Mundgod	33879	6004	8336.84	76073.67
Sirsi	92968	17230	23073.4	210544.78
Siddapur	73573	12258	17901.68	163352.83
Supa	21569	4820	5566.94	50798.33
Yellapur	46341	11874	12355.35	112742.57
District	463729	100786	118950.21	1085420.67

Note: Body weights - Cattle: 200 kgs, Buffalo: 260 kgs

From the discussion on fuel and fodder availability, it is evident that, to bring down further degradation of resource base and hence ecological imbalance, it is necessary to: (1) enhance the total natural resource base, (2) produce and optimally use biomass for various end uses in domestic and agro processing sectors through fuel efficient devices, and (3) produce fuel and fodder biomass from degraded lands. In the next section, the alternative devices, which help in optimal utilisation of bio residues and their technical and economical feasibility in the context of village ecosystem is discussed.

3.8.0 TECHNO-ECONOMIC ANALYSES OF BIOENERGY SYSTEMS

The fundamental forms of bioenergy use are,

(i) Traditional domestic use for household cooking, lighting and water heating (for bath). The efficiency of conversion of the biomass to useful energy is between 5 to 15%.
(ii) Rural industrial use in agro processing, bricks and tile manufacture, and pig iron where the biomass is considered as free energy source. There is generally little incentive to use the biomass efficiently, so conversion of feedstock to useful energy commonly occurs at an efficiency of 15% or less.
(iii) Biological conversion, including anaerobic digestion for biogas production and fermentation for alcohol.

The overall efficiency of biomass utilisation depends on the moisture content of the fuel and type of stoves used. Freshly cut wood contains about 25-60% moisture. The removal of a kilogram of water from wood involves an expenditure of about 620-670 kcals of heat energy. It is noticed that a reduction of 25% moisture in fuel wood causes a saving of nearly 15% of fuel wood. The most common method adopted in Uttara Kannada for removal of moisture is solar drying. Normally wood is lopped and collected from forests during summer and dried in the open for use during monsoon. However, over drying also has its own adverse effects such as fire hazards, etc. It is observed that dried wood with moisture content of 8% releases heat too fast and the whole log tends to burn, bringing the flame out of the stove

3.8.1 Fuel Efficient Stoves

The most commonly used stoves in most households for cooking are either mud stoves or three stone stoves, also referred as traditional cookstoves (TCs). The efficiency of these stoves are less than 10%. Applying the principles of combustion and heat transfer, wood and other biomass burning fuel efficient stoves have been designed by ASTRA (Lokras, 1992), also called as Astra stoves or Improved cookstoves (ICs). In Astra stoves complete combustion of fuel wood takes place with as little excess air as practicable to generate the highest temperature of flue gases. In ICs, combustion of fuel wood is carried out over a grate in an enclosed fuel box with ports of suitable size for entry of air. The grate helps in entry of air below the fuel bed to burn the char as well as for separation of ash from fuel. Air required for burning the volatile matter released as a consequence of heating the fuel (also referred as secondary air), enters through a port at a level slightly above the grate. Heat gets transferred to the pans by the mechanism of conduction, convection and radiation. Fuel efficiency studies conducted in 82 households of a cluster of villages in Sirsimakki microcatchment of Sirsi taluk have shown that the fuel need for cooking is about 1.92(avg)$\pm$1.02(Sd) kgs/person/day in TCs, while in ICs it is about 1.1(avg)$\pm$0.78(Sd) kgs/person/day. This means that there is a saving of about 42% of fuel by switching over to ICs from TCs. The average fuel consumption for water heating in traditional stoves (efficiency 10-16%) is about 1.68(avg)$\pm$0.80 (Sd) kgs/person/day, while in improved stoves (efficiency 35-60%), it is about 1.36(avg)$\pm$0.63(Sd) kgs/person/day. This result based on sample study of 104 households shows a potential of 19 to 24% saving in improved design. All traditional stoves in these villages are without chimney. With the chimney, the efficiency of the stove increases and is about 22%. In improved stoves the fuel is allowed to burn over a cast iron grate. Air required for burning is allowed through the grate so that combustion is controlled. Rate of burning (to a temperature of 45-50°C) in improved stoves is about 30 minutes for 50 litres.

3.8.2 Biogas Technology

Biogas is a product of anaerobic fermentation of organic matters, and consists of about 60-70% methane, 30-40% carbon dioxide, etc. The inputs for biogas digesters are the wastes that are found locally; animal dung, agricultural residues, leaf litter from forests, etc. The residues are introduced into a closed digester, where, without the presence of free oxygen the responsible micro organisms work successively to convert complex organic matter into CH_4, CO_2, H_2, H_2S, etc. The optimum conditions for biogas production are: temperature 30-35°C, pH 6.8-7.5, Carbon/Nitrogen ratio 20-30, solid contents 7-9% and retention time 20-40 days. Among these parameters, temperature is the most difficult or costly to control. The gas formation virtually stops when the temperature drops below 10°C. Retention time decides the rate at which the waste is digested. The longer the time, the larger is the volume of gas produced from a given amount of waste and vice versa. Thus, if the available amount of input materials is limited, a bigger digester can be adopted to exploit the gas potential more fully; and where the waste is abundant, it can be fed at a higher loading rate into a small digester to maximise the gas production per unit volume of the digester. The optimum retention time depends on the temperature. In practice, a longer retention time is usually adopted to cope with cool seasons. There are various designs of biogas digesters such as:

(1) Floating gas holder type designed by Kadhi and Village Industries Commission (Directorate of Gobar Gas Scheme, 1979).
(2) Optimised design developed by Application of Science and Technology to Rural Areas (ASTRA) at Indian Institute of Science (Subramanian, 1984).
(3) Fixed dome type designed by University of Agricultural Sciences - Bhagyalaxmi design.
(4) Raitabandu Biogas Plant - designed by a farmer from Sagar taluk, Shimoga district to suit the needs of Malnad region.

(i) *Biogas Usage*: Biogas can be used for many purposes, mainly cooking and lighting in rural areas. It can be burned with a gas mantle or be converted to electricity using dual mode engine. The per capita requirement of gas for cooking is in the range 0.34-0.43 m^3/day (efficiency of standard burner is about 60%). The gas requirement to generate one unit of electricity (kWh) is about 0.54 m^3. The calorific value of m^3 of gas is about 4713 kcals. Relative values of biogas compared with other energy sources are listed in Table 3.8. The potential of biogas in hilly taluks of Uttara Kannada district is quite appealing. As indicated in Table 3.7.1, biogas can meet cooking requirement of at least 30% of the total district population.

Table 3.8: Biogas Value Compared with Other Energy Sources

Relative Calorific Value	Relative Monetary Value
1 Cubic Meter Biogas	1.00
3.6 kg Fuel wood	0.88
1.5 kg Charcoal	1.28
5 kWh Electricity	3.00
13 kg Cow dung	1.30
0.6 litres Kerosene	2.34

Note: Costs - Kerosene Rs. 3.90 per litre (through Public Distribution System)
Fuel wood Rs. 24.60 per quintal (Price in Government Fuel wood Depots at Kumta), Electricity Rs.0.60 per kWh, Cow dung Rs. 0.10 per kg

(ii) *Biogas as Substitute for Fuel wood*: Socio-economic survey conducted in villages of Kumta, Sirsi, Siddapur and Ankola has shown that per capita requirement of fuel wood is around 0.7-1.1 t/yr. If 30% population opts for biogas, then fuel wood saved in the district would be in the range 2.6-4.10 lakh tonnes/year.

By switching over to biogas, the drudgery involved in collection and transport of fuel wood (which takes about 5-6 hours per day) gets reduced, and also reduces cooking time from 4-6 hours (fuel wood stoves) to 1.5-3 hours (biogas stoves) a day.

(iii) *Electricity Generation*: The decentralised electricity generation system by a community biogas plant was tried out in 1987 in Pura village, Kunigal taluk, Tumkur district for illumination and drinking water facility (Ravindranath and Ramachandra, 1996). The biogas system is accepted and maintained by villagers due to multiple advantages like better illumination facility than kerosene lamps, convenient method of drinking water supply and better fertilizer in the form of sludge without pathogens and weeds (due to anaerobic condition and long retention time in closed digesters). A 5 kW electricity generation unit using biogas-diesel engine-genset system caters the needs of pumping domestic water and illumination in the evening (for 4 hours). Two youths have been trained to maintain and manage the system. This community system provides revenue for the village to the extent that the total revenue received through outputs exceeds the expenses for diesel and dung charges (paid to villagers). Detailed economic analyses carried out based on life cycle costing method show that at 4.3 hours/day capacity utilisation, the cost of electricity is Rs. 2.75/kWh. It is noticed that as capacity utilisation increases to 15 hours/day the cost per kWh reduces to Rs. 1.40. Comparison of decentralised biogas system with nuclear power station (Reddy and Balachandra, 1991) show that in a capital starved situation where the real discount rates are high, the cost/kW of installed capacity is lower for biogas systems compared to nuclear power plants.

This review of bioenergy production from animal residues has shown that potential solutions do exist, but there are very few in practical operation. The economics of such solutions would ultimately depend on local conditions, requirements, subsidies, etc. If farmers are forced/convinced to treat the residues for biogas prior to disposal to compost pits, it would be a valuable way to conserve fuel wood and improve the cooking environment in the kitchen.

3.8.3 Energy Plantations

Technically speaking, energy plantation means growing select species of trees and shrubs which are harvestable in a comparably shorter time and are specifically meant for fuel. The fuel wood may be used either directly in wood burning stoves and boilers or processed into methanol, ethanol and producer gas. These plantations help provide wood either for cooking in homes or for industrial use, so as to satisfy local energy needs in a decentralised manner (Vimal and Tyagi, 1984). The energy plantations provide almost inexhaustible renewable sources (with total time constant of 3-8 years only for each cycle) of energy which are essentially local and independent of unreliable and finite sources of fuel. The attractive features of energy plantations are: (a) heat content of wood is similar to that of Indian coal, (b) wood is low in sulphur and not likely to pollute the atmosphere, (c) ash from burnt wood is a valuable fertiliser, (d) utilisation of erosion prone land for raising these plantations helps to reduce wind and water erosion, thereby minimising hazards from floods, siltation, and loss of nitrogen and minerals from soil and (e) help in rural employment generation - it is estimated that an hectare of energy plantation is estimated to provide employment for at least seven persons regularly. Selection of multipurpose species provides a number of by-products like oils, organic compounds, fruits, edible leaves, forage for livestock, etc. Data collected from Forest Department (plantations in Kumta and Sirsi taluks under Social Forestry Programme) reveals that annual woody biomass available is in the range 11.9 to 21 t/ha/yr. An energy forest raised at Hosalli village in Tumkur district to support a wood gasifier plant has annual yield of 6 t/ha/yr.

3.8.4 Biomass Fired Steam Power Generation System

Biomass fuelled steam power system, providing electricity as well as the process heat, has been utilised for many decades in most countries of the world. Such systems have found widespread applications in paper and pulp, plywood and metallurgical industries. In such a system, the energy in the biomass fuel is utilised through the process of direct combustion to generate steam for driving steam engine or turbine (Tillman, et al., 1981). The efficiency of these devices is around 35%. In a study on biomass fire steam power system (Prasad, 1995), results show that for systems utilising biomass fuels 0 and 60% (wet basis), excess air rates between 0 and 200%, boiler efficiency 84 to 49%, steam engine efficiency 18 to 13%, overall plant efficiency 11 to 8.4% and electricity fuel ratio between 0.7 and 0.2 kWh/kg, the cost per unit of electricity ranging from 24 to 39 paise.

Land requirement for energy plantation (and hence biomass fired thermal system): *Casuarina* plantation with a yield of 6 t/ha/yr and a growth cycle of 10 years is considered to compute land requirement for setting up 1000 MW thermal power station. Assuming calorific value of wood as 4670 kcals/kg, load factor of thermal plant as 0.5 and thermal efficiency as 35%, land requirement can be computed as follows:

Total yield of harvested area	= 60 t/ha/yr
Heat value/ha	= 280.2 x 10^6 kcals/ha/yr
	= 326.43 x 10^3 kWh(th)/ha/yr
	= 1,14,251.55 kWh(elec)/ha/yr
Power generated/ha at 0.5 load factor	= 26.08 kW/ha
Area needed for 1000 MW	= 38,336.46 ha
Total area	= 3,83,364.6 ha
Area to be harvested/day	= 105 ha
Nursery area required	= 1918 ha

The area required for setting up 1000 MW power station for different wood species at different yield rates has been computed and tabulated in Table 3.9. *Casuarina* species with yield of 6-11 t/ha/yr and growth cycle of 10 years requires land of 3833.64-2091.07 sq.km. to support 1000 MW thermal station. Similarly *A. auriculiformis* with yield of 11-36 t/ha/yr requires land of 2091-638.9 sq.km.

Table 3.9: Area Required for a 1000 MW Thermal Power Station

Species	Annual Yield tons/ha	Growth Cycle	Total Yield t/ha/yr	KWh (thermal) kWh/yr/ha	Electrical Energy kWh/yr/ha	Power per hect.	Area Reqd for 1000MW hect.	Total area (growth Cy.) hect.	Area to be harvested per day- hect.	Nursery hect.
Casuarina	6	10	60	326433.0	114251.6	26.08	38336.46	383364.60	105.03	1916.82
Casuarina	11	10	110	598460.5	209461.2	47.82	20910.80	209107.96	57.29	1045.54
Eucalyptus	6.5	8	52	282908.6	99018.0	22.61	44234.38	353875.02	121.19	2211.72
Eucalyptus	13	8	104	565817.2	198036.0	45.21	22117.19	176937.51	60.60	1105.86
Eucalyptus	20	8	160	870488.0	304670.8	69.56	14376.17	115009.38	39.39	718.81
A. auriculi-formis	11	8	88	478768.4	167568.9	38.26	26138.50	209107.96	71.61	1306.92
A. auriculi-formis	21	8	168	914012.4	319904.3	73.04	13691.59	109532.74	37.51	684.58
A. auriculi-formis	36	8	288	1566878.4	548407.4	125.21	7986.76	63894.10	21.88	399.34
Fuel wood	6.5	40	260	1414543.0	495090.1	113.03	8846.88	353875.02	24.24	442.34
Fuel wood	13.5	40	540	2937897.0	1028264.0	234.76	4259.61	170384.27	11.67	212.98
Fuel wood	27	40	1080	5875794.0	2056527.9	469.53	2129.80	85192.13	5.84	106.49

3.8.5 Cost of Wood Based Power Station

Considering life of thermal plants as 30 years, at 10% annual interest rate, capital recovery per year is computed. Cost of thermal station's equipment and civil cost works out to Rs. 4.4 crores/MW. Assuming cost of wood as Rs. 150 per tonne and annual operation and maintenance costs at 1% capital cost of machinery, the cost per unit of electricity comes to Rs. 0.98 (for yield of 6 t/ha/yr) and Rs. 0.69 (for 13.5 t/ha/yr yield).

3.8.6 Scope for Wood Based Power Generation in Uttara Kannada District

The present waste/barren land of 20,068 ha in the district can be converted to energy plantations, the yield of which after 4-5 years would be sufficient to support wood based power stations of about 415 MW.

3.8.7 Wood Gasification

Biomass gasification is basically conversion of solid biomass (wood, wood waste, agricultural residues, etc.) into combustible gas mixture normally called producer or low Btu gas. The process is typically used for woody biomass and involves partial combustion of such biomass. Partial combustion occurs when air supply is less than adequate for complete combustion of biomass. It produces carbon monoxide (CO) as well as hydrogen (H_2), both are combustible gases. Conversion to gas results in loss of up to 25% energy. Use of gas can be highly efficient and hence overall efficiency could be very high. Gas can be fed directly into internal combustion engines (I.C. engines) thereby saving commercial fuels. Also, it can be employed at any scale (above few kilo watts), and hence is ideally suited for decentralised applications - shaft power, electricity or thermal energy. In case of shaft power / electricity, the gas is basically burnt inside an engine (diesel based - compression ignition engine) with pilot diesel injection to start combustion.

Biomass gasification provides a valuable fuel for both mobile and non mobile uses. Producer gas can replace natural gas, gasoline or fuel oils used to make steam for generating electricity and fire boilers, and produce heat for industries and homes. It can also be used as fuel for internal combustion engines for a wide array of purposes. Gasification is an efficient way to extract heat from biomass. It is estimated that for each 100 kcal of potential energy in solid fuels, gasification can extract about 80 kcal in hot raw gas (Dayal, et al., 1989). This is more efficient than many devices that burn biomass directly in a hearth or fire box. Producer gas can be piped short distances and used for industrial purposes like fuel kilns: brick kilns, ceramics, glass pottery, etc., for boilers in rice mills, saw mills, cashew industry or for power generation. In any case, while using producer gas for heating, the burner must be designed for operation on low energy gas.

The function of a gasifier is to convert solid carbonaceous fuel, such as wood into combustible gas by a combination of oxidation, pyrolysis and reduction process (Talib, et al., 1987). The major chemical reaction taking place are both exothermic ($C+O_2 \rightarrow CO_2$, $C+0.5O_2 \rightarrow CO$) and endothermic ($C+H_2O \rightarrow CO+H_2$, $CO_2+C \rightarrow 2CO$). In order to effectively carry out these reactions, the solid fuel passes through the following zones: drying, pyrolysis, combustion and reduction. In drying zone the temperature is around 150°C and moisture in the biomass is driven off. In pyrolysis zone, at 400°C thermal break down of wood / bio residues takes place in the absence of air, resulting in the formation of methanol, acetic acid and heavy hydro carbon including tar. The solid material left after pyrolysis is primarily fixed carbon in the form of charcoal. The pyrolysis material along with the gases and the organic vapors produced passes through the combustion zone. In combustion zone, exothermic reaction takes place and the heat released is used for sustaining both the pyrolysis and the reduction reactions. The temperature in this zone ranges from 1000-1500°C. Combustion zone has controlled introduction of air.

By forcing the vapors through a narrow area directly underneath the combustion zone, organic liquids and tar formed in the pyrolysis zone are cracked. In the cracking process, heat intensity increases. The gases thus formed are drawn to reduction zone where endothermic reaction takes place. The mixture of final product gases (producer gas) consisting of 18-25% CO, 13-15% H_2, 3-5% methane (CH_4), 0.2-0.4% heavy hydro carbon, 5-10% CO_2, 45-54% N_2, 10-15% H_2O and particulate matter is drawn into the clean up system at about 250°C. In clean up system, the gas is cooled and cleaned to remove particulates and later introduced into the engine. Commercially available biomass gasification system for power generation covers the range of 3-500 kW. The major areas where this system is relevant for power generation in a decentralised manner are:

(a) Village electrification in remote hilly terrains which have adequate biomass resources (villages like Yan, Sandolli, Malavalli, Muttolli, Kodambale, Yedtare, Chimmoli, Medine, Morshe, Mudnalli, Kurigadde, Algar, Bangane, Kalve, Hebbail, etc., in Kumta and Mogadde, Audhal, Kodigar, Menshi, etc., in Sirsi taluk),
(b) Energisation of a number of pumpsets located in a cluster,
(c) Captive power for industrial units located in rural areas having extensive land area which can be devoted for energy plantations, and
(d) Captive power for industries that have biomass waste processes such as paper mills, saw mills, rice mills, etc.

3.8.8 Wood Gasifier Based Electricity System

Community wood gasifier installed at Hosalli, a non electrified village in Tumkur district, by I.I.Sc (Ravindranath and Ramachandra, 1996) demonstrates that, these units can be self sustaining and technically and economically feasible decentralised systems if managed properly. Main features of this system are:

(1) The 5 kW wood gasifier-diesel system in operation since 1988, meets the lighting (for 4-5 hours/day) and drinking water requirement of all 43 households.
(2) An energy plantation in 2 ha village land has been raised (in 1987) consisting of *Leucaena leucocephalla, Dalbergia sisso, Eucalyptus* hybrid, *Cassia siamea, Acacia auriculiformis* and *Casuarina equisetifolia* with density of 6600 plants/hect and approximate annual productivity of 6 t/ha/yr.
(3) The annual wood requirement of the gasifier is 5.1 tonnes at the rate of 1.2 kgs/kWh. Households are provided with two electrical lighting points of 40 W and 25 W bulbs. Water is pumped using a 2 kW submersible pump from a borewell, at a depth of 87 m, for drinking and domestic use. Apart from this, the energy is used for 9 street lamps and a flour mill. This system is being managed (technical and financial management) by three trained village youths.
(4) Diesel replacement up to 85% is achieved under favorable operating conditions.

Capital cost of gasifier, engine, generator, accessories, wood chipping machinery, energy forest and building is about Rs. 63,600. Life of gasifier and engine is considered to be 50,000 and 20,000 hours respectively. At an operational level, annual maintenance cost is taken as 5% and 10% of cost for gasifier and engine respectively. Economic analysis carried out using the discounted cash flow techniques (NPV) method (considering total life cost and benefits) shows that at the current level of operation of 4 hours per day, the cost per unit of electricity is about Rs. 3.50/kWh. However, analyses show that cost per kWh can be reduced to Rs. 2.50 by increasing the hours of operation to 20 hours per day.

3.8.9 Scope for Energy Plantations in Uttara Kannada

As per land use statistics, it is seen that about 20,068 hectares of land is barren / degraded due to human pressure, grazing pressure, etc. Assuming productivity of 6 t/ha/yr, the fuel wood available from energy plantations grown in wastelands is about 1,20,408 tonnes. It is seen that 1 kWh per 1.2 kgs of fuel wood can be generated through wood gasification mode of power generation. This means that from wasteland plantation, 100.34 million units of electricity can be expected annually. From this experience, it is evident that the gasification programme can be expanded tremendously by considering the integration of large-scale plantation on sub standard soil / barren wasteland with fast growing tree species. *Kharif* crop is not grown in Uttara Kannada in most of the taluks due to lack of proper irrigation facilities. Irrigation can increase production manifold to that obtained under rain fed conditions. The approach of combining biomass production and gasification will result in many benefits such as wasteland utilisation, energy supply and employment generation. This would help rural population to be energy independent from shortages of oil and fluctuating electricity supply from centralised option. Realising the benefits of the integrated approach, planning machinery of State Government should start implementing energy plantation and power generation through nodal agencies, voluntary sector and universities.

3.9.0 Conclusions

Based on detailed investigation of biomass resource availability and demand, Uttara Kannada district can be categorised into two zones (a) Biomass surplus zone, consisting of taluks mainly from hilly area and (b) Biomass scarce zone, consisting of thickly populated coastal taluks like Bhatkal, Kumta, Ankola, Honnavar and Karwar. Fuel wood is mainly used for cooking and horticulture residues from coconut and arecanut trees, mainly for water heating purposes. The present inefficient fuel consumption could be brought down by using fuel efficient stoves. Availability of animal residues for biogas generation in Sirsi, Siddapur and Yellapur taluks gives a viable alternative for cooking and lighting fuel, and can also be a useful fertiliser. However, to support the present livestock population, fodder from agricultural residues is insufficient in these taluks. There is a need to supplement the fodder availability with fodder crops as successfully tried in Banavasi village by some progressive farmers. Bioenergy experiences all kinds of problems that affect most economic development projects. Lack of capital, skilled labour and service backups are serious impediments to the progress of alternate devices such as fuel efficient stoves, biogas, wood gasifier, etc. A more serious concern is the lack of coordinated effort / approach and integrated planning among various bureaucratic setups and ministries. These are the main hurdles to the successful implementation of biomass cultivation projects and development of bioenergy. Policies are to be formulated to remove the constraints at local/regional level. Policies must engender the communication between the different institutions and government sectors, that is, agricultural, forestry, land planning and energy sectors, involved with the establishment of a significant and sustainable biomass energy programme. Hence, the prudent management practices of biomass production offers the opportunity to address multiple environmental concerns, such as, land degradation, biodiversity, CO_2 emissions, acid rain pollutants, and local and regional health problems.

3.10.0 Acknowledgements

I thank my colleagues Mr. D.M. Bhat and Prof. M.D. Subashchandran for providing useful reference materials and valuable tips on various aspects related to Forestry, Biomass productivity, Minor forest products, etc. I thank Mr. Ramesh and Mr. Bhat, Karnataka State Council for Science and Technology, Bangalore and Ms. Savitha for their kind help in digitising maps and for computation of area under various vegetation types. I thank the Forest Department for its positive support during field studies. Forest Department officials at Bangalore, Sirsi, Kumta and Honnavar provided useful statistical information required for this research. This research was funded by the Ministry of Environment and Forests, Government of India and Indian Institute of Science.

3.11.0 References

1. Hall, D.O., 1991. Biomass energy. *Energy Policy,* 19(8):711-737.
2. Hall, D.O. and Rosillo-Calle, F., 1991. Why biomass matters. *Biomass Newsletter*, 5(4): Special Issue.
3. Hall, D.O., Rosillo-Calle, F., Williams, R.H. and Woods, J., 1992. Biomass for energy: supply prospects. Chapter 14. In: *Renewables for Fuel and Electricity*. Johansson, T.B., Kelly, H., Reddy, A.K.N. and Williams, R.H. (Eds.). Island Press, Washington.

4. Gadgil, M., 1990. Eco development of selected microcatchments in the Bedthi - Aghnashini river basins of the Uttara Kannada District. In: *An Operational Research Programme for Integrated Development of Microcatchments Report.* CES and KSCST, IISc., Bangalore, India.
5. Buchanan, F.D., 1801. *Journey Through the Northern Parts of Kanara*, Volume 1. Higginbothams and Company, Madras, India.
6. Talbot, W.A., 1909. *Forest Flora of the Bombay Presidency and Sind*, Volume 1 and 2. Government Photozincographic Press, Poona, India.
7. Aitchison, P.E., 1910. *The Soppinahosalli High Forest Working Plan.* Block XXXVII. Government Central Press, Bombay, India.
8. Mavinkurve, G.R., 1955. *Working Plan for Inland Coastal Forests of Kanara Western Division.* Government Central Press, Bombay, India.
9. Shanmukhappa, G., 1966. *Working Plan for the Unorganised Forests of Sirsi and Siddapur.* Karnataka Forest Department, Government of Karnataka, India.
10. Shanmukhappa, G., 1977. *Working Plan for the Unorganised Forests of Karwar and Honnavar Divisions.* Karnataka Forest Department, Government of Karnataka, India.
11. Dhareshwar, S.S., 1941. The denuded condition of the minor forest in Kanara coastal tract. *Ind. Forester*, 67: 68-81.
12. Chandran, M.D.S., 1993. *Vegetational Changes in the Evergreen Forest Belt of Uttara Kannada District of Karnataka State*, Ph.D. Thesis, Karnatak University, Dharwad, India.
13. Pascal, J.P., Shyamsundar, S., Meher-Homji, V.M. and Legris, P., 1983. *Vegetation Map of Karnataka, Belgaum, Shimoga.* French Institute, Pondicherry, India.
14. Pascal, J.P., 1986. *Explanatory Booklet on the Forest Map of South India.* Institut Francais de Pondicherry, Travaux de la section Scientifique et Technique, Hors Serie N° 18, Pondicherry, India.
15. Campbell, J.M., 1883. *Kanara-Gazetteer of the Bombay Presidency*, Volume XV, Part-I. Government Central Press, Bombay, India.
16. Kamath, S.U., 1985. *Uttara Kannada District.* Karnataka State Gazetteer, Government of Karnataka, V.B. Soobbaiah & Sons, Bangalore, India.
17. Pascal, J.P., 1988. *Wet Evergreen Forests of the Western Ghats of India*, Institut Francais de Pondicherry, Pondicherry, India.
18. Bhat, D.M., 1990. Litter production and seasonality in tropical moist forest ecosystems of Uttara Kannada District, Karnataka, *Proc. Indian Acad. Sci. (Plant Sci)*, 100(2):139-152.
19. Proctor, J., 1983. Tropical forest litterfall I - Problem of data comparison. In: *Tropical Rain Forest: Ecology and Management.* Sutton, S.L., Whitemore, T.C. and Chadwick, A.C. (Eds.), pp. 267-274. Blackwell, Oxford.
20. Garg, R.K. and Vyas, L.N., 1975. Litter production in deciduous forest near Udaipur (South Rajasthan) in India, In: *Tropical Ecological Systems*, Golley, F.B. and Medina, E. (Eds.), pp. 131-135. Springer-Verlag, Berlin.
21. Rai, S.N. and Proctor, J., 1986a. Ecological studies on four rain forests in Karnataka, India: II. Litterfall. *Jour. Ecol*, 74: 455-463.
22. Rai, S.N. and Proctor, J., 1986b. Ecological Studies on four rainforests in Karnataka: 1. Environment, Structure, Floristics and Biomass. *Journal of Ecology*, 74:439-454.
23. Songwe, N.C., Fasehun, F.W. and Okali, D.W., 1988. Litterfall and productivity in a tropical rain forest, Southern Baakundu forest reserve, Cameroon. *Jour. Trop. Ecol*, 4:25-37.
24. Greenland, D.J. and Kowal, J.M.L., 1960. Nutrient content of the moist tropical forest of Ghana. *Pl. Science*, 12:154-164.
25. Whitemore, T.C., 1975. *Tropical Rain Forests of the Far East*, pp. 238. Clarendon Press, Oxford, London.
26. Rai, S.N., 1984. Above ground biomass in tropical rain forests of Western Ghats, India. *Indian Forester*, 110 (8).
27. Prasad, S.N., Hegde, H.G., Bhat, D.M. and Hegde, M., 1987. *Estimates of Standing Biomass and Productivity of Tropical Moist Forests of Uttara Kannada District, Karnataka, India.* CES Technical Report No. 19, CES, IISc, Bangalore, India.
28. Lele, S.M., 1993. *Degradation, Sustainability or Transformation? A Case Study of Villagers' Use of Forest Lands in the Malnad Region of Uttara Kannada District, India.* CES Technical Report No. 27, CES, IISc, Bangalore, India.

29. Thampan, P.K., 1981. *Handbook on Coconut Palm*, pp. 335. Oxford and IBH Publishing Co. Pvt. Ltd., Bombay, India.
30. Bavappa, K.V.A., Nair, M.K. and Kumar T.P., 1982. *The Arecanut Palm*. Central Plantation Crops Research Institute, Kasargod, Kerala, India.
31. Bidappa, K.G., 1960. Utilisation of Arecanut waste. *Arecanut Jour.*, 11:106-108.
32. Jaiswal, P.L., Wadhwani, A.M., Rajinder, S., Chhabra, N.N. and Sarma, P.S.N., 1980. *Handbook of Agriculture*, Indian Council of Agricultural Research, New Delhi, India.
33. Khandelwal, K.C. and Mahdi, S.S., 1986. *Biogas Technology: A Practical Handbook*, Tata McGraw-Hill Publishing Company Limited, New Delhi, India.
34. Lokras, S.S., 1992. *Fuel Efficient Stoves*. CES Technical Report No. 65, pp.205-208. CES, IISc, Bangalore, India.
35. Directorate of Gobar Gas Scheme, 1979. *Gobar Gas: Retrospect and Prospects*, Khadi and Village Industries Commission, Bombay, India.
36. Subramanian, D.K., 1984. Energy Utilisation in Karnataka. In: *Karnataka State of Environment Report*, Saldanha, C.J. (Ed.), pp. 103-123.
37. Ravindranath, N.H. and Ramachandra, T.V., 1996. Renewable Energy Options for Decentralised Electricity Generation, *Encology,* 11(2):27-32.
38. Reddy, A.K.N. and Balachandra, P., 1991. The economics of electricity generation from the Pura community biogas system. In: *Power Generation Through Renewable Sources of Energy*, Pai, B.R. and Prasad, M.S.R (Eds.). Tata Mc-Grawhill Publishing Co., New Delhi, India.
39. Vimal, O.P. and Tyagi, P.D., 1984. *Energy from Biomass - an Indian Experience*. Gricole Publishing Academy, New Delhi, India.
40. Tillman, D.A., Rossi, A.J. and Kitto, W.D., 1981. *Wood Combustion: Principles, Processes and Economics*. Academic Press, New York, USA.
41. Prasad, S.B., 1995. Biomass - Fired Steam Power Cogeneration System: A Theoretical Study. *Energy Conversion and Management*, 36(1):65 - 77.
42. Dayal, M., Vimal, O.P. and Singh, K.K., 1989. Biomass Gasification in India - DNES activities. *Biomass*, 18:197-204.
43. Talib, A., Gross, J.R., Vigil, J.F., Grover, P.D., Rao, T.R., Gaur, S. and Anuradha, N. 1987. *Development and Field Testing of Small Biomass Gasifier - Engine Systems in India* - A Project Report. The Mitre Corporation, Virginia 22102 - 3481, USA.
44. op cit. Ref. 36.

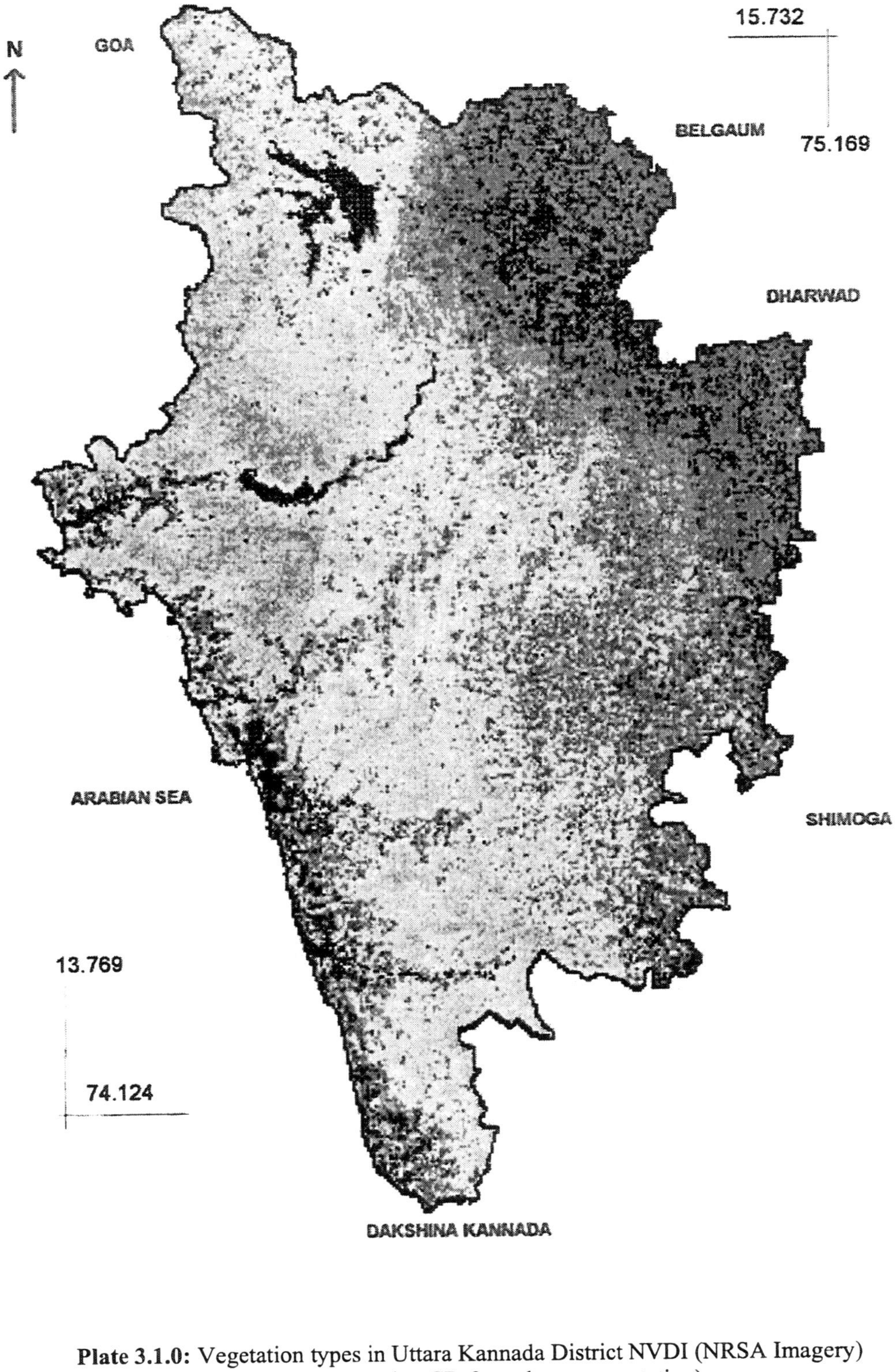

Plate 3.1.0: Vegetation types in Uttara Kannada District NVDI (NRSA Imagery)
(see accompanying CD for color representation)

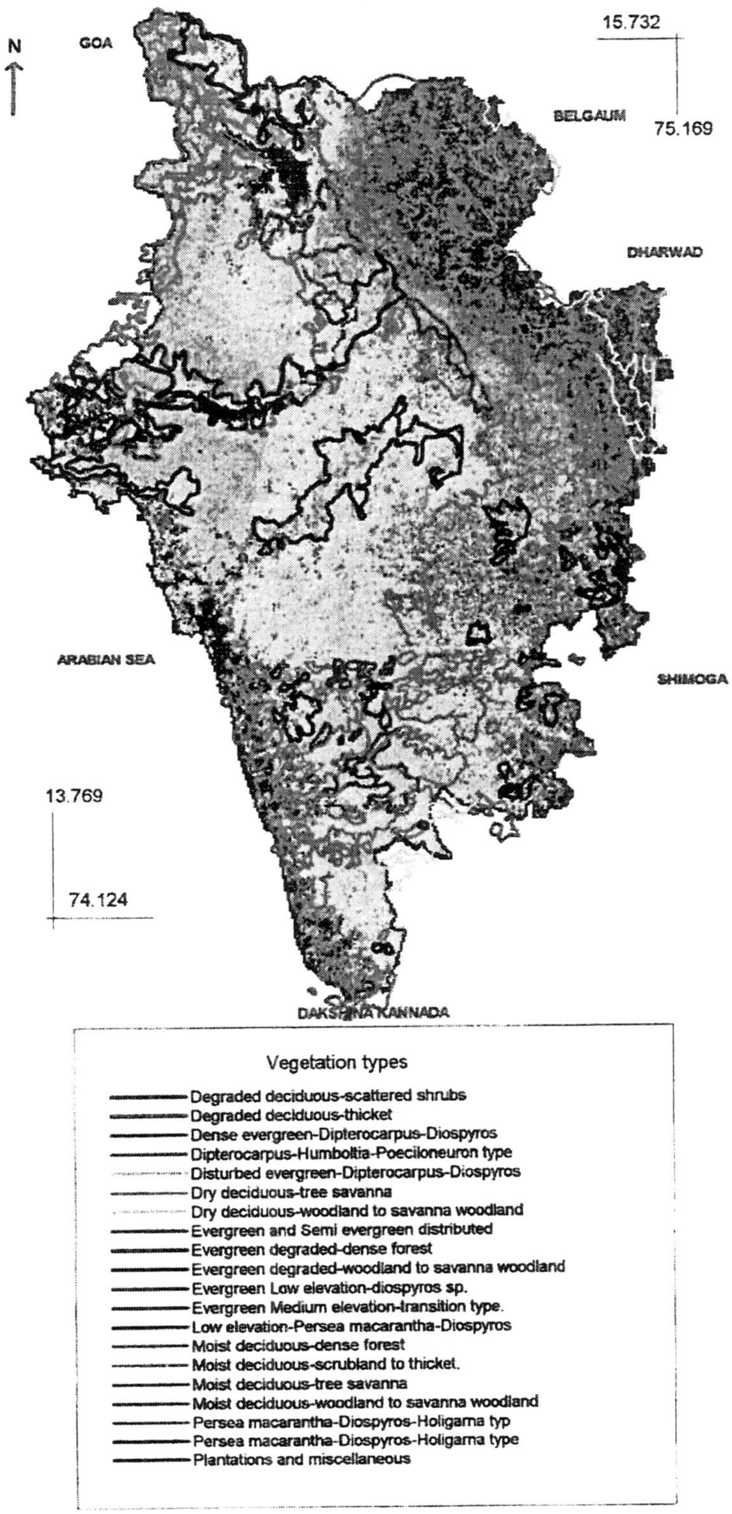

Plate 3.2.0: Vegetation types in Uttara Kannada District
(Based on Forest Map of South India) (see accompanying CD for colcr representation)

Chapter 4

HYDRO ENERGY POTENTIAL AND PROSPECTS

4.0 SUMMARY

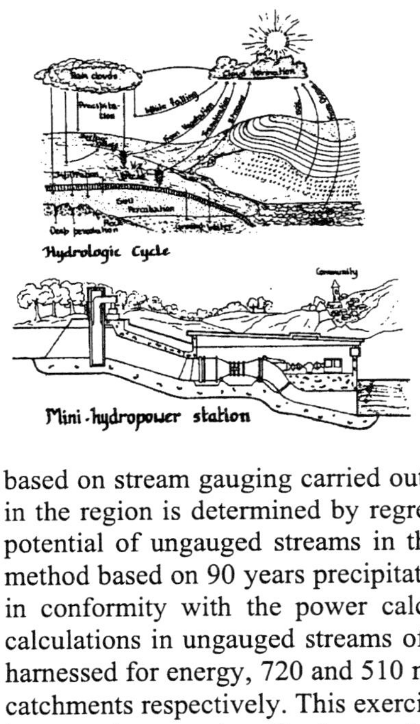

The amount of power available at a given site is decided by the volumetric flow of water and the hydraulic head or water pressure. In hydro schemes the turbines that drive the electricity generators are directly powered either from a reservoir or the "run of the river". Large schemes may include a water storage reservoir, providing daily or seasonal storage, to match the electricity production with the demand for power. Large-scale hydroelectric plants have been producing power in Karnataka for many years, with the first built in 1902. Majority of these are located in Uttara Kannada district. Due to environmental constraints, construction of large storage reservoirs is limited and attention has been focused on developing small-scale environment friendly hydro schemes to cater the needs of the region. In this chapter, potential assessment of streams in Bedthi and Aghnashini river catchments of Uttara Kannada district in Western Ghats is discussed. Hydro potential at five feasible sites is assessed based on stream gauging carried out for 18 months. Relationship between stream run off and precipitation in the region is determined by regression analyses. This relationship is used to find out the hydro energy potential of ungauged streams in the catchments. Computations of discharge on empirical basis/rational method based on 90 years precipitation history, and the subsequent power and energy values computed are in conformity with the power calculations based on stream gauging. This method is used for power calculations in ungauged streams of Bedthi and Aghnashini catchments. It is estimated, if all streams are harnessed for energy, 720 and 510 million kWh of electricity could be generated in Bedthi and Aghnashini catchments respectively. This exercise provides insight to meeting the regional energy requirement through integrated approaches like harnessing hydropower in a decentralised way during seasons and meeting the lean season requirement through solar or other thermal options.

Keywords: Catchment yield, Run off, Precipitation, Micro / mini / small hydel plants.

4.1.0 INTRODUCTION

Hydraulic potential is the combination of the possible flows and distribution of gradients, and hydraulic resource is that fraction of hydraulic potential that is still accessible after economic considerations. Hydropower owes its position as a renewable resource to the varying, but more or less continuous flow of a certain amount of water in the stream. This water, supplied by rain and always moving, constantly flows from the mainland to the sea, where it evaporates into the atmosphere in an unending cycle controlled by two opposing forces, heat of the sun and the earth's gravity. Hydropower is a precipitation dependent resource, subject to the uncertainties which it entails. Water run off can vary within wide limits.

The harnessing of energy of the falling water to provide mechanical power has been one of man's greatest achievements. Water wheels providing mechanical power for grinding and water pumping are still in common use. The industrial revolution created new requirements, which demanded larger power generating capacities, which led to rapid improvement in the design of water wheels, turbines and generators. The capacities of hydropower plants became very large as electricity generation became common place by the end of 19th century. The position has ultimately reached a stage where large-scale hydroelectric plants contribute significantly to the State's and the National energy demand. Unfortunately, every development has its cost and the cost of exploiting hydropower is vigilance in ensuring the environment is not irreparably damaged and the life of the river continues to flourish. This demands considerable care and attention in planning.

A micro, mini or small hydropower station can divert only potential energy of the water which would have dissipated to no benefit in the natural flow along the water course. The domain where these can have potential impact on development is domestic lighting and stationary motive power demand for such diverse productive uses as water pumping, wood and metal work, grain mills, agro processing industries, etc.

Mini, micro and small hydro plants combine the advantages of large hydro plants on one hand and decentralised power supply on the other. The disadvantages associated with large hydropower plants, like high transmission costs, environmental issues like submergence of forests and crop lands, and displacement of families, are not present in small plants. Moreover, the harnessing of local resources like small hydro resources, being of a decentralised nature, lends itself to decentralised utilisation, local implementation and management, making rural development possible based on self-reliance and use of local natural resources. In this section, potential assessment of streams in Bedthi and Aghnashini catchments is discussed.

4.2.0 Study Area

Uttara Kannada extends to about 328 kms north-south and about 160 kms east-west. It is a region of gentle undulating hills, rising rather steeply from a narrow coastal strip bordering the Arabian Sea to a plateau at an altitude of 500 m, with occasional hills rising above 600 to 860 m. The annual precipitation largely confined to monsoon from June to September, ranges between 3500 mm on the coast, rising to 4500-5000 mm on the crestline and declining to 1000 mm on the eastern plateau. According to recent landsat imageries, 67.04% of the geographical area is under forest, 1.94% under paddy and millet cultivation, 1.26% under coconut and areca gardens, 1.94% under rocky outcrops and the remaining 27.82% under habitation and reservoirs.

There are four major rivers, Kalinadi in the north, Bedthi or Gangavalli about 32 kms south, Aghnashini or Tadri, rising far to the south but falling into the sea only about 10 kms south of Gangavalli, and Sharavathi or Gerusoppa about 24 kms south of Tadri, where it reaches the foothills and becomes a tidal creek. Besides these, there are many minor streams flowing in the district. Varada, a tributary of Tungabadra, which originates in the north-eastern part of neighboring Shimoga district and flows north-east, is the only eastward flowing river in the district. The district can be broadly categorised into three distinctive regions - the coast lands, Sahyadrian interior and the eastern margin, where the tableland begins. It has 11 taluks, of which Karwar, Ankola, Kumta, Honnavar and Bhatkal are located in the coastal belt, while Supa, Yellapur, Sirsi and Siddapur are located in the mostly forested Sahyadrian region. Parts of Haliyal, Yellapur and Mundgod are plain country. A range of climatic areas from arid to humid, with physiographic conditions ranging from plains, mountains to coast, is seen here. This large variety of natural conditions provides the basis for generalisation of the theoretical probability distributions for annual precipitation.

Among the four rivers, hydro potential of Kali and Sharavati has been tapped already for power generation. Various developmental projects undertaken or proposed in the district are listed in Table 4.1. The completed large-scale projects have caused serious environmental damage by submerging productive natural virgin forests, and horticultural and agricultural lands. In view of these, potential assessment of streams in Bedthi and Aghnashini catchments is undertaken to explore ecologically sound means of harnessing hydro energy.

Table 4.1: Development Projects in Uttara Kannada District

(A) Hydel Projects		
(i) 810 MW	Nagjhari power house	completed
(ii) 100 MW	Supa dam	project of
		Kali river
(iii) 120 MW	Kodasalli dam	
	Kadra	
(iv) 150 MW	dam	
Proposed projects:		
Bedthi Basin		
Bedthi stage I		210 MW
Bedthi stage II		210 MW
(at Sonda pattanadahalla)		
Kali Basin		
Danelli dam		60 MW
Kali stage III		400 MW
Aghnashini Basin		
Aghnashini		435 MW
Bennehole scheme		
(B) Nuclear Power Plants		
Kaiga (under construction)		2 x 235 MW
Kaiga II (proposed)		4 x 234 MW

4.3.0 Definitions

4.3.1 Catchment Characteristics

(i) *Catchment area*: It is the area (also referred as watershed area) above a specific point on a stream from where water drains into the stream. Catchment boundaries are located using the contour lines on a topographical map and sketched following the ridge tops which appear as downhill pointing V-shaped crenulations. The boundary should be perpendicular to the contour lines it intersects. The mountain tops, often marked as dots on the map and the location of roads which follow the ridges are other clues. Catchment area can be measured directly from the marked maps by: a) planimeter, b) superimposing a grid of squares or a dot grid over the map and counting the number of squares or dots that fall within the catchment area, or c) digitising the boundaries.

(ii) *Stream length*: Stream length will influence the extent of stream habitat in a catchment. It also affects the flowing time of water in a drainage system and availability of sediment for transport. Opisometer (map wheels), which measures distance as it rolls across a map, is used to measure stream length.

(iii) *Mean catchment slope*: The slope of the catchment will influence surface run off rates and is related to drainage density and basin relief.

$$S_b = \frac{(\text{Elevation at } 0.85\text{ L})-(\text{Elevation at } 0.10\text{ L})}{0.75\text{ L}}$$

(iv) *Mean stream slope (Average channel slope)*: This is given by

$$S_c = \frac{(\text{Elevation at source - Elevation at mouth})}{\text{Length of stream}}$$

(v) *The longitudinal profile*: This describes the way in which the stream's elevation changes over distance. The x-axis represents the distance along a stream, as measured from some out-fall point such as a stream junction, a lake or an ocean.

(vi) *Perennial streams*: Perennial streams are those which flow year-round. They are primarily effluent and consist of base flow during dry periods.

(vii) *Intermittent streams*: These flow only certain times of the year, when they receive water from springs and run off. Depending on the season, they are either influent or effluent. During dry years they may cease to flow entirely or they may be reduced to a series of separate pools.

(viii) *Gross head*: The gross head on a hydroelectric plant is the difference between head water and tail water elevation when the plant is not in operation. Only a part of the gross head can be utilised in power generation. There is loss of water head in the penstock, conduits or head-race supplying the turbines, in water passages to the power plant and in the tail-race. The difference in elevation of the water surface in tail-race for no plant discharge, with all water flowing through the plant, may be from 1 to 3 meters. With a low head plant, this loss may be a substantial percentage of the head available.

(ix) *Net head*: The net head or effective head of a hydroelectric plant is the gross head minus all losses above entrance to the scroll case and below an exit from the draft tube. Net head varies with gross head, and in addition, it varies with load on the unit because friction and velocity losses increase approximately as the square of the discharge. The net head at full load for low head developments, with short or no conduits or tail-race, may be nearly as much as the gross head; but for high head developments, requiring long conduits, the net head may be only 90%, or even less, of the gross head. Drawdown for pondage varies with topographical features at the site and nature of the load. For low head developments, the average drawdown required is not often in excess of 5% of the total head, and for very low head developments it is usually negligible.

4.3.2 Classification of Hydropower Plants

Hydropower plants can be classified based on:

(i) *Power and head:* mini, micro and small hydropower plants.

		Head (metres)		
Type of plants	Power (kW)	Low	Medium	High Head
Micro	up to 50	< 15	15 - 50	> 50
Mini	50 - 500	< 20	20 - 100	> 100
Small	500 - 5000	< 25	25 - 130	> 130

The upper and lower head and output limits adopted for any classification is only indicative.

(ii) Intake:
Run-of-river (lateral intake from a main water course).
Run-of-river with reservoir or dam.

(iii) Regulation:
Adjustable flow, either manual or automatic.
Constant load, due to actual nature of the load or through dissipation of excess energy.

(iv) Links to the grid:
Isolated plants.
Plants connected to small electric grids.
Plants connected to major zonal or National grids.

(v) Technological concepts:
Plants using conventional technology - this implies quality civil engineering works for the intake, canal and forebays; silt basin at the intake, steel piping, expensive electro-mechanical equipment constructed to strict material and manufacturing criteria, and fully instrumented switchboards. Plants using non-conventional technology - they often use intakes from existing irrigation canals, which are improved, the forebay installed in line on the canal incorporating the silt basin, electro-mechanical equipment designed and constructed with technologies appropriate to the level of industrial development of the region, and the availability of local materials, standardised equipment and modular switchboards with minimum instrumentation.

4.3.3 Classification of Basins and Sub-Basins

Based on a study of geographic and topographical maps, preliminary approximations were carried out. It included approximate determination of hydrographic and physical parameters of the basins and sub-basins of the region, either on the basis of measurements and studies that have been carried out or by inference from mathematical models. Overall evaluation studies of the resources in the region covers ecology, hydrology, geology, geomorphology and the availability of aggregates.

4.3.4 Criteria for Site Selection

The choice of the site is based on a close interaction between the various conditions like the pattern of the stream, integrity of the site works, environmental integration, etc. It is necessary to establish the inventory of energy demand in various sectors and assess various other sources like solar, biomass, wind, etc. Various factors involved in estimating hydro potential are: (1) head, (2) hydrological pattern; defined from measurements or from interrelationships between effective rain and discharge, (3) usage of water, upstream of the intake to determine the flow which is available and downstream to determine the effects of diverting the water from present and future uses, (4) distance from the intake to the power station and from the power station to the consumer site, and (5) size of the scheme involved and evaluation of their stability depending on various lithological, morphological and topographical conditions.

4.3.5 Energy

Energy is the capacity to do work. Work is the utilisation of energy and is equal to force times distance (that is force*distance). The potential energy of a volume of water is the product of its weight by the head, or the vertical distance it may be lowered. In the transformation of this energy into useful work, part will be lost in the conduit system and part in the apparatus of the development. The hydraulic power, normally available at a given site is defined as

$$P = g*p*Q*H$$

where, P is hydraulic power (in watts), g - acceleration due to gravity (equal to 9.81 m s^{-2}), p - liquid density (in case of water p = 1000 kg m^{-3}), Q - flow of discharge (in kilograms per cubic meter) and H the head or height of the waterfall (in meters). Therefore

$$P = 9.81* Q*H$$

The corresponding electrical production

$$E = P*t*\eta*f$$

where, E is the electrical energy produced (in kilowatt-hours), P - hydraulic power (in kilowatts), t - operating time (in hours) and η- the turbine-generator assembly efficiency, usually between 0.6 and 0.9.

4.3.6 Effect of Number of Units on Plant Efficiency

It is seen that a two-unit plant has decidedly better average efficiency than a single-unit one, but the advantage of a three-unit plant over a two-unit plant is less marked. In a hydro plant, which has pondage and is part of a large interconnected system, the number of units in the plant makes very little difference in overall efficiency. In effect, entire system is a single plant and nearly all the units are operated at point of maximum efficiency or no efficiency at all. Without pondage or with deficient pondage, the efficient utilisation of the available water supply may advocate the use of a number of units, even though the size of these units may be smaller than would otherwise be economical (Creager, W.P. and Justin, J.D., 1950).

4.4.0 OBJECTIVE

The prime objective of this endeavour is to assess the potential of streams in Bedthi and Aghnashini catchments based on eighteen months field experiment, and precipitation and topographic information (including ungauged streams).

4.5.0 METHODOLOGY

The hydrology of the river and streams under consideration were studied by –

(i) *Reconnaissance Study or Pre-feasibility Study:*
Pre-feasibility study was carried out in all streams with water-head greater than three meters in Bedthi and Aghnashini catchments.

(ii) *Feasibility Study:*
This involved the measurement of catchment area and stream discharge for a substantial period of time.

(a) Measurement of catchment area: Catchment boundaries were located using contour lines on a topographical map. They were drawn following the ridge tops which appear on topo maps as downhill pointing V-shaped crenulations. The boundary should be perpendicular to the contour lines it intersects. The mountain tops are often marked as dots on a map, and the location of roads which follow ridges are other clues. Catchment area thus marked/traced is measured directly using a planimeter.

(b) Stream discharge: Both direct and indirect methods were carried out. Indirect method was tried in order to assess the potential of ungauged streams.

(i) Direct estimation of flows at site: Stream discharge is the rate at which a volume of water passes through a cross-section per unit of time, usually expressed in units of cubic meters per second (m^3/s). The velocity - area method using a current meter is used for estimating discharge. The cup type current meter is used in a section of a stream, where water flows smoothly and the velocity is reasonably uniform in the cross-section. As far as possible a cross-section is chosen where the current is reasonably regular over the whole width. This measurement of five readings for three consecutive days every month is carried out for 18 months to account for day to day fluctuations and seasonal variations.

(ii) Indirect estimation of flows at site: Run off is the balance of rain-water, which flows or runs over the natural ground surface after evaporation, interception and infiltration. The yield of a catchment is the net quantity of water available for storage, after all losses, for the purpose of water resources utilisation and planning. The run off from rainfall may be estimated by empirical formula and rational method.

(1) Empirical formula: The relationship between run off and precipitation is determined by regression analyses based on their field data.

(2) Rational method: A rational approach is used to obtain the yield of a catchment by assuming a suitable run off coefficient (Raghunath, 1985).

$$\text{Yield} = C*A*P$$

where, C is run off coefficient, A - catchment area and P - rainfall. The value of "C" varies depending on the soil type, vegetation, geology, etc. as listed below:

Type of Catchment	Value of C
Rocky and impermeable	0.8 - 1.0
Slightly permeable and bare	0.6 - 0.8
Cultivated or covered with vegetation	0.4 - 0.6
Cultivated absorbent soil	0.3 - 0.4
Sandy soil	0.2 - 0.3
Dense forest	0.1 - 0.2

This involved (i) analyses of 90 years precipitation data collected from India Meteorological Department (IMD), (ii) computations of discharge by empirical/rational method based on this precipitation history and comparison of the same with values obtained by actual stream gauging, and (iii) computation of discharge and power in ungauged streams by empirical/rational method in Bedthi and Aghnashini catchments.

(c) Computation of power and total energy available in all streams: The hydraulic power which is naturally available at a given site is defined by $P = 9.81*Q*H$, where Q is the discharge in cubic meters per second and H is height of the waterfall or head in meters.

The corresponding electrical energy produced could be $E = P*t*\eta*f$, where E is the electrical energy produced in kilowatt-hours, P - hydraulic power in kilowatts, t - operating time in hours, η - efficiency of turbine-generator assembly (usually between 0.5 and 0.9, here taken as 0.7), and f- coefficient to allow for seasonal flow variations for run-of-river installations.

4.6.0 Analyses of Data and Discussion

4.6.1 Rainfall

The response of the watershed to precipitation is the most significant relationship in hydrology. The first portion of rainfall wets the surface of soil and vegetation. Next, a thin film is formed on the ground surface. Not all the rainfall moves to the stream by overland flow. Part is lost by evaporation and part infiltrates into the soil, which involves the air-water-soil interface. The movement of water under ground is governed by laws of flow in porous media. One portion percolates directly to the ground-water table, but part moves to the stream without reaching the water table. The response of stream run off to precipitation is a result of interactions among these processes in which the effect of one process gradually overshadows the effects of another (Holland, 1967). The complexity of the rainfall-run off relationship is increased by areal variation like geologic formations, soil conditions and vegetation, and areal and time variations of meteorological conditions. Vegetation influences rainfall-run off relationship not only through interception and surface detention, but also by its effect on the impact energy of rainfall, which may initiate turbulence in overland flow and increase erosion.

The extent to which a stream/river could be developed for power depends not only on stream flow but also details of design and the amount of energy available, etc. "Rainfall or precipitation may be defined as the total condensation of moisture from the atmosphere that reaches the earth, including all forms of rain, ice and snow". Precipitation and river flow are governed by chance phenomena, that is, there are so many causes at work that the influence of each cannot be readily identified. Therefore, statistical and probability methods are applied to describe these hydrologic phenomena (Markovic, 1976). An attempt is made here to find (1) theoretical probability functions of best fit to distributions of annual precipitation and (2) whether there is any relationship in year to year precipitation and to see if the data reveals any significant trend. These analyses were carried out for the data collected from all rain gauge stations in Uttara Kannada district.

The rate of rainfall is expressed in millimeter/centimeter/inches of water during a given period of time. Most rainfall records are obtained by periodic observation of gauges. The usual interval is 24 hours. In Uttara Kannada district, various government agencies are maintaining rainfall stations, namely IMD, Agriculture Department and Public Works Department. IMD has three observatories in coastal belts at Karwar, Honnavar and Shirali, where various parameters like temperature, humidity, rainfall, solar radiation and cloud cover are recorded at regular intervals through automatic weather stations. In 27 rain-gauge stations distributed all over the district, rainfall is recorded manually, using manual rain-gauges (managed by the Tahsildar's office).

All precipitation records collected from various agencies are used with full realization of their limitations, which are summarised as follows:

(a) Rainfall varies with elevation above sea level and the records obtained at these stations (at different elevations) are indicative of average precipitations.

(b) Annual precipitation varies considerably from year to year, and records of considerable length are necessary to indicate accurately the mean rainfall for a given station.

(c) Accuracy of rainfall records varies greatly, and for important hydrological studies careful investigation of accuracy is necessary.

4.6.2 Mean Rainfall of the Region

All annual rainfall records, on or in the vicinity of the drainage area, are tabulated from 90 years rainfall data maintained by IMD, Government of India. The inter annual variability of the yearly and monsoon rainfall has considerable impact on agriculture, water management and energy generation. In view of this, monthwise rainfall data collected all over the district by various agencies has been looked at in greater detail and year to year rainfall variability has been studied. Rainfall variability of Ankola, Bhatkal, Kumta, Karwar, Sirsi, Siddapur and Yellapur taluks was studied for 90 years beginning from 1901, while the same for Haliyal was studied for 22 years, Supa 27 years and Honnavar 62 years. The annual average rainfall and standard deviation computed for each taluk based on these data are listed in Table 4.2.1. This shows that Bhatkal taluk receives highest rainfall of 3942 (avg)± 377 (Sd) mm, while Haliyal receives least rainfall among 11 taluks with 1174 (avg)± 166 (Sd) mm. The coefficient of variation computed to see the relative variability of precipitation data is listed in Table 4.2.2. It ranges from 0.0851 (Karwar) to 0.1105 (Honnavar) in the coastal region, 0.0810 (Yellapur) to 0.1217 (Siddapur) in the Sahyadrian interior and 0.1415 (Haliyal) to 0.2083 (Mundgod) in the plains.

Table 4.2.1: Annual Rainfall (mm) – Talukwise

Taluk	Years	Pre-Monsoon		Southwest Monsoon		Northeast Monsoon		Annual Rainfal			
		Avg	Sd	Avg	Sd	Avg	Sd	Avg	Sd	Min	Max
Ankola	90	125	80	3084	251	204	67	3413	273	2915	4265
Bhatkal	90	162	107	3514	344	257	95	3942	377	3062	4987
Haliyal	22	126	38	907	121	141	35	1174	166	899	1452
Honnavar	62	126	82	3094	313	216	67	3436	380	2520	4090
Karwar	90	140	82	2794	216	198	64	3132	267	2548	3736
Kumta	90	140	84	3445	920	222	83	3566	364	2759	4450
Mundgod	90	145	60	887	203	181	53	1235	257	846	2078
Siddapur	90	110	50	2635	417	188	54	2997	365	2206	3951
Sirsi	90	96	34	2150	242	183	47	2430	263	1743	3037
Supa	27	35	38	2305	302	117	37	2485	219	2047	2803
Yellapur	90	100	30	2200	214	170	49	2470	200	1879	2878

Table 4.2.2: Talukwise Computation of COV (Coefficient of Variation)

Taluk	Region Type	COV
Karwar	Coastal Region	0.0851
Ankola		0.0798
Kumta		0.1021
Honnavar		0.1105
Bhatkal		0.0956
Supa	Sahyadrian Interior	0.0880
Yellapur		0.0810
Sirsi		0.1081
Siddapur		0.1217
Mundgod	Plain Region	0.2083
Haliyal		0.1415

4.6.3 Properties of Observed Data

It is seen that coastal taluks Ankola, Karwar, Kumta, Bhatkal and Honnavar receive highest rainfall in the district. Among these, Bhatkal receives maximum rainfall of 3132 to 3942 mm. The hilly taluks Siddapur, Sirsi, Yellapur and Supa follow coastal taluks with mean annual rainfall of 2470 to 2997 mm, while the plain taluks Haliyal and Mundgod get least rainfall in the district. This variation in rainfall from coastal to plain area is shown pictorially in Figure 4.1. It is seen that the annual precipitation has considerable range, indicating the large variety of climatic and physiographic conditions in the district. The south-west monsoon constitutes 88 to 90% of the total precipitation in these taluks. Various statistical tests are carried out with the annual rainfall data for each taluk to find out whether the rainfall regime in any taluk follows a particular trend and whether there is any relationship among variables and to see if the rainfall during any particular year depends on that of the previous years. Statistical Runs Test (runs above and below mean, and runs up and runs down) conducted for annual rainfall data for each taluk suggests that the year to year annual rainfall variables are independent of each other. The null hypothesis of independence could not be rejected based on these tests for the data from any of the 11 taluks and for the annual rainfall of the district.

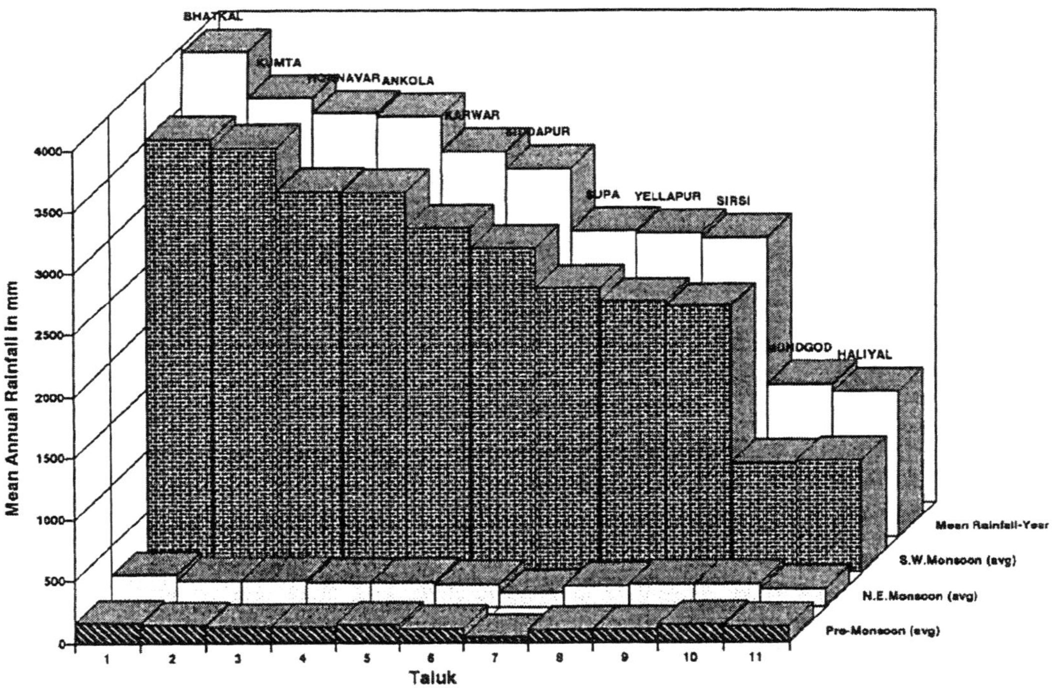

Figure 4.1: Mean Annual Rainfall (Talukwise) – Uttara Kannada District

4.6.4 Statistical Tests - Goodness of Fit Test

According to properties of observed data, the distribution functions of best fit to observed frequency distributions of annual precipitation should be a bell shaped but skewed curve. Screening of the applicable functions with respect to the criteria required, their convenient usage in mass computation and the experience already obtained in applying them in this kind of analysis, as listed by various researchers (Pearson, 1957; Zar, 1984) in literature, lead to the selection of (a) Normal Density function, (b) Log Normal and (c) Gamma.

To test the theoretical probability distribution functions for goodness of fit to observed data, the data is classified into mutually exclusive and exhaustive categories of class intervals. The number of class intervals is chosen in a way that the average expected frequency of any class interval is at least five. Chi-square test is used as a measure of goodness of fit of the theoretical probability distributions to observed ones. If the probability of a hypothesized function is less than the assigned level of significance, then the function would be acceptable as a good approximation to the distribution of a sample considered. The normal and log-normal distributions are fitted to the distribution of annual precipitation data of taluks in Uttara Kannada district. It is found that, the departure between normal and observed distribution would give the probability of Chi-square less than 0.95 for Siddapur, Yellapur, Karwar, Mundgod, Haliyal and Supa taluks. While, for Ankola, Kumta, Sirsi, Bhatkal and Honnavar taluks, annual precipitation data follows log normal distribution as the departure between observed and log-normal distribution, which has the probability of Chi-square 0.32, 0.55, 0.82, 0.44 and 0.79 (all less than 0.95, accepted at 95% level of significance). Smaller the value of probability, smaller is the departure between theoretical and observed distributions, and the better theoretical function fits an observed distribution.

These analyses illustrate that altitude and distance from the sea are not likely to explain the differences between the distributions of annual precipitation of the taluks. For example, two neighboring taluks in the coastal belt, Ankola and Karwar (or Siddapur and Sirsi in the interior) give two different distributions. There may be certain other factors like latitude, temperature, evaporation and prevailing wind direction of moist air masses governing the difference in distribution. However, analyses show that regional characteristics especially do not favour the use of one of the probability functions in fitting the observed distributions of annual precipitation.

4.6.5 Stream Flow and Precipitation

Distributions of annual river and stream flow are affected by physiographic factors of a watershed area apart from its precipitation. The literature regarding watershed response can be classified into two general groups. Majority of the research has dealt with obtaining time distributions of direct surface run off at a point, given the volume and distribution of effective rainfall. The remaining has centered around total rainfall-run off relationship, including the estimation of the volume of effective rainfall considering loss functions experienced by storm rainfall. Studies regarding the conversion of effective rainfall to hydrographs of stream flow at the catchment outlet stem primarily from unit hydrograph theory. Hydrograph is the graph of discharge against time, which refers to the pattern of stream flow that occurs over a season or a year. The theory has been modified, applied, verified and used for analysis and synthesis. The concept of instantaneous unit hydrograph along with various storage and routing ideas has led to the development of numerous theoretical response models. On the other hand, few rainfall-run off models have been investigated with emphasis on conversion of rainfall to effective run off.

Stream flow and ecology are affected by catchment conditions. Changes in stream discharge and sediment loading, caused by the modification of the catchment, are reflected in variations in the rate of sediment transport, channel shape and stream pattern. Responses to change may be immediate, delayed or dependent on a critical factor reaching the threshold level. It is necessary to know the response of a catchment/watershed to rainfall in order to design structures such as overflow spillways on dams, flood-protection works, highway culverts and bridges (Gordon, et al., 1992). The rate at which run off moves towards the stream depends on the drainage efficiency of the hill slopes, which is influenced by the slope and length of an upland surface, its micro topography, permeability and moisture content of the soil, subsurface geology and vegetation cover. The hydro potential of each stream is assessed for establishing micro, mini and small hydropower plants.

4.7.0 Feasibility Study of Mini, Micro ad Small Hydro Sites

Catchment areas for the streams are obtained from the Survey of India toposheets. Stream gauging is done with both direct and indirect methods.

4.7.1 Catchment Area

Catchment area measured from marked toposheets using planimeter for the streams in Bedthi basin is listed in Table 4.3.1, while Table 4.3.2 lists the same for streams in Aghnashini basin. Drops at various points along the Bedthi river course have been identified as: Kalghatgi (80 m), Kaulgi Halla (64 m), major drop at Magod (340 m), and lowest drop of 8.5 m at a place about 129 kms from Kalghatgi (in Ankola taluk). Numerous streams join the river along its course from Kalghatgi. Major streams with good drops (head) are Shivganga (119 m), Handinadi (230.50 to 318.50 m) and Mattigatta (270 m). Aghnashini river has major drops at Unchalli (360 m) and its major streams are Benne (400 m), Bhimavara (290 m), Mudanalli (270 m), etc.

The average channel slope (S_c) is one of the factors controlling water velocity, while slope of the catchment (S_b) influences surface run off rates. These two parameters which give an idea about the nature of a stream are computed for Bedthi and Aghnashini basins and listed in Tables 4.3.1 and 4.3.2 respectively. Magod has a stream slope of 61.34°. Shivganga and Mattigatta streams of Bedthi have slopes of 43.83° and 40.03° respectively, while Muregar and Boosangeri have respective slopes of 6.27° and 2.29°. The Muregar Jog has a catchment of 25.97 sq.km., while Boosangeri has 11.29 sq.km. Stream gauging at regular intervals is carried out in Muregar, Boosangeri, Vanalli and Shivganga.

4.7.2 Catchment Shape

The shape of Boosangeri catchment is short and wide (fan shape), while Muregar, Mattigatta and Shivganga catchments are elongated.

4.7.3 Stream Flow Measurement (Direct Method) and Computation of Power (kW)

Stream gauging is carried out as explained in methodology using a current meter every month. Stream discharge ranges from 1.12 (August) to 0.015 cum/sec (February) for Boosangeri stream, while for Muregar, it ranges from 1.395 to 0.026 cum/sec. The monthly hydraulic power computed for these streams, listed in Table 4.4.1, suggests that streams of this kind are seasonal. Power generated during June-Sept is sufficient to meet the energy needs of the nearby villages. Power computed for these streams is shown in Figure 4.2. The relationship between run off and precipitation determined by regression analyses (Indirect method - Empirical formula) based on the field data is

$$R = 0.85*P + 30.5$$

where, R is run off and P is rainfall in cms.

Table 4.3.1: Catchment Area, Stream Slope and Catchment Slope Computed for Various Streams of Bedthi River

Streams	Latitude		Longitude		Cat area	Height	St. slope	Cat slope
	From	To	From	To	sq.km.	m	degrees	degrees
Magod	14° 45'		74° 45'		2084.46	340.00	61.346	68.459
Shivganga	14° 45'	14° 50'	74° 46'		88.48	119.00	55.713	60.642
Kaulgi Halla	14° 56'		74° 50'		591.45	64.00	3.814	2.545
Kalghatgi	14° 0'	15° 10'	74° 55'	75°	785.74	80.00	5.711	3.814
Arumatti Halla	14° 40'	14° 45'	74° 29'		16.50	145.00	32.800	40.671
Handimadi	14° 40'	14° 45'	74° 28'		27.29	230.50	86.524	15.440
Varangudda	14° 40'	14° 45'	74° 28'		25.60	240.00	37.569	44.556
Talagudde	14° 35'	14° 40'	74° 22'		27.29	95.00	19.058	24.731
Hirehalla	14° 35'	14° 40'	74° 29'		66.82	288.50	23.532	30.141
Motigudda	14° 35'	14° 40'	74° 30'		35.92	345.00	34.606	38.191
Kolgi (Hosakambi)	14° 35'	14° 40'	74° 30'		15.30	37.00	0.025	1.232
Chandmata	14° 35'	14° 40'	74° 22'		28.65	14.00	0.345	24.731
Malgaon	14° 40'	14° 45'	74° 33'		20.25	60.00	23.575	21.200
Heggarni	14° 40'	14° 45'	74° 30'	74° 35'	26.42	220.00	64.537	67.380
Aladbail	14° 40'	14° 45'	74° 30'	74° 35'	16.24	190.00	32.347	40.179
Kotegudda	14° 40'	14° 45'	74° 30'	74° 35'	13.16	240.00	60.191	61.509
Kanchikeri	14° 40'	14° 45'	74° 30'	74° 35'	13.40	14.00	43.025	51.216
Maratikoppa	14° 40'	14° 45'	74° 30'		16.80	30.00	8.531	11.310
Bairekoppa	14° 40'	14° 45'	74° 30'		12.60	40.00	21.801	14.931
Gubargadda	14° 35'	14° 40'	74° 30'	74° 35'	40.26	230.00	83.667	84.644
Vadgar	14° 35'	14° 40'	74° 30'	74° 35'	22.70	230.00	44.081	45.300
Kumbhatt	14° 50'	14° 55'	74° 45'		9.85	180.00	55.222	60.396
Dabguli (Panasguli)	14° 50'	14° 55'	74° 35'	74° 40'	34.68	220.50	14.964	16.007
Mulepal	14° 50'	14° 55'	74° 35'	74° 40'	18.60	180.00	28.980	33.281
Mogadde Halla	14° 50'	14° 55'	74° 35'	74° 40'	39.65	280.00	13.379	16.841
Garge Halla	14° 45'	14° 55'	74° 30'	74° 35'	29.72	230.00	22.694	23.575
Ramanguli	14° 45'	14° 55'	74° 30'	74° 35'	21.03	90.00	16.669	10.081
Barehalla	14° 45'	14° 50'	74° 30'	74° 35'	39.41	215.00	24.353	29.918
Benguli	14° 45'	14° 50'	74° 30'	74° 35'	20.46	330.00	51.499	54.888
Mulemane	14° 45'	14° 50'	74° 30'	74° 35'	13.46	230.00	47.265	48.478
Matti Gatta	14° 35'	14° 45'	74° 35'	74° 40'	32.13	270.00	30.434	30.541
Hasehalla	14° 35'	14° 45'	74° 35'	74° 40'	99.30	235.00	12.495	16.461
Vaidyaheggar	14° 45'	14° 50'	74° 35'	74° 40'	11.43	230.00	56.889	57.995
Kaigadi	14° 45'	14° 50'	74° 40'	74° 45'	12.60	230.00	42.614	49.560
Bili Halla	14° 35'	14° 50'	74° 35'	74° 40'	79.95	241.50	9.146	11.628

Table 4.3.1: Catchment Area, Stream Slope and Catchment Slope Computed for Various Streams of Bedthi River (Cont'd)

	Latitude		Longitude		Cat area	Height	St. slope	Cat slope
Streams	From	To	From	To	sq.km.	m	degrees	degrees
Songi Mane	14° 45'	14° 50'	74° 40'	74° 45'	22.68	200.00	8.498	9.528
Achave Stream	14° 35'	14° 50'	74° 35'	74° 40'	62.55	230.00	78.690	78.690
Vatehalla	14° 35'	14° 40'	74° 40'	74° 45'	45.48	50.00	7.125	7.595
Kalalegadde	14° 35'	14° 40'	74° 35'	74° 45'	40.26	40.00	83.935	83.935
Pattanadahall	14° 40'	14° 50'	74° 40'	74° 45'	92.34	52.50	3.163	3.774
Shalmala	14° 35'	14° 45'	74° 45'	74° 50'	85.38	191.50	10.336	13.668
Hemmadi	14° 40'	14° 45'	74° 50'	74° 55'	16.20	50.00	8.130	8.664
Anegundi	14° 40'	14° 45'	74° 50'	74° 55'	33.60	70.00	7.595	8.664
Manigeri	14° 40'	14° 45'	74° 50'	74° 55'	18.40	70.00	11.310	10.784
Hosahalli	14° 40'	14° 45'	74° 50'	74° 55'	26.40	70.00	11.310	10.784
Neggodu	14° 40'	14° 45'	74° 45'	74° 52'	14.80	30.00	6.226	5.539
Sadashivalli	14° 35'	14° 45'	74° 45'	74° 52'	35.70	90.00	7.883	8.459
Nilkani	14° 35'	14° 45'	74° 45'	74° 52'	28.60	60.00	3.270	3.633
Hunsekopp	14° 35'	14° 40'	74° 45'	74° 50'	24.60	50.00	6.340	6.759
Shigehalla	14° 35'	14° 40'	74° 45'	74° 50'	13.20	30.00	5.711	5.080
Kengrehole	14° 35'	14° 45'	74° 45'	74° 50'	19.70	20.00	2.083	1.389
Gunjavati	14° 55'	15° 0'	74° 50'	74° 55'	98.40	60.00	4.086	5.440
Tandi Halla	14° 35'	14° 45'	74° 45'	74° 50'	34.60	50.00	4.764	5.080
Kaulgi Hall	14° 50'	15° 00'	74° 50'	74° 0'	79.37	130.00	3.787	3.332
Majjige Halla	14° 45'	14° 55'	74° 50'	74° 0'	80.62	47.50	2.862	3.013
Mavinkatte	14° 45'	14° 55'	74° 50'	74° 0'	35.09	56.50	5.865	5.539
Kolemane	14° 45'	14° 55'	74° 50'	74° 0'	70.72	56.50	4.040	4.432
Devigadde	14° 50'	14° 55'	74° 45'	74° 55'	32.71	40.00	9.090	0.090
Koppadgadde	14° 45'	14° 50'	74° 45'	74° 50'	40.24	64.00	12.043	13.496
Onitata	14° 45'	14° 50'	74° 45'	74° 50'	28.20	40.00	6.520	6.520
Bajankoppa	14° 45'	14° 50'	74° 45'	74° 50'	33.20	60.00	3.923	4.357
Bharni	14° 45'	14° 50'	74° 45'	74° 50'	32.40	60.00	5.711	6.340
Heggarni	14° 45'	14° 50'	74° 45'	74° 50'	28.60	50.00	7.125	5.711

Table 4.3.2: Catchment area, Stream slope and Catchment slope computed for various streams of Aghanashini river

Streams	Latitude		Longitude		Cat. Area	Height	St slope	Cat slope
	From	To	From	To	sq.km.	m	degrees	degrees
Unchalli	14° 20'	14° 25'	74° 45'	74° 50'	605.67	360.00	35.538	6.039
Benne	14° 25'	14° 32'	74° 35'	74° 40'	140.75	400.00	23.962	17.874
Melinsalve	14° 30'	14° 37'	74° 35'	74° 40'	20.60	93.75	11.165	12.656
Koregudda	14° 30'	14° 35'	74° 35'	74° 40'	12.62	30.00	11.310	10.081
Kokande	14° 30'	14° 35'	74° 35'	74° 40'	14.60	70 00	17.281	19.573
Mundagar	14° 30'	14° 35'	74° 35'	74° 40'	24.80	60 00	11.310	11.717
Salikoppa	14° 25'	14° 30'	74° 35'	74° 40'	18.90	15 00	36.870	84.289
Ne Morse	14° 25'	14° 30'	74° 38'	74° 45'	12.80	276.50	47.881	50.812
Bhimavara	14° 25'	14° 30'	74° 38'	74° 45'	18.60	290.00	44.029	46.848
Halebail	14° 25'	14° 30'	74° 35'	74° 40'	9.20	280.00	44.029	63.935
Bandal	14° 30'	14° 35'	74° 35'	74° 40'	21.80	210.00	34.992	43.025
Arehalla	14° 30'	14° 35'	74° 50'	74° 55'	12.40	43.00	8.506	3.450
Terakanalli	14° 30'	14° 35'	74° 50'	74° 55'	14.60	60.00	2.923	3.895
Gopinathpuram	14° 30'	14° 35'	74° 50'	74° 55'	12.40	73.00	3.977	3.633
Nigodu-Kilar	14° 20'	14° 30'	74° 45'	74° 55'	28.00	67.25	4.805	22.620
Mattimane	14° 15'	14° 25'	74° 45'	74° 55'	46.30	60.00	2.923	2.209
Nigmandikai	14° 15'	14° 25'	74° 45'	74° 55'	20.00	70.00	3.814	5.080
Markodinalla	14° 15'	14° 25'	74° 45'	74° 55'	28.00	169.50	11.278	6.266
Amminalli	14° 30'	14° 38'	74° 40'	74° 45'	34.50	60.00	6.654	8.842
Keremane	14° 20'	14° 25'	74° 40'	74° 45'	42.50	218.00	19.968	25.017
Balur	14° 20'	14° 25'	74° 40'	74° 45'	36.80	222.00	27.580	33.378
Birlamakki	14° 20'	14° 25'	74° 40'	74° 45'	12.60	226.00	56.427	61.821
Markodi Halla	14° 25'	14° 30'	74° 40'	74° 45'	48.90	85.50	6.100	20.723
Huldevarakodalu	14° 22'	14° 28'	74° 35'	74° 40'	48.90	85.50	6.100	20.723
Huldeva	14° 22'	14° 28'	74° 35'	74° 40'	10.20	170.00	51.033	58.758
Shavemane	14° 22'	14° 28'	74° 35'	74° 40'	18.40	210.00	43.025	58.758
Mudanalli	14° 22'	14° 28'	74° 35'	74° 40'	22.60	270.00	44.474	49.325
Basolli	14° 22'	14° 28'	74° 35'	74° 40'	14.40	14.00	34.992	30.964
Santeguli	14° 22'	14° 28'	74° 30'	74° 35'	33.20	190.00	26.870	26.463
Vadgeri	14° 22'	14° 30'	74° 40'	74° 45'	14.20	209.00	37.235	44.119
Abbigadde	14° 22'	14° 30'	74° 40'	74° 45'	18.40	200.00	45.000	51.710
Hukkalli	14° 22'	14° 30'	74° 40'	74° 45'	20.40	220.00	48.814	54.175
Tingar	14° 25'	14° 30'	74° 30'	74° 40'	16.20	190.00	32.347	30.018

Table 4.3.2: Catchment area, Stream slope and Catchment slope computed for various streams of Aghanashini river (Cont'd)

Streams	Latitude From	Latitude To	Longitude From	Longitude To	Cat. Area sq.km.	Height m	St. slope degrees	Cat slope degrees
Ambiguli	14° 25'	14° 30'	74° 30'	74° 40'	14.80	210.00	30.964	37.304
Holavali	14° 25'	14° 30'	74° 30'	74° 40'	12.60	230.00	53.902	61.327
Kanakale	14° 25'	14° 30'	74° 30'	74° 35'	8.60	14.00	48.240	56.192
Kavaladi	14° 25'	14° 30'	74° 30'	74° 35'	8.80	190.00	59.370	58.924
Korebail	14° 25'	14° 30'	74° 30'	74° 35'	19.60	240.00	34.439	35.897
Shiragunj	14° 25'	14° 30'	74° 30'	74° 35'	12.80	95.00	32.347	21.801
Devangadde	14° 25'	14° 30'	74° 30'	74° 35'	9.90	45.00	14.421	6.520
Chandikanadi	14° 25'	14° 30'	74° 25'	74° 32'	24.80	145.00	21.140	17.745
Bogorbail	14° 25'	14° 30'	74° 25'	74° 32'	11.60	65.00	27.474	25.641
Khaire	14° 25'	14° 30'	74° 25'	74° 32'	22.60	245.00	31.487	32.347
Kudirahalla	14° 25'	14° 35'	74° 22'	74° 28'	50.44	171.50	19.852	22.147
Bargi	14° 25'	14° 35'	74° 22'	74° 28'	9.60	14.00	44.029	40.179

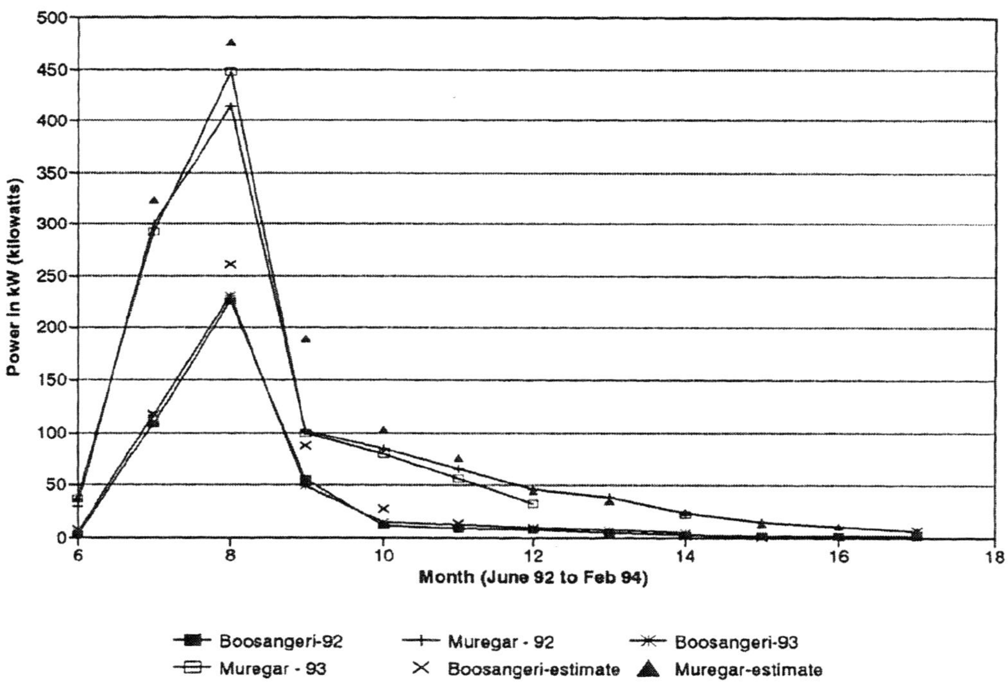

Figure 4.2: Power in Muregar and Boosangeri (Field Data and Estimate)

Table 4.4.1: Estimated Power in kW (monthwise) in Boosangeri Stream and Muregar Jog (based on field measurements)

	Boosangeri	Muregor
June 1992	2.94	30.09
July 1992	108.31	300.12
August 1992	224.99	414.78
September 1992	55.09	101.85
October 1992	12.14	84.76
November 1992	8.27	65.42
December 1992	7.60	45.65
January 1993	4.19	38.54
February 1993	2.68	23.37
March 1993	1.72	14.56
April 1993	1.60	10.75
May 1993	1.49	5.23
June 1993	4.19	36.92
July 1993	116.45	292.12
August 1993	230.00	447.48
September 1993	48.72	99.47
October 1993	15.12	79.45
November 1993	12.54	56.33
December 1993	9.41	32.61
February 1994	4.44	23.04

4.7.4 Rational Method

In order to estimate hydropower potential of ungauged streams, either rational method or empirical relationship of run off and precipitation is used. In rational method, monthly yield is derived by assuming a suitable run off coefficient (which depends on catchment type) - for each site Yield (Y= C*A*I) is computed with the knowledge of catchment area (A), catchment coefficient (C) and precipitation intensity (I). An attempt has been made through this method to compute the monthly yield from these catchments and the subsequent power that could be harvested from the streams. Results of this computation are listed in Table 4.4.2. The estimated monthly power by indirect method, shown pictorially in Figure 4.3, conforms with the power computed by direct method (of gauged streams, Boosangeri and Muregar). Therefore, rational method is used to compute the hydropower of ungauged streams.

This study explores the possibility of harnessing hydro potential in an ecologically sound way (by having run-of-river plants with no storage options) to suit the requirements of the region. Sirsi, Siddapur and Yellapur taluks of Uttara Kannada district, being located in a hilly terrain amidst evergreen forests with large number of streams, are ideally suited for micro, mini or small hydropower plants. Monthly stream gauging at Muregar and Boosangeri has revealed that mini hydropower plants could be set up at these sites. The stream at Muregar is perennial and with a flow of about 0.26 m^3/sec during summer, power of 10 - 20 kW could be generated, while during the seasons, 300-400 kW could be harnessed.

Computations of discharge on empirical basis/rational method, based on precipitation history of last 90 years, and subsequent power calculated are in conformity with the power calculations based on stream gauging. Based on this field experience of gauged sites, an attempt has been made to compute water inflow (using indirect methods), hydraulic power available and energy that could be harnessed monthwise for all ungauged streams in Bedthi and Aghnashini catchments. Hydropower potential of all streams is listed in Table 4.5.1, while energy potential of streams in Bedthi catchment is listed in Table 4.5.2. It is estimated that about 720 million kWh of electricity can be generated from various streams in Bedthi catchment. Similarly, Tables 4.6.1 and 4.6.2 provide information of hydro resource potential in ungauged streams of Aghnashini river catchment. The electrical energy that could be harnessed annually from various streams of Aghnashini basin is about 510 million kWh.

The potential assessment of streams in Bedthi and Aghnashini catchments show that most of the streams are seasonal and cater the needs of the local people in a decentralised way during the seasons. This assures continuous power supply during heavy monsoon, which otherwise gets disrupted due to various problems like dislocation of electric poles due to heavy winds or falling of trees/branches on transmission lines. Detailed household survey of villages in hilly areas of the district shows that people have to spend at least 60 to 65% of the monsoon season without electricity from centralised supply due to environmental problems.

There is demand for alternatives to meet the region's requirement during the lean season as most of these streams are seasonal. This exercise provides insight into the regional energy requirement through integrated approaches like harnessing of hydropower in a decentralised way during seasons and meeting the lean season requirement through solar or other thermal options. In view of this, an ecologically friendly design of a hydroelectric plant is proposed at Magod, which would generate maximum electricity during seasons and have storage options to store enough water to generate sufficient electricity to meet the region's lean season requirement. This design is discussed in detail in the next chapter.

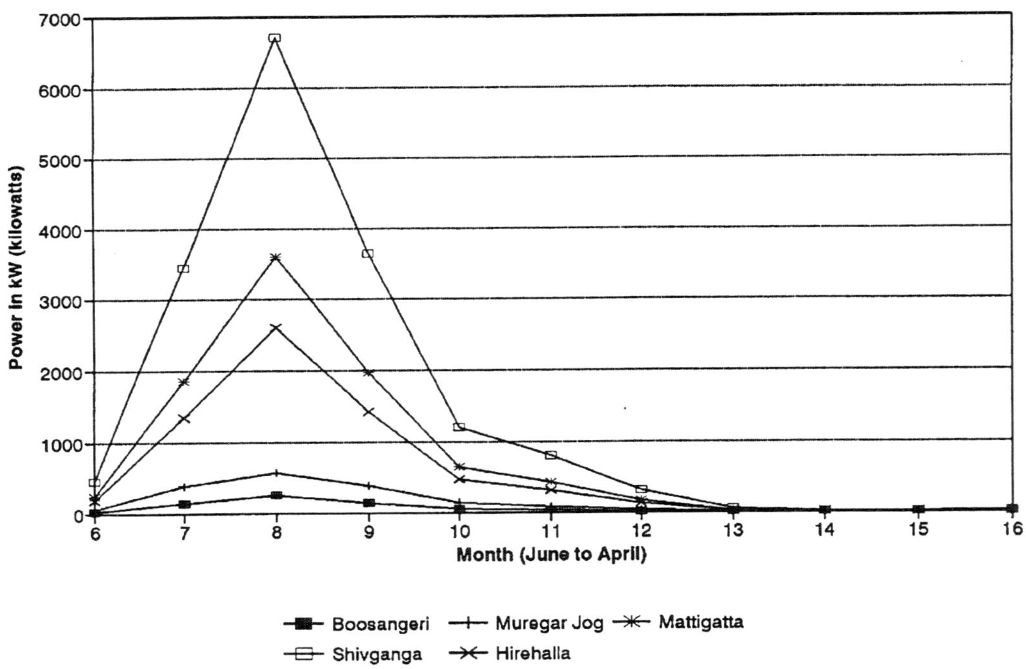

Figure 4.3: Power in Muregar & Boosangeri (Based on Rational Method)

Table 4.4.2: Estimated Power in kW (monthwise) in Boosangeri Stream and Muregar Jog by Rational Method - with the knowledge of Catchment Area and Precipitation (100 years)

Year	Month	Boosangeri	Muregar	Mattigatta	Shivaganga	Hirehalla
1992	June	12.50	48.02	242.23	450.45	174.96
	July	138.38	379.69	1853.68	3447.02	1338.85
	August	260.19	576.83	3601.43	6697.07	2601.19
	September	141.70	388.82	1961.39	3647.32	1416.64
	October	48.23	132.33	646.04	1201.35	466.61
	November	31.17	85.52	431.42	802.22	311.59
	December	12.73	34.92	170.48	317.02	123.13
1993	January	1.85	5.08	25.63	47.66	18.51
	February	0.24	0.65	3.29	6.11	2.37
	March	0.24	0.73	3.34	6.22	2.42
	April	1.04	2.86	14.41	26.80	10.41
	May	6.67	17.08	83.41	155.10	60.24

Table 4.5.1: Estimated Water Discharge and Power (MW) in Ungauged Streams of Bedthi Catchment Area

	Streams	April	May	June	July	August	September	October	November	Avg
1	Arumatti Halla	0.01	0.11	1.00	1.79	1.02	0.38	0.27	0.10	0.58
2	Handinadi	0.20	1.05	8.17	14.44	8.21	3.24	2.34	1.00	4.83
3	Varangudda	0.05	0.30	2.59	4.61	2.63	1.00	0.72	0.28	1.52
4	Talagudde	0.02	0.13	1.10	1.95	1.11	0.43	0.30	0.12	0.64
5	Hirehalla	0.19	1.06	8.23	14.55	8.27	3.25	2.35	1.00	4.86
6	Motigudda	0.11	0.64	5.26	9.33	5.31	2.06	1.48	0.60	3.10
7	Kolgi	0.00	0.02	0.24	0.42	0.24	0.09	0.06	0.02	0.14
8	Malgaon	0.01	0.06	0.51	0.91	0.52	0.20	0.14	0.05	0.30
9	Heggarni	0.04	0.29	2.46	4.36	2.49	0.95	0.68	0.27	1.44
10	Aladbail	0.01	0.13	1.29	2.30	1.32	0.49	0.34	0.12	0.75
11	Kotegudda	0.01	0.13	1.31	2.35	1.34	0.49	0.34	0.12	0.76
12	Kanchikeri	0.01	0.08	0.78	1.40	0.80	0.29	0.20	0.07	0.45
13	Maratikoppa	0.00	0.02	0.21	0.38	0.21	0.08	0.06	0.02	0.12
14	Bairekoppa	0.00	0.02	0.21	0.37	0.21	0.08	0.05	0.02	0.12
15	Gubbargadda	0.52	2.62	19.68	34.75	19.73	7.85	5.71	2.49	11.67
16	Kumbhatt	0.00	0.06	0.73	1.31	0.75	0.27	0.18	0.05	0.42
17	Datguli	0.07	0.39	3.24	5.75	3.27	1.27	0.91	0.37	1.91
18	Mulepal	0.02	0.15	1.40	2.50	1.43	0.54	0.38	0.14	0.82
19	Mogadde Halla	0.10	0.58	4.72	8.36	4.76	1.85	1.33	0.55	2.78
20	Garge Halla	0.05	0.34	2.89	5.14	2.93	1.13	0.80	0.32	1.70

Table 4.5.1: Estimated Water Discharge and Power (MW) in ungauged Streams of Bedthi Catchment Area (Cont'd)

	Streams	April	May	June	July	August	September	October	November	Avg
21	Ramanguli	0.01	0.09	0.80	1.42	0.81	0.31	0.22	0.08	0.47
22	Barehalla	0.08	0.44	3.60	6.38	3.63	1.41	1.01	0.42	2.12
23	Benguli	0.04	0.32	2.84	5.05	2.88	1.09	0.77	0.29	1.66
24	Mulemane	0.01	0.13	1.29	2.30	1.32	0.48	0.34	0.12	0.75
25	Matti Gatta	0.07	0.44	3.68	6.52	3.71	1.43	1.03	0.41	2.16
26	Hasehalla	0.25	1.30	9.98	17.64	10.02	3.96	2.87	1.23	5.91
27	Vaidyaheggar	0.00	0.10	1.09	1.95	1.12	0.40	0.28	0.09	0.63
28	Kaigadi	0.01	0.11	1.20	2.15	1.23	0.45	0.31	0.10	0.70
29	Bili Halla	0.20	1.07	8.25	14.58	8.29	3.27	2.37	1.01	4.88
30	Songi Mane	0.03	0.22	1.91	3.40	1.94	0.74	0.52	0.20	1.12
31	Achave Stream	0.14	0.78	6.14	10.86	6.17	2.43	1.75	0.74	3.63
32	Vatehalla	0.02	0.12	0.97	1.71	0.97	0.38	0.27	0.11	0.57
33	Kalalegadde	0.01	0.08	0.68	1.21	0.69	0.27	0.19	0.08	0.40
34	Pattanadahall	0.05	0.27	2.07	3.66	2.08	0.82	0.60	0.26	1.23
35	Shalmala	0.17	0.91	6.99	12.35	7.02	2.77	2.01	0.86	4.13
36	Hemmadi	0.00	0.04	0.34	0.60	0.35	0.13	0.09	0.03	0.20
37	Anegundi	0.02	0.12	1.00	1.77	1.01	0.39	0.28	0.11	0.59
38	Manigeri	0.01	0.06	0.54	0.96	0.55	0.21	0.15	0.05	0.32
39	Hosahalli	0.01	0.09	0.78	1.39	0.79	0.30	0.22	0.09	0.46
40	Neggodu	0.00	0.02	0.19	0.33	0.19	0.07	0.05	0.02	0.11
41	Sadashivalli	0.03	0.17	1.36	2.42	1.38	0.53	0.38	0.16	0.80
42	Nilkani	0.01	0.09	0.73	1.29	0.73	0.28	0.20	0.08	0.43
43	Hunsekopp	0.01	0.06	0.52	0.92	0.53	0.20	0.14	0.06	0.30
44	Shigehalla	0.00	0.02	0.16	0.29	0.17	0.06	0.04	0.01	0.10
45	Kengrehole	0.00	0.02	0.17	0.29	0.17	0.06	0.04	0.02	0.10
46	Gunjavati	0.06	0.33	2.53	4.46	2.53	1.00	0.73	0.31	1.49
47	Tandi Halla	0.01	0.09	0.73	1.30	0.74	0.29	0.21	0.08	0.43
48	Kaulgi Halla	0.11	0.57	4.41	7.79	4.43	1.75	1.26	0.54	2.61
49	Majjige Halla	0.06	0.30	2.28	4.02	2.28	0.91	0.66	0.28	1.35
50	Mavinkatte	0.06	0.34	2.56	4.52	2.57	1.02	0.74	0.32	1.51
51	Devigadde	0.06	0.31	2.38	4.21	2.39	0.95	0.69	0.29	1.41
52	Koppadgadde	0.02	0.13	1.09	1.94	1.10	0.43	0.31	0.13	0.64
53	Onitata	0.01	0.06	0.48	0.85	0.48	0.19	0.13	0.05	0.28
54	Bajankoppa	0.02	0.10	0.84	1.50	0.85	0.33	0.24	0.10	0.50
55	Bharni	0.02	0.10	0.82	1.46	0.83	0.32	0.23	0.09	0.48
56	Heggarni	0.01	0.07	0.61	1.07	0.61	0.24	0.17	0.07	0.36
57	Magod Falls	6.31	22.09	224.09	520.86	229.31	84.08	58.55	20.31	145.70
58	Chandmata	0.03	0.20	1.70	3.01	1.72	0.66	0.47	0.19	1.00

Table 4.5.2: Estimated Water Discharge and Energy ('000 units) in Ungauged Streams of Bedthi Catchment Area

	Streams	April	May	June	July	August	September	October	November	Total
1	Arumatti Halla	0.60	5.39	49.69	91.62	52.34	18.86	13.69	4.83	237.02
2	Handinadi	5.73	30.75	230.66	421.41	239.47	91.36	68.28	28.14	1115.80
3	Varangudda	1.40	9.33	77.84	142.92	81.45	30.14	22.18	8.46	373.72
4	Talagudde	14.71	95.64	789.15	1448.24	825.13	306.21	225.68	86.72	3791.48
5	Hirehalla	144.66	784.94	5924.46	10827.26	6153.75	2343.31	1749.60	717.87	28645.85
6	Motigudda	79.52	476.41	3786.89	6937.88	3948.90	1481.03	1097.41	433.60	18241.64
7	Kolgi	1.84	18.01	170.11	313.96	179.46	64.28	46.50	16.10	810.26
8	Malgaon	5.59	42.09	367.74	676.56	386.02	141.05	103.11	37.94	1760.09
9	Heggarni	32.38	213.18	1768.28	3245.93	1849.60	685.41	504.76	193.21	8492.75
10	Aladbail	11.01	100.24	928.81	1712.91	978.69	352.23	255.43	89.78	4429.09
11	Kotegudda	7.42	94.56	944.56	1746.96	999.78	353.31	253.67	83.94	4484.21
12	Kanchikeri	4.62	56.62	561.39	1038.00	593.95	210.26	151.11	50.31	2666.26
13	Maratikoppa	1.89	16.56	151.85	279.93	159.90	57.70	41.90	14.84	724.57
14	Bairekoppa	1.04	14.79	150.50	278.53	159.47	56.11	40.19	13.10	713.72
15	Gubbargadda	383.45	1950.83	14172.70	25851.80	14676.60	5654.21	4245.84	1790.20	68725.63
16	Kumbhatt	0.34	45.09	524.11	974.34	559.26	191.15	134.65	39.21	2468.14
17	Datguli	48.43	292.63	2335.73	4280.07	2436.40	912.69	675.87	266.23	11248.05
18	Mulepal	14.15	113.38	1011.34	1862.23	1063.06	386.34	281.63	101.98	4834.11
19	Mogadde Halla	73.69	431.94	3396.50	6219.52	3538.98	1331.43	988.11	393.53	16373.71
20	Garge Halla	40.51	255.78	2083.46	3821.35	2176.46	810.63	598.53	232.24	10018.95
21	Ramanguli	8.99	66.17	573.32	1054.41	601.48	220.27	161.22	59.71	2745.59
22	Barehalla	56.13	329.43	2592.06	4746.61	2700.92	1015.96	753.91	300.12	12495.16
23	Benguli	31.33	234.48	2044.00	3760.12	2145.28	784.36	573.60	211.43	9784.59
24	Mulemane	7.72	93.61	926.55	1713.06	980.20	347.14	249.53	83.20	4401.02
25	Matti Gatta	53.26	328.47	2647.09	4852.77	2763.11	1032.25	763.35	298.56	12738.86
26	Hasehalla	184.77	970.35	7187.03	13122.37	7454.11	2854.72	2137.44	888.94	34799.72
27	Vaidyaheggar	3.62	73.37	782.11	1449.85	830.85	289.28	205.96	64.59	3699.64
28	Kaigadi	5.98	85.04	865.36	1601.56	916.93	322.63	231.08	75.31	4103.88
29	Bili Halla	148.90	794.56	5940.22	1050.94	6165.52	2354.53	1750.47	727.29	28742.43
30	Songi Mane	22.88	161.36	1376.14	2529.16	1442.19	530.40	339.07	145.83	6597.03
31	Achave Stream	106.72	583.19	4419.33	8078.13	4591.79	1746.43	1303.18	533.17	21361.93
32	Vatehalla	15.72	89.77	696.69	1274.96	725.20	273.88	203.65	81.89	3361.77
33	Kalalegadde	10.74	62.76	492.76	902.26	513.37	193.23	143.44	57.19	2375.76

Table 4.5.2: Estimated Water Discharge and Energy ('000 units) in ungauged Streams of Bedthi catchment area (Cont'd)

	Streams	April	May	June	July	August	September	October	November	Total
34	Pattanadahall	38.07	200.94	1492.58	2725.60	1548.40	592.47	443.41	184.03	7225.50
35	Shalmala	127.19	675.14	5032.03	9190.58	5221.63	1995.92	1493.02	618.15	24353.68
36	Hemmadi	2.88	26.29	243.80	449.64	256.91	92.44	67.03	23.55	1162.54
37	Anegundi	14.71	89.62	718.12	1316.13	749.28	280.37	207.51	81.51	3457.24
38	Manigeri	5.38	43.49	388.97	716.31	408.93	148.51	108.22	39.10	1858.90
39	Hosahalli	10.29	67.77	562.20	1032.01	588.06	217.91	160.47	61.42	2700.13
40	Neggodu	1.36	13.95	133.29	246.11	140.71	50.26	36.30	12.45	634.43
41	Sadashivalli	20.57	123.42	981.76	1798.72	1023.81	383.90	284.43	112.32	4728.95
42	Nilkani	9.98	63.81	522.72	958.99	546.28	203.14	149.87	57.91	2512.69
43	Hunsekopp	6.56	44.50	373.73	686.41	391.26	144.50	106.23	40.28	1793.47
44	Shigehalla	0.94	11.87	118.44	219.05	125.36	44.31	31.82	10.54	562.33
45	Kengrehole	1.77	13.55	119.18	219.32	125.15	45.65	33.34	12.21	570.17
46	Gunjavati	46.70	245.41	1818.28	3319.95	1885.90	722.17	540.69	224.81	8803.92
47	Tandi Halla	10.95	66.18	528.41	968.28	551.19	206.46	152.89	60.21	2544.57
48	Kaulgi Halla	79.49	424.44	3174.31	5798.58	3294.79	1258.11	940.63	388.50	15358.85
49	Majjige Halla	42.93	222.86	1639.14	2991.76	1699.11	652.10	488.77	204.29	7940.97
50	Mavinkatte	47.65	249.24	1841.73	3362.31	1909.81	731.93	548.22	228.38	8919.27
51	Devigadde	43.81	231.12	1716.38	3134.25	1780.54	681.35	509.95	211.68	8309.09
52	Koppadgadde	17.18	100.37	788.02	1442.89	820.99	309.01	229.38	91.46	3799.29
53	Onitata	6.51	41.85	343.53	630.31	359.07	133.44	98.42	37.97	1651.09
54	Bajankoppa	12.40	75.78	608.10	1114.58	634.56	237.35	175.62	68.91	2927.30
55	Bharni	11.98	73.70	593.25	1087.52	619.21	231.40	171.14	66.99	2855.19
56	Heggarni	8.32	53.18	435.60	799.16	455.23	169.28	124.89	48.25	2093.91
57	Magod Falls	4695.0	16435.5	161345.0	387522.4	170605.2	60538.7	43563.4	14620.8	859326.1
58	Chandmata	23.34	149.19	1221.85	2241.59	1276.88	474.86	350.34	135.39	5873.46
									Total	1319325

Table 4.6.1: Estimated Water Discharge and Power (in MW) in ungauged streams of Aghnashini Catchment Area

Streams	April	May	June	July	August	September	October	November	Avg
Melinsalve	0.04	0.10	1.22	2.26	1.41	0.39	0.14	0.04	0.70
Koregudda	0.02	1.12	0.24	0.44	0.28	0.07	0.02	0.01	0.27
Kokande	0.02	0.05	0.64	1.19	0.74	0.20	0.07	0.02	0.37
Mundagar	0.03	0.08	0.94	1.75	1.08	0.30	0.11	0.04	0.54
Salikoppa	0.06	0.14	1.79	3.32	2.06	0.57	0.20	0.06	1.03
Ne Morse	0.07	0.16	2.21	4.13	2.57	0.69	0.23	0.05	1.26
Bhimavara gudda	0.12	0.27	3.40	6.32	3.93	1.08	0.38	0.11	1.95
Halebail	0.04	0.10	1.60	2.99	1.87	0.49	0.15	0.02	0.91
Bandal	0.10	0.24	2.89	5.37	3.33	0.93	0.33	0.11	1.66
Arehalla	0.01	0.02	0.33	0.62	0.39	0.10	0.03	0.01	0.19
Terakanalli	0.02	0.04	0.55	1.02	0.64	0.17	0.06	0.02	0.31
Gopinathpuram	0.02	0.04	0.57	1.06	0.66	0.18	0.06	0.01	0.32
Nigodu-Kilar	0.05	0.10	1.19	2.21	1.37	0.39	0.14	0.05	0.69
Mattimane Halla	0.07	0.16	1.77	3.27	2.03	0.58	0.22	0.08	1.02
Nigmandikai	0.03	0.07	0.88	1.64	1.02	0.28	0.10	0.03	0.51
Markodinalla	0.11	0.26	3.01	5.57	3.46	0.97	0.35	0.12	1.73
Amminalli	0.05	0.12	1.31	2.43	1.51	0.43	0.16	0.06	0.76
Keremane	0.24	0.53	5.89	10.90	6.76	1.92	0.72	0.26	3.40
Balur	0.20	0.46	5.19	9.61	5.96	1.69	0.63	0.23	2.99
Birlamakki	0.05	0.13	1.78	3.32	2.07	0.55	0.18	0.04	1.02
Markodi Halla	0.11	0.24	2.66	4.92	3.05	0.87	0.33	0.12	1.54
Huldevarakodalu	0.03	0.07	1.08	2.02	1.26	0.33	0.10	0.02	0.61
Shavemane	0.08	0.19	2.43	4.52	2.81	0.77	0.27	0.08	1.40
Mudanalli	0.14	0.32	3.86	7.16	4.45	1.24	0.44	0.14	2.22
Basoli	0.04	0.09	1.26	2.35	1.47	0.40	0.13	0.03	0.72
Santeguli	0.16	0.35	4.00	7.41	4.60	1.30	0.48	0.17	2.31
Vadgeri	0.06	0.14	1.86	3.47	2.16	0.58	0.20	0.05	1.06
Abbigadde	0.08	0.18	2.32	4.31	2.68	0.74	0.26	0.08	1.33
Hukkalli	0.10	0.23	2.83	5.26	3.27	0.90	0.32	0.10	1.63
Tingar	0.06	0.15	1.93	3.60	2.24	0.61	0.21	0.06	1.11
Ambiguli	0.06	0.15	1.95	3.63	2.26	0.61	0.21	0.06	1.12
Holavali	0.06	0.13	1.81	3.38	2.11	0.56	0.19	0.04	1.03
Kanakale	0.02	0.04	0.74	1.40	0.87	0.22	0.07	0.01	0.42
Kavaladi	0.03	0.06	1.03	1.94	1.21	0.31	0.09	0.01	0.59
Korebail	0.10	0.24	2.97	5.51	3.43	0.95	0.33	0.10	1.70
Shiragunj	0.02	0.05	0.76	1.42	0.88	0.24	0.08	0.02	0.43
Devangadde	0.01	0.02	0.28	0.52	0.32	0.08	0.03	0.00	0.16
Chandkanadi	0.08	0.19	2.27	4.22	2.62	0.73	0.26	0.09	1.31
Bogorbail	0.01	0.03	0.47	0.88	0.55	0.15	0.05	0.01	0.27
Khaire	0.13	0.29	3.50	6.49	4.03	1.12	0.40	0.13	2.01
Kudirahalla	0.22	0.50	5.51	10.18	6.31	1.80	0.68	0.25	3.18
Bargi	0.02	0.05	0.83	1.56	0.97	0.25	0.08	0.01	0.47
Benne	0.71	5.60	70.91	121.26	58.65	17.83	11.37	1.33	35.96

Table 4.6.2: Estimated Water Discharge and Energy (in '000 units) in ungauged Streams of Aghnashini Catchment Area

	Streams	April	May	June	July	August	September	October	November	Total
1	Melinsalve	32.39	73.82	877.53	1683.70	1046.48	280.42	103.11	31.19	4128.66
2	Koregudda	15.18	832.07	170.46	328.44	204.64	53.13	18.23	4.08	1626.24
3	Kokande	15.42	35.47	461.62	888.14	552.90	145.16	51.08	12.94	2162.73
4	Mundagar	25.99	59.04	677.78	1298.97	806.84	218.01	81.54	26.17	3194.34
5	Salikoppa	46.51	106.18	1286.50	2469.87	1535.66	409.69	149.25	43.62	6047.28
6	Ne Morse	50.52	116.80	1593.98	3070.89	1913.20	497.28	171.05	38.87	7452.58
7	Bhimavara gudda	88.10	201.22	2447.14	4698.64	2921.61	778.76	283.18	82.20	11500.83
8	Halebail	30.12	71.10	1149.52	2224.12	1389.07	349.44	111.02	14.53	5338.92
9	Bandal	77.82	177.16	2081.83	3992.89	2481.21	666.69	246.52	76.07	9800.19
10	Arehalla	7.50	17.36	239.96	462.46	288.18	74.70	25.54	5.62	1121.32
11	Terakanalli	13.22	30.40	395.67	761.27	473.91	124.42	43.78	11.09	1853.77
12	Gopinathpuram	12.73	29.47	407.37	785.10	489.23	126.82	43.36	9.54	1903.63
13	Nigodu-Kilar	33.62	76.25	858.88	1645.01	1021.40	277.26	104.67	34.63	4051.71
14	Mattimane Halla	52.92	119.41	1272.41	2432.38	1508.59	415.26	161.14	57.94	6020.05
15	Nigmandikai	23.31	53.16	635.86	1220.26	758.53	202.96	74.40	22.25	2990.73
16	Markodinalla	84.75	192.17	2164.75	4146.16	2574.38	698.81	263.82	87.29	10212.12
17	Amminalli	38.14	86.28	946.06	1810.32	1123.44	307.00	117.45	40.50	4469.20
18	Keremane	174.98	395.09	4241.24	8109.80	5030.56	1382.12	534.35	190.12	20058.27
19	Balur	151.78	343.12	3735.77	7146.82	4434.50	1213.98	466.07	162.44	17654.48
20	Birlamakki	40.35	93.35	1282.02	2470.31	1539.19	399.54	137.02	30.66	5992.43
21	Markodi Halla	80.05	180.56	1915.66	3661.45	2270.67	625.74	243.33	88.05	9065.52
22	Huldevarakodalu	21.83	51.12	776.28	1499.72	935.85	238.15	77.89	13.01	3613.87
23	Shavemane	62.92	143.74	1752.70	3365.56	2092.80	557.51	202.47	58.49	8236.19
24	Mudanalli	104.56	237.88	2776.20	5323.49	3307.63	890.20	330.29	103.13	13073.38
25	Basoli	30.26	69.62	910.33	1751.69	1090.56	286.04	100.43	25.20	4264.13
26	Santeguli	115.62	261.65	2881.99	5515.68	3423.20	934.40	356.69	122.18	13611.41
27	Vadgeri	44.29	101.98	1339.73	2578.30	1605.32	420.62	147.35	36.59	6274.17
28	Abbigadde	59.92	136.90	1669.24	3205.30	1993.14	530.96	192.83	55.70	7843.99
29	Hukkalli	75.10	171.18	2038.99	3912.42	2431.82	651.33	239.26	72.11	9592.20
30	Tingar	48.20	110.49	1393.10	2677.78	1666.10	440.49	157.40	42.62	6536.17
31	Ambiguli	47.14	108.36	1404.22	2701.34	1681.54	441.90	155.82	39.87	6580.19
32	Holavali	41.06	95.00	1304.71	2514.03	1566.43	406.61	139.44	31.20	6098.49
33	Kanakale	13.31	31.62	536.04	1038.26	648.84	161.87	50.33	5.20	2485.46
34	Kavaladi	18.85	44.69	745.00	1442.45	901.24	225.50	70.65	7.99	3456.36
35	Korebail	77.92	177.75	2135.84	4099.40	2548.44	681.19	249.16	73.94	10043.64
36	Shiragunj	17.36	40.13	547.66	1055.10	657.34	170.85	58.77	13.35	2560.56
37	Devangadde	5.50	12.90	199.26	385.12	240.38	60.98	19.79	3.11	927.04
38	Chandkanadi	62.81	142.68	1637.97	3139.19	1949.86	526.85	197.06	63.24	7719.65
39	Bogorbail	10.25	23.81	338.76	653.38	407.33	104.97	35.40	7.22	1581.11
40	Khaire	94.88	215.86	2519.14	4830.58	3001.37	807.77	299.70	93.58	11862.89
41	Kudirahalla	166.08	374.54	3964.26	7576.37	4698.29	1295.53	504.39	183.11	18762.57
42	Bargi	16.23	38.17	600.57	1161.26	725.00	183.28	58.96	8.65	2792.13
43	Benne	510.12	4033.00	51052.70	87307.16	42225.82	12838.77	8188.46	959.20	55085.70

4.8.0 CONCLUSIONS

The potential assessment of streams in Bedthi and Aghnashini catchments shows that most of the streams are seasonal and cater the needs of the local people in a decentralised way during the seasons. This assures continuous power supply during heavy monsoon, which otherwise gets disrupted due to various problems like dislocation of electric poles due to heavy winds or falling of trees/branches on transmission lines. This study explores the possibility of harnessing hydro potential in an ecologically sound way (by having run-of-river plants with no storage options) to suit the requirements of the region. Sirsi, Siddapur and Yellapur taluks of Uttara Kannada district, being located in a hilly terrain amidst evergreen forests with large number of streams, are ideally suited for these micro, mini or small hydropower plants. Monthly stream gauging at Muregar and Boosangeri has revealed that mini hydropower plants could be set up at these sites. Stream at Muregar is perennial and during summer, with a flow of about 0.26 m^3 /sec, 10-20 kW power could be generated, while during seasons power of 300-400 kW could be harnessed. Energy that could be harnessed monthwise is computed for all ungauged streams in Bedthi and Aghnashini catchments based on empirical basis/rational method using precipitation history of last 90 years. The hydro energy potential of streams in Bedthi catchment is estimated to be about 720 million kWh, while in Aghnashini catchment it is 510 million kWh.

4.9.0 ACKNOWLEDGEMENTS

I thank Mr. Sampath and Mr. Dinesh, Information Analysts from Karnataka State Council for Science and Technology, for taking part in an exploratory survey to identify feasible sites and for useful tips during the field survey. I thank my colleagues Mr. Deepak Shetti, Mr. Rosario Furtado, Mr. Kumar and Mr. Raghavendra Rao for their assistance in collection of stream flow data at regular intervals. Mr. Rosario, apart from driving jeep, took keen interest in the experiments along with Mr. Deepak and ensured the work was not hampered due to the extremely hostile environment, especially during the monsoon season. I appreciate Mr. Rosario and Mr. Deepak for their commitment and promptness in work.

4.10.0 REFERENCES

1. Creager, W.P. and Justin, J.D., 1950. *Hydroelectric Handbook*, John Wiley and Sons, Inc., England.
2. Raghunath, H.M., 1985. *Hydrology*, Wiley Eastern Ltd., New Delhi.
3. Holland, M.E., 1967. *An Experimental Rainfall-Run off Facility,* In Hydrology Papers, Colorado State University, Colorado.
4. Markovic, R.D., 1976. *Probability Functions of Best Fit to Distributions of Annual Precipitations and Run off*, In Hydrology Papers, Colorado State University, Fort Collins, Colorado.
5. Pearson, K., 1957. *Table of Incomplete Gamma Function*, Cambridge, The University Press.
6. Zar, J.H., 1984. *Biostatistical Analysis*, Prentice Hall, Inc., Englewood Cliff, N.J. 07632.
7. Gordon, N.D., McMahon, T.A. and Finlayson, B.L., 1992. *Stream Hydrology: An Introduction for Ecologists*, John Wiley and Sons, Ltd., England.

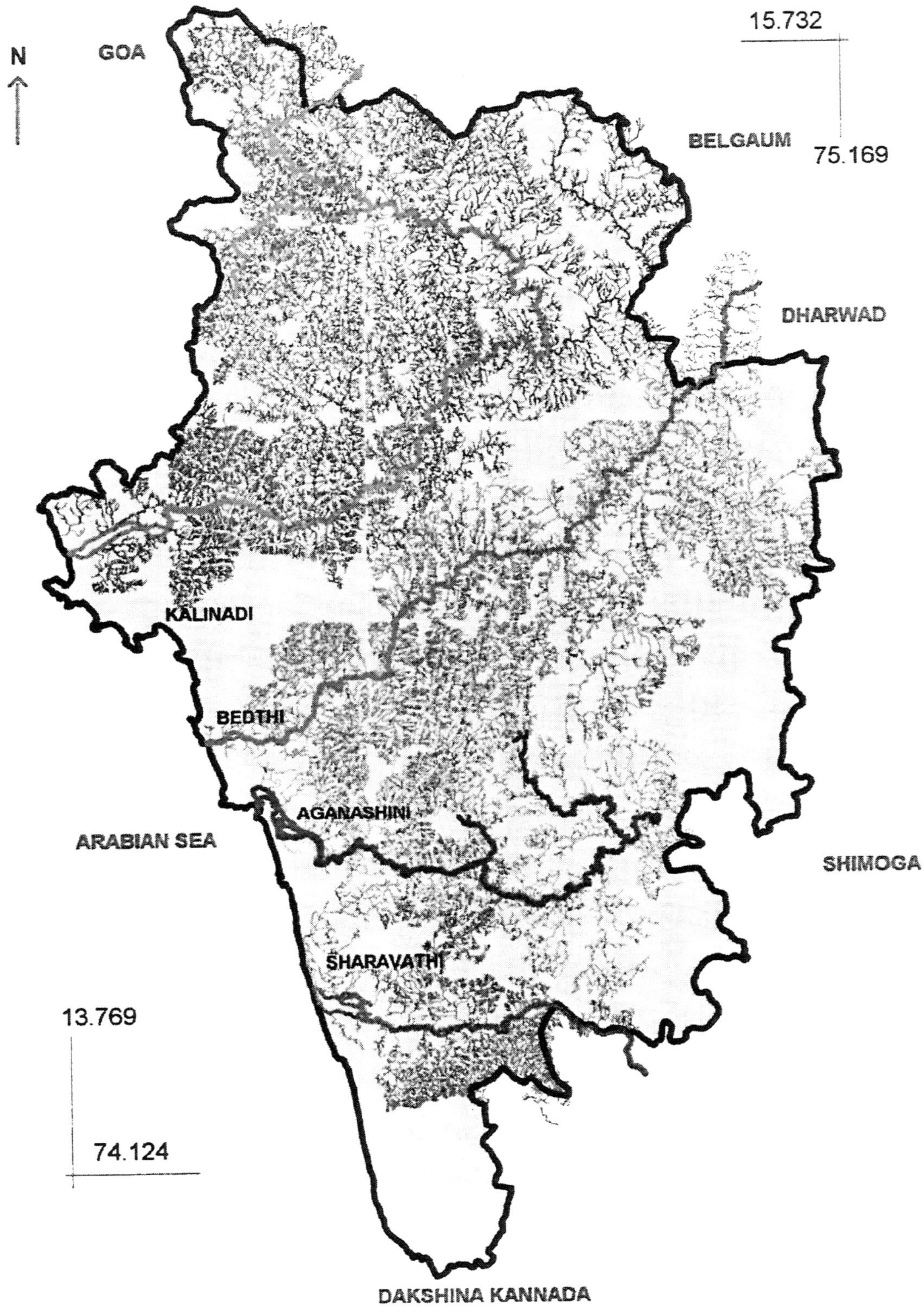

Plate 4.1.0: Rivers and Streams in Uttara Kannada District (see accompanying CD for color representation)

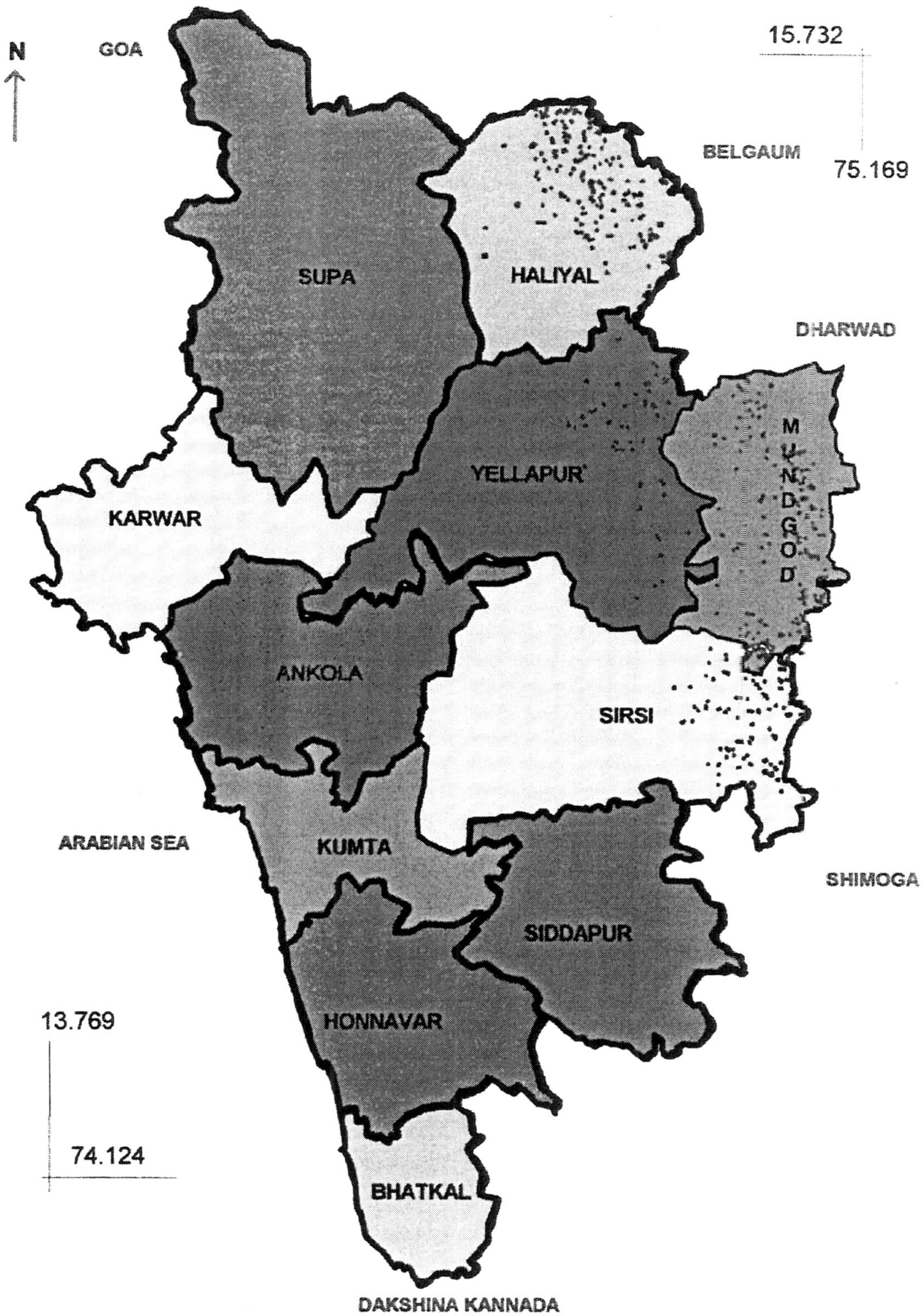

Plate 4.2.0: Distribution of Tanks in Uttara Kannada District
(see accompanying CD for color representation)

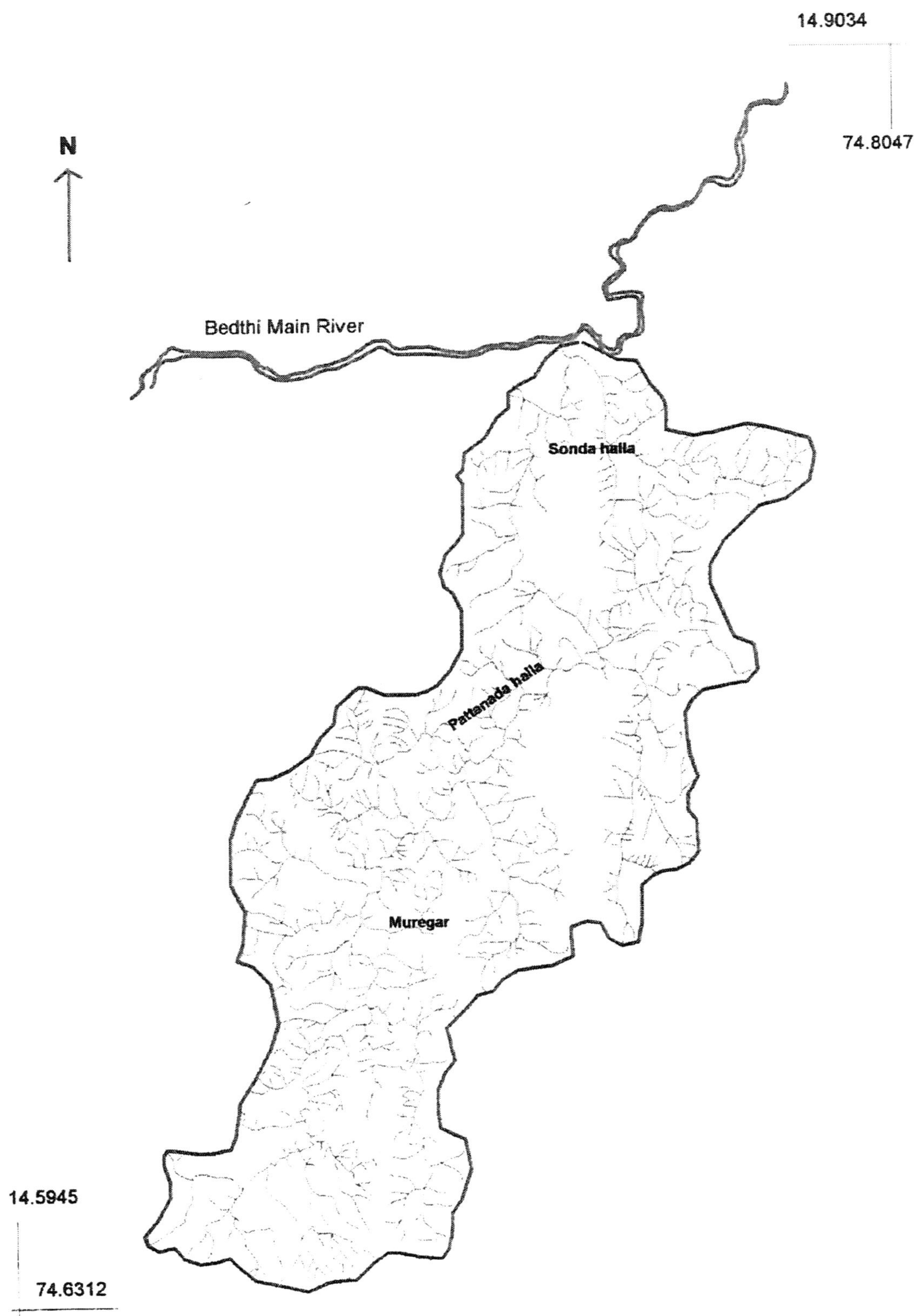

Plate 4.3.0: Catchment Area and Streams at Muregar Jog
(see accompanying CD for color representation)

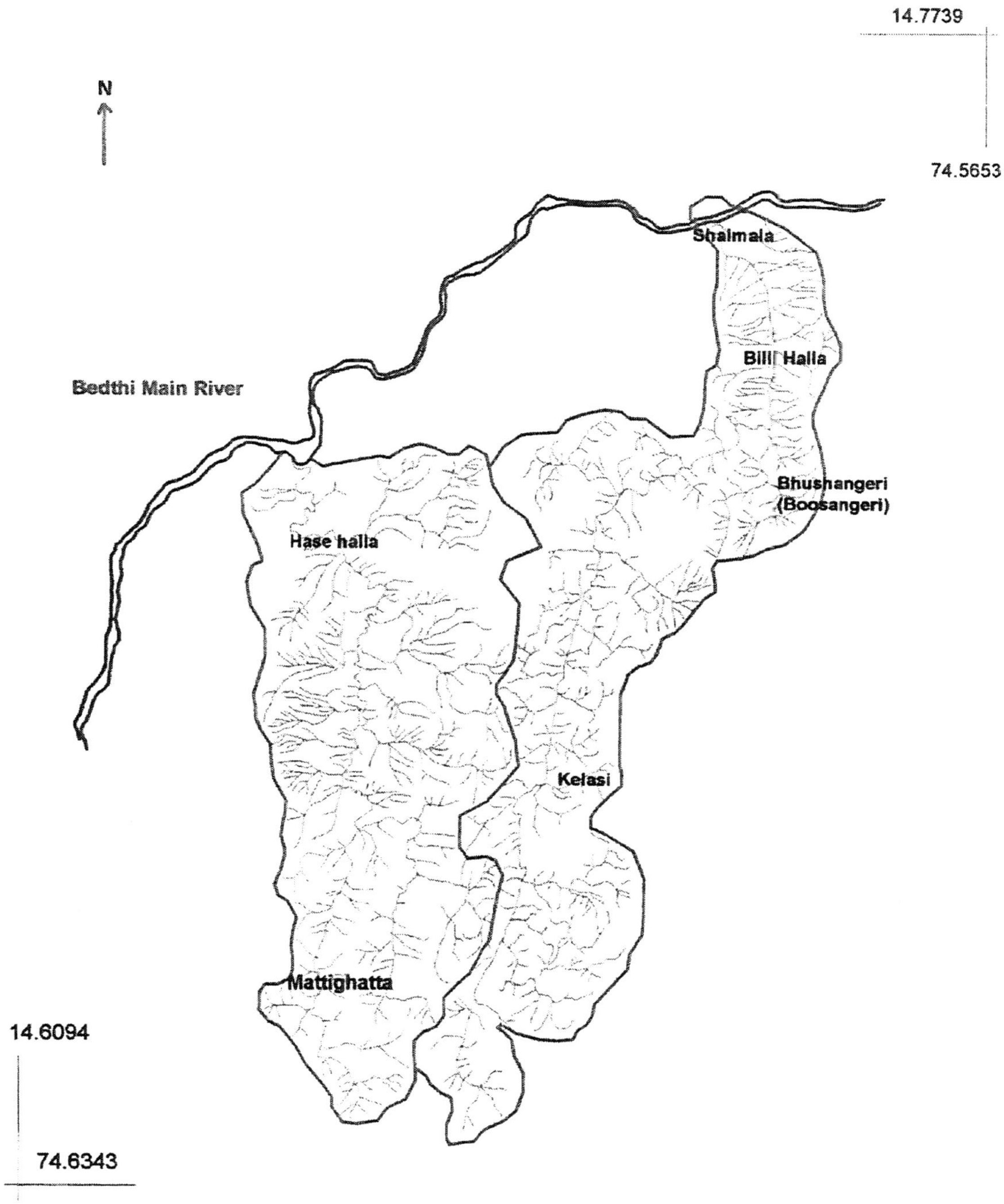

Plate 4.4.0: Catchment Area and Streams at Bhushangeri Jog (Boosangeri)
(see accompanying CD for color representation)

Chapter 5

AN ECOLOGICALLY FRIENDLY OPTIMAL DESIGN OF HYDROELECTRIC PROJECTS

5.0 SUMMARY

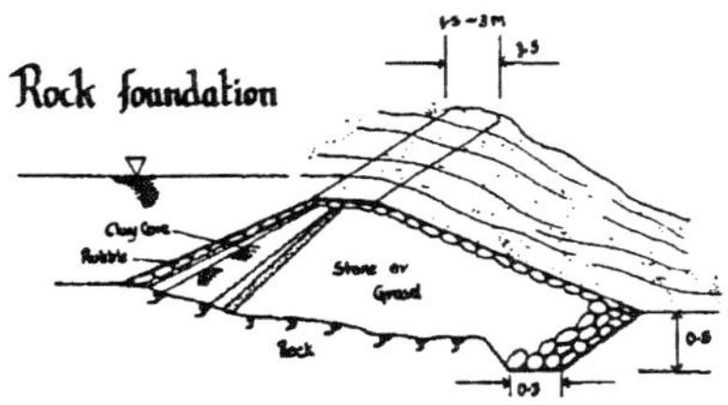

The planning of water resources depends on type and size of projects, the ecological factors involved, etc. Emphasis is placed on presenting an overview of water resources in Uttara Kannada district through meteorological, hydrological, ecological and economic data. Economic data includes all costs and benefits, specifically hitherto under estimated environmental and social costs and benefits. This study is carried out in the catchment areas of Bedthi and Aghnashini rivers, out of the four rivers flowing in the district. The potential of the other two rivers, Kali and Sharavati, has been tapped already through a storage reservoir to even out seasonal and yearly fluctuations in the river's flow. This has submerged vast tracts of natural forests and dislocated a large number of families. The submergence of forest and agricultural lands has affected the availability of fuel wood in the region. These two hydroelectric projects are supposed to generate electricity throughout the year in an average rainfall year. Focussing on land submergence impact, alternatives are proposed: sites have been identified to set up small, mini or micro hydropower plants, and a model is proposed to minimise submergence and maximise net energy in a region with seasonal power generation, reservoir storage capacity and installed generation capacity as the decision variables.

Net energy analyses incorporating biomass energy lost in submergence at Magod (Bedthi river) and Unchalli (Aghnashini river) show that savings in land submergence could be achieved and net energy at a site maximised if the hydroelectric generation capacity is adjusted according to seasonal variations in the river's discharge/run off. However, there are already two river projects in the region designed to supply power throughout the year that can cater the lean season needs (if there are any shortages). A proposal for seasonal run-of-river plants, based on net energy analyses of Bedthi and Aghnashini river projects, is discussed in the following section.

Keywords: Hydro Energy, Bioenergy, Net Energy Analyses, Parametric Optimisation, Cost-Benefit Analyses, Rehabilitation Costs, Submersion Costs, Land use Analyses, Ecologically Sound Design, Optimal Design.

5.1.0 INTRODUCTION

Hydraulic potential is the combination of the possible flow and distribution of gradients, and hydraulic resource is that fraction of hydraulic potential which is still accessible after economic considerations. Hydropower owes its position as a renewable resource to the varying but more or less continuous flow of a certain amount of water in the stream. This water, supplied by rain and always moving, constantly moves from the mainland to the sea, where it evaporates into the atmosphere in an unending cycle controlled by two opposing forces, the heat of the sun and earth's gravity. The river is thus very dynamic with fluctuation due to gains from precipitation, losses through evaporation and consumption, and reduced flow caused by obstacles, which include turbulence in a torrent or soil permeability leading to entrapment of water that continually tries to reach its lowest gravity potential, the sea. The various time-lags introduced while traversing this geological maze modulate the initial variability of the rains and determine the different patterns of the streams, which form an image of rain fallen on the catchment area after filtering through complex geological and topographical environments.

5.1.1 Hydroelectric Development - Present Status and Approaches in Design

Installed electrical energy capacity in Karnataka State is 3005 MW, out of which 79% is through hydropower source (2375 MW) and the balance 21% by coal based thermal power stations. At present, hydroelectric development work is undertaken on ad hoc project-by-project basis with capacity fixed on the basis of 90% availability of water annually, estimated over a 10-day period. The current planning approach is characterised by an attempt to provide constant power throughout the year maximising the electrical energy output through supply expansion options without considering environmental impacts. The seasonal and yearly fluctuations of rainfall/stream flow being high, planners try to achieve their goals opting for large storage reservoirs having greater submergence area, ignoring energy (in terms of bioresources) and economic costs of the submerged area. Also, this approach ignores the complimentarity of the different options like (a) run-of-river hydro plants, (b) seasonal peak power generation and (c) other options, especially thermal option due to available biomass in the region on renewable basis. There is a need to consider 90% reliability of the system in the region, and not of the individual projects, to prevent gross under utilisation of the natural resources. Besides, the integrated systems planning of supply and demand, which would help in optimal development of the potential, should be considered.

Hydroelectric projects are decided on the basis of two criteria in India. The Planning Commission specifies that the Benefit-Cost (B-C) ratio should be greater than 1.5 for clearing any hydroelectric project, while the State agencies such as Karnataka Power Corporation Ltd. (KPCL), choose least cost of generation between the alternatives. These approaches do not take into account energy/economic costs of submergence and alternate designs involving seasonal variation in generation. The land area submerged can be viewed either specifically as a source of bioenergy in the form of fuel wood, etc., or as a source of economic benefits in general. In view of these, the following alternatives are suggested for calculating the benefits:

(i) Computation of Benefit-Cost strictly on energy terms: This involves calculation of net-energy from the project, defined as the annual electrical energy generated minus the energy lost from the potential annual increment of wood biomass in the submerged land.

(ii) Assigning monetary value to the submerged land and including it in the construction cost of the project itself.

The main objective of economic analysis in developmental projects is to see whether the resources are put to effective use. This analysis has to be carried out by incorporating a number of socio-economic and environmental variables. This incorporation is different from the conventional economic analysis in which cost minimisation and profit maximisation are the primary objectives.

Economic feasibility consists of examining the total benefits that result from a project, finding out whether they would exceed those which would occur without the project by an amount in excess of the project cost. It depends on engineering feasibility that assures benefits and environmental feasibility, minimum submergence. In 1958, the Planning Commission while reviewing five major projects showed that, large benefits accrued from irrigation in terms of double cropping, diversification and better quality crops, higher yields, larger income and greater employment opportunities. The indirect benefits were the establishment of processing industries, expansion of consumer industries, retail trade, transport and communications. The committee recommended that economic benefit criteria be adopted instead of financial criteria for sanctioning hydropower projects for irrigation or power. In 1966, the Ministry of Irrigation and Power, Government of India, laid down that, a project has to be considered worthwhile only if the B-C ratio is not less than 1.5. However, there is special consideration for scarcity and backward areas.

The main objective of Cost-Benefit Analysis (CBA) is to look for economic viability of a project. It tries to include all technical, physical, socio-cultural and environmental variables affected by the project in terms of economic value fixed for the pre-determined year. Cost includes any expenditure incurred by the project authorities or the society directly or indirectly over the economic life of the project. This also includes construction cost, and operation and maintenance cost till the end of the life of the project. If the life of a dam is "n" years, counted from the beginning of construction, the cost could be indicated as C1, C2, ...Cn. Benefit includes all the advantages that accrue to the society, directly or indirectly attributable to the project, which can be expressed in monetary terms. Once the project is completed, benefits begin to flow and gradually grow and remain constant once the full capacity utilisation level is achieved. These are indicated by B1, B2, ...Bn.

Hence, the Present Value of Cost (PVC) at discount rate "r" is given by

$$PVC = \frac{C_t}{(1+r)^{t-1}} \quad \text{where } t = 1, 2, \ldots\ldots n.$$

The Present Value of Benefit (PVB) at discount rate "r" is given by

$$PVB = \frac{B_t}{(1+r)^{t-1}} \quad \text{where } t = 1, 2, \ldots\ldots n.$$

The Benefit-Cost ratio is

$$\text{B-C ratio} = \frac{PVB}{PVC} = \frac{\dfrac{B_t}{(1+r)^{t-1}}}{\dfrac{C_t}{(1+r)^{t-1}}} \quad \text{where } t=1, 2, \ldots\ldots n.$$

The project is considered economically viable if B-C ratio is greater than 1, which implies that for every rupee invested, the return is greater than one rupee. However, this technique has certain limitations as follows.

(1) This analysis presumes that the prices given by the market forces are not distorted because of skewed income distribution. However, even if shadow prices are used, the influence of skewness of income distribution in an economy cannot be overcome.

(2) This technique does not take into account the income redistribution and regional imbalances caused by the project.

(3) The compensatory payment done to oustees is usually arbitrary as used in CBA. The CBA technique is unable to take into account quantification of suffering of oustees which involves interpersonal comparison of utilities.

(4) CBA analysis presupposes accurate information and absence of risk and uncertainties, which are unrealistic. Ignoring uncertainties arising from price and cost movement and risk of dam failure makes the CBA unreliable.

(5) A complete dependence on CBA will not give us the picture of alternate investment opportunities. In the absence of alternate techniques, CBA continues to be used extensively in assessing economic viability of projects.

This chapter, focusses on design of a hydroelectric project with emphasis on economic efficiency and engineering soundness subject to social, political, environmental, ecological, cultural, institutional and legal constraints. The complexity is compounded by lack of methodologies that can quantify these constraints or express them in tangible terms. Some of these constraints are not mutually exclusive and are even conflicting. Clearly, these involve trade-offs between different constraints that may be qualitative. Thus, satisfying these constraints plays a dominant role in achieving the objectives in water management. Optimisation (programming) and Descriptive (predictive) model approaches are usually adopted to accomplish this complicated task (Falkenmark, M. 1989). The optimisation models aid in decision-making attempts to derive the optimal water management scheme for a set of objectives. The descriptive models attempt to predict the consequences of a management scheme for a known quantity of water. The steps involved in developing a water management scheme are: evaluation of water availability, specification of constraints, specification of objectives and their expression into quantitative criteria, selection of appropriate model scenarios, derivation of alternate schemes, evaluation of each scheme and comparative evaluation of schemes in measurable terms.

5.1.2 Definitions

(i) *Types of Hydro Plants:* Hydro plants may be classified on the basis of their functions as follows:

(1) Run-of-river plant: A run-of-river plant has no storage capacity and must either use or spill whatever water is flowing in the river at any given time. This flow can vary considerably over a given period of time. A run-of-river plant often has little, if any, firm capacity. Its principal value is associated with its ability to displace other, more expensive, sources of energy from thermo-electric stations whenever sufficient water is available.

(2) Run-of-river plant with pondage: Pondage is meant for sufficient storage at the plant to take care of hourly fluctuations in the load on the plant throughout a week. In the usual run-of-river plants with pondage, installation is on a high basis in relation to minimum stream flow. In some plants, discharge is about 10 to 20 times minimum of stream flows. When plenty of water is available, these plants operate on the base of load curve. With seasonal decrease in stream flow, they gradually change their position on the load curve until, under minimum stream flow conditions, they operate entirely on the peaks of load curve. The annual capacity factors of these plants vary from 45 to 65%.

(3) Base-load hydropower plants: These are run-of-river plants without pondage that are capable of substantially continuous operation in the base of load curve throughout the year. Consequently, their full plant discharge is seldom greater than the minimum flow of the river. The annual capacity factor of the base-load plants varies from 70 to 100%.

(4) Reservoir plants: While hydroelectric projects that take their flow directly or indirectly from large storage reservoirs are suitable for development as peak-load plants, provided the required conduit is not too long or expensive, they are not necessarily peak-load plants. This is because, market conditions may not require all the peak-load hydropower available at such a site. Consequently, many plants of this type operate on a daily capacity factor of 30 to 100%.

(5) Peak-load hydropower plants: A peak-load hydropower plant is designed and constructed primarily to take care of the peak-loads of a power system. Run-of-river plants with pondage operate both as peak-load and base-load plants as the river's flow permits. A large pond or reservoir is essential, and extensive seasonal storage is usually provided. An important characteristic of these plants is high basis of installation in relation to stream flow. They frequently have large seasonal storage at relatively high heads and are likely to be located on small watersheds. They store the water and clip the peaks off the top of the load curve.

(6) Pumped storage plants for peak-loads: Pumped storage plants are peak-load plants that pump all or a portion of their own water supply. They essentially consist of a tail water pond, which may be replaced by a river or a natural lake, and a head-water pond. During peak-load, water is drawn from the head-water through the penstocks to operate the hydroelectricity generating units over the peak of the load curve. During the off-peak hours, pumps are operated to shunt back water from the tail water pond to the head-water pond. Power for operating the pumps is furnished by off-peak steam generated energy, sometime supplemented by secondary hydro energy. For heads up to 300 feet, it is feasible to use the same unit for both pumping water and generating power. Reservoirs for such developments are large enough to provide for the plant's operation over the top of the peak-load of the system it is designed to serve, with some margin to permit the plant's usage for short-time breakdown service. A reservoir capacity that will permit full-capacity operation for four to ten hours is usual. Pumped storage peak-load hydro plants have one peculiar advantage over the usual peak-load hydro plant on a reservoir outlet. In the latter type, if the reservoir is drawn too low, firm capacity will be sacrificed, whereas, with the pumped storage type of plant under such conditions, firm capacity can be maintained by additional off-peak pumping.

Peak-load hydroelectric plants are especially useful in systems subject to sharp and unexpected short time peaks. A load curve of this sort necessitates a larger percentage of reserve capacity than is required by most power companies. If generation is 100% steam, a large part of the reserve must be kept hot and turning over, ready to pick up any unusual load. This in turn means that one reserve boiler capacity is to be kept hot to furnish steam in the event of such a demand. Consequently, to serve and to be ready to serve such unusual peaks by means of steam, electric power becomes expensive. Peak-load hydroelectric plants on the other hand, are particularly suited to perform such service and can frequently do it at much less cost.

Even if the entire expense of the reservoir must be borne by the peak-load hydroelectric plant, the low capacity factor and consequent high basis of installation in relation to average annual stream flow would frequently assure a low capital cost per kilowatt. The prime functions of such plants are to carry short time and unusual loads, and to serve in case of need as an instantly available reserve capacity for the system; but, when operating as reserve, such plants may operate further down on the load curve, drawing down the reservoir for this purpose. Such peak-load points permit the better steam plants to operate at more efficient capacity factors: they obviate the necessity for retaining in service so many antiquated high production cost steam plants, for carrying so many boilers hot and so many steam units in hot reserve.

(ii) *Load Factor*: The load factor of any system is the ratio obtained by dividing the number of kilowatt-hours over a given length of time by the product of the peak-load and the number of hours in this period (that is the ratio of the average load to the peak-load).

(iii) *Capacity Factor:* It is the ratio of the average load to installed capacity. The capacity factor at which a power plant operates has a major influence on the cost of power. The capacity factor at any given period of time (day, week, month or year) may be defined as the ratio of energy that the plant has actually produced to the energy that it might have produced if operated at full capacity throughout the period.

5.1.3 Past and Present Planning Approaches in Hydroelectric Projects

The utilisation of hydropower as a motive mechanical energy using water wheels has been in practice for several centuries. The development of hydraulic turbine and the subsequent improvements in its efficiency and flexibility in usage for varying discharges and heads, and finally coupling it with electric generators enabled the use of hydropower to produce electric energy. As a result, several small hydroelectric power stations were established in various parts of the country. Subsequently, improvement in technology and demand for electric energy with industrial revolution shifted the attention towards the development of large hydroelectric stations in the early part of this century. The planning and design of hydroelectric projects has become a well developed subject all over the world. Karnataka State has the distinction of being one of the pioneering States to harness hydropower by installing a 42 MW hydroelectric power station, commissioned near Shivanasamudra across the Cauvery river.

Various literatures are available regarding design methods of all structures necessary for hydroelectric development. A publication entitled "Hydroelectric Handbook" by William P. Creager and Joel Justin [1950] attempts to present a compendium of all phases of hydroelectric practice. This provides mainly hydrological and geological investigations, techno-economic aspects of hydropower projects, and broad aspects of planning and development of hydroelectric power. Complete exposition of design methods for various accessories, civil structures and electrical components are also provided in this handbook.

By mid 1980's, decentralisation of hydropower generation became one of the accepted means of developing energy resources to meet rural energy requirements and support programs involving rural industrialisation and decentralisation of industries in developing countries like China, Nepal, Sri Lanka and India. In many regions, hydropower remains an abundant and largely untapped resource that could be tapped in a cost effective, environmental problem-free and renewable way. The manuals published by United Nations Industrial Development Organisation and Economic and Social Commission for Asia and the Pacific (UNIDO, 1986; ESCAP, 1982) provide practical and useful reference for design of mini, micro and small hydropower plants. In 1985, Central Electricity Authority (agency involved with formulation of power projects in India) published guidelines for development of hydroelectric projects. This publication contains information on design details, major components of hydropower scheme, cost-benefits and economic considerations. In 1988, realising the advantages of small hydroelectric power stations, KPCL undertook a pilot survey in various river basins and canal drops in Karnataka to assess hydro potential. It has identified 223 stations worth 215 MW for setting up mini and micro hydroelectric schemes (KPCL, 1988). However, this report leaves out potential feasible sites in Uttara Kannada district of Western Ghats region and concentrates mainly on canal drops in Mandya, Raichur, Shimoga, Mysore, Belgaum, Bijapur and Gulbarga districts. In this connection, this Chapter and the previous one provide information on feasible sites in Bedthi and Aghnashini basins in Uttara Kannada district for setting up mini, micro and small hydropower plants to cater the needs of remote rural areas in a decentralised way.

The present planning of hydroelectric projects in Karnataka is purely viewed from power generation aspects only. That is, to try to use every cubic centimetre (cc) of water in the rainfall area and every metre of head available. Many factors, such as: (a) cost of the land, (b) siltation problem, (c) effect of reduced vegetation cover and therefore reduction in bioresources, (d) effect of impounding large quantity of water, (e) social costs involved in rehabilitation of inhabitants, changes in their life style, etc., and (f) cost of other alternatives are ignored. Conventional selection criteria is normally based on the cost of energy generation as a major yardstick in proposing projects. The cost of generation is computed purely based on fixed costs of construction and machinery, and a component that linearly varies with the installed capacity. The importance of carrying out extensive and detailed analyses of the varying costs of hydroelectric projects has been given least importance. However, in today's scenario, there is a need to develop integrated and optimal energy plan taking into consideration many alternatives, incorporating all environmental costs and analysing many competing factors.

This necessitated an integrated approach taking into account bioenergy available in the submerged area and detailed engineering design of the power station, while formulating alternate schemes to the major hydropower scheme through a large reservoir proposed at Magod site of Bedthi river and Unchalli site of Aghnashini river by a government agency.

5.2.0 OBJECTIVE

The prime objective of this endeavor is to design a hydroelectric plant for optimal utilisation of energy in the water, with minimum submergence area and economic costs, taking into account inter-seasonal variation in generation capability so as to meet the region's demand during all seasons.

5.3.0 METHODOLOGY

The various stages of achieving the aforementioned objectives are as follows:

5.3.1 Collection of Secondary Data

Meteorological, land use, forest products and population data were compiled from various agencies such as India Meteorological Department (Bangalore and Pune), Tahsildar's Office at each taluk, Village Accountant Offices, Block Development Offices, Forest Department, Agriculture Department, Revenue Department, Uttara Kannada Jilla Parishad at Karwar and District Statistical Office, Karwar. Schedules of rates for civil and electrical work were collected from Karnataka Power Corporation Limited (KPCL) and Public Works Department (PWD). Sectorwise consumption of electricity and power tariffs were collected from Karnataka Electricity Board Zonal Office at Karwar. Soil type and ground-water table data were collected from Department of Geology and Mines at Karwar.

5.3.2 Primary Data

The hydrology of the river under consideration was studied by

(i) *Feasibility Study*: This involved the measurement of catchment area and stream discharges for a substantial period of time. Relationship between run off and rainfall is determined by regression analysis.

(ii) *Measurement of Catchment Area*: Catchment boundaries were located using contour lines on a topographical map. They were drawn by following the ridge tops that appear on topo maps as downhill pointing V-shaped crenulations. The boundary should be perpendicular to the contour lines it intersects. The mountain tops are often marked as dots on a map, and the location of roads which follow ridges are other clues. Catchment area thus marked/traced is measured directly from the marked maps using a planimeter.

5.3.3 Potential Assessment at Selected Sites

(1) *Hydrology*: The hydrology of Bedthi and Aghnashini rivers was studied at Magod, Shivganga, Unchalli and Benne sites. The stream discharges at these sites were analysed on daily and weekly basis using daily precipitation data of five years.

(2) *Engineering design aspects*: Combining hydrological aspects, topographic mapping (a contour map), geological feasibility studies, best dam site, tunnel alignment and power house location information are obtained from the Department of Geology at Karwar and from reports available at the KPCL library.

(3) *Stream discharge data analyses*: In the second stage, inflow data at Magod, Shivganga (Bedthi river), Unchalli and Benne (Aghnashini river) was analysed; average, standard deviation, maximum, minimum and mode values were computed on monthly basis using 90 years precipitation data.

(4) *Mapping of contours*: Various contours at dam sites were traced and area determined using planimeter.

(5) *Calculation of submergence area*: Submergence area was determined for various heights of the dam and their corresponding full reservoir level (FRL).

(6) *Computation of reservoir volume for various heights of the dam*: A certain amount of "dead" storage capacity is added to take into account the sedimentation in the reservoir.

(7) *Parametric analyses*: Computation of annual electrical energy for various heights of the dam and seasonal power drafts.

(8) *Net energy analyses:* Net energy is computed by determining the total electrical energy generated minus the energy lost from the potential annual increment of wood biomass (for various productivities documented in the region).

(9) *Design and cost estimates*: The preliminary designs and cost estimates for all the components of the project - dam, pressure shaft, power house, intake and outlet structures (civil and electrical components and other accessories) - are also determined according to the present costs.

(10) *Economic analyses:* It was done based on Benefit-Cost analysis, by incorporating environmental costs of submergence, rehabilitation, etc., for various heights of the dam at each site.

5.4.0 HYDROELECTRIC PROJECTS CONSIDERED - BEDTHI AND AGHNASHINI: HISTORICAL PERSPECTIVE

The Karnataka Power Corporation Limited proposed a hydropower project of 210 MW (2 x 105 MW) on the river Bedthi (Stage I) in 1969, with a dam at FRL of 480 m (1570 feet) and submergence area of 9780 hectares (MPCL, 1977; KPCL, 1973). This proposal raised strong protests and agitation from academicians and locals in 1981 (Sharma, L.T. and Sharma, R. 1981). Centre for Ecological Sciences, I.I.Sc., and local NGO's jointly organised a three day seminar (17-19, Jan 1981) at Sirsi on "Implications of hydroelectric projects for Uttara Kannada district". Papers on environmental impacts, alternatives and biological diversity were presented at the seminar.

Subramanian's paper (Subramanian, 1981) in this seminar gave alternate energy strategies in planning of major hydroelectric projects. Various options suggested were (a) a single plant of maximum generating capacity possible, (b) two plants, one peaking type and the other average power producing type, (c) large number of plants of various sizes, to reduce storage and submersion area, and (d) hydro-thermal mixed system to generate equivalent amount of power and energy during lean period. The saved submergence area could grow biomass that would give energy used for domestic (cooking and water heating), industrial and other end uses. Hydro-thermal mix is based on the assumption that only when the water-head is very high and the reservoir profile a deep valley with steep walls at its sides, hydroelectric energy would become competitive compared to firewood and other bioresources. With reduction in head and slow slope of the terrain, the smaller depths of the reservoir make firewood a more attractive, and below certain depths, bioresources (trees) more efficient storage devices than water.

Focussing on land submergence impact, Lele, S.M. (1986) proposed an optimisation model for the Koyna hydroelectric project in Maharashtra and Bedthi project in Uttara Kannada district. Two alternative approaches, net energy and net economic costs approach, were adopted to formulate a mathematical model. His calculation shows that by allowing variations in power generation to exploit the seasonal variations in the inflow, there is reduction in submergence area.

In response to the outcome of the seminar at Sirsi in 1981, the Government stopped the project and appointed a five member expert committee to give its views on the feasibility of the project, taking into account various alternatives proposed in the seminar. This committee submitted its report in 1986, proposing a dam height at FRL of 470 m (1540') with submersion of 6800 hectares. This proposal, cleared by the Government in 1992 (KPCL, 1993), was not acceptable and drew protest from local people and academicians. In response to the views of the public, the Ministry of Environment and Forests in its letter dated October 1992, requested KPCL to submit a revised project proposal along with the environmental management plan and also instructed it not to undertake any work on the project, preliminary or otherwise.

Meanwhile, an alternate plan was mooted by Chelvaraj (1994) which consisted of 16 diversion weirs involving a total submergence of 4372 hectares. The proposed scheme is a staircase of dams 20 to 60 m high. Due to these, this proposal will be expensive and will not be feasible or sustainable due to siltation and would not reduce submergence of forest area. The economic analyses in this report are based on externalized costs, overestimated benefits, incomplete technical analyses, lack of realistic assessment of reservoir sedimentation and inadequate environmental assessment. Sedimentation reduces the storage capacity and therefore reduction in the yield over a given time. Deposition at the head of the reservoir leads to an increase in flood levels in the contributing rivers upstream. Also, it increases evaporation losses for a given storage volume. All these approaches (of KPCL, Expert Committee and Mr. Chelvaraj) are ad hoc and heuristic without considering ecological and economic aspects from an optimal point of view. In this chapter, the design is carried out as a parametric optimisation problem with an objective to maximise energy, subject to ecological constraints (minimisation of submergence area).

In this context, the hydro potential in the catchment that can be harnessed in an ecologically sound way for electricity generation is assessed and a hydroelectric project designed (to ensure electricity in all seasons to meet the region's demand) based on net energy approach at Magod. The biomass energy lost in the submerged land is compared to the electrical energy generated for various heights of the storage dam. Economic analyses are carried out incorporating environmental costs. Results of these analyses show that submergence area reduces considerably and the Benefit-Cost ratio increases by allowing seasonal variations in the hydroelectric generation capacity.

5.5.0 Analysis of Water Resources of a River with Constraints on Energy and Ecology

A storage reservoir has become the most important technique for regulating surface run off to match supply and demand in recent years. Nature does not meet the demand for water at all places at all times. Due to the environmental impact associated with a large storage reservoir, it has become necessary to estimate the proper size of a dam or reservoir to meet the target demand of a region. Therefore, planning, decision on size, and management of water resources and storage capacities represent a very important research topic both from practical and theoretical points of view. In the practical engineering approach the historic water input data is basically used to determine the storage capacity, assuming the operation of the reservoir to have started at the beginning of data collection and lasted till the end of the available data series (Mutreja, 1976). Storage problems are studied with the premise that a given water draft regime, satisfying a given water demand or solving complex water resource's problems by regulation, is accomplished by that capacity during the historic data period. On the other hand, theoretical approaches have been developed for studying various storage approaches as governed by the law of chance. Based on 90 years precipitation data and 18 months field data of water run off, a methodology is proposed to design a storage reservoir taking into account ecological constraints to meet the region's demand during all seasons.

Design of storage systems for power generation in hilly terrain is constrained mainly by ecological factors these days. Submergence of land is a major impact of hydroelectric power projects apart from other problems like siltation, delays in construction and escalating costs due to delayed implementation of the projects. The environmental impact of large-scale hydroelectric projects are (i) loss of forests (flora and fauna), agricultural lands and human habitats, (ii) upstream and downstream ecosystem disruption, (iii) impoundment impacts like health hazard, flood hazard, increased water loss, micro-climatic changes, etc., and (iv) influx of labour population leading to further loss of forests and land degradation. The submergence of forest and agricultural lands is a major concern today due to the rapidly declining forest cover. These negative aspects of hydroelectric projects necessitate the minimising of submergence area subject to criteria such as: (1) cost should not be very high, (2) minimum or no wastage of water and (3) seasonal constraints, as the region receives maximum rainfall during the south-west monsoon.

5.5.1 Design Assumptions and Parameters - Scenario Creation

(i) *Assumptions*: As already mentioned, KPCL has proposed to build storage reservoirs at Magod (Bedthi river) and Unchalli (Aghnashini river) sites. Detailed geological and topographical investigations have been undertaken to determine the best sites for the dam, pressure shaft alignment, power house location, etc. The same location can be used for implementing this proposed alternate design.

(ii) *Parameters*: Input data consists of the following types.
(a) Site Specific Data: Includes data on water discharge, water yields in a catchment, generation head, evaporation rate, seepage rate, catchment type, area spread for each contours, volume of water stored, submersion area for various heights of the dam and dead storage capacity of a dam.
(b) Technical Data: Data consists of potential biomass productivity of submerged land, allowable inter-seasonal variation in power generation, load factor, generator and turbine efficiency, project life span (of the dam), electrical machinery, dependability norm for storage capacity, and sectorwise electricity and fuel wood demand in the region.
(c) Economic Data: Consists data on costs of civil, electrical and other accessories, environmental cost of submergence, rehabilitation cost, etc., electricity tariff, fuel wood cost, interest rate, project life span, construction time, inflation rate, and operation and maintenance costs.

(iii) *Decision Variables:* These are used to determine the optimum value of the storage capacity of a reservoir, installed generation capacity and seasonal power drafts, so as to maximise the net energy available in the region (objective function) subject to hydrological constraints (seasonal variation in water inflow, etc.) and minimise submergible area and costs, meeting the energy requirement of the region during all seasons.

5.5.2 Background

The broad outline of an optimisation model for exploiting a river for hydropower and irrigation, incorporating energy cost and net incremental yield due to irrigation with constraints on irrigable area and crop water requirement and seasonal variation in precipitation, was formulated earlier by Subramanian (1985). On the same lines, with detailed engineering design and quantification of data, the parametric optimisation approach is used to find the alternatives to get maximum energy subject to ecological constraints.

5.5.3 Net Energy Model - Parametric Optimisation

This section introduces a model for optimal utilisation of energy in the water and thermal energy in the region. The proposed model is a parametric one, solved for various scenarios, with the objective to maximise the net energy available in the region given by

$$Enet = Eh - Ebi \quad (1)$$

The model also includes an equation that computes the potential monthly hydropower production as a function of the (a) volume of water discharged, (b) gross head of this water and (c) efficiency of the turbine-generator couple.

Hydropower is given by $P(\text{in kW}) = 9.81*Q\ (\text{in } m^3/\text{sec})*H\ (\text{in metres})$ (2)

The corresponding electrical production $Eh= P*t*\eta$, where E is the electrical energy produced (in kilowatt-hours), P the hydraulic power (in kilowatts), t operating time (in hours) and η the turbine-generator assembly efficiency usually between 0.7 and 0.85. Because of the estimates of the turbine-generator couple and gross height, this is only an approximate characterization of the energy harnessed for the purpose of illustration.

Eh_t is the monthly hydro energy generated in million kWh, written as

$$Eh_t = \Sigma(9.81*D_t*H*\eta)/3600 \qquad t=1.......12 \quad (3)$$

where H = average net generation head, η = efficiency of the turbine-generator couple and D_t = power draft from reservoir during a month (million cubic metres, Mm^3).

$$Eh_t = \Sigma k1*D_t$$

That is, a sum of the energy generated from the power drafts in all months. This equation may further be decomposed to indicate seasonal draft as

$$Eh_t = \Sigma k1*D_{tm} + \Sigma k1*D_{td} \quad (4)$$

where, tm = 1.......4, representing water drawn during monsoon months and
td = 5.......12, water drawn during non-monsoon months (dry period)

$$E_{bio} = A_{sub} * Gr * (CV) * \eta_c \quad (5)$$

where, E_{bio} is energy loss due to submergence, A_{sub} - area to be submerged, Gr - annual rate of growth or productivity, CV - energy equivalence factor (for example, one tonne of dry matter has the heating value of 18.6 Giga Joules, which is equal to 590 watts) and η_c is energy conversion efficiency.

A_{sub} is further classified based on land type as forest, agriculture, gardens, etc. The primary productivity of biomass in forests is estimated to be 6.5-27.5 t/ha/yr. The estimated paddy yield is 18-24 quintals/ha/yr and areca and coconut residues from gardens, 30-45 quintals/ha/yr The thermal content of biomass is computed in terms of their primary energy contents. For example, a ton of dry fuel wood with a calorific value of 4400 kcal/kg has a thermal content of 5112.8 kWh, which is considered equal to the same amount of electric energy.

This parametric model is subject to the following constraints.

(i) *Hydrological constraints*: These operate on monthly basis and consist mainly of the following continuity equation, which has been a standard tool for hydrologists and civil engineers for many years (Maass, A., et al., 1962).

$$V_{t+1} = V_t+I_t-S_t-E_t-D_t \quad (6)$$

where V_t- Volume of the reservoir at the beginning of month t, I_t - Inflows to the reservoir during month t, S_t - Seepage loss, E_t - Evaporation loss and D_t - Discharge during the same month from the reservoir. Several inputs required to solve this equation are as follows:

(a) functional relationship between volume, surface area and level of water in the reservoir,
(b) relationship between seepage and volume,
(c) relationship between evaporation and loss,
(d) a sequence of monthly inflows to the reservoir,
(e) a policy for determining the discharges from the reservoir and
(f) volume of the reservoir when simulation begins.

(ii) *Dependability*: Dependability/reliability is recognised as an indispensable element in the reliable design of a reservoir. The storage capacity (V) of any reservoir is the function of both the targeted drafts (D) and the reliability (R), which can be expressed as

$$V = f(D,R) \quad (7)$$

It is seen that the required reliability of the targeted draft has a direct relation to the effective storage capacity which has to be provided. For a given draft, particularly the one approaching a mean flow, the required storage is extremely sensitive to reliability. Similarly, for a given level of reliability, increase in targeted drafts would result in an increasingly large storage. Thus, for a given draft, storage would increase substantially with increased level of reliability. For the purpose of hydro planning, in case of generation schemes, a dependability criterion of 90% is normally adopted. Failure in one month in a period of ten years would mean a reliability of 99.2% on monthly basis, but 90% on yearly basis, since failure in any one month is considered failure for the whole year.

(iii) *Constraint of seasonal variation on generation capacity:*

If no variation is allowed, $P_{tm} = P_{td}$ or $P_1=P_2$ (8)

That is, hydroelectric generation capacity during monsoon is same as dry season (non-monsoon period) capacity. This region receives maximum rainfall during the south-west monsoon (June to September is considered as monsoon here).

If variation of "r" is allowed, $P_{tm} \geq P_{td}$

That is, $P_{tm} \geq r * P_{td}$ (9)

where P_{tm} is generation capacity during monsoon and P_{td}, non-monsoon/dry season capacity.

$$P_{tm} = k1*D_{tm}/LF_{tm} \text{ and } P_{td} = k1*D_{td}/LF_{td}$$

where D_{tm} is power draft from a reservoir during monsoon (in Mm^3), LF_{tm} - average load factor for the monsoon months (assumed as 0.5), D_{td} - power draft during non-monsoon months and LF_{td} the average load factor during dry months. Therefore,

$$D_{tm}/LF_{tm} = D_{td}/LF_{td} \quad (10)$$
$$D_{tm}/LF_{tm} \geq r*D_{td}/LF_{td} \quad (11)$$

Due to same load factors for monsoon and non-monsoon/dry months, the seasonal variation in generation capacity could also be written in terms of power draft during respective seasons as

$$D_{tm} = D_{td} \quad (12)$$
$$D_{tm} \geq r*D_{td} \quad (13)$$

(iv) *Constraint of minimum storage*:

Active storage capacity $Ka \geq V_t$ (14)

where t = 1...........12

(v) *Operating policy of the reservoir:*

The feasible operating policy considering seasonal variation in water inflow could be summarised as

$$\{r*S1*P_{tm} + S2*P_{td}\}*30*24*LF*(\text{amount of water/million kWh}) = \text{Total quantity of water available at a site}$$

where S1 = monsoon (4 months in a year is considered as monsoon season).

Therefore S1 = 4 and S2 = 8.

By substituting for $P_{tm} = k1*D_{tm}/LF_{tm}$ and $P_{td} = k1*D_{td}/LF_{td}$,

and when $LF_{tm} = LF_{td} = LF$,

the constraint reduces to

$$4*D_{tm} + 8*D_{td} = V_{t+1}$$

where $D_{tm} = D_{td}$ (if no variation is allowed)

$D_{tm} \geq r*D_{td}$ (if variation 'r' is allowed)

Value of "r" ranges from 1.............. ∞

and $V_{t+1} = V_t + I_t - S_t - E_t - D_t$

With change in dam height, the corresponding submergence area and storage volume capability of the reservoir changes (submergence area and volume computation are discussed later in case studies).

The regulation through storage could be shown as follows:

If $V_t+I_t-S_t-E_t-D_t \leq$ storage volume of a reservoir (V_s)

then $D_t = V_t+I_t-S_t-E_t$ (15)

If $V_t+I_t-S_t-E_t-D_t \geq$ storage volume of a reservoir (V_s)

then $D_t + d = V_t+I_t-S_t-E_t$ (16)

where d = excess quantity available for generation.

(vi) *Positivity constraints*:

Decision variables are positive.

$D_t \geq 0$, for all t = 1, 2 . . . 12

and $V_t \geq 0$, for all t = 1, 2 . . . 12

Therefore, $Ka \geq 0$ (17)

5.6.0 Case Studies

This design is used to assess energy benefits from storage based hydroelectric schemes at Magod and Unchalli sites of Bedthi and Aghnashini rivers in Uttara Kannada.

5.6.1 The Bedthi or Gangavali River

The Bedthi or Gangavali is formed by the confluence of two streams near Kalghatgi (Dharwad district), namely, Shalmala from Dharwad south and Bedthi originating from Hubli taluk. After flowing westward for about 25 kms, it enters Uttara Kannada district and flows in a fairly south-westerly course for about 32 kms before falling into the sea, about 32 km south of Kalinadi. The united stream flows about 8 km south-east to the border of Uttara Kannada, and during the course of 96 kms (out of the total length of 161 kms) across the district, it receives no feeder of any size. At Magod village, about 40 kms from where it enters the district, the Bedthi dashes over the western face of Sahyadris in a cataract known as the Magod falls. The Sonda river, one of its tributaries, joins it beyond Harigadde village. About 16 kms further, near the village of Gundbale, it meets the tide, and for the remaining 24 kms of its course, it is navigable to boats of one to five tonnes (Kamath, S.U., 1985). The river course drops at various places, and contours are shown in the photographs (Plates 5.1, 5.3, 5.4, 5.5 and 5.6 respectively).

The Bedthi river, flowing mostly in hilly tracts and deep gorges, cannot be used for irrigation development. On account of the available water potential and the many waterfalls, the topography of the course offers ideal sources for hydropower development. Suitable dam sites have been identified near Magod falls in the upstream for hydropower development. Based on the model explained earlier, water would be drawn on the run-of-river basis (during the monsoon period) and a small storage would meet the non-monsoon requirement of the region.

5.6.2 The Aghnashini or Tadri

The Aghnashini or Tadri or Donihalla river (total length 121 kms) originates at Manjguni near Sirsi and after a winding westerly course of about 70 kms falls into the sea about 10 km south of the Gangavali river. It receives no feeder of any size throughout its course. It has two sources, the Bakurhole rising in a pond at Manjguni about 25 km west of Sirsi and the Donihalla whose source is close to Sirsi. The streams meet near Mutthalli, about 16 km south of Sirsi, under the name of Donihalla and flow about 24 km south of Sirsi in a winding westerly course to the western face of the Sahyadris down which, about 13 km north of Bilgi at Unchalli, leap over as the Lushington or Unchalli Falls. At Uppinapattana, about 16 km west of the falls, Donihalla meets the tide. For the remaining 24 kms, during which it is navigable to craft of four to ten tonnes, the river is known as the Tadri or Aghnashini, deriving the name from two towns on the right and left banks of its mouth. From Uppinapattana, it winds south-west and north-west together for about 13 kms to Mirjan village, from where it becomes a backwater or lagoon 13 km long and 2 to 6 km wide running parallel to the coast, cut off from the sea by a belt of land with a nearly uniform breadth of about 1.5 km. The outlet to the sea is about 5 kms from the north end of the lagoon, between two hills of 91 and 122 m altitude respectively with a depth of about 6 m at high tide. In the interior, there is as much as eight metres of water, where vessels of 20 tonnes could be docked near the shore. Unfortunately, the entrance is narrow and blocked by a rocky reef, and not to be tried during the south-west monsoon (Campbell, J.M., 1883). The river course drops at various places, and contours are shown through photographs (Plates 5.2, 5.7 and 5.8 respectively).

5.6.3 Catchment Area

The Bedthi catchment is bounded by the watersheds of Kalinadi in the north, Aghnashini in the south and Tungabadra and Malaprabha on the east. The river drains a total area of 4060 sq.kms of which about 2084 sq.kms comes under the command of the proposed dam site above Magod falls. River Sonda (Shivganga falls) drains an area of 1077 sq.kms, while Bedthi drains a total of 1976 sq.kms below Magod falls, flowing mostly in the hilly tracts and coastal plains receiving rainfall mainly from south-west monsoon. The shape of the Bedthi basin is elongated from its origin up to the Arabian sea. The maximum length of the basin is 160 kms and the maximum width, 40 kms. At the proposed dam site, the shape of the catchment is elongated, the maximum length and width being 80 and 26 kms respectively. The perimeter of the catchment is 240 kms. The level of the river-bed at the proposed dam site is at an elevation of 373.38 m. The maximum altitude of the catchment is about 782 m above mean sea level. The shape of Aghnashini catchment is also elongated from its origin to the Arabian sea. At the proposed dam site, above Unchalli falls, it commands an area of about 605.67 sq.kms.

5.6.4 Rain-gauge Stations

There are about 27 rain-gauge stations in Uttara Kannada of which nineteen are located in and around Bedthi catchment and the rest in and around Aghnashini catchment.

5.6.5 River Discharge

River discharge is the rate at which a volume of water passes through a cross-section per unit of time. It is usually expressed in cubic metres per second or Mm^3 per second depending on the quantity. The water yield at the dam site is computed based on the relationship R=0.85*P+30.5, where R is run off and P catchment precipitation. An attempt is also made to compute the same by rational method assuming an appropriate run off coefficient. The yield is given by Y=C*A*I, where C is run off coefficient, A- catchment area and I- rainfall intensity in metres/hour. It is seen that the average annual yield at Magod is 1125 million cubic metres by rational method, compared to 1105 Mm^3 by empirical method. A dependable water yield of 90% is estimated to be 994.55 Mm^3. The tributaries, like Shalmala which joins near Kalghatgi, contribute a lower yield because of scanty precipitation in the region, while Koulgi Nalla, Nalgounda Nalla and Suremane Nalla, which join Bedthi in the upstream of Magod falls, command a sizeable catchment and receive substantial precipitation during the south-west monsoon. This part of the catchment is hilly and Ghat type.

Based on precipitation data of 90 years collected from India Meteorological Department at Pune and Bangalore, and using empirical method water yield is computed at Magod, Shivganga, Unchalli and Benne identified for storage dam construction. Inflow data for these sites were analysed; average, standard deviation, maximum, minimum and mode were computed.

Table 5.1 lists water discharge details at Magod site, which shows that water quantity varies from 0.25(avg.)±1.25(Sd) during January to 364.67(avg.)±136.18(Sd) in July. Mode value computed based on 90 year's data, ranges from 16.80 (frequency of occurrence: 61) in May, 200.70 (frequency of occurrence: 73) in June and 339.16 (number of occurrences: 61) in July. Similarly, Table 5.2 lists water discharge at Shivganga (of Bedthi river), which shows variation from 0.15(avg.)±0.59(Sd) for January to 167.43(avg.)±65.72(Sd) during peak monsoon month - July. Table 5.3 lists the same for Unchalli site, the variation ranging from 0.19(avg.)±0.74(Sd) for February to 442.02(avg.)±91.19(Sd) for July. Table 5.4 lists monthly values of water discharge at Benne (of Aghnashini river). From these Tables, it is evident that all the sites receive good flow only during four months in a year.

Table 5.1: Water Discharge (million cubic meters) at Magod (Bedthi river)

Month	Avg	Sd	Var	Max	Min	Mode	Mode (frequency)
Jan	0.25	1.25	1.57	10.73	0.00	0.000	0(77), 0.47(11)
Feb	0.28	1.47	2.17	11.13	0.00	0.000	0(98)
Mar	2.21	5.83	33.99	35.25	0.00	0.000	0(60), 1.88(18)
Apr	22.72	47.62	2267.53	278.91	0.00	0.000	0(20),1.040 (8)
May	71.14	118.72	14094.77	586.88	0.00	16.800	16.80(61)
June	243.73	113.01	12770.57	645.86	70.51	200.700	200.70(73)
July	364.67	136.18	18545.73	713.64	86.50	339.160	339.16(61)
Aug	216.27	111.56	12444.74	576.58	48.28	156.310	156.31(57)
Sept	77.83	59.87	3585.01	357.33	0.00	66.230	66.23(84)
Oct	47.65	34.83	1213.19	157.20	0.00	36.080	36.08(68)
Nov	15.50	24.29	589.77	157.20	0.00	11.130	11.13(60)
Dec	4.24	12.40	153.78	68.99	0.00	0.000	0 (58), 0.1513 (13), 0.7406(10)

Table 5.2: Water Discharge (million cubic meters) at Shivganga–Sonda (Bedthi River)

Mon	Avg	Sd	Var	Max	Min	Mode	Mode (frequency)
Jan	0.15	0.59	0.35	4.76	0.00	0.00	0(71), 0.541(12)
Feb	0.18	1.18	1.38	9.69	0.00	0.00	0(81)
Mar	0.76	1.86	3.48	10.01	0.00	0.00	0(59), 0.64947(15)
Apr	4.59	5.39	29.10	21.06	0.00	1.81	1.81(58)
May	11.44	13.23	174.99	77.99	0.00	9.02	9.02(75)
June	91.49	38.69	1497.15	214.20	34.69	84.10	84.096(79)
July	167.43	65.72	4319.46	409.36	36.46	162.52	162.52(63)
Aug	90.41	38.14	1454.96	227.88	28.73	75.85	75.85(82)
Sep	31.56	16.34	266.91	79.06	4.91	29.13	29.13(81)
Oct	22.28	12.79	163.64	55.50	1.23	20.25	20.25(80)
Nov	8.93	10.77	116.01	52.82	0.00	6.18	6.18(52),0(17),21.09(13)
Dec	1.08	2.76	7.61	12.94	0.00	0.00	0(60),0.667(11)

Table 5.3: Water Discharge (million cubic meters) at Unchalli (Aghnashini river)

Mon	Avg	Sd	Var	Max	Min	Mode	Mode (frequency)
Jan	0.21	1.10	1.21	9.54	0.00	0.00	0(76)
Feb	0.19	0.74	0.55	4.68	0.00	0.00	0(80)
Mar	0.73	2.02	4.08	14.11	0.00	0.00	0(61),1.89(27)
Apr	4.60	5.47	29.94	24.08	0.00	2.82	2.82(52)
May	88.14	20.15	406.17	130.21	0.00	18.12	18.12(80)
June	283.49	49.78	2478.27	356.94	35.48	219.47	219.47(81)
July	442.02	91.19	8316.30	603.30	51.35	414.12	414.12(73)
Aug	246.18	50.59	2558.88	263.92	30.90	184.12	184.12(77)
Sept	89.14	24.50	600.11	115.93	4.04	50.12	50.12(75)
Oct	25.06	16.02	256.52	75.28	0.15	16.47	16.47(62)
Nov	9.69	12.69	161.09	65.63	0.00	5.77	5.77(64),21.12(10)
Dec	1.70	4.98	24.76	40.61	0.00	0.00	0(59),2.499(26)

Table 5.4: Water Discharge (million cubic meters) at Benne (Aghanashini river)

Mon	Avg	Sd	Var	Max	Min	Mode	Mode (frequency)
Jan	0.08	0.21	0.04	1.79	0.00	0.00	0(77),0.056(11)
Feb	0.07	0.14	0.02	0.88	0.00	0.00	0(87),0.056(6)
Mar	0.27	0.38	0.14	2.65	0.00	0.00	0(60),0.029(18)
Apr	1.71	1.03	1.07	4.53	0.00	1.24	1.2445(42), 0.8331(9)
May	5.70	3.81	14.55	24.48	0.00	5.87	5.87(75)
June	41.23	8.14	66.31	44.24	6.67	43.81	43.81(81)
July	83.74	15.31	234.54	84.54	9.66	84.51	84.51(52)
Aug	45.95	9.44	89.18	49.62	5.81	41.90	41.90(65)
Sept	14.73	4.54	20.62	19.71	0.76	14.43	14.43(82)
Oct	9.17	2.78	7.75	13.69	0.03	9.32	9.32(86)
Nov	3.44	2.12	4.50	11.67	0.00	3.48	3.48(34)
Dec	0.68	0.95	0.91	7.64	0.00	0.00	0(58),0.15(13),0.678(10)

5.6.6 Evaporation and Seepage Loss

Total evaporation loss from water surfaces, soil, vegetation and transpiration by plants and seepage loss through percolation of precipitation into soil and underground are estimated as 99.14 Mm^3 per annum for an area of 100 sq.kms in the region.

5.6.7 Silting Capacity

About 48% of the total catchment area is plain with partial vegetation cover receiving moderate rainfall, while the remaining is hilly with evergreen vegetation. The siltation rate is computed using empirical method (Ingli's formula). The silt rate per annum is given by $S = C(A)^{3/4}$ where, C is catchment coefficient and A catchment area in sq.m. For the catchment with plain and forested tracts, C is assumed to be 4.5. For Magod site, the siltation rate is about 0.82875 Mm^3 per year. This computation helps in assessing the life of a reservoir. At this rate of siltation, the life of a reservoir at FRL 450-455 m is approximately 50 years.

5.6.8 Dam Site

The major portion of the water yield in the hilly region is contributed by Suremane Nalla, Koulgi Nalla and Nalgounda Nalla. In order to use the maximum water yield, it is advantageous to construct the storage dam between Magod falls and the confluence of Suremane Nalla. The site for the proposed design is same as the one proposed earlier (MPCL, 1977) at about 0.91 km upstream of Magod falls at longitude 74°45'28"E and latitude 14°51'41"N. This site commands a catchment of 2084 sq.kms with exposed rocky bed at the flanks on either side. The river flows in a deep and well defined gorge. At this site, the river-bed level is 373 m and the river is 36.58 m wide, while at different elevations the length of the dam would be: 392 m at 450 m contour, 436 m at 460 m contour and 579 m at 480 m contour.

5.6.9 Dam Height and Submergence Area

The maximum quantity of water available at this site is computed from rainfall-run off relationship. Hydroelectric energy could be generated by constructing a reservoir here. But, the submergence area has different habitats like forests, agricultural lands, horticultural lands, grasslands, houses, etc. It also has bioresources like firewood, twigs, etc., and bio residues from agricultural and horticultural lands, used for domestic, commercial and other end uses. This energy is not insignificant. Only when the water-head is very high and the reservoir profile, a deep valley with steep walls at its sides, can hydroelectric energy be very competitive compared to bioenergy from submergence area (Subramanian, 1985). When the water-head is not much and the terrain has a slope less than 25°, then the smaller depths of the reservoir and hence less submergence area, enabling the use of firewood, becomes an attractive option. In order to find the submergence area for certain dam heights, the corresponding contours are traced from Survey of India's 1:50,000 scale toposheets. Contours in 1:50,000 scale toposheets are at 20 m interval (Plates 5.3, 5.5, 5.6 and 5.8). The submergence area is computed using planimeter. Volume of water stored in the dam is also calculated assuming the volume at any two consecutive contours as trapezoids, which is given by

$$V_{12} = (a_1+a_2)*0.5*h_{12} \qquad (18)$$

where V_{12} is the volume between contours 1 and 2,

a_1 - area of spread of contour 1, a_2 - area of spread of contour 2 and

h_{12} - height difference between contours 1 and 2.

Similarly, to compute volume between contours 1 and 3,

$$V_{13} = V_{12}+V_{23} = [(a_1+a_2)*0.5*h_{12}] + [(a_2+a_3)*0.5*h_{23}]$$

where V_{13} is the volume between contours 1 and 3,
V_{23} - volume between contours 2 and 3,
V_{12} - volume between contours 1 and 2,
a_2 - area of spread of contour 2,
a_3 - area of spread of contour 3 and
h_{23} - height difference between contours 2 and 3.

To compute volume at dam height 87 m, that is at 460 m contour, volume between contours 373 and 380, 380 and 400,......, 440 and 460 have to be computed. The volume of the reservoir at 460 m FRL is given by the cumulative volume of all the consecutive contours. The generalised form could be written as

$$V_{ij} = \sum_{k=1}^{k=j-1} V_{k,k+1} \quad \text{where } i=1 \text{ and } j=i+1,i+2,......, \qquad (19)$$

The relationship between dam height and submergence area, computed by regression method, is used to get the area of spread for intermediate contours such as 470, 475 or 485 m. Dam height and quantity/volume of water that could be stored is also computed by the same method. Submergence area and corresponding volume computed for different dam heights at Magod is listed in Table 5.5.1. It shows that, when the dam height is 67 m, the corresponding submergence area is 5.7 sq.kms and volume 106.35 Mm^3. With increase in height of the dam to 70 m, submergence area increases to 9.98 sq.kms. Beyond 87 m, there is a steep increase in submergence area as evident from area of 95.02 sq.kms for height of 107 m. Tables 5.6.1, 5.7.1 and 5.8.1 list submergence area and volume computed for various dam heights at Shivaganga, Unchalli and Benne sites.

Table 5.5.1: Submergence Area (sq. kms) for Various Heights of the Dam (in metres) at Magod (Bedthi River)

Height of dam (metres)	Corresponding contour (metres)	Submergence area in sq.kms	Volume (million cubic metres)
0	373	0.45	
7	380	0.68	6.48
27	400	1.84	24.64
47	420	2.86	71.64
67	440	5.70	106.35
70	443	9.98	185.72
87	460	28.59	500.14
107	480	95.02	1736.34
127	500	195.85	4645.14

Table 5.6.1: Submergence Area (sq. kms) for Various Heights of the Dam (in metres) at Shivganga (Bedthi River)

Height of dam (metres)	Corresponding contour (metres)	Submergence area in sq.kms	Volume (million cubic metres)
0	372	0.71	
8	380	1.02	12 48
28	400	2.45	37.28
48	420	4.13	103.04
68	440	5.30	197.30
88	460	30.51	555.45
108	480	52.54	1486.01
128	500	194.88	4645.14

Table 5.7.1: Submergence Area (sq. kms) for Various Heights of the Dam (in metres) at Unchalli (Aghnashini River)

Height of dam (metres)	Corresponding contour (metres)	Submergence area in sq.kms.	Volume (million cubic metres)
0	420	0.12	
20	440	0.40	5.19
40	460	1.53	24.52
60	480	17.99	291.83
80	500	65.92	1059.04

Table 5.8.1: Submergence area (sq. kms) for Various Heights of the Dam (in metres) at Benne (Aghnashini River)

Height of dam (metres)	Corresponding contour (metres)	Submergence area in sq.kms.	Volume (million cubic metres)
0	360	0.37	
20	380	0.63	10.17
40	400	1.86	35.01
60	420	4.88	102.49
80	440	13.06	231.94
100	460	20.94	621.99

Both linear and non-linear regression analyses were carried out of the variables dam height, submergence area and volume. Table 5.5.2 lists the relationship between area and height, and volume and height based on data of Table 5.5.1. The relationship between submergence area and dam height is found to be either powerlaw or exponential. The correlation coefficient (r) for this relationship is 0.99 and percentage error is 0.45. Similarly, the probable relationship between volume of water and dam height is exponential. This relationship of submergence area and height, and volume and height for Magod is shown in Figures 5.1.1 and 5.1.2 respectively. Actual and predicted area and volume for different dam heights are also shown in the Figures. The best fit relationship which helps in predicting submergence area and volume of water for various heights of the dam is used to predict the same for dam heights 77 and 82 m at FRL 450 and 455 m respectively.

Table 5.5.2: Probable Relationship between Submergence Area versus Height and Volume versus Dam Height at Magod

X variable	Y variable	Corr.coef (r)	% error	Best fit relationship
Height (H_{dam})	Submergence area (A_{sub})	0.99	0.45	A_{sub} =0.38*e^(0.048*H_{dam})
Height (H_{dam})	Volume (V_{dam})	0.9982	0.22	V_{dam} =5.03*e^(0.054*H_{dam})
Submergence area (A_{sub})	Volume (V_{dam})	0.9994	0.18	V_{dam} =34.79+13.15*A_{sub}+0. 0529* (A_{sub})*(A_{sub})

Similar exercise was carried out for Shivganga, Unchalli and Benne sites. The corresponding submergence areas are listed in Tables 5.6.1, 5.7.1 and 5.8.1 respectively, while volume of water is listed in Tables 5.6.2, 5.7.2 and 5.8.2 respectively.

Table 5.6.2: Probable Relationship between Submergence Area versus Height and Volume versus Dam Height at Shivganga

X variable	Y variable	Corr.coef. r)	% error	Best fit relationship
Height (H_{dam})	Submergence area (A_{sub})	0.9875	0.485	A_{sub} =0.58*e^(0.044*H_{dam})
Height (H_{dam})	Volume (V_{dam})	0.7962	1.82	V_{dam} =0.11*e^(0.096*H_{dam})

Table 5.7.2: Probable Relationship between Submergence Area versus Height and Volume versus Dam Height at Unchalli

X variable	Y variable	Corr.coef (r)	% error	Best fit relationship
Height (H_{dam})	Submergence area (A_{sub})	0.9892	0.445	A_{sub} =0.091*e^(.082*H_{dam})
Height (H_{dam})	Volume (V_{dam})	0.8574	1.44	V_{dam} =0.0036*e^(.18*H_{dam})

Table 5.8.2: Probable Relationship between Submergence Area versus Height and Volume versus Dam Height at Benne

X variable	Y variable	Corr.coef (r)	% error	Best fit relationship
Height (H_{dam})	Submergence area (A_{sub})	0.9927	0.325	A_{sub} =0.337*e^(.043*H_{dam})
Height (H_{dam})	Volume (V_{dam})	0.7738	3.38	V_{dam} =0.0158*e^(.13*H_{dam})

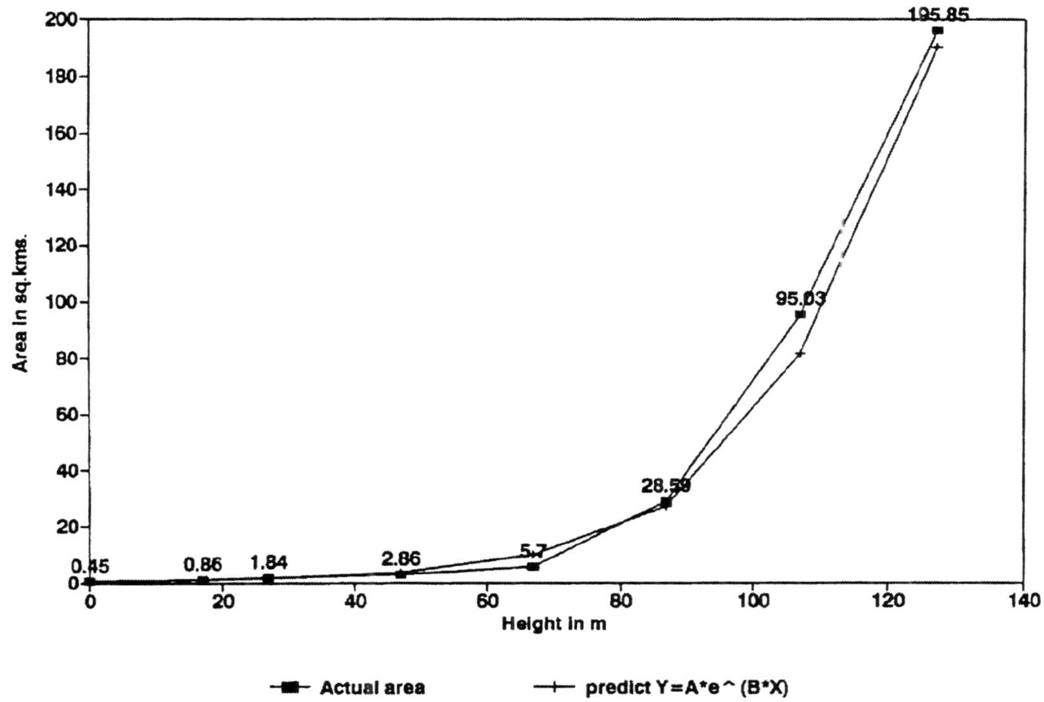

Figure 5.1.1: Height versus Submergence Dam at Magod (Bedthi River)

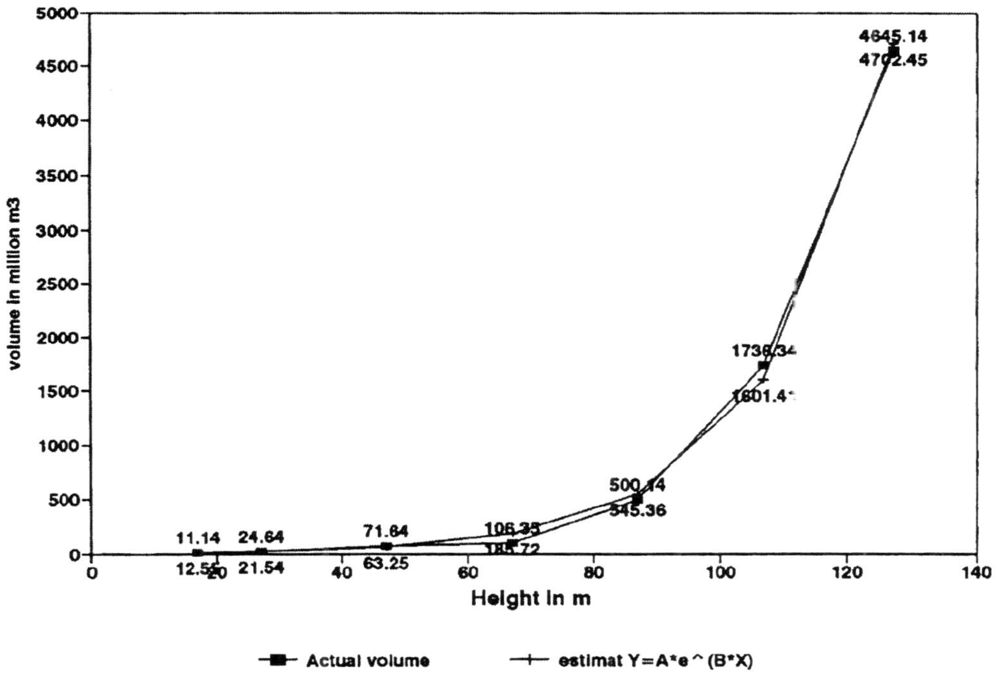

Figure 5.1.2: Height versus Storage Volume Dam at Magod (Bedthi River)

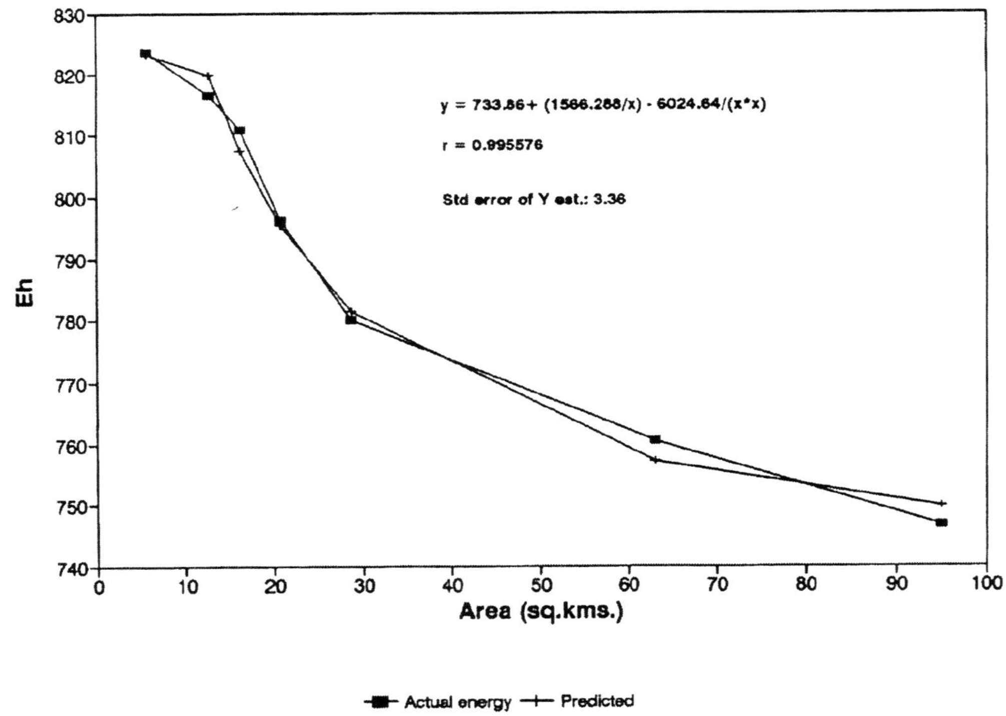

Figure 5.2: Area vs. Energy Generated

5.7.0 NET ENERGY ANALYSES

Hydropower equivalent has been computed for each site before maximising the net energy function. Energy from hydro source computed based on parametric optimisation technique for identified sites of Bedthi and Aghnashini rivers is listed in Tables 5.9 and 5.10 respectively. If variation is allowed between P_{tm} (P1) and P_{td} (P2), the storage capacity requirement is reduced. More of the inflow during monsoon could be used directly without storing. By allowing P1 to P2 ratio as 3, the submergence area saved is about 69.97%. That is, with P1 = P2, area required is about 95.03 sq.kms, which comes down to 28.59 sq.kms when P1 = 3*P2. With this, height of the dam comes down to 87 m from 107 m, which results in reduction of civil costs of the project.

Reduction in dam height from 107 to 77 m changes the submergence area from 95.05 to 16.27 sq.kms. Further reduction of dam height by 10 m (height 67 m) would bring down the submergence area to 5.70 sq.kms. More hydroelectric energy could be harnessed by drawing the monsoon inflow on run-of-river basis and storing small quantity to meet the lean season (non-monsoon) requirement. The saved area would also help in meeting the thermal energy requirement of the region like firewood, etc., and leaf manure for agriculture and horticulture lands. By allowing variation in P_{tm} and P_{td} ratio (results in reduction in dam height and corresponding submergence area), there is increase in electric energy generated. For example, at Magod for a dam height of 107 m (when P1 = P2) energy generated is 746.66 million kWh per year, while for dam height of 77 m (when P1 = 3*P2) it is 810.92 million kWh and for 67 m it is 823.55 million kWh. This is mainly due to lower submergence area for smaller dam heights resulting in reduced evaporation and seepage losses.

Similarly, as listed in Table 5.10, at Unchalli for dam heights of 60 and 35 m, the energy generated is 1814.40 and 1864.40 million kWh respectively. Figure 5.2 gives a graph of submerged area versus energy generated. Regression analysis of these variables gives hyperbolic relationship, given by

En = 733.86+[(1566.28)/Area]-[(6024.64)/(area)(area)], with r (regression coeff.)=0.9966 and % error=1.04.

Table 5.9: Power (MW) and Energy (mkWh) Available at Major Streams of Bedthi River Basin

						Hydropower			@Turbine-Generator η of 70%
Site	Height metres	Contour m	Sub area sq.kms	Volume million M^3	P1/P2 Ratio	P1 MW	P2	Energy mkWh	Energy mkWh
Magod	107	480	95.03	1736.35	1.00	85.00	85.00	746.66	522.66
	97	470	63.23	1134.53	1.00	90.00	90.00	790.55	553.39
	87	460	28.59	545.36	3.00	157.51	52.50	768.62	538.03
	87	460	28.59	545.36	2.50	146.20	58.48	770.53	539.37
	87	460	28.59	545.36	2.00	129.99	65.00	761.24	532.87
	87	460	28.59	545.36	4.00	138.04	34.51	768.65	538.06
	87	460	28.59	545.36	6.00	199.82	33.30	780.08	546.05
	82	455	20.77	416.61	6.00	203.98	34.00	796.36	557.45
	82	455	20.77	416.61	4.00	180.00	45.00	790.55	553.39
	82	455	20.77	416.61	3.00	163.37	54.46	796.27	557.39
	77	450	16.27	416.61	7(for 2 mon)	245.01	35.00	771.14	539.80
					4(for 2 mon)	140.00			
	77	450	16.27	416.61	6.00	210.00	35.00	819.87	573.91
	77	450	16.27	416.61	4.00	184.63	46.16	810.90	567.63
	77	450	16.27	416.61	3.00	166.15	55.38	810.82	567.57
	72	445	12.74	243.12	10.00	250.03	25.00	806.48	564.54
					& 8	200.02			
	70	443	9.98	185.72	16.00	319.95	20.00	823.55	576.49
					& 8	159.98			
	67	440	5.70	106.35	32.00	320.00	10.00	823.55	576.49
					& 16	160.00			
Shivganga									
	48	420	4.13	103.04	16.00	59.45	3.72	152.93	107.05
					8	29.73			
	68	440	5.30	197.50	3.00	31.08	10.36	152.38	106.67
Mattigatta									
	30	400	1.45	7.50	2.00	2.04	1.02	11.97	8.38
Bilihalla									
	20	420	0.68	17.50	3.00	8.22	2.74	40.09	28.06
Barehalla									
	10	390	0.88	19.50	3.00	3.90	1.30	19.02	13.31
@ Magod if dam is up to 460 m contour then Kaulgi Halla is also included									
	10	460	8.28	82.80	7.00	2.31	0.33	8.23	5.76
						1.65			
Kaulgi Halla (near Bommanalli side)									
	10	460	3.32	33.20	16.00	2.52	0.16	8.29	5.80
Kaulgi Halla (left side) 48i/6									
	10	460	8.28	82.80	5.00	3.92	0.78	16.05	11.24

Table 5.10: Power (MW) and Energy (mkWh) Available at Major Streams of Bedthi River Basin

						Hydropower		Power		Power		Power			@Turb-Gen efficiency of 70%
Site	Height metres	Contour m	Sub. Area sq.kms.	Volume million M^3	P1/P2 Ratio	P1 MW	P2	P1 MW	P2	P1 MW	P2	P1 MW	P2	Energy mkWh	Energy mkWh
Unchalli	60	500	65.92	1059	1	210.00	210.00	420.00	420.00	350.00	350.00	323.08	323.08	1814.40	1270.08
	45	485	28.71	451.14	4	420.00	105.00	840.00	210.00	700.00	175.00	646.15	161.54	1814.40	1270.08
	40	480	18	220	12	700.00	58.33	1400.00	116.67	1166.67	97.22	1076.92	89.74	1840.80	1288.56
	40				24	345.00		690.00		575.00		530.77			
	35	475	8.38	86.9	40	600.00	15.00	1200.00	30.00	1000.00	25.00	923.08	23.08	1864.40	1305.08
	30	470	5.55	52.27	70	600.00	7.17	1200.00	14.34	1000.00	11.95	923.08	11.03	1829.30	1280.51
Benne	40	400	1.86	35.01	8	52.93	6.62	105.86	13.23	88.22	11.03	81.43	10.18	190.55	133.38
	50	410	2.9	57	5	50.11	10.02	100.22	20.04	83.52	16.70	77.09	15.42	202.04	141.43
	55	415	4.08	88.06	3	45.19	15.06	90.38	30.13	75.32	25.11	69.52	23.17	216.91	151.84

5.7.1 Biomass Energy from Lands to be Submerged for Hydroelectric Generation

Reduction in submergence area implies reduction in wood energy loss and hence thermal energy. From the submerged area, the energy that would be lost due to submersion of forests and agricultural lands (in terms of biomass) is

$$E_{biomass} = A_{sub} * Gr * (CV\text{-}) * \eta_{c\text{-}}$$

where, $E_{biomass}$ - is energy loss due to submergence, A_{sub} - area to be submerged, Gr - annual rate of growth / productivity, CV energy equivalence factor and η_c - energy conversion efficiency. Then, net energy is given by

$$E_{net} = E_{hydro} - E_{biomass}$$

The land use patterns in the submergence area for various heights of the dam at Magod are listed in Table 5.11.1. It shows the area under natural forest as the major constituent of the land submerged in all scenarios. Figure 5.3.1 gives submergence area (in sq.kms) for various dam heights at Magod, while Figure 5.3.2 provides the land use patterns of these submergence areas, which consist of evergreen, semi-evergreen and deciduous forests where primary production of biomass is estimated to range from 6.5 to 27.5 t/ha/yr (Bioresource productivity, etc., is discussed in Chapter 3).

Table 5.11.2 lists thermal equivalent of energy (in Mega Watts) in this area for biomass productivities 6.5, 13.5, 20 and 27.5 t/ha/yr. It is seen that at productivity of 6.5 t/ha/yr, dam height of 107 m submerges land worth 34.21 MW. With reduction in dam height, this value comes down. For dam height of 67 m, the thermal energy lost is about 3.38 MW.

Table 5.11.3 provides energy equivalent in million kWh for each category of land and total land for respective dam heights. Thermal energy for each land use type (forest, garden, agriculture, etc.) is computed. Total bioenergy is computed by considering various biomass productivities of 6.5, 13.5, 20 and 27.5 t/ha/yr for areas under forests. The total bioenergy in an area of 95.09 sq.kms ranges from 329.77 (biomass productivity of 6.5 t/ha/yr) to 1302.60 million kWh (biomass productivity of 27.5 t/ha/yr).

Net energy ($E_{net} = E_{hydro} - E_{biomass}$) computed considering thermal value of bio residues in the region is given in Table 5.12.1, which shows that the net energy available increases with decrease in dam height. At Magod, the net energy increases from 416.89 (dam height of 107 m) to 787.81 million kWh (height of 67 m). Similarly, at biomass productivity of 13.5 t/ha/yr, net energy available ranges from 61.76 (height 107 m) to 749.31 million kWh (height 67 m). Figure 5.4.1 gives hydroelectric energy generated and bioenergy lost for various submergence areas (that is, for various dam heights). Figure 5.4.2 gives net energy available with various designs. The efficiency of the hydropower stations is the combined efficiency of turbine, generators, etc., which is estimated to be around 70%. For this scenario, the net energy function becomes

Enet = 0.7*Eh - Ebio (last)
Enet = Eh'- Ebio

Table 5.11.1: Landuse Pattern in Submergence Area for Various Heights of Dam

			<--Land use Classification by category-->					
	Dam Height	Submer. Area	Forest	Garden	Paddy	Cultivable waste	NA for Cultivation	Total
Magod (Bedthi)	107	95.03	88.53	2.66	1.49	0.13	2.22	95.03
	97	63.23	60.27	1.21	0.68	0.06	1.01	63.23
	87	28.59	25.93	1.32	0.69	0.00	0.65	28.59
	82	20.77	18.84	0.96	0.50	0.00	0.47	20.77
	77	16.27	14.06	1.30	0.57	0.00	0.34	16.27
	72	12.74	11.01	1.02	0.45	0.00	0.27	12.74
	70	9.98	8.62	0.81	0.35	0.00	0.21	9.98
	67	5.70	4.49	0.68	0.32	0.00	0.21	5.70
Shivganga (Bedthi)	48	4.13	3.61	0.00	0.25	0.14	0.14	4.13
	68	5.30	4.45	0.07	0.49	0.15	0.13	5.30
Unchalli (Aghnashini)	60	65.92	52.41	0.48	6.72	1.70	4.61	65.92
	45	28.71	22.83	0.21	2.93	0.74	2.01	28.71
	40	18.00	13.80	0.15	1.94	0.54	1.57	18.00
	35	8.38	6.43	0.07	0.90	0.25	0.73	8.38
	30	5.55	4.26	0.05	0.60	0.17	0.48	5.55
Benne (Aghnashini)	55	4.08	3.64	0.08	0.15	0.01	0.20	4.08
	50	2.90	2.58	0.06	0.11	0.01	0.14	2.90
	40	1.86	1.66	0.04	0.07	0.00	0.09	1.86

Table 5.11.2: Thermal power computation based on biomass productivity in each category (in Megawatts) (Net primary production of biomass in evergreen and deciduous forests is 6.50, 13.5, 20 to 27.5 t/ha/yr)

			<-----Power in MW----->					Total from all categories (F+G+P+CW+NA.C) Thermal Power in Megawatts			
Place	Dam Height	Area sq.kms.	Forest	Garden	Paddy	Cultivable. Waste	NA for Cultivation	@27.5 t/ha/yr	@20.0 t/ha/yr	@13.5 t/ha/yr	@6.5 t/ha/yr
Magod (Bedthi)	107	95.03	141.03	0.35	0.14	0.01	0.03	141.56	105.00	71.05	34.21
	97	63.23	93.84	0.16	0.06	0.00	0.01	94.19	69.86	47.27	22.76
	87	28.59	41.31	0.18	0.07	0.00	0.01	41.56	30.85	20.91	10.07
	82	20.77	30.01	0.13	0.05	0.00	0.01	30.19	22.41	15.19	7.31
	77	16.27	22.40	0.17	0.05	0.00	0.00	22.63	16.82	11.43	5.50
	72	12.74	17.54	0.14	0.04	0.00	0.00	17.72	13.17	8.95	4.31
	70	9.98	13.74	0.11	0.03	0.00	0.00	13.88	10.32	7.01	3.38
	67	5.70	7.85	0.11	0.03	0.00	0.00	7.93	5.89	4.00	1.93
Shivganga (Bedthi)	48	4.13	5.75	0.00	0.02	0.01	0.00	5.78	4.29	2.91	1.40
	68	5.30	7.09	0.01	0.05	0.01	0.00	7.16	5.32	3.61	1.74
Unchalli (Aghnashini)	60	65.92	83.49	0.06	0.64	0.08	0.06	84.33	62.69	42.59	20.50
	45	28.71	36.36	0.03	0.28	0.03	0.03	36.73	27.30	18.55	8.93
	40	18.00	21.99	0.02	0.18	0.02	0.02	22.24	16.54	11.24	5.41
	35	8.38	10.24	0.01	0.09	0.01	0.01	10.35	7.70	5.23	2.52
	30	5.55	6.78	0.01	0.06	0.01	0.01	6.86	5.10	3.47	1.67
Benne (Aghnashini)	55	4.08	5.79	0.01	0.01	0.00	0.00	2.92	4.32	5.82	2.80
	50	2.90	4.12	0.01	0.01	0.00	0.00	2.08	3.07	4.14	1.99
	40	1.86	2.64	0.00	0.01	0.00	0.00	1.33	1.97	2.65	1.28

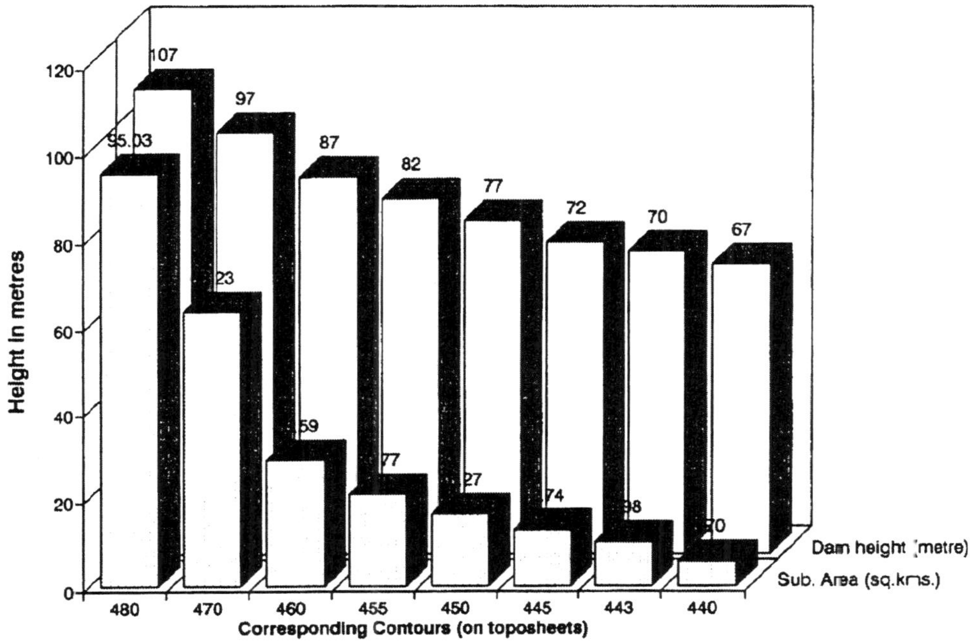

Figure 5.3.1: Total Submergence Area (for various dam heights)

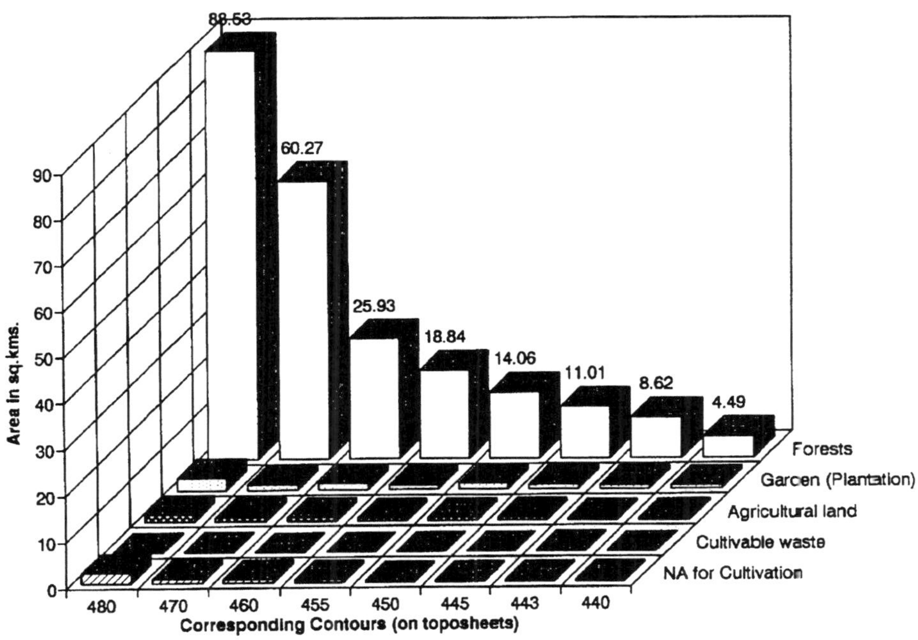

Figure 5.3.2: Land-use of Submerged Area (for various dam heights)

Table 5.11.3: Energy from biomass (mkWh) from various categories of land for various productivity rates. (Net primary production of biomass in evergreen and deciduous forests in 6.50, 13.5, 20 to 27.5 t/ha/yr)

			<----------Energy in mkWh----------> from various categories					<-----Total Energy in mkWh -----> for biomass productivity (t/ha/yr)			
	Dam Height	Subm. Area	Forest	Garden	Paddy	Culti. waste	NA for culti.	Ebio''' 27.50	Ebio'' 20.00	Ebio' 13.5	Ebio 6.50
Magod (Bedthi)	107	95.03	1235.40	27.50	15.40	1.39	22.90	1302.60	982.31	684.90	329.77
	97	63.23	822.00	12.51	7.00	0.63	10.42	866.71	653.60	455.71	219.42
	87	28.59	361.87	13.64	7.17	0.00	6.69	389.37	295.55	208.44	100.36
	82	20.77	262.89	9.92	5.21	0.00	4.86	282.88	214.72	151.44	72.91
	77	16.27	196.18	13.44	5.92	0.00	3.54	219.07	168.21	120.98	58.25
	72	12.74	153.62	10.54	4.63	0.00	2.77	171.56	131.74	94.75	45.62
	70	9.98	120.34	8.27	3.63	0.00	2.17	134.41	103.21	74.24	35.74
	67	5.70	68.73	8.27	3.63	0.00	2.17	76.77	58.95	42.40	20.41
Shivganga (Bedthi)	48	4.13	50.34	0.00	2.61	1.41	1.42	55.76	42.71	30.60	14.73
	68	5.30	62.15	0.72	5.09	1.54	1.39	70.89	54.78	39.82	19.17
Unchalli (Aghnashini)	60	65.92	731.35	4.96	69.48	17.55	47.70	871.04	681.43	505.36	243.32
	45	28.71	318.52	2.17	30.26	7.64	20.77	379.37	296.79	220.11	105.98
	40	18.00	192.62	1.54	20.05	5.59	16.25	236.05	186.11	139.74	67.28
	35	8.38	89.67	0.72	9.30	2.60	7.57	109.86	86.61	65.02	31.31
	30	5.55	59.39	0.48	6.20	1.72	5.01	72.80	57.40	43.11	20.75
Benne (Aghnashini)	55	4.08	50.73	0.84	1.55	0.09	2.11	55.31	42.16	29.95	14.42
	50	2.90	36.06	0.60	1.14	0.06	1.50	39.35	30.00	21.32	10.27
	40	1.86	23.12	0.38	0.72	0.04	0.96	25.23	19.24	13.67	6.58

Table 5.12.1: Hydro and Thermal, and hence net energy from submerged area

	Dam Ht. m	Subm. Area sq.kms	Eh Hydro mkWh	Eh @0.7Eh mkWh	<----------Ebiomass---------> in mkWh @6.5t/h/yr	@13.5/t/h/yr	@20t/h/yr	@27.5t/h/yr	Net Energy mkWh Eh-Ebio (6.5t)	Eh-Ebio' (13.5t)	Eh-Ebio'' (20t)	Eh-Ebio''' (27.5t)
Magod (Bedthi)	107	95.03	746.66	522.66	329.77	684.90	982.31	1302.60	416.89	61.76	-235.65	-555.94
	97	63.23	760.55	532.39	219.42	455.71	653.60	866.71	541.13	304.84	106.95	-106.16
	87	28.59	780.08	546.06	100.36	208.44	295.55	389.37	679.72	571.64	484.53	390.71
	82	20.77	796.27	557.39	72.91	151.44	214.72	282.88	723.36	644.83	581.55	513.39
	77	16.27	810.82	567.57	58.25	120.98	168.21	219.07	752.57	689.84	642.61	591.75
	72	12.74	816.48	571.54	45.62	94.75	131.74	171.56	770.86	721.73	684.74	644.92
	70	9.98	823.55	576.49	35.74	74.24	103.21	134.41	787.81	749.31	720.34	689.14
	67	5.70	823.55	576.49	20.41	42.40	58.95	76.77	803.14	781.15	764.60	746.78
Shivganga (Bedthi)	48	4.13	152.93	107.05	14.73	30.60	42.71	55.76	138.20	122.33	110.22	97.17
	68	5.30	152.38	106.67	19.17	39.82	54.78	70.89	133.21	112.56	97.60	81.49
Unchalli (Aghnashini)	60	65.92	1814.40	1270.08	243.32	505.36	681.43	871.04	1571.08	1309.04	1132.97	943.36
	45	28.71	1814.40	1270.08	105.98	220.11	296.79	379.37	1708.42	1594.29	1517.61	1435.03
	40	18.00	1840.80	1288.56	67.28	139.74	186.11	236.05	1773.52	1701.06	1654.69	1604.75
	35	8.38	1864.40	1305.08	31.31	65.02	86.61	109.86	1833.09	1799.38	1777.79	1754.54
Benne (Aghnashini)	55	4.08	96.41	67.49	14.42	29.95	42.16	55.31	81.99	66.46	54.25	41.10
	50	2.90	89.62	62.73	10.27	21.32	30.00	39.35	79.35	68.30	59.62	50.27
	40	1.86	70.13	49.09	6.58	13.67	19.24	25.23	59.86	56.46	50.89	44.90

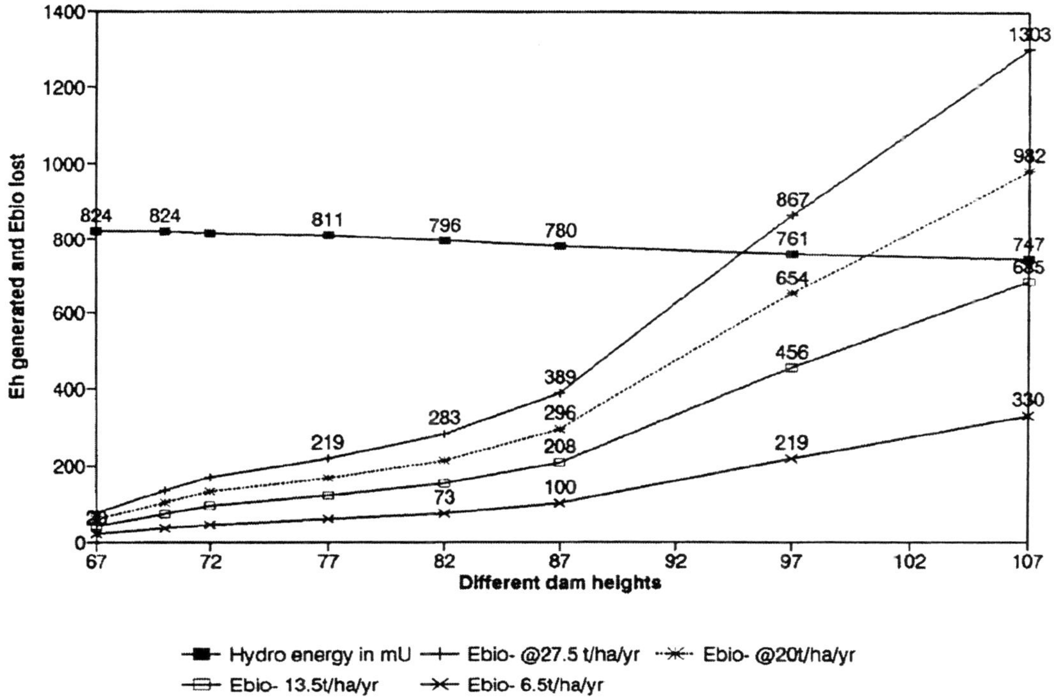

Figure 5.4.1: Hydro Energy Generated and Bioenergy Lost – mkWh (for Various Dam Heights)

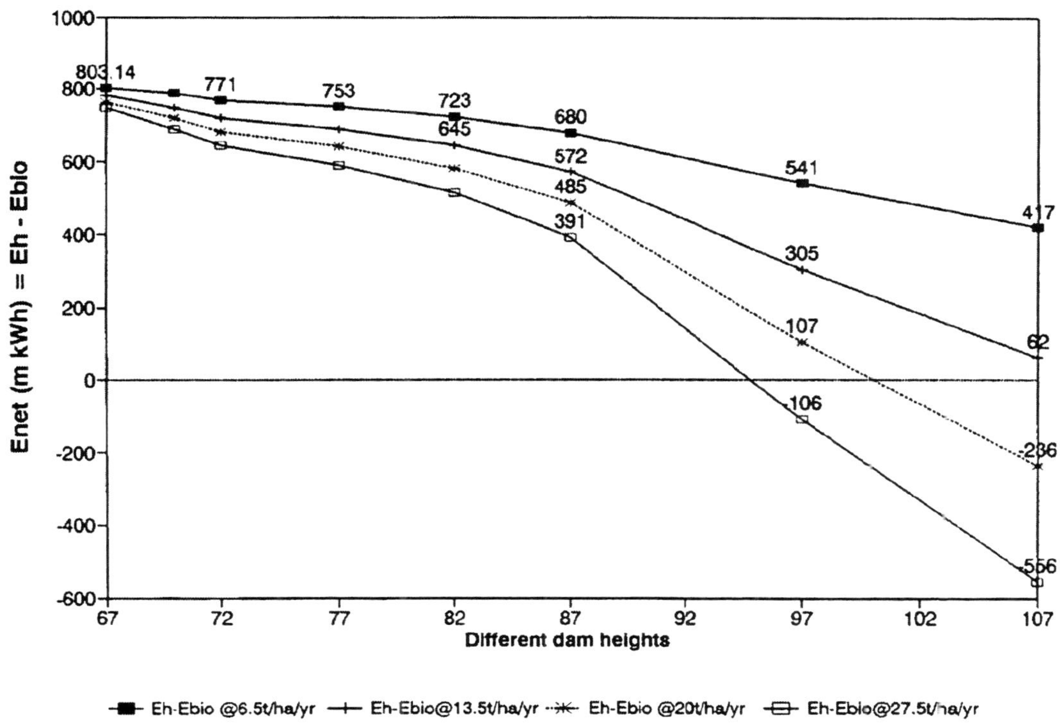

Figure 5.4.2: Net Energy available – mkWh for Various Dam Heights (Eh – Ebio)

Table 5.12.2: Net Energy Computations for Various Heights of Dam at each Site considering Electrical Machinery's Efficiency (as 70%)

	Dam Height	Sub area sq.kms	Hydro Energy (mkWh) Eh Eh	Eh' 0.7Eh	Energy from biomass (mkWh) Eb Ebio(6.5)	Eb' Ebio(13.5)	Eb'' Ebio(20)	Eb''' Ebio(27.5)	Net Energy (mkWh) Eh'-Eb	Eh'-Ebio'	Eh'-Ebio''	Eh'-Ebio'''
Magod (Bedthi)	107	95.03	746.66	522.66	329.77	684.90	982.31	1325.48	192.90	-162.23	-459.65	-802.82
	97	63.23	760.55	532.39	219.42	455.71	653.60	866.71	312.97	76.68	-121.22	-334.33
	87	28.59	780.08	546.06	100.36	208.44	295.55	396.07	445.70	337.62	250.50	149.98
	82	20.77	796.27	557.39	72.91	151.44	214.72	287.75	484.48	405.95	342.66	269.64
	77	16.27	810.82	567.57	58.25	120.98	168.21	222.71	509.32	446.59	399.36	344.87
	72	12.74	816.48	571.54	45.62	94.75	131.74	174.41	525.91	476.78	439.80	397.13
	70	9.98	823.55	576.49	35.74	74.24	103.21	136.64	540.74	502.25	473.28	439.85
	67	5.70	823.55	576.49	20.41	42.40	58.95	76.77	556.08	534.09	517.54	499.72
Shivganga (Bedthi)	48	4.13	152.93	107.05	14.73	30.60	42.71	56.70	92.32	76.45	64.34	50.35
	68	5.3	152.38	106.67	19.17	39.82	54.78	72.04	87.49	66.85	51.89	34.62
Unchalli (Aghnashini)	60	65.92	1814.40	1270.08	243.32	505.36	681.43	884.58	1026.76	764.72	588.65	385.50
	45	28.71	1814.40	1270.08	105.98	220.11	296.79	385.27	1164.10	1049.97	973.29	884.81
	40	18.00	1840.80	1288.56	67.28	139.74	186.11	239.62	1221.28	1148.82	1102.45	1048.94
	35	8.38	1864.40	1305.08	31.31	65.02	86.61	111.52	1273.77	1240.06	1218.47	1193.56
	30	5.55	1829.30	1280.51	20.75	43.11	57.40	73.90	1259.76	1237.40	1223.11	1206.61
Benne (Aghnashini)	55	4.08	96.41	67.49	14.42	29.95	42.16	56.25	53.07	37.54	25.33	11.24
	50	2.9	89.62	62.73	10.27	21.32	30.00	40.02	52.47	41.41	32.73	22.72
	40	1.86	70.13	49.09	6.58	13.67	19.24	25.66	42.51	35.42	29.85	23.43

Table 5.12.3: Net Energy Computation Considering Conversion Efficiency of Biomass as 35%

	Dam Height	Sub area sq.kms	Hydro Energy (mkWh) Eh Eh	Eh' 0.7Eh	0.35*Eb	Energy from biomass (mkWh) 0.35 *Eb' Ebio (13.5)	0.35 *Eb'' Ebio (20)	0.35 *Eb''' Ebio (27.5)	Net Energy mkWh Eh'-0.35 *Eb	Eh'-035 Ebio'	Eh'-0.35 Ebio''	Eh'-035 Ebio'''
Magod (Bedthi)	107	95.03	746.66	522.66	115.42	239.71	343.81	463.92	407.24	282.95	178.85	58.74
	97	63.23	760.55	532.39	76.80	159.50	228.76	303.35	455.59	372.89	303.63	229.04
	87	28.59	780.08	546.06	35.13	72.95	103.44	138.63	510.93	473.10	442.61	407.43
	82	20.77	796.27	557.39	25.52	53.00	75.15	100.71	531.87	504.39	482.24	456.68
	77	16.27	810.82	567.57	20.39	42.34	58.87	77.95	547.19	525.23	508.70	489.63
	72	12.74	816.48	571.54	15.97	33.16	46.11	61.04	555.57	538.37	525.43	510.49
	70	9.98	823.55	576.49	12.51	25.98	36.12	47.82	563.97	550.50	540.36	528.66
	67	5.70	823.55	576.49	7.14	14.84	20.63	26.87	569.34	561.65	555.85	549.62
Shivganga (Bedthi)	48	4.13	152.93	107.05	5.16	10.71	14.95	19.84	101.89	96.34	92.10	87.21
	68	5.3	152.38	106.67	6.71	13.94	19.17	25.21	99.96	92.73	87.49	81.45
Unchalli (Aghnashini)	60	65.92	1814.40	1270.08	85.16	176.88	238.50	309.60	1184.92	1093.20	1031.58	960.48
	45	28.71	1814.40	1270.08	37.09	77.04	103.88	134.84	1232.99	1193.04	1166.20	1135.24
	40	18.00	1840.80	1288.56	23.55	48.91	65.14	83.87	1265.01	1239.65	1223.42	1204.69
	35	8.38	1864.40	1305.08	10.96	22.76	30.31	39.03	1294.12	1282.32	1274.77	1266.05
	30	5.55	1829.30	1280.51	7.26	15.09	20.09	25.87	1273.25	1265.42	1260.42	1254.64
Benne (Aghnashini)	55	4.08	96.41	67.49	5.05	10.48	14.76	19.69	62.44	57.01	52.73	47.80
	50	2.9	89.62	62.73	3.59	7.46	10.50	14.01	59.14	55.27	52.23	48.73
	40	1.86	70.13	49.09	2.30	4.78	6.73	8.98	46.79	44.31	42.36	40.11

Table 5.12.2 gives the net energy computed for various dam heights at Magod considering the efficiency of turbine-generator sets, which shows variation of 192.90 to 540.74 million kWh for biomass productivity of 6.5 t/ha/yr. Net energy ranges from 407.24 (height 107 m) to 563.47 million kWh (height 67 m) at productivity of 13.5 t/ha/yr.

To account for only the final amount of electricity that a thermal source can produce, the thermal energy content of that source is discounted by the conversion efficiency of a thermal power plant based on that source. The conversion efficiency of wood-based thermal power station ranges from 35% to 50%. With this, the net energy function becomes

$$Enet = Eh' - 0.35*Ebio \text{ (where Eh' is } 0.7*Eh)$$

This is calculated for various biomass productivities such as Ebio (6.5 t/ha/yr), Ebio' (13.5 t/ha/yr), Ebio" (20 t/ha/yr) and Ebio'"(27.5 t/ha/yr). The results of this computation are given in Table 5.12.3. With this, variation in net energy available ranges from 407.24 to 569.34 million kWh (biomass productivity of 6.5 t/ha/yr). For productivity of 13.5 t/ha/yr, it ranges from 282.95 to 561.64 million kWh.

The domestic fuel wood consumption survey carried out in this region reveals that 82 to 90% of the households still depend on fuel wood and agro residues like coconut husks, areca husks, etc., to meet their domestic cooking and water heating requirements. The annual fuel wood energy requirement of the region, based on primary household surveys, is estimated to be 312.05 million kWh. This is computed taking into account per capita per day fuel wood consumption of 2.5 kgs for cooking and water heating, which shows clearly that fuel wood is as important as electricity to meet the basic needs of the region. System efficiency, peak power and socio-economic considerations rule out the possibility of electricity entirely substituting fuel wood in the region as a domestic energy source. Therefore, only a fraction of wood energy is converted to electrical energy, which is equal to the fraction of purely non-thermal consumption of electric energy, in a total wood and electric energy consumption. Dam height of 107 m at Magod submerges about 95.03 sq.kms, about 88.53 sq.kms of which is under evergreen and semi-evergreen forests, rich in biodiversity. This demands for eco-friendly options to reduce submergence of prime forest lands.

Considering this area of 95.03 as a region of energy sources, which can cater both thermal and electric energy requirements, analyses were carried out for various dam heights and corresponding submergence areas. The net energy gain for various scenarios is computed considering the following three cases to calculate the total energy available in the area.

Case I. Enet (gain) = Eh' - Ebio (lost).

Case II. Considering the area of 95.03 sq.kms as an energy resource region consisting of (a) water storage area and (b) area under forests - to meet the thermal requirement of the region. Therefore, $Area = Area_1$ (submerged for hydroelectric generation) + $Area_2$ (forests - bioresource available to meet thermal requirements)

Net energy in this case would be

$$Enet = Eh' \text{ (from } Area_1) + Ebio \text{ (from } Area_2)$$
$$\text{where } Eh' = 0.7 * Eh$$

Taking into account conversion efficiency of thermal energy (bioenergy to electricity), total energy from this area is given by

$$Enet = Eh' + 0.35 * Ebio \text{ (from } Area_2)$$

Case III. Taking into account fuel wood requirements of the region for domestic purposes.

$$\text{(a) } Enet = Eh' + Efw + 0.35 * (Ebio - Efw), \text{ if } Ebio \geq Efw$$
$$\text{(b) } Enet = Eh' + Ebio, \text{ if } Efw \geq Ebio$$

In Cases II and III, the objective is to maximise the total energy generation from both hydro and thermal options.

In Table 5.13.1 (Case I), Ebio is computed based on earlier findings of net primary productivity in the region as 6.5 t/ha/yr. Analysis is carried out to maximise the net energy for different dam heights and productivities. It is seen from the Table that net energy increases from 192.89 to 556.08 million kWh with reduction in the dam height by 40 m.

Table 5.13.2 (Case II) shows net energy gain of 63.9% for a reduction of 37.3 % in dam height. That is, energy increases from 522.66 to 896.35 million kWh by hydro-thermal combined option in a land area of 95.03 sq.kms. If efficiency of thermal conversion is taken into account, then the hydro-thermal combined scheme would generate 688.44 as against 522.66 million kWh (case of pure hydro scheme with dam height of 107m and submergence area of 95.03 sq.kms.) of energy.

Table 5.13.3 lists the net energy gain from hydro and thermal combination by considering thermal conversion (bioresource to electricity) efficiency. With decrease in submergence area from 95.03 to 5.7 sq.kms., the net energy available in the area increases from 522.66 (height of 107 m) to 688.44 million kWh (height of 67 m).

Table 5.13.4 lists the net energy gain under [Enet = Eh' + Efw + 0.35* (Ebio - Efw), if Ebio ≥ Efw] and [Enet = Eh' + Ebio, if Efw ≥ Ebio] scenarios. The fuel wood requirement in the region is about 312.05 million kWh for domestic purposes. Net energy computed for various heights of the dam indicates that dam height of 67 m stores enough water to meet the lean season electricity requirement. This area with natural vegetation and bioresource of 319.85 million kWh also caters the thermal energy demand of 312.05 million kWh in the domestic sector.

Similar analyses carried out for productivities of 13.5, 20 and 27.5 t/ha/yr are listed in Tables 5.14.1 to 5.14.4, Tables 5.15.1 to 5.15.4 and Tables 5.16.1 to 5.16.4 respectively. These are depicted pictorially in Figures 5.5.1 to 5.5.4 respectively. Figure 5.5.1 gives biomass available for various productivities and submergence areas saved due to reduced dam heights. Figure 5.5.2 shows net energy available in the region (thermal+hydro option) for biomass productivity of 6.5 t/ha/yr.

Table 5.14.4 (Case III) shows that it is possible to cater the fuel wood needs of the domestic sector and generate electricity of 823.55 million kWh by having minimum submergence. With increase in productivities, the net energy from hydro+thermal source increases considerably. For productivity of 13.5 t/ha/yr, net energy gain from 95.03 sq.kms is 1240.97 million kWh compared to 522.66 million kWh of a pure hydro scheme.

Figure 5.5.3 depicts net energy available (hydro+thermal = Eh+Ebio) for various productivities. By taking turbine-generator efficiency into account, the net energy available from this region is given in Figure 5.5.4 (Eh'+Ebio). This option of hydro+thermal combination has various advantages like: (1) reliability of mixed system is higher than pure hydro system since hydroelectric system is more prone to the vagaries of climate than the productivity of forests and (2) significant reduction in dam height and size brings down economic costs of the project. In case the forest land is already degraded due to various reasons, an energy plantation of mono/poly cultures could be tried in the area, which would help in having a hydro+thermal system. An energy plantation is seen as an intensive form of land use providing adequate opportunities for employment and other benefits like fodder, leaf manure for mulching, soil conservation, etc. The model and subsequent quantitative analyses demonstrate that land submergence could be reduced substantially if the hydroelectric generation capacity is adjusted according to seasonal variations in the river's run off. The viability of a mixed hydro and biomass generation system is shown in energy terms, which leads to a significant reduction in the total area used for power generation. At Magod, with a large storage reservoir, energy generated is 522.66 million kWh (taking into account efficiency of generation). Adjusting capacity of storage reservoir according to the seasonal variation and with hydro-biomass combination, the net energy generated is about 809.06 million kWh. Apart from this, there is scope for generating hydroelectric energy from streams in a decentralised way. Similar analyses were carried out for biomass productivities of 20 and 27.5 t/ha/yr. Results are listed in Tables 5.15.1 to 5.15.4 and Tables 5.16.1 to 5.16.4.

Table 5.13.1: Case 1 – Net Energy Gain for Various Designs and Biomass Productivity of 6.5 T/Ha/Yr from Forests

	Dam Height	Area	Hydro energy		Thermal energy		Fuel wood	Net Energy Gain	
				Eh'				Net Energy	
	m	sq.kms	Eh	0.7*Eh	Ebio	0.35*Ebio	Efw	Eh'-Ebio	Eh'-0.35 *Ebio
Magod	107	95.03	746.66	522.662	329.77	115.42	312.05	192.89	407.24
	97	63.23	760.55	532.385	219.42	76.80		312.97	455.59
	87	28.59	780.08	546.056	100.36	35.13		445.70	510.93
	82	20.77	796.27	557.389	72.91	25.52		484.48	531.87
	77	16.27	810.82	567.574	58.25	20.39		509.32	547.19
	72	12.74	816.48	571.536	45.62	15.97		525.92	555.57
	70	9.98	823.55	576.485	35.74	12.51		540.75	563.98
	67	5.7	823.55	576.485	20.41	7.14		556.08	569.34

Table 5.13.2: Case 2 – Considering the Area of 95.03 Sq.Kms. divided into Forests/Fuel Wood and Water Storage

	Dam height	Area under (in sq.kms)			Eh'		Net Energy Gain
		Submerge	Forests/fw	Eh	0.7*Eh	Ebio	Eh'+Ebio
Magod	107	95.03	0	746.66	522.662	0.00	522.66
	97	63.23	31.8	760.55	532.385	110.35	642.74
	87	28.59	66.44	780.08	546.056	219.14	765.20
	82	20.77	74.26	796.27	557.389	244.94	802.33
	77	16.27	78.76	810.82	567.574	259.78	827.35
	72	12.74	82.29	816.48	571.536	271.42	842.96
	70	9.98	85.05	823.55	576.485	304.58	881.06
	67	5.7	89.33	823.55	576.485	319.86	896.35

Table 5.13.3: Net Energy: Thermal Efficiency into Consideration (If Wood is Converted to Electricity)

	Dam Height m	Area (in sq.kms)			Eh'		0.35*	Net Energy Gain
		Submerge	Forests/fw	Eh	0.7*Eh	Ebio	Ebio	Eh'+0.35* Ebio
Magod	107	95.03	0	746.66	522.662	0.00	0.00	522.66
	97	63.23	51.8	760.55	532.385	110.35	38.62	571.01
	87	28.59	66.44	780.08	546.056	219.14	76.70	622.76
	82	20.77	74.26	796.27	557.389	244.94	85.73	643.12
	77	16.27	78.76	810.82	567.574	259.78	90.92	658.50
	72	12.74	82.29	816.48	571.536	271.42	95.00	666.53
	70	9.98	85.05	823.55	576.485	304.58	106.60	683.09
	67	5.7	89.33	823.55	576.485	319.86	111.95	688.44

Table 5.13.4: Case 3 – Enet = Eh'+Efw+0.35(Ebio-Efw), if Ebio > Efw and Eh'+Ebio if Efw > Ebio

									Net Energy Gain	
	Dam Height m	Area sq.kms	Forests /fw	Eh	Eh' 0.7*Eh	Ebio	0.35 *Ebio	Taluk fuel wood Required (Efw)	Case 3a Eh'+Efw+ 0.35 (Ebio-Efw)	Case 3b Eh' +Ebio
Magod	107	95.03	0	746.66	522.662	0.00	0.00	312.05		522.66
	97	63.23	31.8	760.55	532.385	110.35	38.62			642.74
	87	28.59	66.44	780.08	546.056	219.14	76.70			765.20
	82	20.77	74.26	796.27	557.389	244.94	85.73			802.33
	77	16.27	78.76	810.82	567.574	259.78	90.92			827.35
	72	12.74	82.29	816.48	571.536	271.42	95.00			842.96
	67	9.98	85.05	823.55	576.485	304.58	106.60			881.06
	67	5.7	89.93	823.55	576.485	319.86	111.95		891.27	

Table 5.14.1: Case 1 – Net Energy Gain for Various Designs and Biomass Productivity of 13.5 T/Ha/Yr from Forests

	Dam Height m	Area sq.kms	Eh	Eh' 0.7Eh	Ebiomass'	0.35*Ebio'	Taluk wood energy requirement mkWh	Net Energy Eh'-Ebio'
Magod	107	95.03	746.66	522.66	684.90	239.71	312.05	-162.23
	97	63.23	760.55	532.39	455.71	159.50		76.68
	87	28.59	780.08	546.06	208.44	72.95		337.62
	82	20.77	796.27	557.39	151.44	53.00		405.95
	77	16.27	810.82	567.57	120.98	42.34		446.59
	72	12.74	816.48	571.54	94.75	33.16		476.78
	70	9.98	823.55	576.49	74.24	25.98		502.25
	67	5.7	823.55	576.49	42.40	14.84		534.09

Table 5.14.2: Case 2 – Considering Area of 95.03 sq.km divided into Forests/Fuel Wood (Fw) and Water Storage

	height	Area under (in sq.kms.)		Forest	Eh	Eh'		
	m	Submerge	Forests /fw	saved		0.7*Eh	Ebio'	Eh'+Ebio'
Magod	107	95.03	0.00	0.00	746.66	522.66	0.00	522.66
	97	63.23	31.80	31.80	760.55	532.39	229.19	761.57
	87	28.59	66.44	66.44	780.08	546.06	484.39	1030.44
	82	20.77	74.26	74.26	796.27	557.39	541.44	1098.83
	77	16.27	78.76	78.76	810.82	567.57	585.65	1153.23
	72	12.74	82.29	82.29	816.48	571.54	612.04	1183.57
	70	9.98	85.05	85.05	823.55	576.49	632.65	1209.14
	67	5.7	89.33	89.33	823.55	576.49	664.49	1240.97

Table 5.14.3: Net Energy considering Thermal Efficiency (if all Wood is Converted to Electricity)

	Height m	Area under (in sq.kms.) submerge	Forests/ fw	Forest Saved	Eh	Eh' 0.7*Eh	Ebio'	0.35*Ebio	Eh'+ +0.35 *Ebio'
Magod	107	95.03	0.00	0.00	746.66	522.66	0.00	0.00	522.66
	97	63.23	51.80	51.80	760.55	532.39	229.19	80.22	612.60
	87	28.59	66.44	66.44	780.08	546.06	484.39	169.54	715.59
	82	20.77	74.26	74.26	796.27	557.39	541.44	189.50	746.89
	77	16.27	78.76	78.76	810.82	567.57	585.65	204.98	772.55
	72	12.74	82.29	82.29	816.48	571.54	612.04	214.21	785.75
	70	9.98	85.05	85.05	823.55	576.49	632.65	221.43	797.91
	67	5.7	89.33	89.33	823.55	576.49	664.49	232.57	809.06

Table 5.14.4: Case 3 – Enet = Eh'+Efw+0.35(Ebio-Efw), If Ebio > Efw and Eh'+Ebio if Efw > Ebio

	Height m	Area under (in sq.kms.) submerge	Forests/ fw	Forest Saved	Eh	Eh' 0.7*Eh	Ebio'	0.35*Ebio'	Fuel wood Efw	Eh'+Efw+ +0.35 *Ebio'	Eh'+ +Ebio' *0.35
Magod	107	95.03	0.00	0.00	746.66	522.66	0.00	0.00	312.05		522.66
	97	63.23	51.80	51.80	760.55	532.39	229.19	80.22			612.60
	87	28.59	66.44	66.44	780.08	546.06	484.39	169.54		918.42	
	82	20.77	74.26	74.26	796.27	557.39	541.44	189.50		949.72	
	77	16.27	78.76	78.76	810.82	567.57	585.65	204.98		975.38	
	72	12.74	82.29	82.29	816.48	571.54	612.04	214.21		988.58	
	70	9.98	85.05	85.05	823.55	576.49	632.65	221.43		1000.75	
	67	5.7	89.33	89.33	823.55	576.49	664.49	232.57		1011.89	

Table 5.15.1: Case 1 – Net Energy Gain for Various Designs and Biomass Productivity of 20.0 T/Ha/Yr from Forests

	Dam Height m	Area sq.kms	Eh	Eh' 0.7*Eh	Ebio"	0.35*Ebio"	Fuel wood Efw	Net energy gain Net Energy Eh'-Ebio"
Magod	107	95.03	746.66	522.662	982.31	343.81	312.05	-459.65
	97	63.23	760.55	532.385	653.60	228.76		-121.22
	87	28.59	780.08	546.056	295.55	103.44		250.50
	82	20.77	796.27	557.389	214.72	75.15		342.66
	77	16.27	810.82	567.574	168.21	58.87		399.36
	72	12.74	816.48	571.536	131.74	46.11		439.80
	70	9.98	823.55	576.485	103.21	36.12		473.28
	67	5.7	823.55	576.485	58.95	20.63		517.54

Table 5.15.2: Case 2 – Considering Area of 95.03 sq.kms. divided into Forests/Fuel Wood (fw) and Water Storage

		Area under (sq.kms)				0.7 Eh		Total Energy
		Submerge	Forests/fw	Forest saved	Eh	Eh'	Ebio"	Eh'+Ebio"
Magod	107	95.03	0	0	746.66	522.662	0.00	522.66
	97	63.23	31.8	31.8	760.55	532.385	328.71	861.10
	87	28.59	66.44	66.44	780.08	546.056	686.84	1232.89
	82	20.77	74.26	74.26	796.27	557.389	767.72	1325.10
	77	16.27	78.76	78.76	810.82	567.574	814.28	1381.86
	72	12.74	82.29	82.29	816.48	571.536	850.91	1422.45
	70	9.98	85.05	85.05	823.55	576.485	879.54	1456.02
	67	5.7	89.33	89.33	823.55	576.485	923.86	1500.35

Table 5.15.3: Case 3: Net Energy considering Thermal Efficiency (if all Wood is used for Electricity Generation)

		Area (sq.kms.)		Forest		Eh'			Eh'+
	Height	Submerge	Forests/fw	saved	Eh	0.7Eh	Ebio"	0.35*Ebio"	+0.35*Ebio"
Magod	107	95.03	0	0	746.66	522.662	0.00	0.00	522.66
	97	63.23	31.8	51.8	760.55	532.385	328.71	115.05	647.43
	87	28.59	66.44	66.44	780.08	546.056	686.84	240.39	786.45
	82	20.77	74.26	74.26	796.27	557.389	767.72	268.70	826.09
	77	16.27	78.76	78.76	810.82	567.574	814.28	285.00	852.57
	72	12.74	82.29	82.29	816.48	571.536	850.91	297.82	869.36
	70	9.98	85.05	85.05	823.55	576.485	879.54	307.84	884.32
	67	5.7	89.33	89.33	823.55	576.485	923.86	323.35	899.84

Table 5.15.4: Case 4 - Enet = Eh'+Efw+0.35(Ebio-Efw), if Ebio > Efw and Eh'+Ebio if Efw > Ebio

	Height	Area (sq.kms.) submerge	Area Under Forests /fw	Forest saved	Eh	Eh' 0.7*Eh	Ebio"	0.35 *Ebio"	Taluk fuel wood reqd(Efw)	Case 3a Eh'+Efw+ +.35(Ebio "-Efw)	Case 3b Eh'+ 0.35* Ebio"
Magod	107	95.03	0	0	746.66	522.662	0.00	0.00	312.05		522.66
	97	63.23	31.8	51.8	760.55	532.385	328.71	115.05		850.27	
	87	28.59	66.44	66.44	780.08	546.056	686.84	240.39		989.28	
	82	20.77	74.26	74.26	796.27	557.389	767.72	268.70		1028.92	
	77	16.27	78.76	78.76	810.82	567.574	814.28	285.00		1055.40	
	72	12.74	82.29	82.29	816.48	571.536	850.91	297.82		1072.19	
	70	9.98	85.05	85.05	823.55	576.485	879.54	307.84		1087.16	
	67	5.7	89.33	89.33	823.55	576.485	923.86	323.35		1102.67	

Table 5.16.1: Case 1 – Net Energy Gain for Various Designs and Biomass Productivity of 27.5 T/Ha/Yr from Forests

	Dam Height m	Area sq.kms.	Area Under Forests/fw	Eh	Eh' 0.7*Eh	Ebio'''	0.35*Ebio'''	Fuel wood Efw	Net energy gain Net energy Eh'-Ebio'''
Magod	107	95.03	0	746.66	522.662	1302.60	455.91	312.05	-779.94
	97	63.23	31.8	760.55	532.385	866.71	303.35		-334.33
	87	28.59	66.44	780.08	546.056	389.37	136.28		156.68
	82	20.77	74.26	796.27	557.389	282.88	99.01		274.51
	77	16.27	78.76	810.82	567.574	219.07	76.68		348.50
	72	12.74	82.29	816.48	571.536	171.56	60.05		399.97
	70	9.98	85.05	823.55	576.485	134.41	47.04		442.08
	67	5.7	89.33	823.55	576.485	76.77	26.87		499.72

Table 5.16.2: Case 2 – Considering Area of 95.03 sq. kms. divided into Forests/Fuel Wood (Fw) and Water Storage

	Dam Height	Area under (in sq.kms) Submerge	Forests/fw		Eh	Eh' 0.7*Eh	Ebio'''	Net Energy Gain Eh'+Ebio'''
Magod	107	95.03	0	0	746.66	522.662	0.00	522.66
	97	63.23	31.8	31.8	760.55	532.385	435.89	968.28
	87	28.59	66.44	66.44	780.08	546.056	904.86	1450.91
	82	20.77	74.26	74.26	796.27	557.389	1011.40	1568.79
	77	16.27	78.76	78.76	810.82	567.574	1060.50	1628.07
	72	12.74	82.29	82.29	816.48	571.536	1108.17	1679.70
	70	9.98	85.05	85.05	823.55	576.485	1145.42	1721.90
	67	5.7	89.33	89.33	823.55	576.485	1203.13	1779.62

Table 5.16.3: Net Energy Considering Thermal Efficiency (if all Wood is Converted to Electricity)

	Height m	Submerge	Forests/fw		Eh	Eh' 0.7*Eh	Ebio'''	0.35*Ebio'''	Net Energy Gain Eh'+0.35*Ebio'''
Magod	107	95.03	0	0	746.66	522.662	0.00	0.00	522.66
	97	63.23	31.8	31.8	760.55	532.385	435.89	152.56	684.95
	87	28.59	66.44	66.44	780.08	546.056	904.86	316.70	862.76
	82	20.77	74.26	74.26	796.27	557.389	1011.40	353.99	911.38
	77	16.27	78.76	78.76	810.82	567.574	1060.50	371.17	938.75
	72	12.74	82.29	82.29	816.48	571.536	1108.17	387.86	959.39
	70	9.98	85.05	85.05	823.55	576.485	1145.42	400.90	977.38
	67	5.7	89.33	89.33	823.55	576.485	1203.13	421.10	997.58

Table 5.16.4: Case 3 – Enet = Eh'+Efw+0.35(Ebio-Efw), if Ebio > Efw and Eh'+Ebio if Efw > Ebio

	Dam Height m	Submerge	Forests/fw		Eh	Eh' 0.7*Eh	Ebio'''	0.35 *Ebio'''	Taluk fuel wood req(Efw)	Net Energy Gain Eh'+Efw+ +0.35 (Ebio'''-Efw)	Case 3b Eh'+0.35 *Ebio'''
Magod	107	95.03	0	0	746.66	522.662	0.00	0.00	312.05		522.66
	97	63.23	31.8	31.8	760.55	532.385	435.89	152.56		887.78	
	87	28.59	66.44	66.44	780.08	546.056	904.86	316.70		1065.59	
	82	20.77	74.26	74.26	796.27	557.389	1011.40	353.99		1114.21	
	77	16.27	78.76	78.76	810.82	567.574	1060.50	371.17		1141.58	
	72	12.74	82.29	82.29	816.48	571.536	1108.17	387.86		1162.23	
	70	9.98	85.05	85.05	823.55	576.485	1145.42	400.90		1180.21	
	67	5.7	89.33	89.33	823.55	576.485	1203.13	421.10		1200.41	

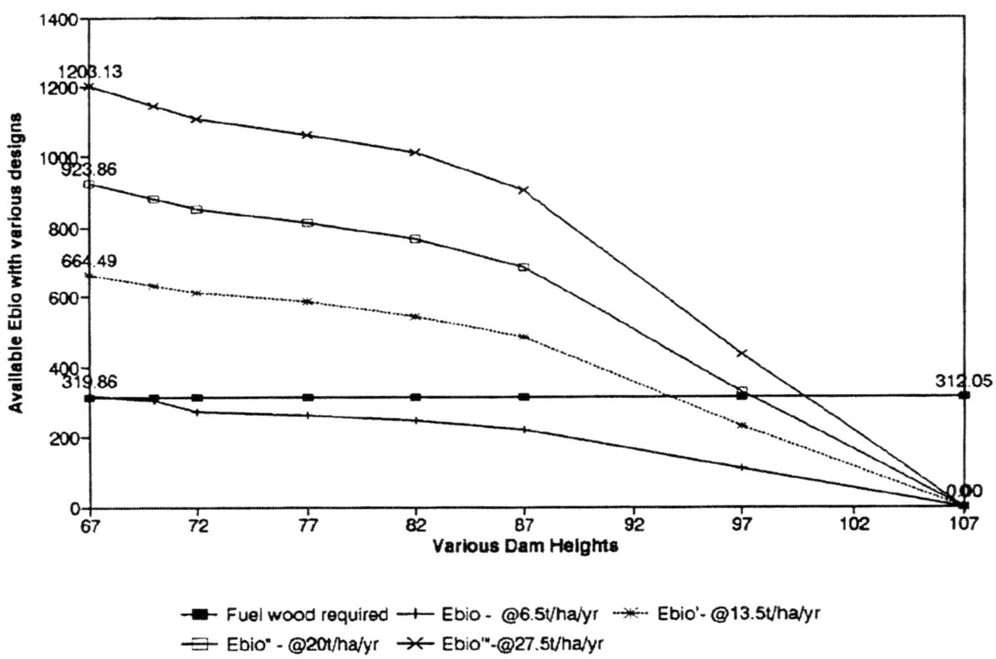

Figure 5.5.1: Fuel Wood Requirement and Ebio Available – mkWh

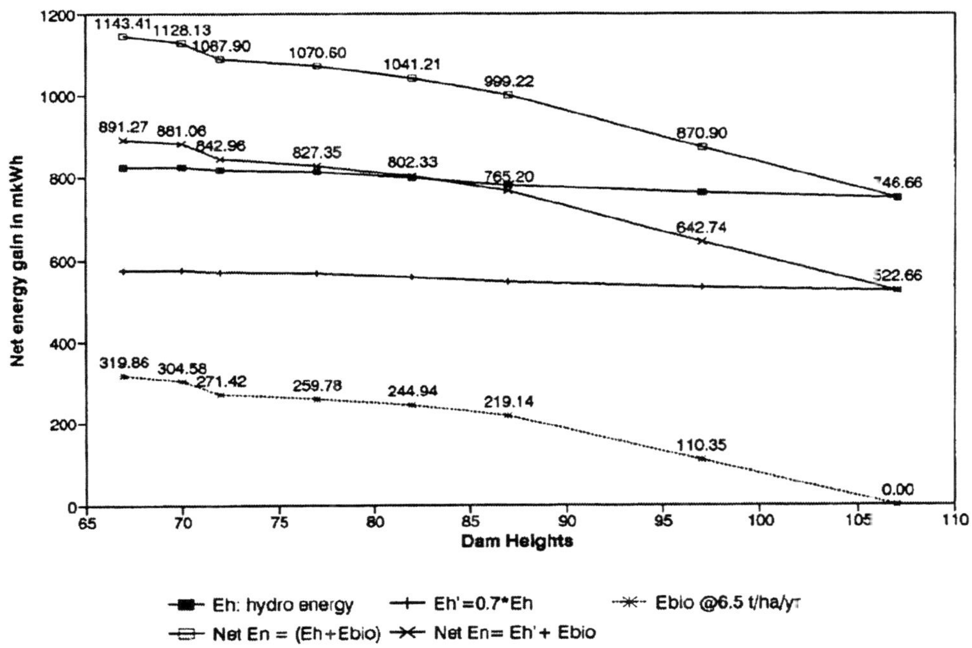

Figure 5.5.2: Net Energy from 95.03 sq.kms Hydro + Thermal (Ebio @6.5 t/ha/yr)

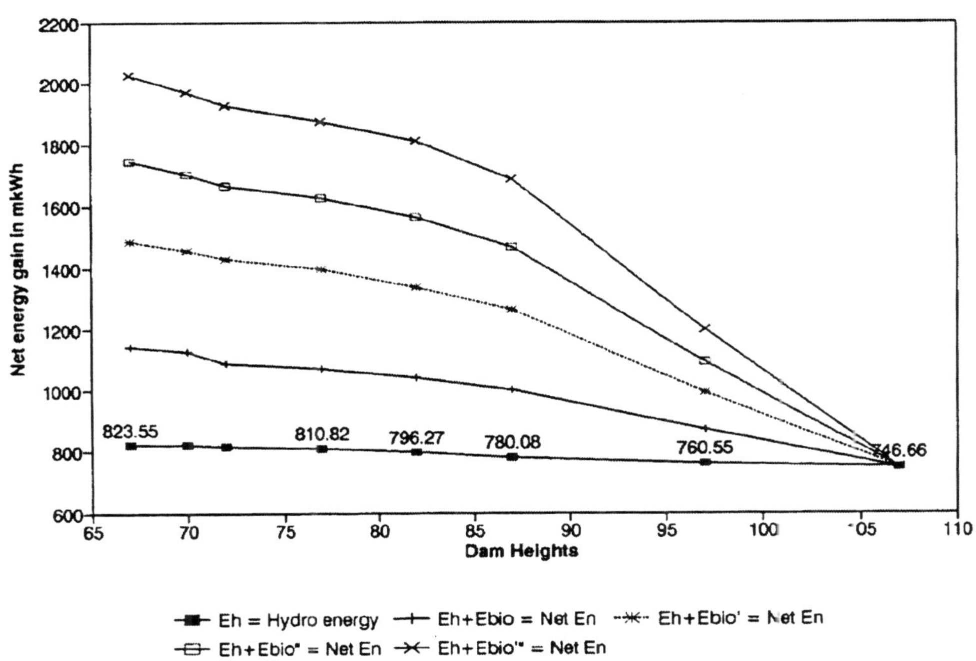

Figure 5.5.3: Net Energy from 95.03 sq.kms Hydro (Eh) + Thermal (different Ebio)

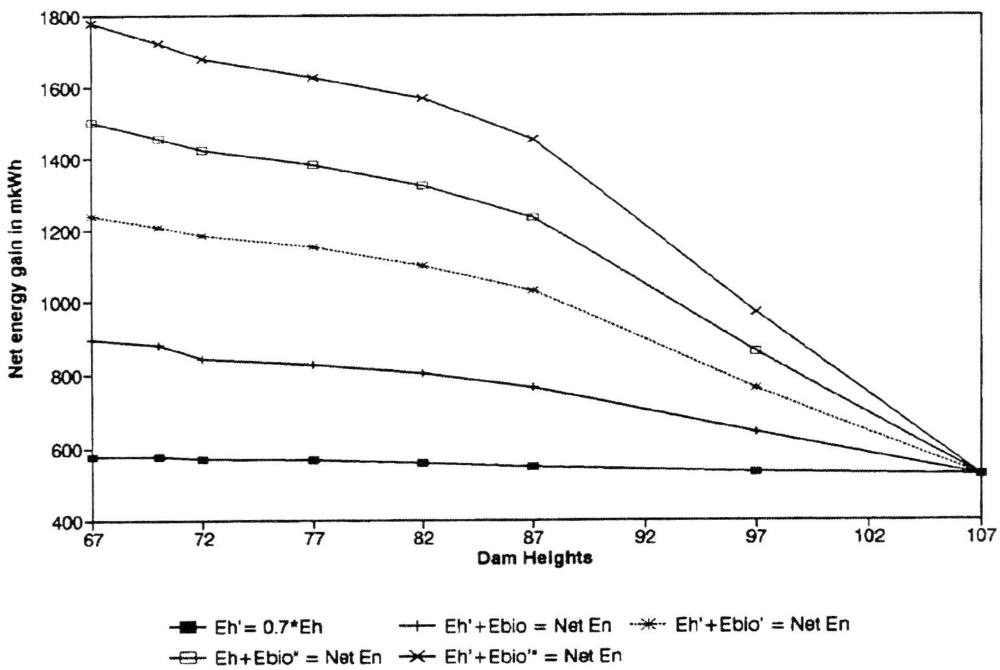

Figure 5.5.4: Net Energy from 95.03 sq.kms Hydro (0.7*Eh) + Thermal (different Ebio)

5.8.0 ECONOMIC ANALYSES

5.8.1 Computation of Costs: Net Loss Due to Forest Submersion

With submersion of forests, benefits such as (a) fuel wood, (b) timber, (c) grass and other biomass material, (d) forest products such as cane, bamboo, tendu leaves, gums and resins, drugs and spices, tans and dyes, fibres and flosses, essential oils, non-edible oil seeds, etc., (e) biodiversity, (f) recreation and (g) environmental benefits such as oxygen production, soil conservation, control of air pollution, recycling of water, control of humidity, etc., are lost. Forests play a vital role not only in the social and economic well being of the society, but also in the maintenance of the ecological balance. Direct costs of submergence are assessed by considering standing biomass in the area, based on species diversity studies carried out by the Centre for Ecological Sciences, I.I.Sc., in the sample plots at Sonda, Kallabe, etc. Cost of timber, fuel wood and minor forest produce works out to be Rs. 2.25 to 4.55 million per hectare depending on the extent of vegetation cover in the area. In this computation, the price of forest land is taken as Rs.1,11,200/ha. Some attempts have been made earlier to quantify the environmental cost of the loss of a forest, assessed to be Rs. 14.13 million per hectare for tropical forests and Rs. 12.87 million per hectare for sub-tropical forests (Maudgal, S. and Kakkar, M., 1992). Details of the environmental benefits derived from a medium sized tree of 50 tonnes during its 50 years life span, excluding value of timber, fruit and flowers, are listed in Table 5.17.

5.8.2 Loss Due to Submersion of Agriculture and Horticulture Lands

The cost involved in the submersion of agricultural lands, and arecanut and coconut plantations is based on the market value of the land in the region. The value of arecanut garden is Rs. 8,90,000 per hect, paddy field Rs. 99,000 per hect and coconut plantation, Rs. 2,18,000/ha. The land use details at Magod are compiled from the records available at Tahsildar's Offices at Yellapur, Sirsi and Siddapur. (Land use particulars for submergence area at Magod, Shivganga, Unchalli and Benne sites are already listed in Tables 5.11.1 and 5.11.2).

5.8.3 Cost of Displacement and Rehabilitation

Details of villages submerged for various dam heights and the subsequent number of households affected are obtained from respective government agencies like Tahsildar's Office at taluk level, Village Accountant's Office, etc., (Table 5.18).

5.8.4 Annual Charges on Capital Costs

Capital cost depends on (a) civil construction costs, which in turn depends on the size and type of dam, and (b) cost of generating units, which depends on the capacity. The capacity is calculated using a normal load factor of 0.5. The schedule of rates approved by the government recently (for the year 1994-95) has been used in computing civil and electrical costs. This exercise is carried out for Magod and listed in Table 5.18.

The annual capital recovery factor (annuity factor) is calculated for the total cost (civil+electrical+environmental+rehabilitation) at i% interest for n years of satisfactory functioning of the hydroelectric power station (over all life of power station is taken as 50 years, considering siltation rate, life of the dam, life of various electrical machineries) by the following relationship.

Annual capital recovery $\quad 'A' = \left(\frac{i}{(1+i)\ \ -1}+i\right) * \text{ total cost}$

Table 5.17: Environment Benefits from One Hectare of Forest (During 50 Year Life Span)

	Environmental Benefits	Tropical Vegetation million Rs. per hect.	Sub-tropical Vegetation million Rs. per hect.
1	Oxygen production	2.25	2.05
2	Conservation of animal protein	0.18	0.16
3	Control of soil erosion	2.25	2.05
4	Recycling of water and control of humidity	2.70	2.46
5	Shelter for birds, reptiles, insects and fauna	2.25	2.05
6	Control of air pollution	4.50	4.10
	Total	14.13	12.87

Table 5.18: Civil, Electrical (El), Environmental (Env) and Rehabilitation costs (million rupees) for Various Heights of Dam

Site	Ht	Cost Civil	Electrical	Civil+El	Env (land)	Rehabilitation	Env cost	Total cost
Magod	107	2800.77	1101.98	3902.75	1282.99	173.25	20.00	5378.99
	97	2177.87	1140.34	3318.21	583.64	138.60	9.10	4040.45
	87	1988.41	1508.66	3497.07	427.76	23.85	3.03	3948.68
	82	1761.24	1830.94	3592.17	339.01	1.12	0.80	3931.19
	77	1657.04	2061.13	3718.18	287.29	0.09	0.18	4005.74
	72	1504.59	2368.06	3872.65	225.15	0.02	0.03	4097.85
	67	1487.68	2541.56	4029.24	218.60	0.02	0.02	4247.88

5.8.5 Operation and Maintenance (O & M) Charges

Annual O & M charge is taken as 1% of the total cost of the project. The depreciation taking into account life of dam as 70 years, surge shaft and penstock as 50 years, power house building as 50 years, power house equipment as 50 years, substation equipment as 50 years, roads, etc., as 35 years, works out to be 1.80%. Hence, annual cost would be C = Annuity factor*Total cost+O&M charges+Depreciation cost.

With this cost and energy information, cost per kWh of energy (net energy = electric energy from hydro + electrical energy from thermal option through bio residues) has been computed.

Table 5.19.1 lists cost per kWh for various designs of the dam and for biomass productivity of 6.5 t/ha/yr. Figure 5.6.1 shows 40.5% reduction in cost (Rs. 1.53 to Rs. 0.92 per kWh) with dam height reduction of 32.75%. If costs of bioresources is excluded considering electricity only from hydro resource, then cost reduces from Rs. 1.53 to Rs. 1.09 per kWh.

Cost per kWh has also been computed without taking into account conversion efficiency. In this case (Table 5.19.1), cost varies from Rs. 1.07 to 0.56 per kWh. Figure 5.6.2 gives cost per kWh of net energy generated for various designs, taking into account various biomass productivities. Analysis for cost per kWh is repeated for biomass productivity of 13.5 t/ha/yr and the results listed in Table 5.19.2. In this case, cost varies from Rs. 1.53 to 0.78 per kWh of energy as a consequence of height reduction from 107 to 67 m.

5.8.6 Benefits

Benefit would be the revenue accrued by electricity consumption in various sectors. The percentage share of each sector is calculated by looking into the annual consumption pattern in Uttara Kannada district. The average power tariff for each sector is used to compute the revenue from each sector. These are given in Table 5.20.1, which shows the industrial sector with a share of 86.38% as the major consumer followed by the domestic sector with a share of 10.20%. Table 5.16.3 lists net electrical energy generated for various heights of the dam. Based on consumption patterns of previous years and values in Table 5.20.2, electric energy requirement in various sectors is estimated and listed in Table 5.20.3. The percentage share of each sector is used to compute the values for various heights of the dam. Based on tariff information provided by the Karnataka Electricity Board at Sirsi, Uttara Kannada district, the sector wise revenue is computed and listed in Table 5.20.4.

Benefit from the power project is computed by

$$B = \sum e_j * t_j$$

where j =1.......6 (representing various sectors),

e_j = electricity consumption in sector "j" and

t_j = tariff in Rs./kWh for sector "j".

Hence, the Benefit-Cost (B-C) ratio = B/C

Table 5.19.1: Case 1 - Cost per kWh of Electricity (net) considering Biomass Productivity of 6.5 t/ha/yr

	Dam Height	Total cost- million Rs.		El energy gen Eh	Eh' 0.7*Eh	Thermal energy (th)	0.35*th	Eh'+0.35 th Net en (mkWh)	Cost/kWh Rs./Eh'	
Magod	107	5378.99		746.66	522.66	0.00	0.00	522.66	1.53	
	97	4040.45		760.55	532.39	110.35	38.62	571.01	1.13	
	87	3948.68		780.06	546.04	219.14	76.70	622.74	1.07	
	82	3931.19		796.36	557.45	244.94	85.73	643.18	1.05	
	77	4005.74		810.82	567.57	259.78	90.92	658.50	1.05	
	72	4097.85		816.48	571.54	271.42	95.00	666.53	1.06	
	67	4247.87		823.55	576.49	319.86	111.95	688.44	1.09	
		Total cost- million Rs.	Int (12%, 50yrs)	Dep (1.806%)	Oper.ch(1%)	Total ex	Net en (mkWh) Eh'+0.35th	Net energy Eh+th	Cost/kWh Eh'+0.35th	Cost/kWh Eh+th
Magod	107	5378.99	647.74	97.14	53.79	798.67	522.66	746.66	1.53	1.07
	97	4040.45	486.55	72.97	40.40	599.93	571.01	870.90	1.05	0.69
	87	3948.68	475.50	71.31	39.49	586.30	622.74	999.20	0.94	0.59
	82	3931.19	473.39	71.00	39.31	583.70	643.18	1041.30	0.91	0.56
	77	4005.74	482.34	72.34	40.05	594.73	658.50	1070.60	0.90	0.56
	72	4097.85	493.46	74.01	40.98	608.44	666.53	1087.90	0.91	0.56
	67	4247.87	511.52	76.72	42.48	630.72	682.65	1126.88	0.92	0.56

Table 5.19.2: Case 2 – Cost per kWh of Electricity (net) considering Biomass Productivity of 13.5 t/ha/yr

		Total cost- million Rs.		El energy gen Eh	Eh' 0.7*Eh	Thermal energy (th)	0.35*th	Net en (mkWh)	Cost/kWh Rs./Eh'	
Magod	107	5378.99		746.66	522.66	0.00	0.00	522.66	1.53	
	97	4040.45		760.55	532.39	229.19	80.22	612.60	1.13	
	87	3948.68		780.06	546.04	484.39	169.54	715.58	1.07	
	82	3931.19		796.36	557.45	541.44	189.50	746.96	1.05	
	77	4005.74		810.82	567.57	585.65	204.98	772.55	1.05	
	72	4097.85		816.48	571.54	612.04	214.21	785.75	1.06	
	67	4247.87		823.55	576.49	664.49	232.57	809.06	1.09	
		Total cost- million Rs.	Int (12%, 50yrs)	Dep (1.806%)	Oper.ch(1%)	Total ex	Net en (mkWh) Eh'+0.35th	Net energy Eh+th	Cost/kWh Eh'+0.35th	Cost/kWh Eh+th
Magod	107	5378.99	647.74	97.14	53.79	798.67	522.66	746.66	1.53	1.07
	97	4040.45	486.55	72.97	40.40	599.93	612.60	989.74	0.98	0.61
	87	3948.68	475.50	71.31	39.49	586.30	715.58	1264.45	0.82	0.46
	82	3931.19	473.39	71.00	39.31	583.70	746.96	1337.80	0.78	0.44
	77	4005.74	482.34	72.34	40.05	594.73	772.55	1396.47	0.77	0.43
	72	4097.85	493.46	74.01	40.98	608.44	785.75	1428.52	0.77	0.43
	67	4247.87	511.52	76.72	42.48	630.72	809.06	1488.04	0.78	0.42

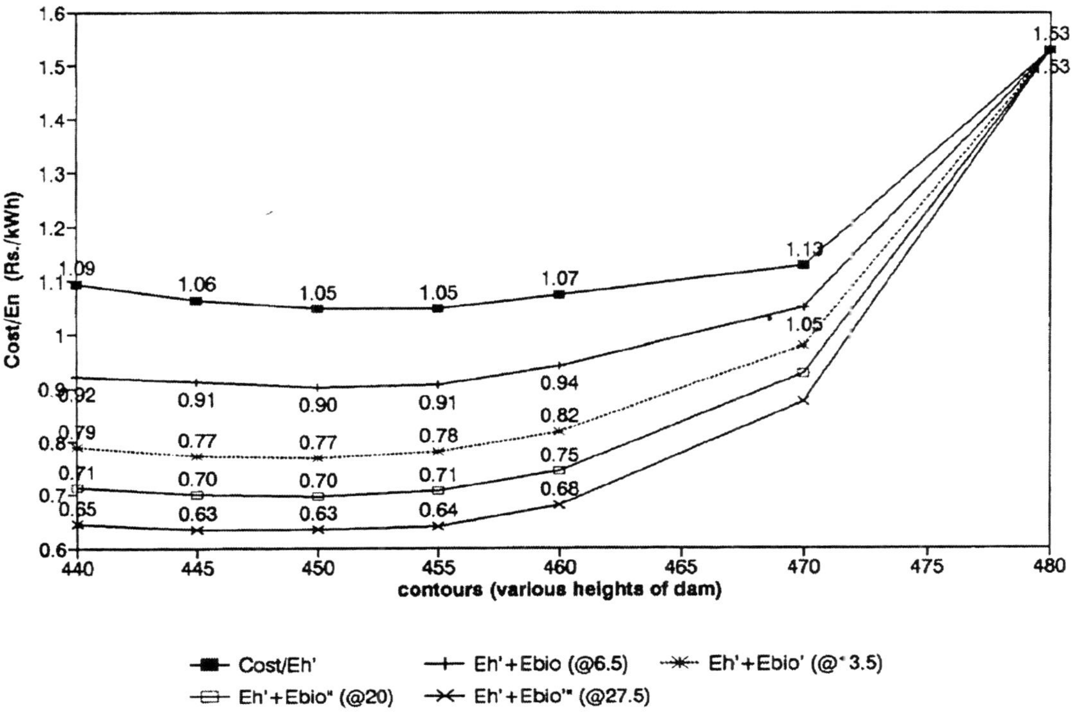

Figure 5.6.1: Cost/unit of Electricity – En=Eh'+0.35*Ebio

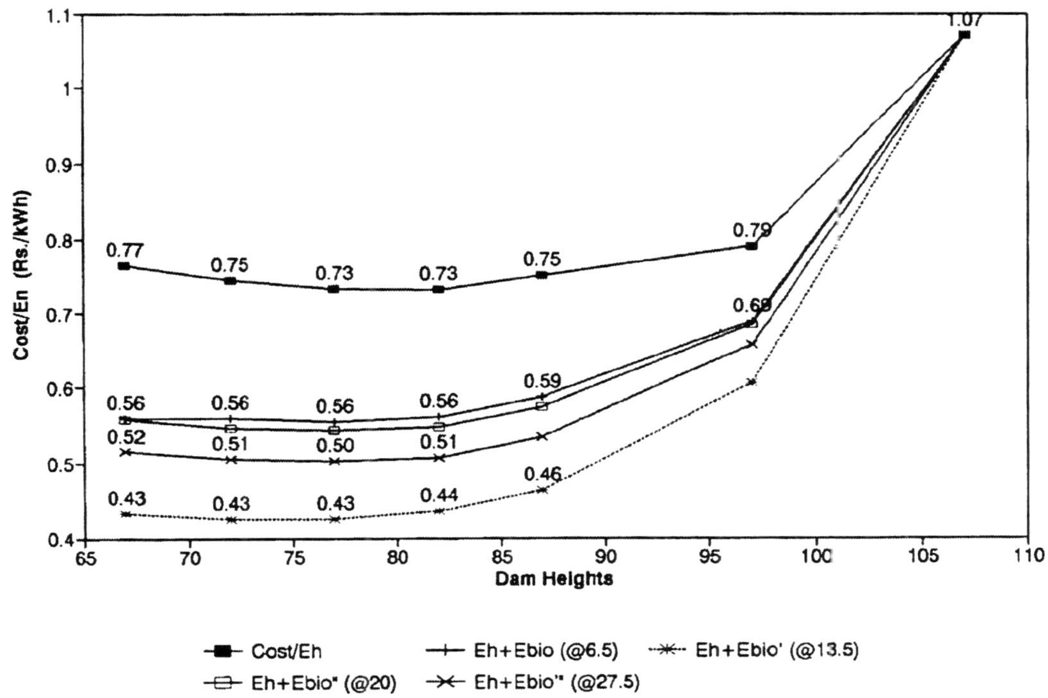

Figure 5.6.2: Cost/unit of Electricity - En=Eh + Ebio

The Benefit-Cost ratio, computed thus for various heights of the dam (H1 - 107 m, H2 - 97 m, H3 - 87 m, H4 - 82 m, H5 - 77 m, H6 - 72 m and H7 - 67 m), is computed for the following cases and are listed in Tables 5.21.1 to 5.21.6.

Case I: Considering value of forest land as Rs. 37,065/ha.
Case II: Considering standing biomass + value of land as Rs. 1,11,200/ha.
Case III: Considering standing biomass (Minor Forest Produce, Timber, Fuel wood, etc.) as Rs. 12,000/yr + value of land as Rs. 1,11,200/ha.
Case IV: Considering standing biomass + value of land as Rs. 2.25 million/ha.
Case V: Considering standing biomass + value of land as Rs. 4.55 million/ha.
Case VI: Considering standing biomass + value of land as Rs. 2.25 million/ha. + environmental value as Rs. 14.13 million/ha.

B-C computed for Case I (Table 5.21.1) is less than 1.5 for heights H1, while for heights H4, H5 and H6 it is greater than 1.5. B-C ratio computed considering Case II is less than 1.5 for heights H1 and H2, while for other heights it is greater than 1.5 (Table 5.21.2). Data from various forest documents (Timber, Minor Forest Produce and other products) and fuel wood depots at Kumta and Ankola show that value of products per hectare of forest land annually is about Rs. 12,000. Here, fuel wood rate of depots to the local residents is taken as Rs. 250/tonne (heavily subsidised price for ration card holders), for decentralised electricity generation Rs. 150/tonne (wood gasifier unit) and to temples and hostels Rs. 750/tonne. Quantity supplied to households, for cremation (through depots), commercial organisations, temples and hostels was obtained from Forest Department (Canara circle). Incorporating this value of Rs. 12,000/ha/year and land value of Rs. 1,11,200/ha (Case III, Table 5.21.3), B-C ratio for heights H4, H5 and H6 is greater than 1.5. Benefit-Cost ratios computed for various designs and environmental values are shown pictorially in Figure 5.7.1, while Figure 5.7.2 gives a plot of annual costs and electricity generated for various dam heights. For Cases IV (Table 5.21.4), V (Table 5.21.5) and VI (Table 5.21.6), B-C is less than 1.5 for all heights of the dam. That means, by assigning environmental values taking all components into consideration, the hydroelectric generation through storage (for minimum submergence also) becomes a less attractive option.

Table 5.20.1: Sectorwise Electricity (El) Consumption ('000 units) and Average Tariff in Uttara Kannada District

Sector	# users	El. cons	% share	Tariff/kWh
Domestic	98117	29994	10.20	0.79
Industrial	3269	254010	86.38	1.31
Agriculture	13633	689	0.23	0.07
Commercial	13089	5811	1.98	1.16
Public lighting	598	788	0.27	0.76
Others	270	2757	0.94	1.92
Total	128976	294049	100	

Table 5.20.2: Net Electrical Energy Generated for Different Submergence Areas (in million kWh) at Magod Site of Bedthi River (Ebio based on 6.5 t/ha/yr)

	H1	H2	H3	H4	H5	H6	H7
Dam height	107	97	87	82	77	72	67
Corres. contour	480	470	460	455	450	445	440
Electricity gen. -mkWh i.e. Eh' + 0.35*Ebio	522.66	612.60	715.59	746.89	772.55	785.75	809.06

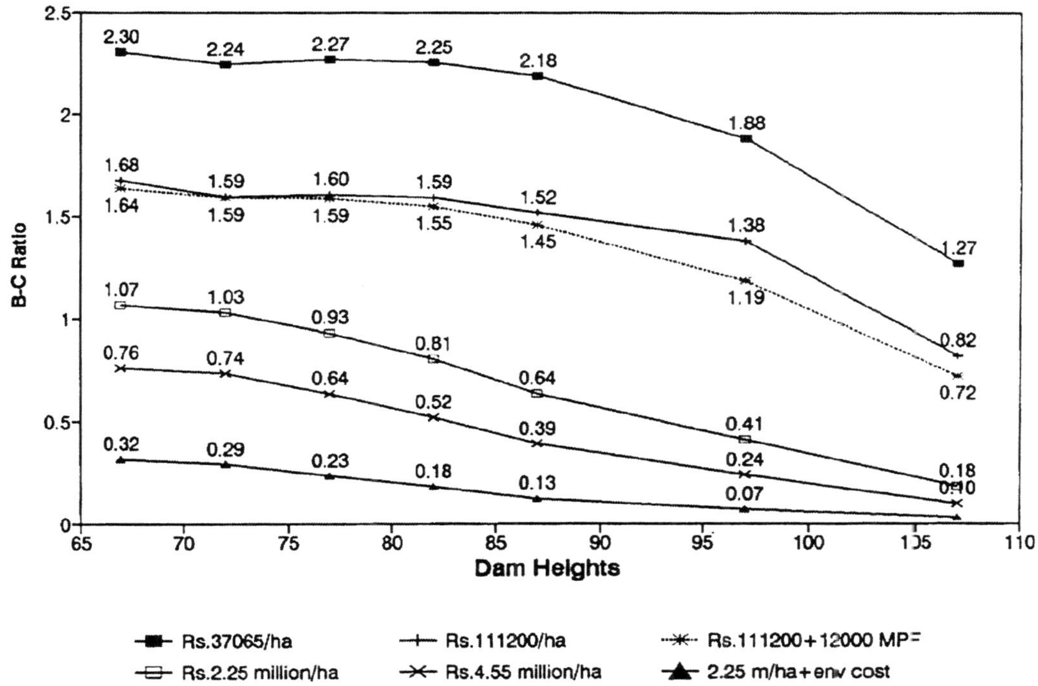

Figure 5.7.1: Benefit Cost Ratio (for various environmental costs)

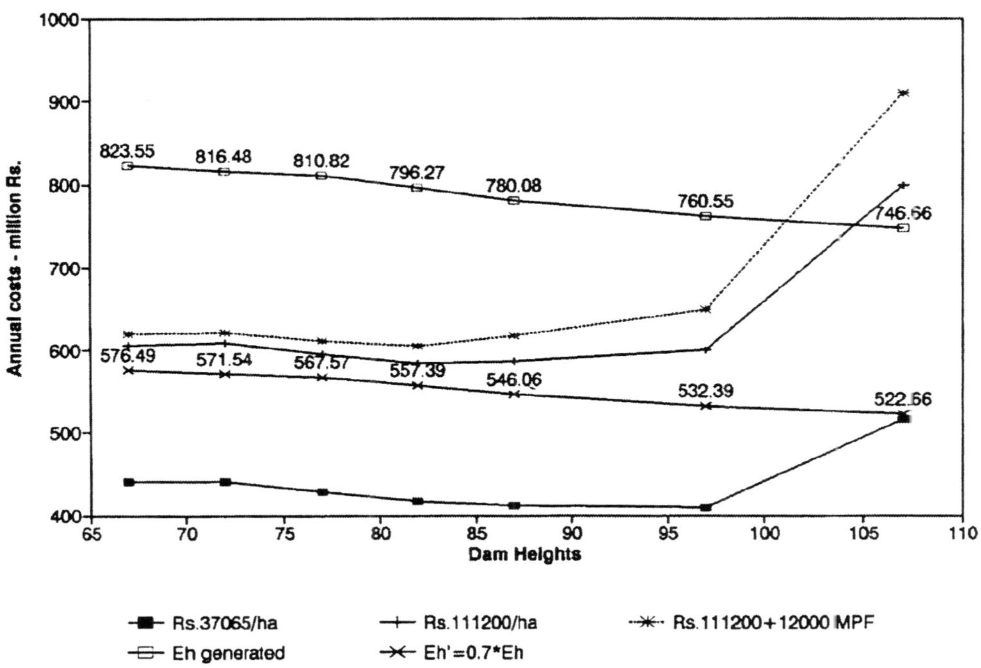

Figure 5.7.2: Annual Costs & Eh generated (for various environmental costs)

Table 5.20.3: Anticipated Sectorwise Consumption of Electricity (million kWh) based on Present Consumption Patterns

Sector	H1	H2	H3	H4	H5	H6	H7
Domestic	53.31	62.49	72.99	76.19	78.80	80.15	82.53
Industrial	451.49	529.19	618.15	645.19	667.36	678.76	698.89
Agriculture	1.22	1.44	1.68	1.75	1.81	1.84	1.90
Commercial	10.33	12.11	14.14	14.76	15.27	15.53	15.99
Public lighting	1.40	1.64	1.92	2.00	2.07	2.11	2.17
Others	4.90	5.74	6.71	7.00	7.24	7.37	7.59
Total	522.66	612.60	715.59	746.89	772.55	785.75	809.06

Table 5.20.4: Sectorwise Revenue (computed as per KEB Tariff) in million Rupees

Sector	H1	H2	H3	H4	H5	H6	H7
Domestic	42.12	49.36	57.66	60.19	62.25	63.32	65.20
Industrial	591.45	693.23	809.78	845.20	874.24	889.17	915.55
Agriculture	0.09	0.10	0.12	0.12	0.13	0.13	0.13
Commercial	11.98	14.04	16.40	17.12	17.71	18.01	18.55
Public lighting	1.06	1.25	1.46	1.52	1.57	1.60	1.65
Others	9.41	11.03	12.88	13.45	13.91	14.14	14.56
Total	656.11	769.02	898.30	937.60	969.81	986.38	1015.64

Table 5.21.1 Benefit to Cost ratio for Various Scenarios considering value of Forest Land as Rs.37,065/ha.

	H1	H2	H3	H4	H5	H6	H7
Annual cost in million Rs.	516.54	409.95	411.87	416.35	428.04	440.73	440.71
Benefit - Cost	1.27	1.88	2.18	2.25	2.27	2.24	2.30

Table 5.21.2: Benefit to Cost ratio for Various Scenarios considering value of Forest Land as Rs.1,11,200/ha.

	H1	H2	H3	H4	H5	H6	H7
Annual cost in million Rs.	798.67	599.93	586.30	583.70	594.73	608.44	606.32
Benefit - Cost	0.82	1.38	1.52	1.59	1.60	1.59	1.68

Table 5.21.3: Benefit to Cost ratio for Various Scenarios considering value of Forest Land as Rs.1,11,200/ha. + biomass, MFP and Timber as Rs. 12,000/year/hect.

	H1	H2	H3	H4	H5	H6	H7
Annual cost in million Rs.	909.91	648.25	617.42	605.64	611.60	621.65	619.88
Benefit - Cost	0.72	1.19	1.45	1.55	1.59	1.59	1.64

Table 5.21.4: Benefit to Cost ratio considering Standing Biomass + Value of Land as Rs.2.25 million/ha.

	H1	H2	H3	H4	H5	H6	H7
Annual cost in million Rs.	3606.05	1877.03	1408.61	1163.38	1040.54	957.53	952.50
Benefit - Cost	0.18	0.41	0.64	0.81	0.93	1.03	1.07

Table 5.21.5: Benefit to Cost ratio considering Standing Biomass + Value of Land as Rs.4.55 million/ha.

	H1	H2	H3	H4	H5	H6	H7
Annual cost in million Rs.	6629.41	3252.38	2294.19	1787.65	1520.65	1333.47	1330.45
Benefit - Cost	0.10	0.24	0.39	0.52	0.64	0.74	0.76

Table 5.21.6: Benefit to Cost ratio considering Standing Biomass + Value of Land as Rs.2.25 million/ha. and Environmental Value as Rs.14.3 million/hect.

	H1	H2	H3	H4	H5	H6	H7
Annual cost in million Rs.	23330.15	10849.70	7186.03	5236.06	4172.73	3410.14	3215.14
Benefit – Cost	0.03	0.07	0.13	0.18	0.23	0.29	0.32

5.9.0 CONCLUSIONS

In order to reduce the submergence of prime forest lands, it is necessary to look for optimal range of reservoir size and installed generation capacity based on certain site specific factors Quantification of energy cost of land submergence reveals that, net energy from a region by hydroelectric energy and thermal energy (due to potential biomass increment in the region) could be maximised by optimal project dimensions. While assessing the thermal energy of biomass, the following aspects have also been considered:

(1) Direct use of fuel wood for various thermal end uses is actually more efficient than the use of electricity.

(2) Substantial domestic demand of fuel wood for cooking and water heating (for bathing), which is as high as 89-92%.

This analysis explains the upper bound/limit on the height of the dam and therefore the area of the reservoir for the project to yield a positive net energy. It is also noticed that, savings in land submergence could be achieved by adjusting hydroelectric generation capacity according to seasonal variation in the river's run off.

Parametric optimisation technique is used to compute energy from a hydro source at Magod site of Bedthi river. If variation is allowed between P_{tm} (P1) and P_{td} (P2), the storage capacity requirement is reduced. More of inflow during monsoon could be used directly without storing. By allowing P1 to P2 ratio as 3, the submergence area saved is about 69.97%, that is with P1 = P2, area required is about 95.03 sq.kms. This comes down to 28.59 sq.kms when P1 = 3*P2. Consequently, the dam height comes down to 87 m from 107 m, resulting in reduction of civil costs of the project.

Reduction in height of the dam from 107 to 77 m changes the submergence area from 95.05 to 16.27 sq.kms. Further reduction by 10 m (height of 67 m) would bring down the submergence area to 5.70 sq.kms. More hydroelectric energy could be harnessed by drawing the monsoon inflow on run-of-river basis, and storing sufficient quantity to meet the lean season (non-monsoon) requirement. The saved area would also help in meeting the thermal energy requirement of the region, like firewood, etc., and leaf manure for agriculture and horticulture lands. By allowing variation in P_{tm} and P_{td} ratio (resulting in reduction in dam height and submergence area), there is increase in electric energy generated. For example, at Magod for a dam height of 107 m (when P1 = P2) energy generated in a year is 746.66 million kWh, while for dam height of 77 m (when P1 = 3*P2) it is 810.82 million kWh and for height of 67 m, 823.55 million kWh. This is mainly due to lower submergence area for smaller heights of the dam, resulting in reduction in evaporation and seepage losses.

Net energy analyses carried out incorporating biomass energy lost in land submergence at Magod show that, savings in land submergence could be achieved and net energy at a site maximised if the hydroelectric generation capacity is adjusted according to the seasonal variations in the river's water discharge/run off. However, there are already two river projects in the region designed to supply power throughout the year that can cater the lean season needs (if there are any shortages).

Ebio is computed for various productivities such as 6.5, 13.5, 20 and 27.5 t/ha/yr (based on the earlier findings of net primary productivity in the region). Analyses carried out to maximise the net energy for different heights of the dam and for productivity of 6.5 t/ha/yr show that, net energy increases from 192.89 to 556.08 million kWh with reduction in dam height by 40 m.

The net energy gain is 63.9% for reduction of 37.3 % in dam height. That is, there is increase in net energy from 522.66 to 896.35 million kWh by hydro-thermal combined option in a land area of 95.03 sq.kms. If efficiency of thermal conversion is taken into account, then hydro-thermal combined scheme would generate 688.44 as against 522.66 million kWh (case of pure hydro scheme, with dam of 107 m height submerging 95.03 sq.kms.). Apart from the distinct reduction in submergence area, the over all reliability of hydro-thermal combined system is much higher than that of pure hydro systems. Pure hydro systems are also very sensitive to fluctuations in rainfall.

Fuel wood requirement in the region is about 312.05 million kWh for domestic purposes. Net energy computed for various dam heights show that, a dam of 67 m height would store enough water to meet the lean season electricity requirement, while area saved with this design (with natural vegetation, etc.) has bioresource potential of 319.85 million kWh. Which means, the thermal energy demand of 312.05 million kWh in the domestic sector can also be met by this design.

Analysis of cost per kWh for various designs of the dam for biomass productivity of 6.5 t/ha/yr, shows 40.5% reduction in cost (Rs. 1.53 to Rs. 0.92 per kWh) for dam height reduction of 32.75%. If the cost of bioresources is excluded considering electricity from hydro resource only, then cost would reduce from Rs. 1.53 to Rs. 1.09 per kWh.

The Benefit-Cost ratio computed, considering forest land value as Rs. 1,11,200 per hectare, is less than 1.5 for dam height of 107 m, while for heights 87, 77 and 67 metres, it is greater than 1.5. B-C ratio increases from 0.82 to 1.68 by reducing the submergence area from 95.03 sq.kms (107 m height) to 5.70 sq.kms (67 m height). Data from various forest documents (Timber, Minor Forest Produce and other products) and fuel wood depots at Kumta and Ankola show that value of products per hectare of forest land is about Rs. 12,000/year. Incorporating this value and land value of Rs. 1,11,200/ha., B-C ratio for heights 87, 77, and 67 m is found to be greater than 1.5.

By incorporating tangible and intangible environmental factors in Benefit-Cost analyses, B-C ratio is found to be less than 1.5 for all heights of the dam. Which means that, by assigning environmental values taking all components into consideration, the hydroelectric generation through storage (for minimum submergence also) becomes a less attractive option.

5.10.0 ACKNOWLEDGEMENTS

I am grateful to Prof. C. J. Saldanha, Dr. P.P. Mujumdar, Dr. Dilip Ahuja and Mr. K.M. Hegde for useful suggestions. I thank Dr. R. Prabhakar for providing useful tips on cartography. I thank the villagers for their kind help during my field work. I thank Mr. Narendra Hegde, Mr. Gopalkrishna and Mr. Sridar Hegde for their help during the field work and collection of socio-economic data in submergence area of Aghnashini. My sincere thanks to Mr. P.R. Bhat, Mr. Deepak and Mr. Rosario for coordinating the monthly discussion meetings with local people in Siddapur taluk. I thank the authorities at Survey of India for providing toposheets required for this work. Techno-economic data required for this analysis was provided by M/s Kirloskar Electricals Ltd., M/s BHEL, M/s Jyothi Limited, Public Works Department (Sirsi, Kumta and Siddapur taluks) and Karnataka Electricity Board (Bangalore, all divisions in Uttara Kannada district). I thank all these organisations for extending their help.

I thank the District Statistical Officer, Karwar and officials of the Block Development Office, Forest Department, Agriculture Department, Revenue Department and Uttara Kannada Jilla Parishad (at Karwar) for providing statistical information. I am also grateful to Dr. Metzler (editor, Hydronet), Electricite de France (National Hydro Engineering Centre), Indian National Trust for Art and Cultural Heritage and International River Network for providing the relevant literatures.

5.11.0 References

1. Falkenmark, M., 1989. *Hydrological Phenomena in Geosphere-Biosphere Interactions - Outlooks to Past, Present and Future*. Monographs and Reports No 1, IAHS, Wallingford, U.K.
2. Creager, W.P. and Justin, J.D., 1950. *Hydroelectric Handbook*. John Wiley and Sons, New York.
3. UNIDO, 1986. *Small Hydropower Series -No 1 (Mini Hydropower Stations), No. 3 (Chinese Experiences in Mini Hydropower)*. United Nations Industrial Development Organisation, Vienna.
4. ESCAP, 1982. *Renewable Sources of Energy. Volume IV - Small Hydropower Development*. Economic and Social Commission for Asia and the Pacific, United Nations Industrial Development Organisation, Vienna.
5. KPCL, 1988. *Micro/Mini Hydel Stations in Karnataka State - A Report*. Karnataka Power Corporation Limited, Office of the Chief Engineer Investigation, Bangalore, India.
6. MPCL, 1977. *Gangavali Stage I (Bedthi) Hydroelectric Project. Detailed Project Report, Volume I*, Mysore Power Corporation Ltd., Bangalore, India.
7. KPCL, 1973. *Bedthi Hydroelectric Project - Supplement to March 1969 Report: Modified Estimates and Financial Aspects*, Karnataka Power Corporation Limited, Government of Karnataka, Bangalore, India.
8. Sharma, L.T. and Sharma, R., 1981. *Major Dams-A Second Look*, Gandhi Peace Foundation, New Delhi, India.
9. Subramanian, D.K., 1981. *Alternative Energy Planning Strategies in Major Hydroelectric Projects*. Paper presented in the seminar - Implications of Hydroelectric Projects for Uttara Kannada, January 17 - 19, Sirsi, Karnataka, India.
10. Lele, S.M., 1986. *An Optimisation Model for Macro Hydel Projects Incorporating Energy / Economic Cost of Land Submergence*. M.Sc. Thesis, I.I.Sc, Bangalore, India.
11. KPCL, 1993. *Draft Policy Paper on Implementation of Power Projects by KPCL and Private Sector*. Karnataka Power Corporation Ltd., Bangalore, India.
12. Chelvaraj, A.S., 1994. *Bedthi Hydroelectric Project*. Proposal submitted to The Deputy Chief Minister, Govt. of Karnataka, Bangalore, India.
13. Mutreja, K.M., 1976. *Reservoir Capacity for Periodic-Stochastic Input and Periodic Output*. Hydrology Paper No. 86, Colorado State University, Fort Collins.
14. Subramanian, D.K., 1985. *An Optimal Energy and Water Model for Design and Analysis of Water Resources of a River with Constraints on Energy and Ecology*. Technical Report No 8, Centre for Ecological Sciences, Indian Institute of Science, Bangalore, India.
15. Maass, A., Hufschmidt, M., Dorfman, R., Thomas, H.A., Marglin, S.A. and Fair, G.M., 1962. *Design of Water Resources Systems - New Techniques for Relating Economic Objectives*. Engineering Analysis and Government Planning. Harvard University Press.
16. Kamath, S.U., 1985. *Karnataka State Gazetteer: Uttara Kannada District*, Government of Karnataka, Bangalore, India.
17. Campbell, J.M., 1883. *Gazetteer of the Bombay Presidency, Volume XV, Part I-Kanara*, Government Central Press, Bombay, India.
18. op. cit. Ref. 6.
19. op. cit. Ref. 14.
20. Maudgal, S. and Kakkar, M., 1992. Evaluation of forests for impact assessment of development project. In: *The Price of Forests*, pp.56-60, Agarwal, A. (Ed.), Centre for Science and Environment, New Delhi, India.

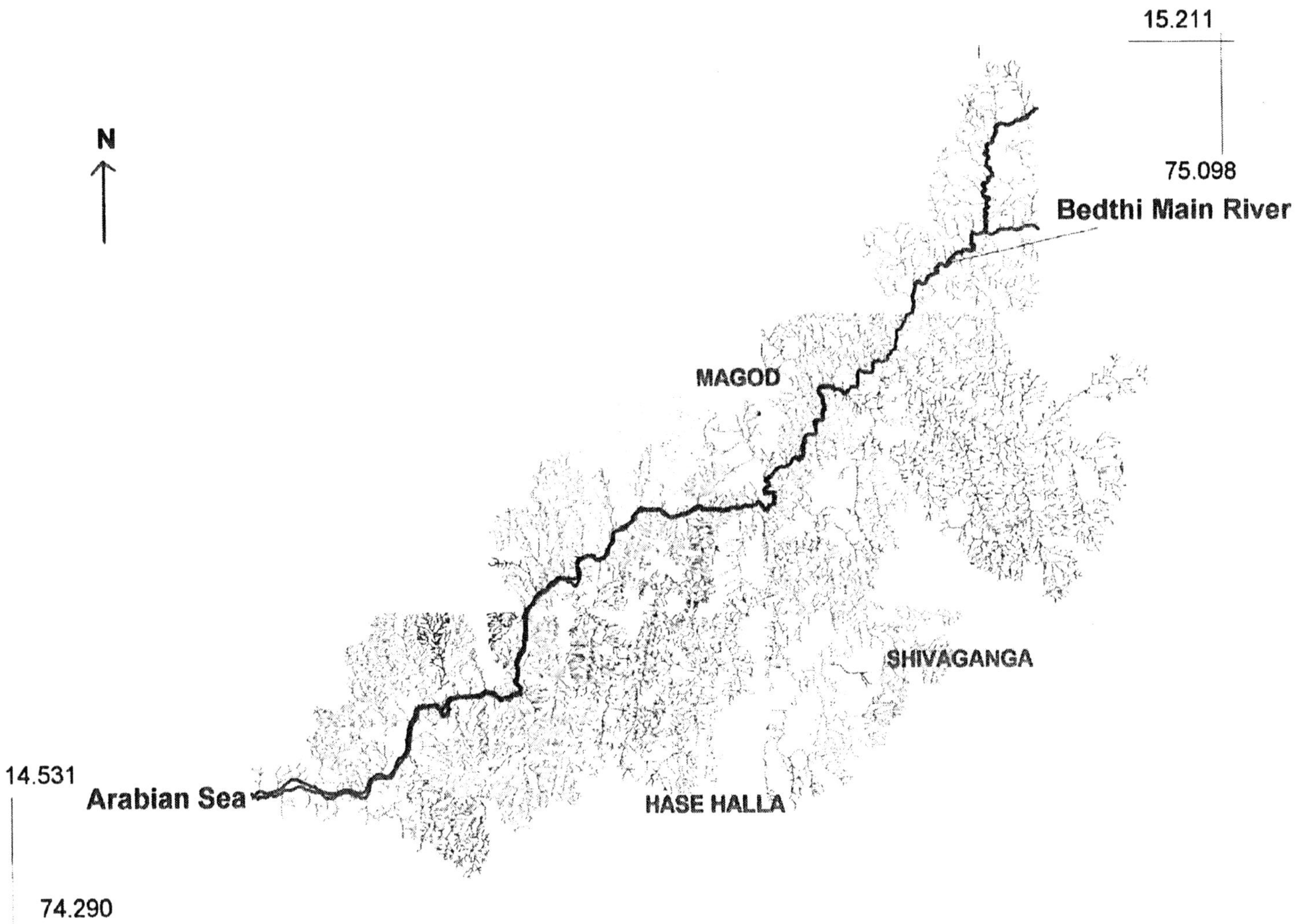

Plate 5.1.0: Tributaries and Streams in Bedthi River Basin
(see accompanying CD for color representation)

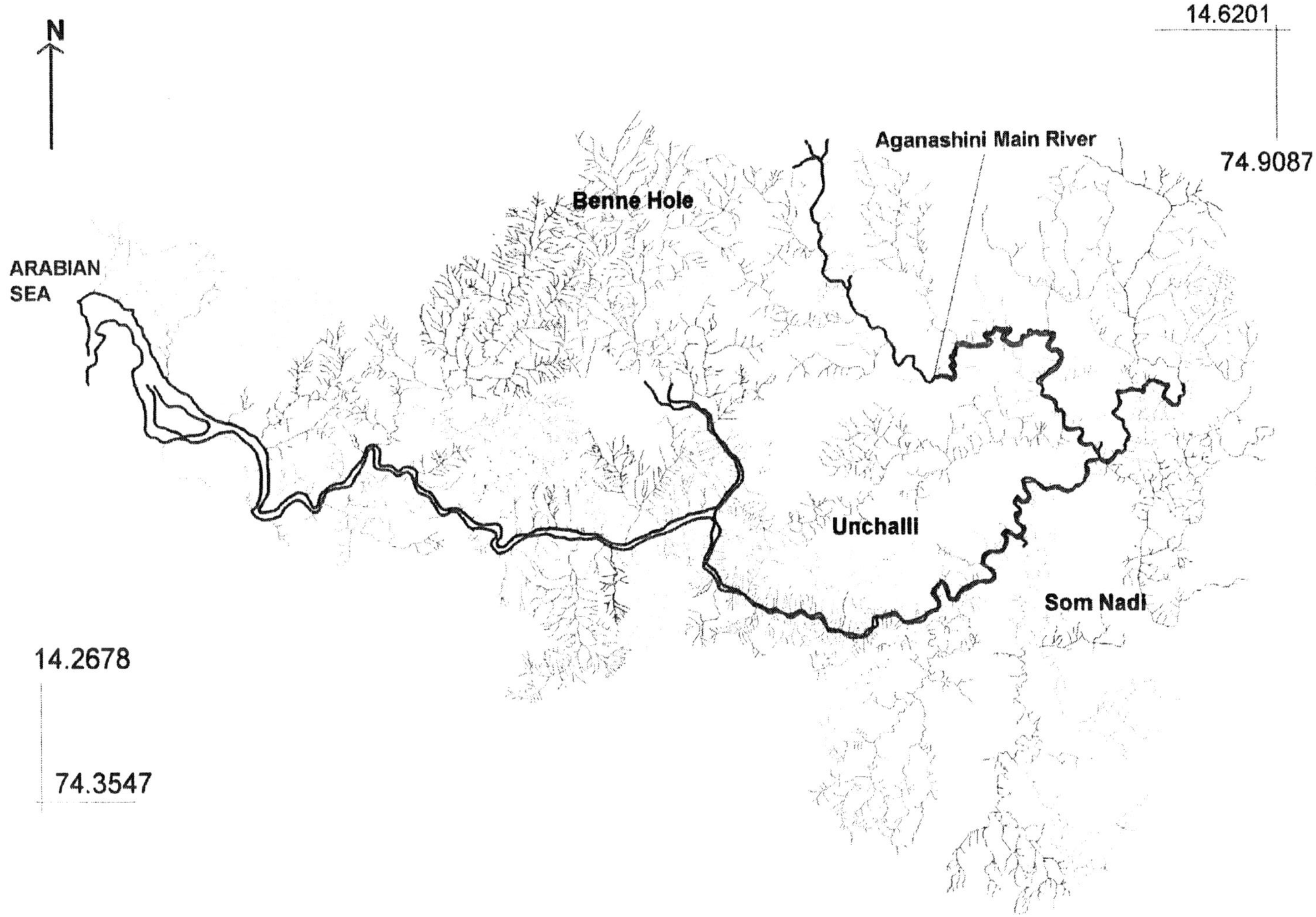

Plate 5.2.0: Tributaries and Streams of Aganashini River Basin (see accompanying CD for color representation)

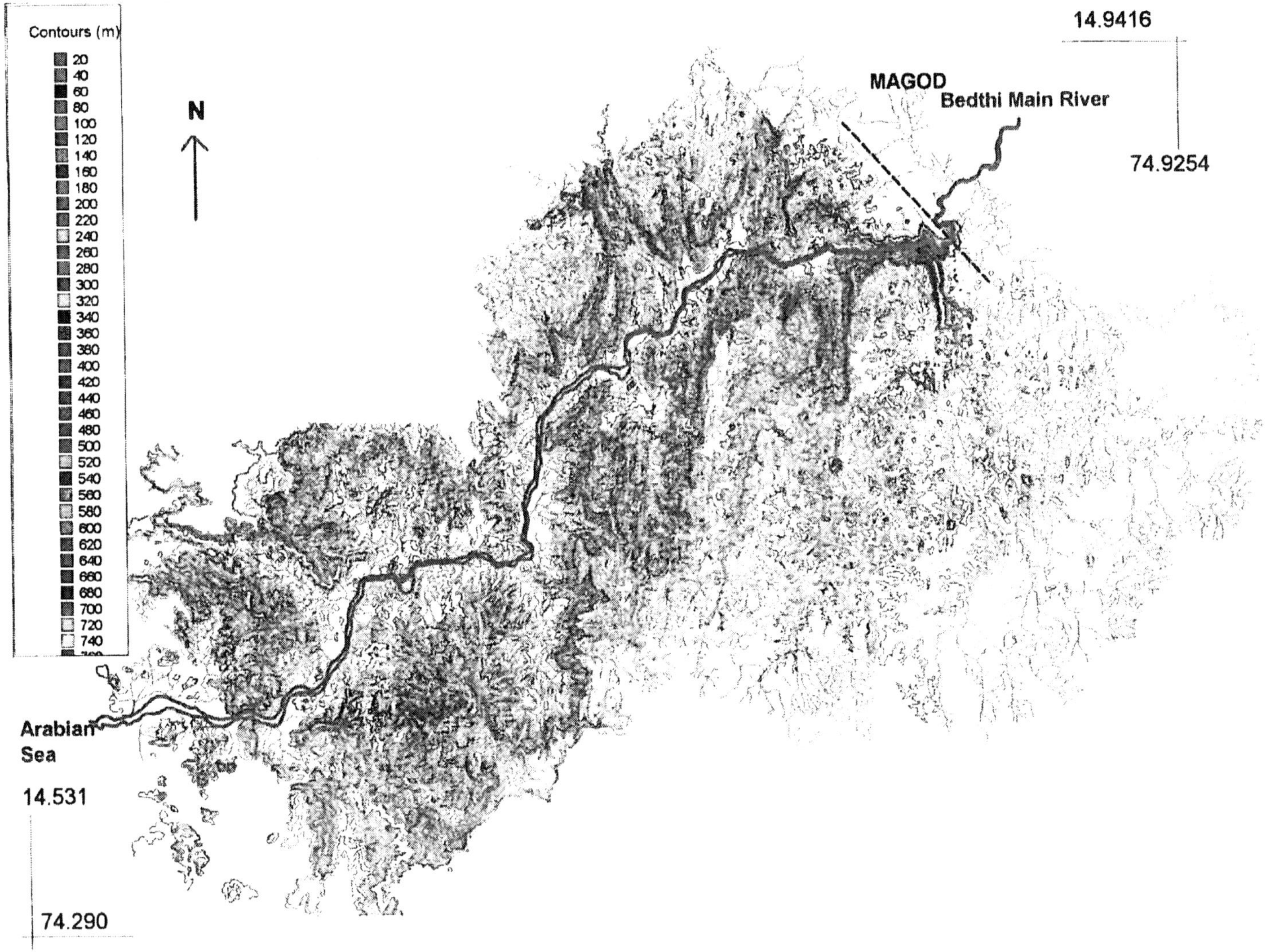

Plate 5.3.0: Contours of Bedthi River (Downstream of Magod Falls) (see accompanying CD for color representation)

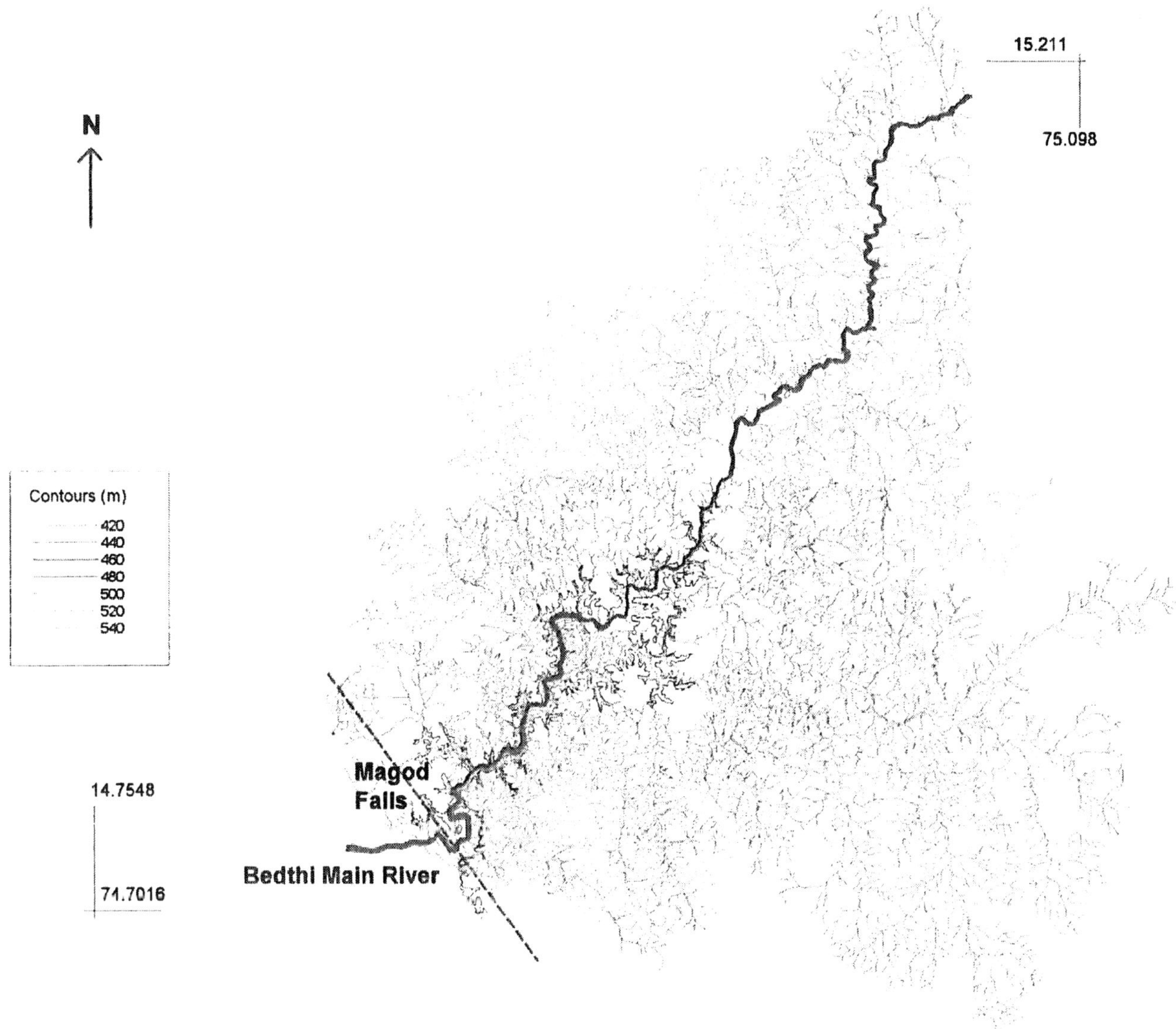

Plate 5.4.0: Contours of Bedthi River (Upstream of Magod Falls) (see accompanying CD for color representation)

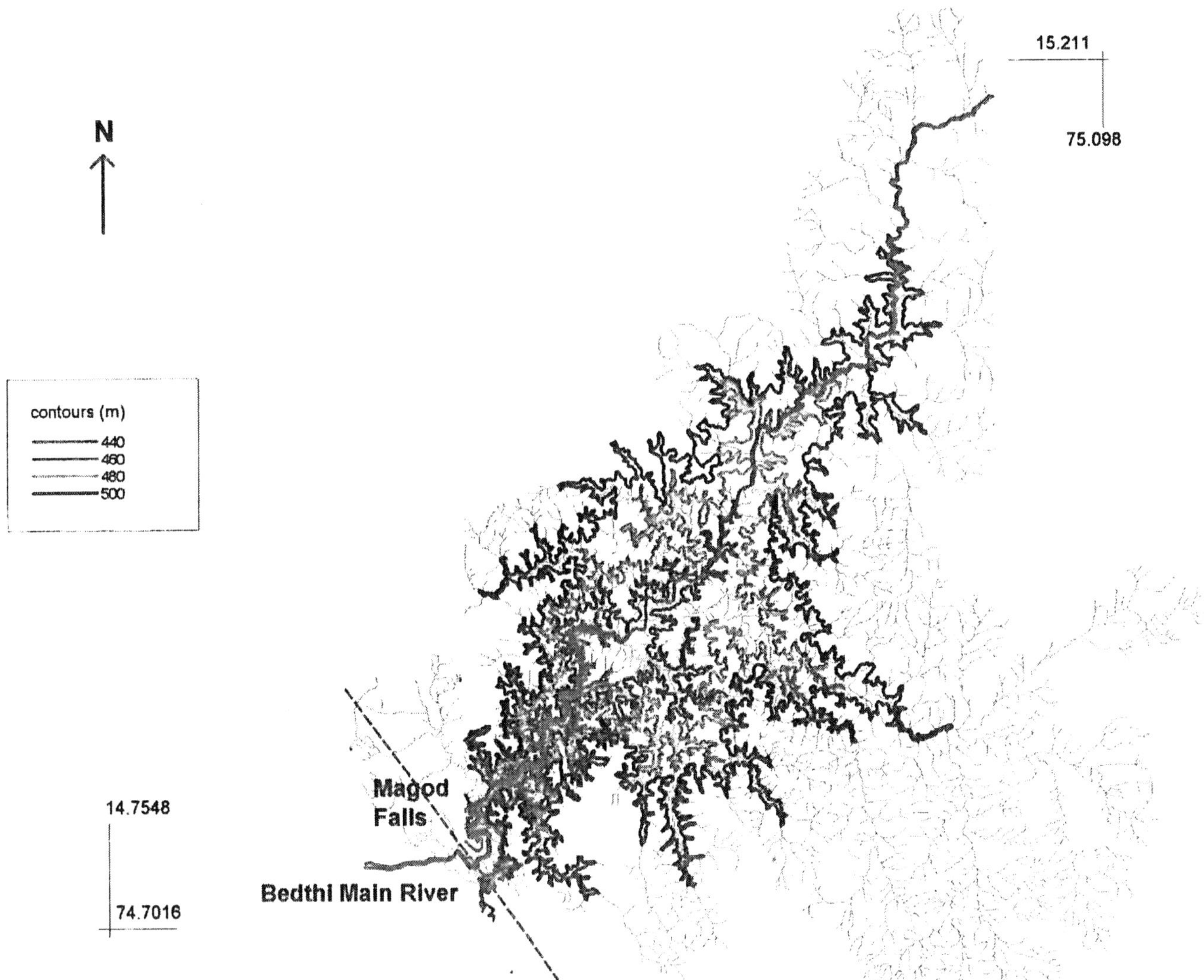

Plate 5.5.0: Various Contours (at 440, 460, 480, 500m) at Magod, Proposed Dam site (see accompanying CD for color representation)

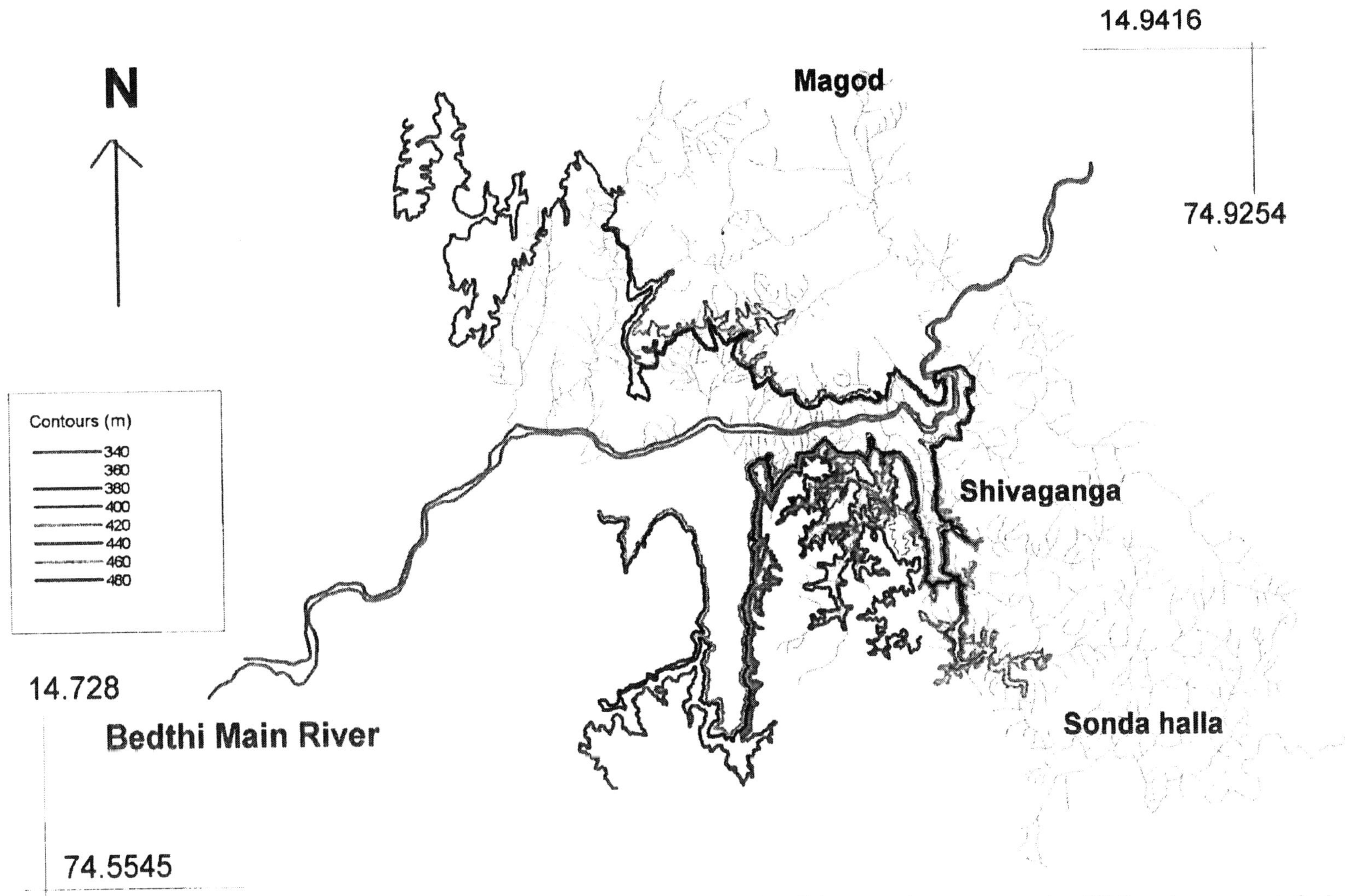

Plate 5.6.0: Various Contours (at 340, 360, 380, 400, 420, 440, 460 and 480m) at Shivaganga (see accompanying CD for color representation)

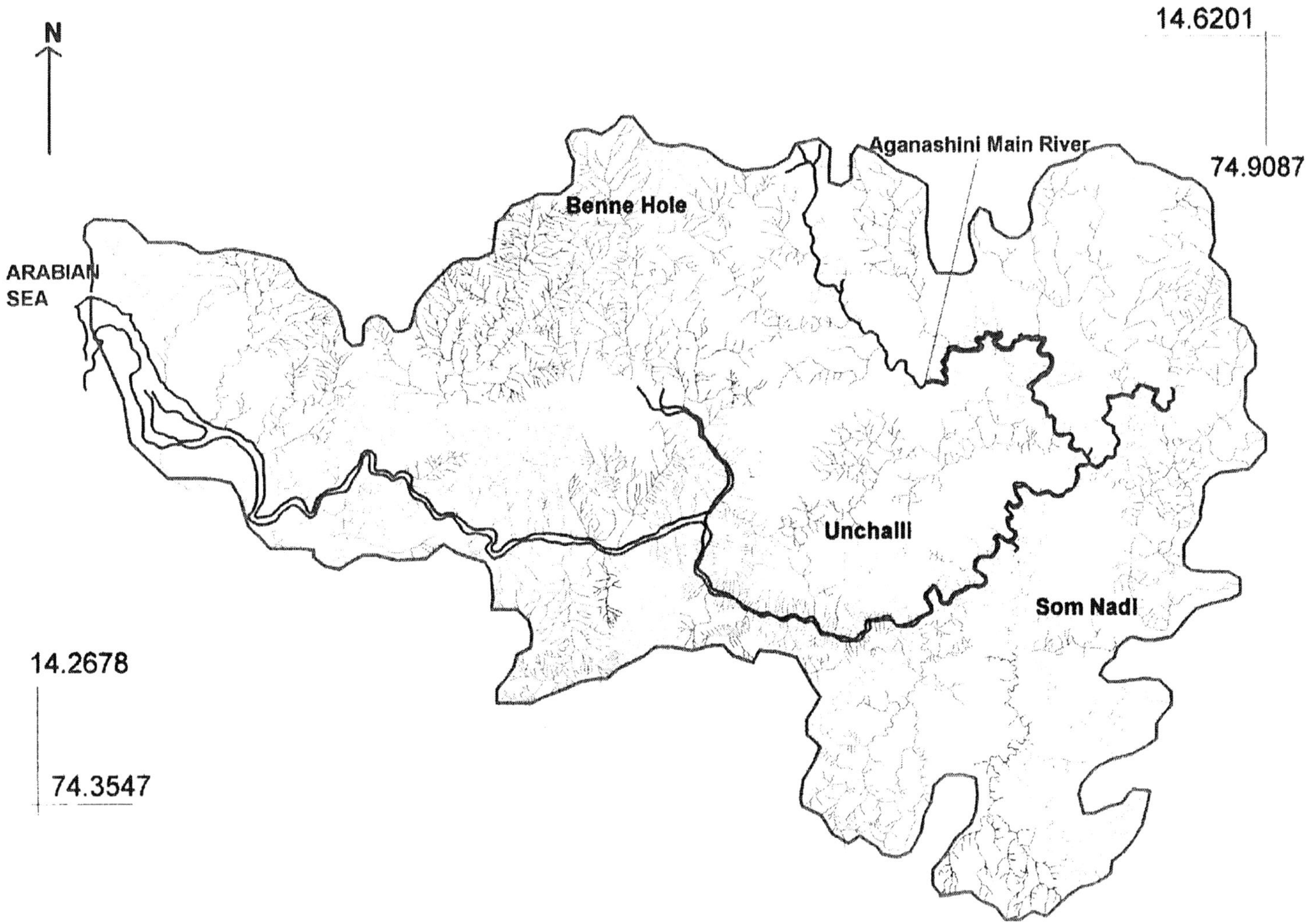

Plate 5.7.0: Catchment Area, Tributaries and Streams in Aganashini River Basin (see accompanying CD for color representation)

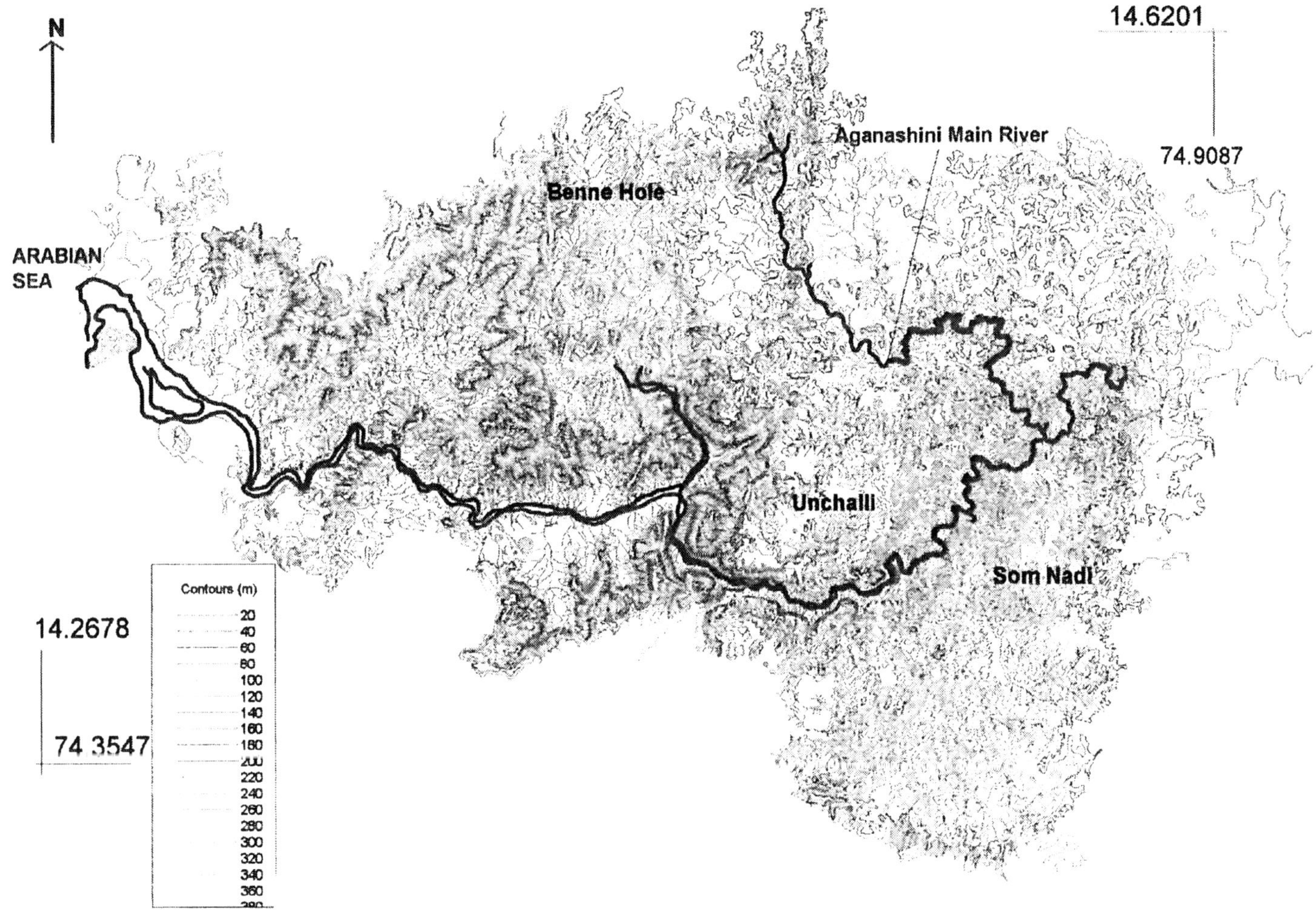

Plate 5.8.0: Contours of Aganashini River (see accompanying CD for color representation)

Chapter 6

SOLAR ENERGY POTENTIAL AND PROSPECTS

6.0 SUMMARY

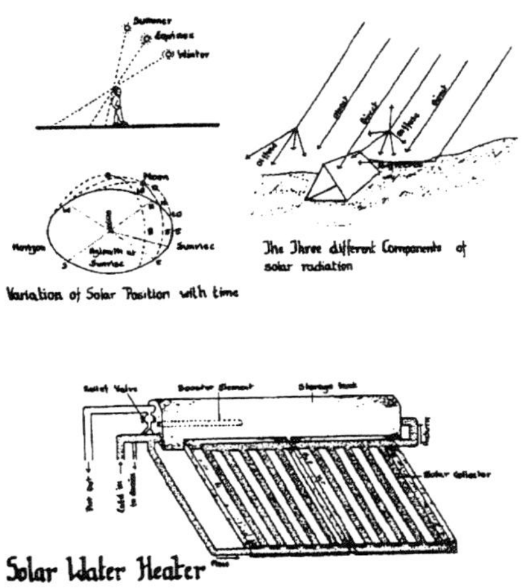

Solar energy can meet a large variety of small-scale decentralised energy needs of a region. The design of solar devices depends on global and diffuse solar radiation data. Estimation of solar radiation for Uttara Kannada district is done on the basis of solar and other climatological data available at stations located in Karwar, Honnavar, Shirali, Mangalore and Goa. A variety of empirical equations for estimating monthly global solar radiation exists in the solar energy literature. The most commonly used empirical formula is the one that relates sunshine duration and global radiation (GR), also referred to as Angstrom's equation. This relationship is used to estimate global radiation using 20 years data at Mangalore and 25 years data at Goa provided by the India Meteorological Department. It is seen that computed and measured values (actual values) of global radiation agree within the range of 2-5% for most months. Data at Goa and Mangalore have been analysed in order to improve the accuracy and establish the role of other climatological parameters, such as, mean daily temperature, relative humidity, specific humidity, minimum and maximum temperature, and rainfall. Based on R^2 value and least value of the standard error of Y-estimate, the empirical formula consisting of mean temperature, sunshine duration and specific humidity is the best relationship to estimate GR with the available climatological information. The computed and estimated values are within the range of $\pm 5\%$. With this empirical relationship, GR (kWh/m^2) is estimated using climatological data of many years for Karwar (37 years), Honnavar (50 years) and Bhatkal (Shirali, 15 years). Karwar has a GR range of 5.5-6.5 for January-May and 4-5 during the monsoon months July-September, while at Honnavar, the GR during January - May is 5.47-6.5 and is minimum during the monsoon months. Thus, this study covering five locations in the coastal belt of Western Ghats, demonstrates that good solar energy potential is available in this region during most months of the year. The amount of solar energy that could be harnessed by utilising 5% of the present wasteland in coastal taluks as a solar collector area, is found to be around 95.72 million units (mkWh) annually. Which means, solar energy can meet at least 32.55% of the present electricity demand of Uttara Kannada from the coastal belt alone. This study suggests that solar conversion technologies have considerable potential for application, provided questions of storage and monsoon season requirement are adequately addressed through other options. The abundance of the solar resource can be illustrated by comparing the land requirements of solar projects (thermal or PV) with those of hydro or energy plantation projects. Except for run-of-river projects and high head sites in deep gorges, the land requirements for hydel projects at today's conversion efficiencies, average around 25 to 45 times. This means that solar energy is capable of supplying 5 to 10 times the electricity demand of Karnataka state, while occupying land area less than that currently used by hydel projects.

Keywords: Solar Energy, Radiation Climatology, Global Solar Radiation, Direct solar radiation, Diffuse solar radiation, Techno-Economic Analyses, Solar Thermal Devices, Photovoltaic Devices, Solar Water Heating.

6.1.0 INTRODUCTION

The sun is the only significant source of energy to the earth. Received entirely as radiation, this energy is subsequently converted, so that at any given time and place, an original quantum may appear in any one of the several diverse forms. Thus, with respect to the energy encompassed in the day to day climate at a given place and time, solar energy may appear as the latent heat of vapor, the advected heat of air masses, radiant energy from sky and terrestrial objects, and unconverted solar radiation arriving as direct beam. The unconverted solar radiation in the form of direct beams is a very important form that is dominant in the hydrologic processes of evaporation, transpiration and snow ablation, as well as, in the growth and succession of vegetation. The total flux of energy to a site at any given time is difficult to measure, and the flux of radiant energy alone can be measured readily at a point. However, this flux varies so widely over most land surfaces that point measurements are of limited value. The spatial heterogeneity in the radiant flux is due to the variation in solar beam irradiation caused by variation in the inclination of the earth's surface with respect to beam direction. Within a geographic region subject to uniform atmospheric conditions, irradiation from the sky is fairly uniform, regardless of surface inclination. However, the total sun and sky irradiation varies widely with surface orientation and slope. The variation of direct beam radiation varies in time with atmospheric condition and path length, the combined effect of which can be estimated.

Systematic long-term measurements of solar and terrestrial radiation in India using calibrated and maintained instruments as per international standards was started in the year 1957-58. The radiation laboratory of the India Meteorological Department, Pune, serves as the radiation centre for India and monitors 25 specifically selected sites, which form the National Radiation Network. Of these 25 stations, 13 (principal stations) are equipped to monitor continuous records of global and diffuse solar radiation, and measure direct solar radiation at regular intervals in addition to recording sunshine. In the remaining 12 stations (ordinary stations), a minimum program is carried out of continuous recording of only global solar radiation in addition to recording sunshine. Apart from these parameters, all stations maintain continuous records of other meteorological parameters such as temperature, relative humidity, rainfall, wind speed and wind direction.

Goa and Mangalore represent a principal (records global radiation, diffuse radiation, direct solar radiation, sunshine and other climatological parameters) and an ordinary station (records global radiation, sunshine and other climatological parameters), respectively. Meteorological observatories at Karwar, Honnavar and Shirali record only climatological parameters. An empirical method is adopted to assess the solar potential in the coastal belt of Uttara Kannada. With this empirical relationship (based on radiation and climatological data at Goa and Mangalore), global radiation (GR in kWh/m^2) is estimated for Karwar (37 years of climatological data), Honnavar (50 years of climatological data) and Bhatkal (Shirali, with 15 years of climatological data).

6.2.0 OBJECTIVE

The objective is to assess the solar energy potential in Uttara Kannada district. Since Uttara Kannada does not have a radiation station, solar climatological data collected at radiation stations set up by the India Meteorological Department closer to Uttara Kannada district, namely, Goa (15° 29' N latitude, 73° 49' E longitude, elevation 55 m) and Mangalore (12° 55' N latitude, 74° 53' E longitude, elevation 102 m) were observed

6.3.0 METHODOLOGY

Radiation and climatological data for radiation stations at Goa and Mangalore and climatological data of the meteorological observatories at Karwar, Honnavar and Shirali were obtained from the India Meteorological Department at Pune and Bangalore and detailed investigations were carried out as discussed below.

6.4.0 ANALYSES, RESULTS AND DISCUSSION

6.4.1 Factors Influencing Radiation Climatology

The quantity of solar radiation reaching the earth's surface during the day is known to be influenced by (a) solar elevation (at noon), (b) duration of the day as determined by astronomical and geographic factors, (c) turbidity of the air, (d) total amount of water vapour in the air, and (e) type and amount of clouds. Of all these factors, solar elevation and duration of the day are important. Turbidity and amount of clouds cause relatively small variation in the quantity of solar radiation. Altitude also influences global, direct and diffuse solar radiation (the explanation for these terminologies and abbreviations used by climatologists are listed in Appendix) received on the ground. The solar radiation climatology of any region is determined by all six of these factors, particularly, the cloud type and amount. The gross solar radiation climatology follows closely the pattern of climatology of low clouds. Based on long-term data collected since 1963 in radiation stations at Goa and Mangalore, mean values of hourly global, direct, and diffuse solar radiation are computed. Table 6.1.1 lists monthly mean hourly global radiation at Goa, while, Table 6.1.2 and Table 6.1.3 list the highest value recorded during the hour, and the standard deviation of hourly value respectively. It is seen that the hourly global radiation is minimum during June-August at Goa. Higher values of standard deviation for May-August are due to large day-to-day variability (usually caused by variable cloudiness). Table 6.1.4 lists monthly mean solar elevation (in degrees) at Goa. Higher values are seen during the May-August period. Monthly mean hourly values of direct solar radiation (I_N) are given in Table 6.1.5 and monthly hourly relative air mass is given in Tables 6.1.6. Tables 6.1.7, 6.1.8 and 6.1.9 list monthly hourly maximum, average and standard deviation values of diffuse solar radiation respectively.

The mean values of direct solar radiation were measured at 0830, 1130, 1430 and 1730 I.S.T. (Indian Standard Time) with an Angstrom pyreheliometer. The pyreheliometer observations are recorded only when the sun's disc and surrounding sky are free of clouds.

Monthly hourly maximum, mean, and standard deviation values of global radiation for Mangalore are listed in Table 6.2.1, 6.2.2 and 6.2.3 respectively. Total global radiation ranges from 5.17 (November) to 6.42 (April), while during the monsoon months, it ranges from 3.032 (July) to 4.93 (October). Higher values of standard deviation for May-August are due to large day-to-day variability (usually caused by variable cloudiness). Monthly elevation of direct solar radiation for Mangalore is given in Table 6.2.4. Monthly variations in mean hourly global radiation at Goa and Mangalore are illustrated in Figure 6.1.1 and Figure 6.1.2 respectively.

Table 6.1.1: Monthwise Mean Hourly Global Solar Radiation (kWh/sq.m.) at Goa

Month	Hours (local apparent time) ending at														Total
	6	7	8	9	10	11	12	13	14	15	16	17	18	19	
January	0.00	0.03	0.21	0.43	0.62	0.75	0.82	0.81	0.74	0.61	0.42	0.21	0.03	0.00	5.69
February	0.00	0.05	0.25	0.48	0.68	0.82	0.90	0.90	0.83	0.68	0.48	0.25	0.05	0.00	6.38
March	0.00	0.07	0.28	0.50	0.69	0.85	0.94	0.95	0.87	0.72	0.52	0.28	0.07	0.00	6.76
April	0.00	0.10	0.30	0.49	0.68	0.86	0.95	0.96	0.88	0.74	0.54	0.31	0.10	0.00	6.93
May	0.01	0.12	0.30	0.47	0.65	0.81	0.90	0.92	0.85	0.72	0.53	0.33	0.13	0.01	6.74
June	0.01	0.09	0.22	0.35	0.48	0.57	0.64	0.64	0.59	0.50	0.37	0.22	0.09	0.01	4.79
July	0.01	0.07	0.18	0.31	0.42	0.50	0.54	0.54	0.48	0.40	0.29	0.17	0.07	0.01	3.99
August	0.00	0.07	0.21	0.35	0.49	0.61	0.67	0.66	0.61	0.51	0.37	0.21	0.07	0.00	4.76
September	0.00	0.06	0.22	0.40	0.53	0.66	0.76	0.77	0.71	0.59	0.42	0.24	0.07	0.00	5.31
October	0.00	0.05	0.21	0.42	0.59	0.71	0.80	0.82	0.75	0.62	0.43	0.22	0.05	0.00	5.61
November	0.00	0.04	0.21	0.42	0.61	0.73	0.79	0.79	0.72	0.59	0.42	0.21	0.03	0.00	5.62
December	0.00	0.03	0.19	0.41	0.59	0.72	0.78	0.77	0.70	0.57	0.40	0.19	0.03	0.00	5.39

Table 6.1.2: Monthwise Max. Value - (highest value recorded during the hour) of Global Radiation at Goa

Month	6	7	8	9	10	11	12	13	14	15	16	17	18	19	Total
January	0.00	0.07	0.30	0.64	0.72	0.86	0.94	0.94	0.87	0.72	0.53	0.30	0.08	0.00	6.67
February	0.00	0.09	0.33	0.60	0.78	0.95	1.03	1.04	0.96	0.81	0.59	0.36	0.10	0.00	7.61
March	0.01	0.13	0.37	0.62	0.82	0.98	1.06	1.06	0.99	0.84	0.64	0.40	0.16	0.01	7.75
April	0.02	0.18	0.45	0.65	0.89	1.00	1.08	1.12	0.99	0.84	0.74	0.40	0.20	0.03	8.13
May	0.03	0.23	0.49	0.68	0.87	1.09	1.09	1.10	1.07	0.93	0.70	0.48	0.24	0.05	8.56
June	0.03	0.23	0.45	0.69	0.85	0.99	1.13	1.12	1.00	0.90	0.67	0.50	0.22	0.10	7.81
July	0.03	0.19	0.42	0.62	0.84	0.97	1.04	1.07	9.65	0.83	0.69	0.45	0.17	0.03	7.50
August	0.02	0.16	0.42	0.65	0.84	0.99	1.10	1.07	1.15	0.89	0.72	0.43	0.18	0.02	7.42
September	0.01	0.14	0.39	0.65	0.91	1.02	1.10	1.09	1.02	0.87	0.71	0.45	0.22	0.02	7.73
October	0.00	0.11	0.37	0.62	0.80	0.96	1.04	1.08	0.97	0.87	0.66	0.45	0.16	0.03	7.13
November	0.00	0.09	0.42	0.58	0.79	0.89	0.96	1.15	0.94	0.78	0.62	0.40	0.10	0.00	6.75
December	0.00	0.09	0.32	0.56	0.78	0.98	1.08	1.11	0.80	0.73	0.54	0.35	0.14	0.00	6.30

Table 6.1.3: Monthwise Standard Deviation of the hourly values of Global Radiation at Goa

Month	6	7	8	9	10	11	12	13	14	15	16	17	18	19	Total
January	0.00	0.02	0.05	0.08	0.10	0.09	0.07	0.06	0.06	0.05	0.05	0.04	0.02	0.00	0.51
February	0.00	0.01	0.03	0.05	0.06	0.07	0.07	0.07	0.07	0.06	0.05	0.03	0.01	0.00	0.45
March	0.00	0.01	0.03	0.05	0.06	0.06	0.06	0.06	0.05	0.05	0.04	0.03	0.01	0.00	0.43
April	0.00	0.03	0.07	0.10	0.12	0.10	0.10	0.09	0.08	0.07	0.06	0.49	0.03	0.00	0.64
May	0.01	0.04	0.09	0.12	0.15	0.17	0.17	0.17	0.15	0.14	0.12	0.08	0.04	0.01	1.11
June	0.01	0.05	0.11	0.17	0.22	0.26	0.29	0.29	0.27	0.24	0.18	0.12	0.05	0.01	1.89
July	0.01	0.03	0.09	0.15	0.20	0.24	0.27	0.27	0.24	0.21	0.16	0.10	0.04	0.00	1.60
August	0.00	0.03	0.09	0.14	0.19	0.23	0.24	0.24	0.24	0.19	0.15	0.09	0.03	0.00	1.47
September	0.00	0.03	0.09	0.14	0.18	0.22	0.25	0.25	0.23	0.20	0.16	0.10	0.04	0.00	1.58
October	0.00	0.02	0.07	0.11	0.15	0.17	0.17	0.17	0.16	0.15	0.13	0.08	0.03	0.00	1.15
November	0.00	0.01	0.05	0.08	0.11	0.12	0.13	0.12	0.12	0.11	0.08	0.05	0.01	0.00	0.78
December	0.00	0.01	0.04	0.06	0.07	0.08	0.09	0.09	0.09	0.08	0.06	0.04	0.01	0.00	0.53

Table 6.1.4: Monthwise Mean Solar Elevation (Degree) at Goa

Month	7	8	9	10	11	12	13	14	15	16	17	18
January	4.00	14.50	26.90	38.30	47.50	52.90	52.90	47.50	38.30	26.90	14.50	4.00
February	5.40	17.60	31.00	43.50	54.20	61.00	61.00	54.20	43.50	31.00	17.60	5.40
March	6.90	21.10	35.30	49.00	61.70	71.00	71.00	61.70	49.00	35.30	21.10	6.90
April	9.70	24.10	38.50	53.00	67.30	80.60	80.60	67.30	53.00	38.50	24.10	9.70
May	11.80	25.80	39.90	54.10	68.30	82.10	82.10	68.30	54.10	39.90	25.80	11.80
June	12.70	26.30	40.10	53.90	67.50	79.70	79.70	67.50	53.90	40.10	26.30	12.70
July	12.40	26.10	40.10	54.00	67.90	80.80	80.80	67.90	54.00	40.10	26.10	12.40
August	10.70	25.00	39.30	53.70	68.20	82.60	82.60	68.20	53.70	39.30	25.00	10.70
September	8.10	22.50	36.90	51.10	64.70	75.70	75.70	64.70	51.10	36.90	22.50	8.10
October	6.00	19.10	32.80	45.90	57.40	65.10	65.10	57.40	45.90	32.80	19.10	6.00
November	4.50	15.50	28.30	40.10	49.80	55.60	55.60	49.80	40.10	28.30	15.50	4.50
December	3.60	13.70	25.90	36.90	45.80	51.00	51.00	45.80	36.90	25.90	13.70	3.60

Table 6.1.5: Monthwise Mean Values of Direct Solar Radiation - (kWh/sq.m.) at Goa

Month	Hours (local apparent time) ending at											
	7	8	9	10	11	12	13	14	15	16	17	18
January	0.16	0.54	0.73	0.81	0.84	0.86	0.84	0.83	0.79	0.71	0.53	0.17
February	0.19	0.55	0.72	0.79	0.83	0.85	0.86	0.84	0.79	0.71	0.54	0.19
March	0.20	0.43	0.56	0.62	0.71	0.77	0.78	0.77	0.72	0.62	0.45	0.21
April	0.14	0.34	0.43	0.53	0.64	0.70	0.72	0.70	0.65	0.56	0.41	0.21
May	0.14	0.28	0.34	0.44	0.55	0.60	0.63	0.62	0.57	0.48	0.38	0.21
June	0.04	0.11	0.15	0.20	0.23	0.26	0.26	0.25	0.24	0.20	0.15	0.07
July	0.05	0.07	0.11	0.13	0.14	0.15	0.15	0.14	0.12	0.10	0.06	0.04
August	0.02	0.09	0.14	0.18	0.23	0.24	0.23	0.22	0.20	0.17	0.11	0.04
September	0.07	0.16	0.25	0.29	0.33	0.40	0.42	0.42	0.38	0.32	0.23	0.13
October	0.10	0.28	0.43	0.50	0.54	0.59	0.62	0.62	0.58	0.48	0.33	0.15
November	0.17	0.47	0.64	0.71	0.73	0.74	0.74	0.72	0.67	0.60	0.45	0.17
December	0.17	0.52	0.71	0.80	0.83	0.83	0.82	0.79	0.74	0.66	0.50	0.16

Table 6.1.6: Monthwise Relative Airmass - (The Total Mass of the Atmosphere Traversed by a Beam of Solar Radiation) at Goa

Month	7	8	9	10	11	12	13	14	15	16	17	18
January	9.74	3.28	1.94	1.45	1.23	1.14	1.14	1.23	1.45	1.94	3.28	9.74
February	12.44	3.94	2.20	1.61	1.35	1.25	1.25	1.35	1.61	2.20	3.94	12.44
March	7.87	2.76	1.73	1.32	1.14	1.06	1.06	1.14	1.32	1.73	2.76	7.87
April	5.76	2.44	1.60	1.25	1.08	1.01	1.01	1.08	1.25	1.60	2.44	5.76
May	4.79	2.29	1.56	1.23	1.08	1.01	1.01	1.08	1.23	1.56	2.29	4.79
June	4.47	2.25	1.55	1.23	1.08	1.01	1.01	1.08	2.25	1.55	2.25	4.47
July	4.57	2.27	1.55	1.23	1.08	1.01	1.01	1.08	1.23	1.55	2.27	4.57
August	5.26	2.36	1.57	1.24	1.08	1.01	1.01	1.08	1.24	1.57	2.36	5.26
September	6.81	2.60	1.65	1.28	1.11	1.03	1.03	1.11	1.28	1.65	2.60	6.81
October	8.90	3.03	1.84	1.40	1.19	1.10	1.10	1.19	1.40	1.84	3.03	8.90
November	11.33	3.70	2.10	1.55	1.31	1.21	1.21	1.31	1.55	2.10	3.70	11.33
December	13.48	4.16	2.28	1.65	1.39	1.29	1.29	1.39	1.65	2.28	4.16	13.48

Table 6.1.7: Monthwise Highest Value - (recorded during the hour) of Diffuse Solar Radiation at Goa

Month	6	7	8	9	10	11	12	13	14	15	16	17	18	19	Total
January	0.00	0.04	0.18	0.28	0.37	0.47	0.50	0.53	0.49	0.38	0.29	0.16	0.04	0.00	3.16
February	0.00	0.07	0.21	0.30	0.39	0.52	0.57	0.52	0.52	0.42	0.28	0.18	0.06	0.00	3.70
March	0.01	0.10	0.26	0.33	0.59	0.51	0.52	0.49	0.46	0.36	0.31	0.20	0.09	0.01	3.68
April	0.02	0.14	0.33	0.40	0.47	0.55	0.57	0.57	0.54	0.47	0.41	0.28	0.17	0.03	4.04
May	0.03	0.16	0.34	0.41	0.56	0.76	0.82	0.79	0.68	0.62	0.50	0.39	0.17	0.03	5.38
June	0.03	0.20	0.36	0.53	0.67	0.81	0.83	0.83	0.78	0.55	0.47	0.30	0.24	0.04	5.21
July	0.02	0.14	0.34	0.45	0.58	0.75	0.70	0.70	0.62	0.58	0.44	0.31	0.16	0.02	4.70
August	0.01	0.14	0.36	0.54	0.83	0.91	0.77	0.91	0.92	0.58	0.56	0.32	0.13	0.01	5.80
September	0.01	0.14	0.30	0.46	0.64	0.66	0.81	0.76	0.76	0.68	0.59	0.36	0.14	0.02	4.60
October	0.00	0.10	0.26	0.35	0.48	0.73	0.90	0.88	0.61	0.69	0.43	0.29	0.15	0.03	5.12
November	0.00	0.07	0.19	0.31	0.49	0.55	0.73	0.68	0.55	0.43	0.31	0.23	0.10	0.11	3.52
December	0.00	0.08	0.14	0.26	0.39	0.46	0.48	0.54	0.42	0.39	0.29	0.16	0.11	0.00	3.00

Table 6.1.8: Monthwise Mean Value of Diffuse Solar Radiation at Goa - in kWh/sq.m.

Month	6	7	8	9	10	11	12	13	14	15	16	17	18	19	Total
January	0.00	0.02	0.07	0.10	0.12	0.13	0.14	0.14	0.13	0.12	0.10	0.07	0.02	0.00	1.15
February	0.00	0.03	0.09	0.11	0.14	0.15	0.15	0.15	0.15	0.14	0.12	0.08	0.03	0.00	1.34
March	0.01	0.09	0.18	0.25	0.29	0.31	0.31	0.29	0.27	0.25	0.22	0.16	0.08	0.01	2.69
April	0.01	0.08	0.17	0.25	0.32	0.36	0.38	0.38	0.35	0.31	0.24	0.16	0.07	0.01	3.04
May	0.00	0.05	0.14	0.24	0.32	0.37	0.39	0.39	0.35	0.30	0.23	0.15	0.06	0.00	3.04
June	0.00	0.05	0.16	0.25	0.31	0.36	0.37	0.36	0.33	0.29	0.23	0.15	0.05	0.00	2.94
July	0.00	0.04	0.12	0.19	0.23	0.26	0.27	0.25	0.23	0.20	0.17	0.11	0.04	0.00	2.11
August	0.00	0.02	0.08	0.12	0.15	0.17	0.18	0.18	0.17	0.16	0.13	0.09	0.02	0.00	1.47
September	0.00	0.02	0.07	0.10	0.11	0.12	0.13	0.14	0.14	0.12	0.11	0.07	0.02	0.00	1.12
October	0.00	0.05	0.13	0.18	0.22	0.22	0.22	0.21	0.20	0.18	0.16	0.12	0.04	0.00	1.88
November	0.00	0.07	0.16	0.22	0.26	0.27	0.26	0.25	0.23	0.22	0.19	0.14	0.07	0.00	2.32
December	0.00	0.06	0.17	0.27	0.35	0.40	0.43	0.43	0.41	0.35	0.26	0.17	0.06	0.00	3.34

Table 6.1.9: Monthwise Standard Deviation of Hourly Diffuse Radiation Values at Goa

Month	6	7	8	9	10	11	12	13	14	15	16	17	18	19	Total
January	0.00	0.01	0.02	0.04	0.05	0.06	0.07	0.07	0.06	0.05	0.04	0.02	0.01	0.00	0.41
February	0.00	0.03	0.03	0.04	0.05	0.06	0.06	0.06	0.06	0.05	0.04	0.02	0.01	0.00	0.41
March	0.00	0.01	0.04	0.06	0.08	0.08	0.07	0.07	0.06	0.05	0.04	0.03	0.01	0.00	0.51
April	0.00	0.02	0.04	0.06	0.08	0.09	0.09	0.09	0.07	0.07	0.05	0.04	0.02	0.00	0.58
May	0.00	0.02	0.04	0.06	0.09	0.11	0.12	0.12	0.10	0.09	0.07	0.05	0.03	0.01	0.67
June	0.01	0.04	0.07	0.10	0.12	0.14	0.15	0.14	0.14	0.12	0.10	0.07	0.04	0.01	0.83
July	0.00	0.03	0.07	0.10	0.13	0.14	0.16	0.15	0.15	0.13	0.11	0.07	0.03	0.00	0.89
August	0.00	0.03	0.06	0.08	0.10	0.12	0.13	0.14	0.13	0.11	0.09	0.06	0.03	0.00	0.79
September	0.00	0.02	0.05	0.07	0.09	0.11	0.14	0.15	0.14	0.12	0.09	0.06	0.02	0.00	0.73
October	0.00	0.02	0.04	0.08	0.10	0.13	0.14	0.14	0.12	0.10	0.08	0.05	0.02	0.00	0.77
November	0.00	0.01	0.03	0.06	0.08	0.11	0.12	0.11	0.09	0.08	0.06	0.03	0.01	0.01	0.67
December	0.00	0.01	0.02	0.04	0.06	0.07	0.08	0.08	0.07	0.06	0.05	0.02	0.01	0.00	0.46

Table 6.2.1: Monthwise Maximum Values - (highest value recorded during the hour) of Global Radiation at Mangalore

Month	Hours (local apparent time) ending at														Total
	6	7	8	9	10	11	12	13	14	15	16	17	18	19	
January	0.01	0.09	0.31	0.54	0.71	0.91	0.98	0.98	0.95	0.70	0.60	0.31	0.15	0.08	6.51
February	0.00	0.09	0.31	0.56	0.78	0.93	1.02	1.03	0.95	0.82	0.72	0.55	0.30	0.07	7.16
March	0.02	0.18	0.41	0.62	0.80	1.01	1.08	1.10	1.00	0.87	0.78	0.55	0.32	0.10	7.54
April	0.03	0.23	0.49	0.71	0.88	1.01	1.18	1.13	1.00	0.87	0.69	0.46	0.22	0.07	7.68
May	0.09	0.35	0.39	0.64	0.83	0.99	1.09	1.11	1.00	0.86	0.71	0.52	0.33	0.10	7.77
June	0.04	0.20	0.53	0.68	0.84	0.92	1.03	1.02	0.95	0.80	0.68	0.40	0.17	0.03	7.05
July	0.03	0.16	0.55	0.65	0.75	0.97	1.01	1.27	1.36	0.79	0.63	0.47	0.22	0.10	6.88
August	0.01	0.11	0.35	0.57	0.91	0.91	1.04	1.04	0.96	0.83	0.69	0.39	0.25	0.05	7.11
September	0.03	0.21	0.46	0.58	0.83	1.00	1.07	1.07	0.99	0.87	0.69	0.58	0.39	0.11	7.48
October	0.01	0.11	0.34	0.61	0.77	0.93	1.16	1.07	1.03	0.83	0.61	0.38	0.14	0.01	6.85
November	0.03	0.21	0.28	0.49	0.80	0.92	0.99	1.11	0.90	0.78	0.60	0.39	0.20	0.03	6.45
December	0.00	0.05	0.26	0.49	0.66	0.83	1.00	0.90	0.85	0.77	0.62	0.43	0.21	0.06	6.05

Table 6.2.2: Monthwise Mean Hourly Values of Global Radiation at Mangalore

Month	6	7	8	9	10	11	12	13	14	15	16	17	18	19	Total
January	0.00	0.02	0.16	0.37	0.56	0.70	0.77	0.78	0.73	0.60	0.41	0.19	0.03	0.00	5.33
February	0.00	0.03	0.20	0.43	0.63	0.78	0.85	0.88	0.82	0.67	0.47	0.23	0.04	0.00	6.02
March	0.00	0.04	0.23	0.46	0.65	0.80	0.90	0.94	0.86	0.71	0.50	0.25	0.06	0.00	6.38
April	0.00	0.06	0.26	0.46	0.62	0.78	0.89	0.93	0.86	0.71	0.51	0.27	0.08	0.00	6.42
May	0.00	0.07	0.23	0.39	0.54	0.67	0.77	0.81	0.75	0.63	0.45	0.25	0.09	0.01	5.73
June	0.00	0.05	0.17	0.30	0.40	0.48	0.53	0.53	0.50	0.40	0.29	0.17	0.06	0.00	3.86
July	0.00	0.03	0.12	0.23	0.31	0.39	0.43	0.45	0.41	0.31	0.21	0.12	0.04	0.00	3.03
August	0.00	0.03	0.15	0.28	0.42	0.52	0.58	0.59	0.53	0.43	0.31	0.16	0.05	0.00	4.07
September	0.00	0.03	0.17	0.34	0.50	0.61	0.69	0.73	0.68	0.55	0.38	0.20	0.05	0.00	4.60
October	0.00	0.02	0.16	0.35	0.52	0.63	0.70	0.73	0.70	0.56	0.36	0.17	0.04	0.00	4.93
November	0.00	0.02	0.16	0.37	0.56	0.69	0.76	0.75	0.69	0.57	0.38	0.19	0.03	0.00	5.17
December	0.00	0.01	0.14	0.35	0.54	0.67	0.76	0.76	0.70	0.57	0.38	0.18	0.03	0.00	5.10

Table 6.2.3: Monthwise Standard Deviation Values of Global Radiation at Mangalore

Month	6	7	8	9	10	11	12	13	14	15	16	17	18	19	Total
January	0.00	0.01	0.07	0.09	0.10	0.10	0.11	0.12	0.08	0.06	0.05	0.04	0.02	0.01	0.65
February	0.00	0.02	0.05	0.07	0.09	0.10	0.11	0.09	0.07	0.06	0.06	0.05	0.03	0.01	0.58
March	0.00	0.02	0.05	0.08	0.10	0.12	0.11	0.09	0.08	0.07	0.07	0.05	0.03	0.01	0.61
April	0.00	0.03	0.08	0.11	0.14	0.14	0.14	0.12	0.10	0.10	0.08	0.07	0.03	0.01	0.73
May	0.01	0.04	0.10	0.14	0.17	0.20	0.22	0.22	0.21	0.19	0.15	0.10	0.04	0.01	1.36
June	0.01	0.04	0.10	0.16	0.21	0.24	0.27	0.28	0.25	0.22	0.16	0.10	0.04	0.01	1.65
July	0.00	0.03	0.09	0.15	0.19	0.23	0.24	0.27	0.24	0.19	0.14	0.09	0.36	0.01	1.50
August	0.00	0.03	0.08	0.13	0.18	0.22	0.24	0.25	0.23	0.20	0.15	0.09	0.03	0.01	1.45
September	0.00	0.03	0.09	0.14	0.17	0.20	0.22	0.23	0.22	0.20	0.15	0.09	0.04	0.01	1.34
October	0.00	0.02	0.07	0.12	0.15	0.19	0.21	0.20	0.19	0.17	0.14	0.09	0.03	0.00	1.10
November	0.00	0.02	0.06	0.10	0.13	0.16	0.17	0.19	0.17	0.14	0.11	0.06	0.02	0.00	1.08
December	0.00	0.01	0.05	0.07	0.09	0.10	0.10	0.10	0.09	0.08	0.07	0.04	0.02	0.01	0.58

Table 6.2.4: Monthwise Elevation of Direct Solar Radiation (Degree) at Mangalore

Month	7	8	9	10	11	12	13	14	15	16	17	18
January	4.50	15.60	28.30	40.00	49.60	55.40	55.40	49.60	40.00	28.30	15.60	4.50
February	5.70	18.40	32.10	44.90	56.20	63.50	63.50	56.20	44.90	32.10	18.40	5.70
March	7.10	21.40	35.80	49.90	63.20	73.40	73.40	63.20	49.90	35.80	21.40	7.10
April	9.30	23.90	38.50	53.10	67.70	81.90	81.90	67.70	53.10	38.50	23.90	9.30
May	11.10	25.10	39.30	53.50	67.60	80.70	80.70	67.60	53.50	39.30	25.10	11.10
June	11.80	25.50	39.30	53.00	66.40	77.60	77.60	66.40	53.00	39.30	25.50	11.80
July	11.50	25.40	39.30	53.30	66.90	79.00	79.00	66.90	53.30	39.30	25.40	11.50
August	10.20	24.60	39.00	53.50	68.10	82.60	82.60	68.10	53.50	39.00	24.60	10.20
September	8.00	22.60	37.20	51.70	65.70	77.80	77.80	65.70	51.70	37.20	22.60	8.00
October	6.30	19.70	33.70	47.10	59.20	67.50	67.50	59.20	47.10	33.70	19.70	6.30
November	4.90	16.60	29.60	41.70	51.90	58.10	58.10	51.90	41.70	29.60	16.60	4.90
December	4.20	14.90	27.40	38.70	48.00	53.50	53.50	48.00	38.70	27.40	14.90	4.20

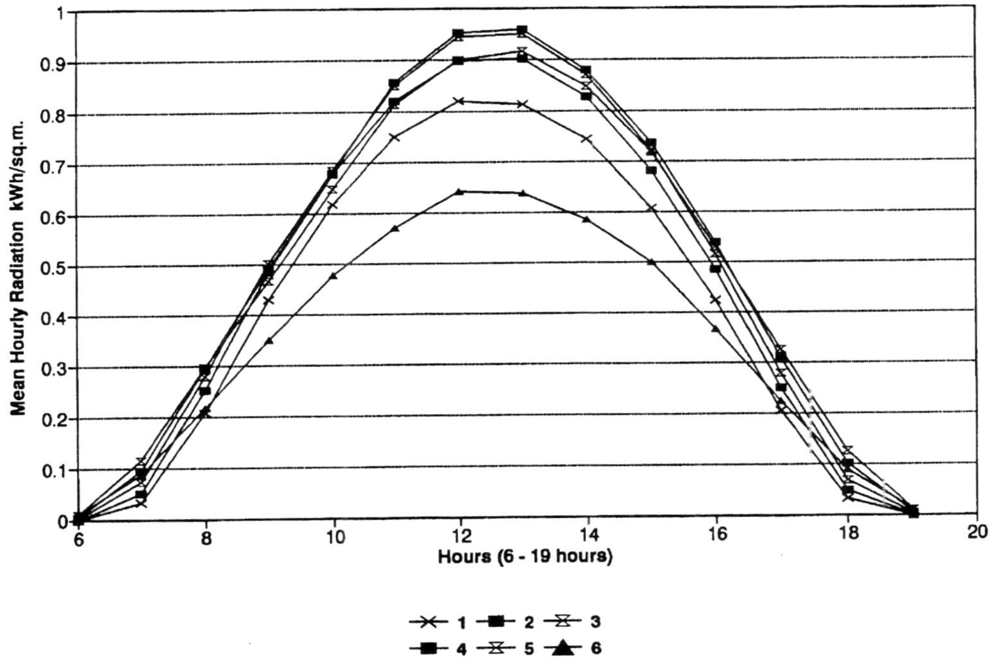

Figure 6.1.1: Mean Hourly Global Solar Radiation at Goa

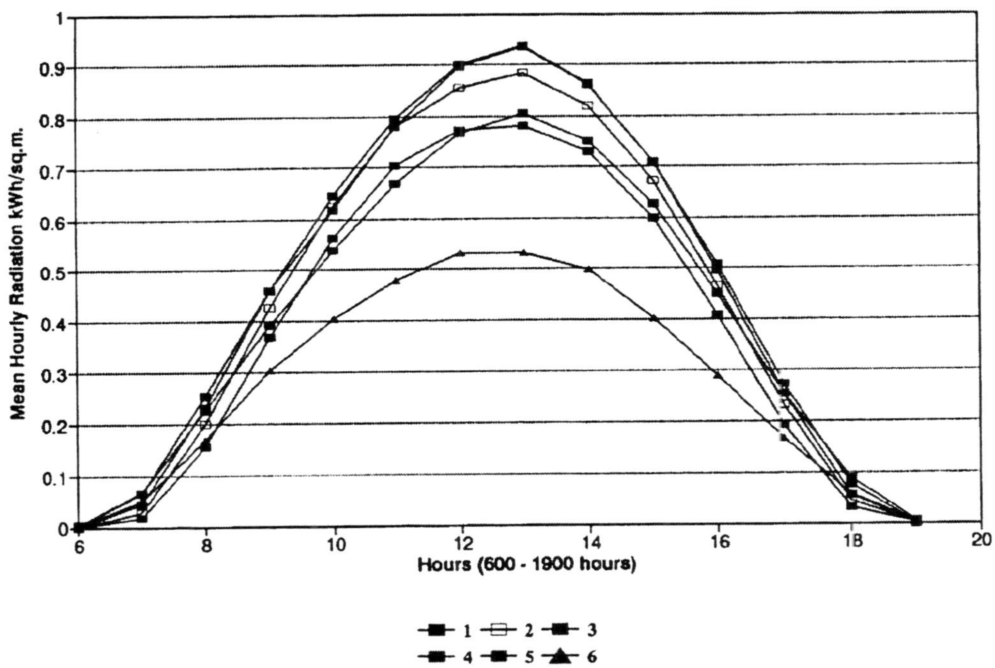

Figure 6.1.2: Mean Hourly Global Solar Radiation at Mangalore

6.4.2 Analyses of Annual and Seasonal Global Solar Radiation

(i) Annual Global Solar Radiation:

Annual global solar radiation received at Goa is about 2064.7 kWh/m^2, while at Mangalore it is about 1813 kWh/m^2. Global solar radiation (in kWh/m^2) received at these stations is listed in Table 6.3.1. Progressive increase in global solar radiation takes place during January and February. During March-May, the coastal belt receives maximum radiation. A marked decrease takes place in June with the onset of the southwest monsoon.

(ii) Sunshine:

At Goa and Mangalore, the average number of hours of sunshine is about 2665 and 2737 respectively, the monthly distribution of which is listed in Table 6.3.2. From these values, it is evident that the duration of sunshine is longest from December to May. During this period, mostly clear sky prevails in the region. With the onset of the southwest monsoon during June to September, a significant reduction in sunshine occurs, dropping to a low of 70-80 hours. From October, an increase in sunshine takes place in the coastal belt of Western Ghats region.

Mean hourly values of the duration of bright sunshine are given in Table 6.3.3 (Goa) and Table 6.3.4 (Mangalore). The daily total columns list the mean duration of sunshine. This is also expressed as a percentage of the maximum possible hours of sunshine. Mean hourly values of the duration of sunshine are illustrated graphically for Goa and Mangalore in Figure 6.2.1 and Figure 6.2.2 respectively. Both graphs concur with inferences drawn earlier from monthly values.

(iii) Diffuse Solar Radiation:

The annual pattern shows diffused solar radiation of 809.90 kWh/m^2 at Goa and 850.00 kWh/m^2 at Mangalore, which increases southward to 920 kWh/m^2 towards the extreme south of the peninsula. This large gradient in diffuse solar radiation is associated with gradients in mean annual cloudiness. The monthly distribution of diffused solar radiation is given in Table 6.3.5. In December and January, diffuse solar radiation is minimum (35-40 kWh/m^2). A general increase takes place during June-August as a result of increased cloudiness and the advance of the monsoon. It reaches maximum (100-105 kWh/m^2) during August. Mean hourly values of diffuse solar radiation and standard deviation for the hourly values are listed in Table 6.1.7 and Table 6.1.8 for Goa. It is seen that diffuse radiation is high between 1000-1600 hours during May-July.

The proportion of diffuse radiation (D) to global radiation (G) varies widely, depending on the cloudiness of the sky. The maximum occurs during June-August and the minimum (20-25%) during November-February. During November-February, D/G values undergo a characteristic diurnal variation under clear-sky conditions with minimum occurrence during noon and maximum in the morning and evening. Monthly D/G values are listed in Table 6.3.6.

Table 6.3.1: Global Solar Radiation (in kWh/sq.m.) at Goa and Mangalore

Months	Goa	Mangalore
January	176.30	165.30
February	178.60	168.60
March	210.00	200.00
April	207.80	200.00
May	208.80	180.00
June	140.00	115.90
July	123.80	91.00
August	147.50	140.00
September	159.20	150.00
October	174.00	170.00
November	168.50	180.00
December	167.10	155.10

Table 6.3.2: Sunshine Duration at Goa and Mangalore

Months	Goa	Mangalore
January	298	298
February	272	288
March	295	295
April	291	288
May	279	276
June	132	126
July	77	80
August	127	136
September	160	171
October	245	229
November	260	258
December	280	279

Table 6.3.3: Mean Duration of Sunshine per Hour at Goa

	Hours (I.A.T.) ending at													
Month	7	8	9	10	11	12	13	14	15	16	17	18	Daily Total	% of Possible hours of sunshine
January	0.0	0.5	0.9	1.0	0.9	0.9	0.9	0.9	0.9	0.9	0.7	0.1	9.0	80
February	0.1	0.8	1.0	1.0	1.0	1.0	1.0	1.0	1.0	1.0	0.8	0.1	9.7	84
March	0.1	0.8	0.9	0.9	0.9	1.0	1.0	1.0	1.0	1.0	0.8	0.1	9.5	79
April	0.3	0.8	0.8	0.9	1.0	1.0	1.0	1.0	1.0	0.9	0.8	0.3	9.7	78
May	0.3	0.6	0.8	0.8	0.9	0.9	0.9	0.9	0.9	0.9	0.8	0.4	9.0	70
June	0.1	0.3	0.3	0.4	0.5	0.5	0.5	0.5	0.4	0.4	0.3	0.2	4.4	34
July	0.0	0.1	0.3	0.3	0.3	0.3	0.3	0.3	0.2	0.1	0.1	0.0	2.5	19
August	0.1	0.3	0.4	0.5	0.5	0.5	0.5	0.4	0.4	0.3	0.2	0.1	4.1	33
September	0.0	0.3	0.5	0.6	0.6	0.7	0.7	0.7	0.7	0.5	0.3	0.0	5.7	47
October	0.0	0.5	0.8	0.8	0.9	0.9	0.9	0.9	0.9	0.7	0.6	0.1	7.9	67
November	0.1	0.6	0.8	0.9	0.9	0.9	0.9	0.9	0.8	0.8	0.6	0.1	8.3	73
December	0.1	0.6	0.9	0.9	0.9	0.9	0.9	0.9	0.9	0.8	0.6	0.1	8.3	75
Annual	0.1	0.5	0.7	0.7	0.8	0.8	0.8	0.8	0.8	0.7	0.5	0.1	7.3	62

Table 6.3.4: Mean Duration of Sunshine per Hour at Mangalore

Month	Hours (I.A.T.) ending at													
	7	8	9	10	11	12	13	14	15	16	17	18	Daily Total	% of Possible hours of sunshine
January	0.3	0.9	1.0	1.0	1.0	0.9	0.9	0.9	1.0	1.0	0.9	0.3	10.1	89
February	0.3	0.9	1.0	1.0	1.0	1.0	1.0	1.0	1.0	1.0	0.9	0.3	10.3	88
March	0.0	0.8	0.9	0.9	1.0	1.0	1.0	1.0	1.0	1.0	0.8	0.0	9.5	79
April	0.3	0.8	0.9	0.9	0.9	0.9	0.9	1.0	1.0	0.9	0.7	0.3	9.6	77
May	0.3	0.6	0.8	0.8	0.8	0.9	0.9	0.9	0.9	0.9	0.8	0.5	8.9	71
June	0.1	0.3	0.4	0.4	0.5	0.5	0.5	0.5	0.4	0.3	0.3	0.2	4.2	33
July	0.1	0.2	0.3	0.3	0.3	0.3	0.3	0.3	0.3	0.2	0.1	0.1	2.9	23
August	0.0	0.2	0.5	0.5	0.5	0.5	0.5	0.5	0.5	0.4	0.3	0.1	4.4	35
September	0.0	0.3	0.6	0.6	0.7	0.6	0.7	0.7	0.7	0.5	0.4	0.1	5.7	47
October	0.1	0.5	0.7	0.7	0.8	0.8	0.8	0.8	0.8	0.7	0.5	0.1	7.4	63
November	0.2	0.7	0.9	0.9	0.9	0.9	0.9	0.9	0.8	0.8	0.7	0.3	8.6	75
December	0.1	0.8	0.9	0.9	0.9	0.9	0.9	0.9	0.9	0.9	0.8	0.3	9.0	80
Annual	0.1	0.6	0.7	0.7	0.8	0.8	0.8	0.8	0.8	0.7	0.6	0.2	7.5	63

Table 6.3.5: Diffuse Solar Radiation (kWh/sq.m.) at Goa and Mangalore

Months	Goa	Mangalore
January	35.8	40.0
February	41.5	45.0
March	58.4	80.0
April	69.7	75.0
May	83.3	85.0
June	91.2	90.0
July	94.1	95.0
August	104.0	100.0
September	88.1	85.0
October	85.5	70.0
November	44.2	44.2
December	34.8	40.0

Table 6.3.6: Monthwise D/G values at Goa and Mangalore

Months	Goa	Mangalore
January	20.3	24.2
February	23.2	20.7
March	27.8	50.0
April	33.5	40.0
May	40.0	47.2
June	65.1	77.0
July	76.0	104.0
August	70.7	71.4
September	55.3	80.0
October	37.8	41.2
November	26.2	27.0
December	20.5	25.0

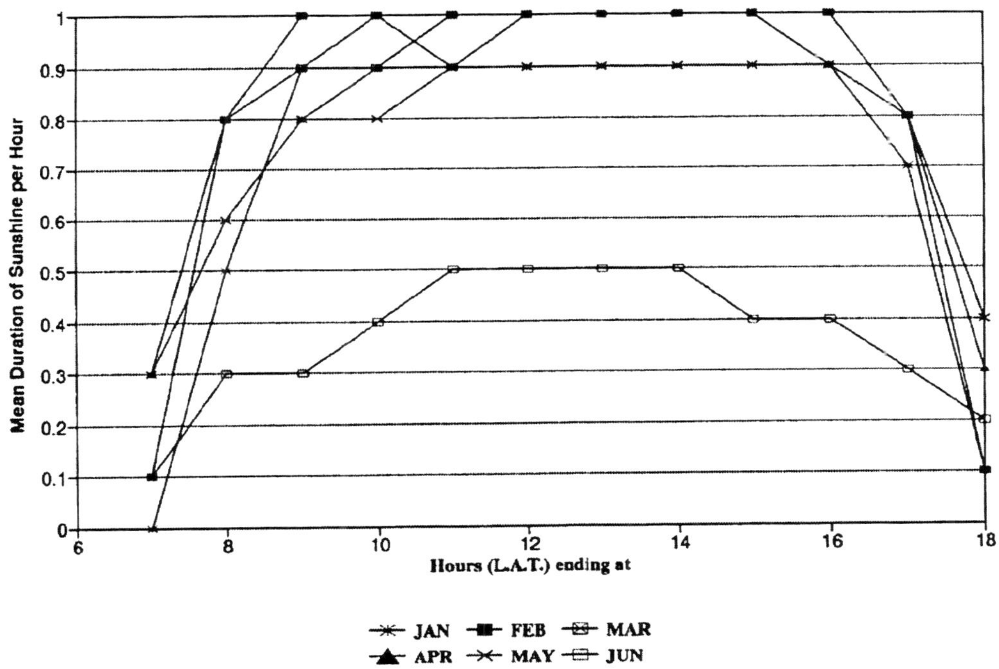

Figure 6.2.1: Duration of Sunshine per Hour at Goa

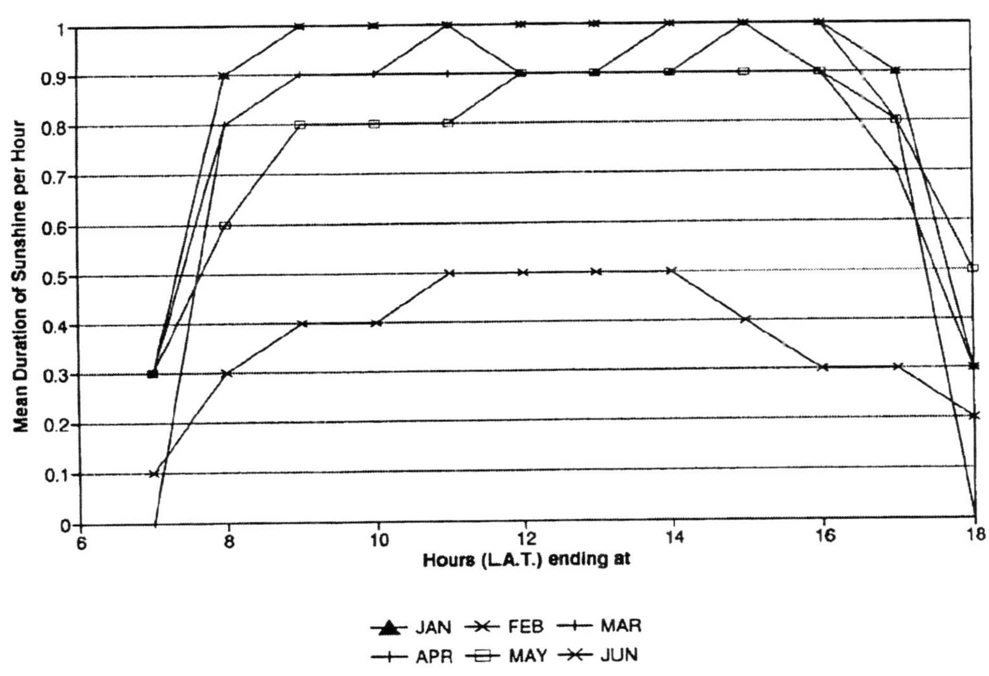

Figure 6.2.2: Duration of Sunshine per Hour at Mangalore

6.4.3 Computation of Global Radiation

In order to estimate solar radiation at places where radiation data are not available, it is computed using regression techniques from parameters (a) sunshine and cloudiness and (b) extra terrestrial radiation, allowing for its depletion by absorption and scattering in the atmosphere. There is an obvious relationship between sunshine duration and the amount of solar energy received on the earth's surface. Hence, earlier researchers like Kimball (1919), Angstrom (1924) and many others have used sunshine data to derive empirical formulas for solar radiation estimates (Angstrom, 1956; Page, 1980; WMO, 1981; Page, 1984; Tasdemiroglu and Sever, 1989; Page, 1989). The statistical relationship formulated between the daily duration of sunshine 'n' and the daily total global radiation 'G' is of the form

$$G/G_0 = a + b\,(n/N) \qquad (1)$$

where G_0 is the daily global radiation with a cloud free atmosphere and N, the maximum possible duration of sunshine. In order to compute n/N at places where only cloud cover data are available, without mean sunshine duration data, the inverse relationship between sunshine (n/N) and cloud cover (C) is used, which is given by 1-n/N = C. Since n/N' (where N' is the maximum possible sunshine with solar elevation $\geq 5°$, in hours and tenths) is used for deriving G (Hoyt, 1977), the relationship between n/N' and C is given by

$$1-(n/N') = 1*C + 0.310*(C^2) + 0.476*(C^3) + 0.100\,(C^4) \qquad (2.1)$$

Due to difficulties in the precise evaluation of G_0 in equation 1, G_0 is replaced by the extra terrestrial radiation (ETR) on a horizontal surface and the relationship becomes

$$(G/ETR) = a + b\,(n/N) \qquad (2.2)$$

extra terrestrial radiation on a horizontal surface for any place for any day / month can be precisely estimated from the relationship

$$ETR\ (kWh/m^2/day) = 10.39\,.k.\,(\cos\phi\ \cos\delta\ \sin\omega_s + \omega_s \sin\phi\ \sin\delta) \qquad (3)$$

where 10.39 represents the solar constant (assumed equal to 1.36 kW/m^2 multiplied by $24/\pi$), k the correction factor for varying earth-sun distance, ϕ latitude, δ mean declination for month and ω_s the sunset hour angle in radians.

The coefficient b in equation 2.2 expresses the rate of increase of G with increase in n/N. For a clear sky, n/N is unity and thus the parameter (a+b) is the fraction of ETR that actually reaches the earth's surface. The magnitude of 'a' generally depends on thickness of prevailing cloud and 'b' depends on the transmission characteristics of the cloud free atmosphere. In the analyses of solar radiation for Canada (Hay, 1979), it was found that the surface albedo (R) plays an important role in determining the magnitude of G in addition to the multiple reflections between the ground and the cloud base or the clear atmosphere. Modified duration of day length N' was also considered instead of N, to take into account the sunshine recorder sensitised card which gets burned only when the sun is at an elevation of 5 degrees or above the horizon. Hence, instead of N, N' is computed as follows.

$$N = \arccos(-\tan\varphi\ \tan\delta)/7.5 \qquad (4)$$
$$N' = \arccos((\cos 85 - \sin\varphi\ \sin\delta)/\cos\varphi\ \cos\delta)/7.5 \qquad (5)$$

And the relationship used to estimate global radiation at a place is

$$G'/ETR = a + b\,(n/N') \qquad (6)$$

where G' is the global solar radiation that first strikes the ground before undergoing multiple reflections with the clear atmosphere and the cloud base. The relationship between G and G', assuming albedo of the cloud free atmosphere as 0.25 and that of cloud base as 0.60 is given by (Hay, 1979),

$$G-G' = G.R.[(0.25\ n/N') + 0.60\ (1-n/N')] \quad (7.1)$$

where R is surface albedo. This relationship (Equation 6) is tried for Goa and Mangalore, to see how far the observed and calculated values agree with the data of Goa and Mangalore. This exercise would help in estimating global radiation (G) at other locations in Uttara Kannada district where radiation data are not available. The regression constants 'a' and 'b' computed for (G'/ETR) and (n/N') are listed in Table 6.4.1. While, Table 6.4.2 gives the calculated and the measured values of G for Goa and Mangalore. It is seen that computed and measured values of G agree within the range of 2-5% for most of the months. This is shown pictorially in Figure 6.3.1.

Table 6.4.1: List of Regression Constants 'a' and 'b' Computed for (G'/ETR) and (n/N')

	Goa	Mangalore
a	0.26	0.21
b	0.58	0.51
r	0.97	0.99
r2	0.95	0.97
Std. Err.	0.03	0.02

Table 6.4.2: Comparison of Computed and Observed Values of Global Solar Radiation

Station	Goa			Mangalore		
Month	Calc	Obs	% diff	Calc	Obs	% diff
January	5.62	5.59	1.20	5.82	5.33	9.20
February	6.33	6.38	0.80	6.37	6.02	5.80
March	6.57	6.76	2.80	6.37	6.38	0.20
April	6.90	6.93	0.30	6.45	6.42	0.50
May	6.73	6.74	0.10	6.02	5.73	5.10
June	4.94	4.79	3.10	4.24	3.86	9.80
July	4.51	3.99	13.10	3.70	3.07	20.50
August	4.93	4.76	3.60	4.33	4.07	6.30
September	5.37	5.31	1.10	4.99	4.96	0.60
October	5.67	5.61	1.10	5.26	4.93	6.70
November	5.72	5.61	2.00	5.53	5.17	7.00
December	5.37	5.39	0.40	5.36	5.10	5.10

6.4.4 Computation of Diffuse Solar Radiation

As in the case of global solar radiation, diffuse solar radiation is given by (Collares-Pereira and Rabl, 1979)

$$D'/G' = c + d\,(G'/ETR) \qquad (7.2)$$

D is computed by $D = D' + GR\,[(0.25(n/N')) + 0.60\,(1-(n/N'))]$ (7.3)

The regression constants c and d computed for (D'/G') and (G'/ETR) are given in Table 6.4.3. Based on this relationship, computed diffuse solar radiation and that actually measured at Goa are given in Table 6.4.4. Similarly, computed and observed values of clear sky noon global solar radiation values are listed in Table 6.4.5, and Table 6.4.6 lists clear sky noon values of diffuse radiation.

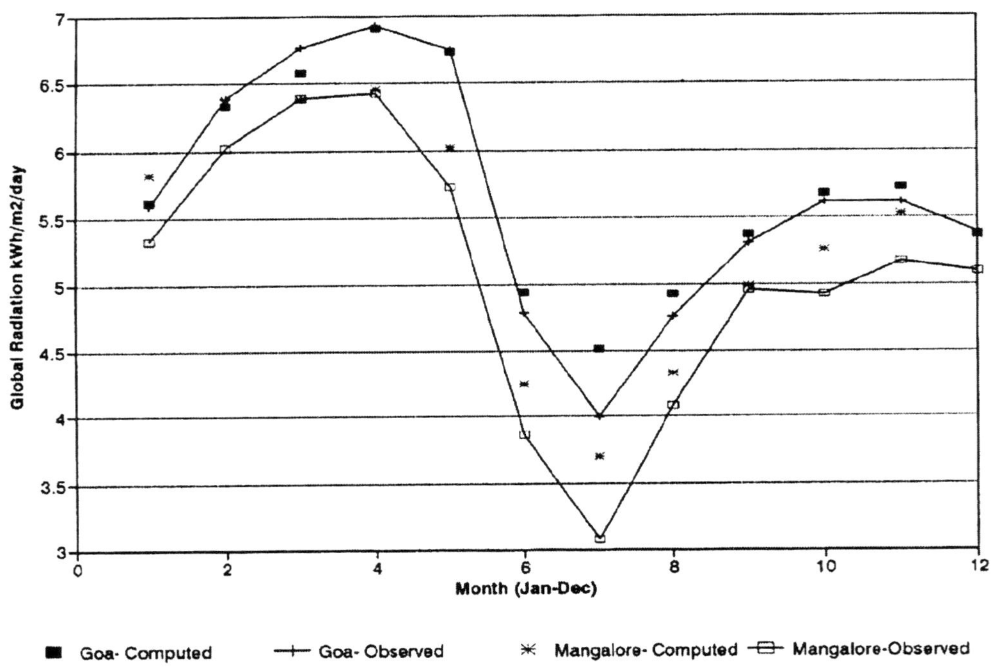

Figure 6.3.1: Computed and Observed Values of Global Solar Radiation

Table 6.4.3: List of Regression Constants 'c' and 'd' Computed for (D'/G') and (G'/ETR)

	Goa
c	1.29
d	-1.66
r	-0.94
r2	0.89
Std. Err.	0.07

Table 6.4.4: Comparison of Computed and Observed Values of Diffuse Solar Radiation

Station	Goa		
Month	Calc	Obs	%diff
January	1.39	1.15	20.70
February	1.50	1.34	12.10
March	2.01	1.88	6.80
April	2.18	2.32	6.30
May	2.43	2.69	9.70
June	3.12	3.04	2.60
July	3.12	3.03	2.70
August	3.08	3.34	7.60
September	2.84	2.94	3.20
October	2.28	2.11	7.90
November	1.54	1.47	4.60
December	1.41	1.12	26.10

Table 6.4.5: Comparison of Computed and Observed Values of Clear Sky Noon Global Solar Radiation

Station	Goa		
Month	Calc	Obs	%diff
January	0.84	0.83	1.30
February	0.92	0.90	1.90
March	0.98	0.96	1.50
April	0.99	0.99	0.30
May	0.97	0.99	2.20
June	0.95		
July	0.95		
August	0.97		
September	0.96		
October	0.91		
November	0.85		
December	0.81		

Table 6.4.6: Comparison of Computed and Observed Clear Sky Noon Values of Diffuse Radiation

Station	Goa		
Month	Calc	Obs	%diff
January	0.12	0.12	1.60
February	0.15	0.14	0.70
March	0.18	0.18	0.00
April	0.19	0.20	4.60
May	0.18	0.19	5.20
June	0.17		
July	0.15		
August	0.14		
September	0.13		
October	0.12	0.12	0.00
November	0.11	0.11	0.90
December	0.10	0.10	0.00

Table 6.5.1 (Goa) and Table 6.5.2 (Mangalore) list monthly mean values of G, D, I(H) [Daily sum of direct solar radiation, given by I(H) = G-D], ETR, n, N' and C.I. (Clearness Index = G/ETR). The Clearness Index gives good measure of the over all transparency of the atmosphere, taking into account the clouding as well as the turbidity and water vapor content.

Table 6.5.1: Mean Values of Solar Radiation and Associated Parameters (computed) for Goa, Lat: 15..48N, Long: 73.82E, Elev: 55 MASL

Month	Mean Daily Radiation and Sunshine Parameters								
	ETR	N	N'	I (H)	D	G	G (C)	C.I.	WS
January	8.10	9.50	10.44	4.23	1.39	5.62	6.00	0.69	83.90
February	9.04	10.00	10.81	4.38	1.50	6.33	4.77	0.70	86.40
March	9.94	9.30	11.23	4.56	2.01	6.58	7.43	0.66	89.40
April	10.52	9.50	11.65	4.73	2.18	6.90	7.74	0.66	92.70
May	10.67	9.00	11.98	4.30	2.43	6.73	7.70	0.63	95.40
June	10.64	4.30	12.14	1.32	3.12	4.94	7.61	0.46	96.80
July	10.61	3.20	12.08	1.40	3.12	4.51	7.56	0.43	96.20
August	10.53	4.30	11.81	1.85	3.08	4.93	7.61	0.47	94.00
September	10.15	5.80	11.42	2.53	2.84	5.37	7.36	0.53	90.90
October	9.36	7.60	10.99	3.39	2.28	5.67	6.78	0.61	87.70
November	8.39	9.30	10.57	4.18	1.54	5.72	6.18	0.68	84.80
December	7.83	9.20	10.34	3.96	1.41	5.37	5.74	0.69	83.30

Table 6.5.2: Values of Solar Radiation and Associated Parameters (computed) for Mangalore, Lat: 12.92N, Long: 74.88E

Month	Mean Daily Radiation and Sunshine								
	ETR	N	N'	I (H)	D	G	G (C)	C.I.	WS
January	8.46	10.10	10.59	4.37	1.45	5.82	6.25	0.69	85.00
February	9.32	10.30	10.90	4.74	1.63	6.37	6.92	0.68	87.10
March	10.09	9.30	11.25	4.17	2.20	6.37	7.49	0.63	89.50
April	10.51	9.10	11.60	4.02	2.43	6.45	7.68	0.61	92.20
May	10.54	8.20	11.87	3.33	2.69	6.02	7.55	0.57	94.50
June	10.45	4.00	12.00	1.25	3.00	4.24	7.47	0.41	95.60
July	10.45	2.70	11.94	0.81	2.89	3.70	7.43	0.35	95.20
August	10.47	4.10	11.72	1.32	3.01	4.33	7.55	0.41	93.30
September	10.22	5.80	11.41	2.09	2.90	4.99	7.40	0.49	90.80
October	9.58	7.10	11.05	2.72	2.54	5.26	6.96	0.55	88.10
November	8.72	8.90	10.70	3.64	1.89	5.53	6.34	0.63	85.70
December	8.21	9.20	10.51	3.71	1.65	5.36	6.02	0.65	84.40

Monthly variation of G, D and I(H) are also illustrated in Figure 6.3.2 for Mangalore and Goa. It is seen that Goa receives higher G than Mangalore for most months. It peaks in April and reaches its lowest value during July. Figure 6.3.3 gives monthly variation of ETR, N and N' for Goa and Mangalore. ETR, N and N' are higher for Goa than Mangalore for most months. Clearness Index is in the range of 0.5-0.7 during October - April, while it is less than 0.5 during the monsoon period (June-August). WS (sunset hour angle) when divided by 7.5 gives duration of the day in hours. Global radiation tilt factors computed for surfaces facing south for various tilt angles of latitude ϕ, $\phi + 15°$, $\phi - 15°$, 22.5° and 90° are listed in Table 6.5.3 (Goa) and Table 6.5.4 (Mangalore). This is also illustrated in Figure 6.3.4 for Goa and Mangalore. The tilt factors help in estimating global radiation on a sloping surface (by multiplying tilt factor with global radiation on horizontal surface).

Table 6.5.3: Global Radiation Tilt Factors (at Goa)

Tilt	LAT-15	LAT	LAT+15	22.5	90	90	90
Azimuth	0	0	0	0	0	+/- 90	180
Month							
January	1.01	1.20	1.31	1.26	1.02	0.64	0.22
February	1.01	1.13	1.19	1.16	0.76	0.63	0.22
March	1.00	1.04	1.03	1.04	0.47	0.62	0.25
April	1.00	0.97	0.90	0.94	0.27	0.60	0.26
May	1.00	0.93	0.82	0.88	0.24	0.60	0.44
June	1.00	0.93	0.84	0.89	0.38	0.60	0.55
July	1.00	0.95	0.86	0.91	0.42	0.60	0.54
August	1.00	0.96	0.89	0.93	0.39	0.60	0.46
September	1.00	1.01	0.96	0.99	0.42	0.61	0.36
October	1.00	1.08	1.10	1.09	0.63	0.62	0.30
November	1.01	1.17	1.26	1.22	0.93	0.63	0.24
December	1.01	1.21	1.35	1.28	1.09	0.64	0.23

Table 6.5.4: Global Radiation Tilt Factors (at Mangalore)

Tilt	LAT-15	LAT	LAT+15	22.5	90	90	90
Azimuth	0	0	0	0	0	+/- 90	180
Month							
January	0.97	1.15	1.27	1.23	0.95	0.63	0.22
February	0.98	1.10	1.15	1.14	0.71	0.62	0.23
March	0.99	1.03	1.01	1.03	0.43	0.61	0.27
April	1.00	0.97	0.90	0.93	0.27	0.60	0.32
May	1.01	0.94	0.84	0.88	0.28	0.60	0.49
June	1.01	0.95	0.87	0.90	0.43	0.60	0.58
July	1.00	0.96	0.89	0.92	0.47	0.60	0.57
August	1.00	0.97	0.90	0.93	0.42	0.60	0.50
September	1.00	1.00	0.96	0.98	0.41	0.60	0.39
October	0.99	1.05	1.07	1.07	0.59	0.61	0.34
November	0.98	1.12	1.21	1.18	0.85	0.62	0.27
December	0.97	1.16	1.28	1.25	1.00	0.63	0.25

6.4.5 Statistical Analyses of Global Radiation Data at Goa and Mangalore

In order to get a complete picture of the solar radiation distribution, median and frequency distribution was observed. "Skewness" factor was used for precise estimation of the degree of departure of a given frequency distribution from a symmetrical one. Also, "Kurtosis" was computed to determine the distortion in the intermediate ranges between the mean and extreme values. Normal distribution has a Kurtosis of 3. The distribution having Kurtosis < 3 is called platykurtic, which has a broader modal part than the normal curve. Kurtosis > 3 is called leptokurtic, where distribution is more sharply peaked than the normal distribution. Skewness and Kurtosis computed for distribution of daily global solar radiation is given by

$$\text{Skewness} = \mu_3/(\mu_2)^{3/2} \text{ and Kurtosis} = \mu_4/(\mu_2)^2$$

where μ_2 = second moment = $(\Sigma x_i^2)/N$, μ_3 = third moment= $(\Sigma x_i^3)/N$, μ_4 = fourth moment = $(\Sigma x_i^4)/N$ and x_i are deviations of the individual values in the series consisting of N items from their arithmetic mean.

Skewness and Kurtosis computed for each month for Goa and Mangalore are listed in Table 6.6.1. It is seen that global radiation values at Goa for June - September have a broader modal distribution, while during November-May with Kurtosis > 3, the values are more sharply peaked than normal distribution. It is also seen that, except for the month of July the global radiation values lower than the mode occur more frequently at Goa and Mangalore.

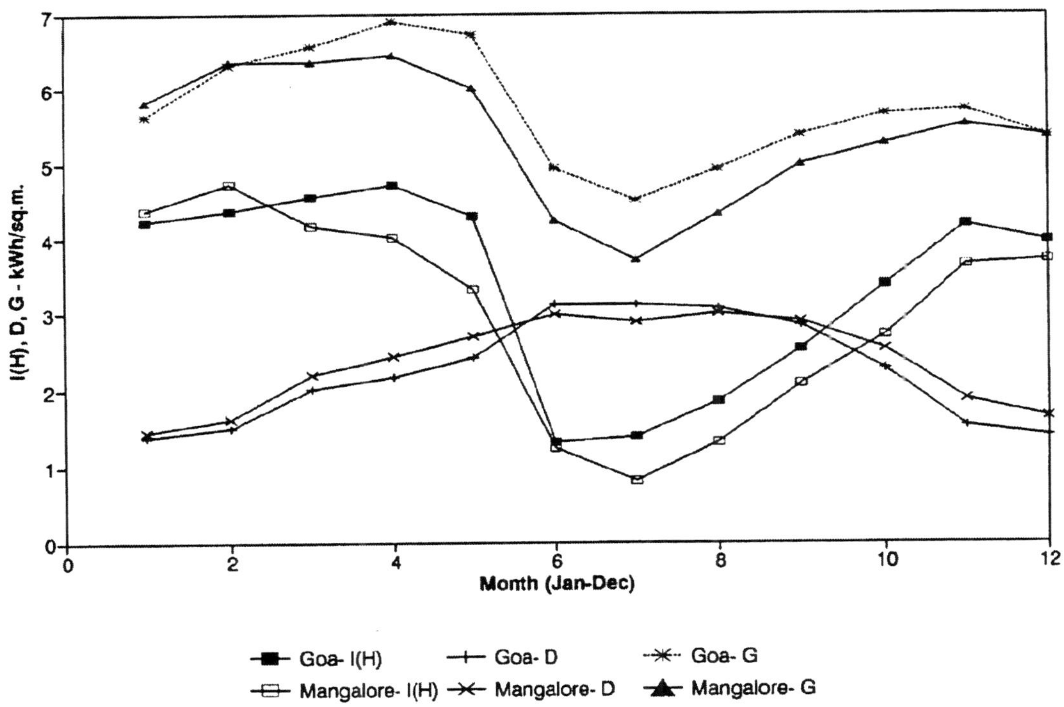

Figure 6.3.2: I(H), Direct Solar Radiation - D: diffuse, G: Global Solar Radiation

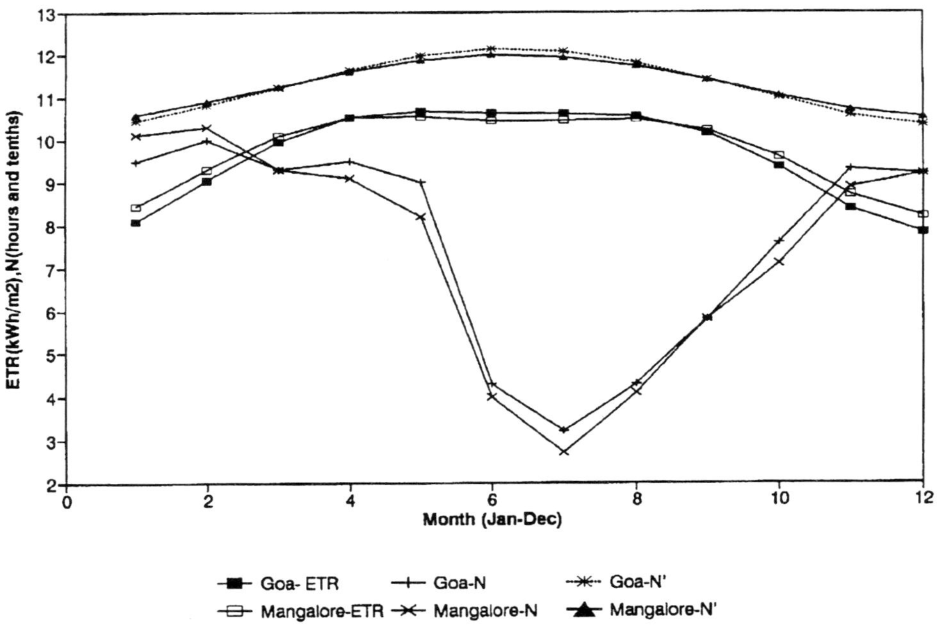

Figure 6.3.3: Extra Terrestrial Radiation - N: Actual sunshine, N': maximum possible

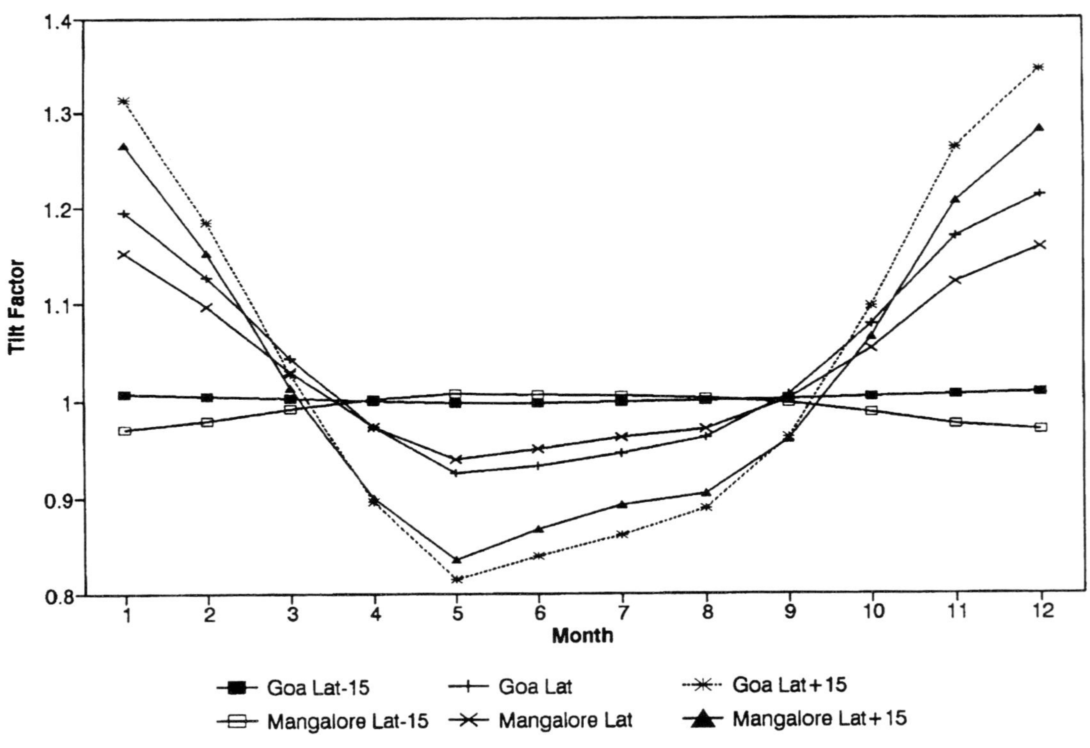

Figure 6.3.4: Daily Global Radiation Tilt Factors at Goa and Mangalore

Median values of hourly and daily global solar radiation are listed in Table 6.6.2 (Goa) and Table 6.6.3 (Mangalore), and are illustrated in Figure 6.4, which depicts that the median values of hourly and daily global radiation is higher during March-May for Goa and Mangalore. It appears from Figure 6.4 that Goa has higher median values of hourly and daily global solar radiation for all months in a year.

The monthly percentage frequency distributions of global solar radiation at Goa and Mangalore are given in Table 6.6.4 and Table 6.6.5. This range gives an idea of maximum solar radiation possible in a month and the design of proper backup devices to store the energy for the lean period requirement in a day. Table 6.6.1 and Table 6.6.2 also list cumulative frequency distribution of solar radiation in the range 0-9 kWh/m^2/day in steps of 0.5. It is seen that percentage of occurrence in the interval 5.5-7 kWh/m^2/day is more during November-May (65 - 85%) for Goa and Mangalore. For June-August, it is 55-70% in the range 2.5-5.5 kWh/m^2/day. The cumulative percentage frequency distribution of daily global solar radiation is illustrated for December-May and June-November for Goa in Figure 6.5.1 and Figure 6.5.2, and for Mangalore in Figure 6.6.1 and Figure 6.6.2 respectively. These illustrations confirm the earlier conclusions that Goa and Mangalore receive global Solar radiation in the range 5.5-7.5 kWh/m^2/day during November-May. During June-August and September-October, the solar radiation received is in the range 2.5-5 and 3.5-6.5 kWh/m^2/day respectively. These analyses show that solar potential at Goa and Mangalore is 5-7 kWh/m^2/day for about nine months a year and 3.5-5 kWh/m^2/day during the monsoon season. Dense cloud cover for about 10-12 days on average in July and 5-6 days in August are the lean periods for solar energy in this region.

Table 6.6.1: Skewness and Kurtosis (monthwise) for Goa and Mangalore

Month	Station		
		Goa	Mangalore
January	Skew	-1.43	-0.99
	Kurt	6.39	4.40
February	Skew	-0.74	-0.65
	Kurt	4.07	3.25
March	Skew	-1.03	-0.66
	Kurt	4.61	3 66
April	Skew	-1.73	-0.82
	Kurt	9.55	3.82
May	Skew	-2.06	-1.45
	Kurt	8.75	4.91
June	Skew	-0.25	-0.17
	Kurt	1.95	2.08
July	Skew	0.00	0.22
	Kurt	1.94	2.25
August	Skew	-0.60	-0.33
	Kurt	2.73	2.38
September	Skew	-0.80	-0.71
	Kurt	2.68	3.01
October	Skew	-1.67	-0.70
	Kurt	6.10	2.88
November	Skew	-1.98	-2.10
	Kurt	8.02	3.14
December	Skew	-2.07	-1.71
	Kurt	10.17	8.77

With these detailed investigations of daily global solar radiation data collected at Goa and Mangalore, estimation of solar potential is attempted in the district. Uttara Kannada consists of three distinct zones, namely, the coastal belt (consisting of Karwar, Ankola, Kumta, Honnavar and Bhatkal), the Sahyadrian Interior (consisting of Supa, Yellapur, Sirsi, Siddapur and part of Haliyal), and the plains (consisting of Mundgod and part of Haliyal). Radiation stations at Goa and Mangalore are also in the coastal belt of the Western Ghats region and are located adjacent to Uttara Kannada (Goa is located in the north, while Mangalore is in the south of Uttara Kannada). The coastal line, covering a distance of 114 km, stretches from Machale near Karwar to Gorte near Bhatkal. The distance between Goa radiation station and the Karwar climatological station (moving from north to south) is about 60 km, between Karwar and Honnavar about 72 km, and between Honnavar and Bhatkal about 40 km. The Mangalore radiation station is about 120 km from Bhatkal. The distances between various stations along the coastal strip of Western Ghats region is given below.

Table 6.6.2: Mean Values of Hourly and Daily Global Solar Radiation in kWh/sq.m. (at Goa)

Month	Hours L.A.T. ending at													Daily
	6	7	8	9	10	11	12	13	14	15	16	17	18	Total
January	0.00	0.00	0.19	0.43	0.65	0.79	0.81	0.81	0.78	0.62	0.43	0.19	0.00	5.75
February	0.00	0.00	0.27	0.48	0.68	0.82	0.92	0.92	0.82	0.68	0.47	0.25	0.00	6.41
March	0.00	0.00	0.29	0.52	0.70	0.89	0.94	0.94	0.89	0.71	0.54	0.30	0.00	6.83
April	0.00	0.00	0.30	0.51	0.71	0.89	0.97	0.97	0.90	0.74	0.55	0.31	0.00	7.03
May	0.00	0.00	0.30	0.49	0.68	0.87	0.95	0.96	0.89	0.76	0.55	0.33	0.14	7.00
June	0.00	0.00	0.22	0.38	0.51	0.61	0.71	0.70	0.60	0.52	0.38	0.22	0.00	5.04
July	0.00	0.00	0.18	0.30	0.42	0.52	0.55	0.53	0.49	0.39	0.29	0.17	0.00	3.93
August	0.00	0.00	0.21	0.36	0.53	0.64	0.71	0.70	0.64	0.54	0.37	0.21	0.00	5.05
September	0.00	0.00	0.22	0.42	0.56	0.70	0.83	0.86	0.79	0.64	0.47	0.26	0.00	5.81
October	0.00	0.00	0.22	0.43	0.63	0.76	0.85	0.88	0.79	0.66	0.45	0.24	0.00	5.91
November	0.00	0.00	0.20	0.43	0.65	0.78	0.82	0.81	0.77	0.63	0.43	0.19	0.00	5.85
December	0.00	0.00	0.18	0.42	0.59	0.73	0.79	0.79	0.70	0.57	0.42	0.18	0.00	5.58

Table 6.6.3: Median Values of Hourly and Daily Global solar radiation in kWh/sq.m. (at Mangalore)

Month	6	7	8	9	10	11	12	13	14	15	16	17	18	Daily Total
January	0.00	0.00	0.17	0.39	0.57	0.72	0.79	0.80	0.73	0.59	0.43	0.19	0.00	5.44
February	0.00	0.00	0.19	0.43	0.65	0.79	0.87	0.90	0.82	0.68	0.45	0.21	0.00	6.12
March	0.00	0.00	0.23	0.46	0.66	0.81	0.92	0.94	0.87	0.71	0.51	0.26	0.00	6.44
April	0.00	0.00	0.27	0.47	0.64	0.80	0.92	0.94	0.89	0.72	0.53	0.29	0.00	6.55
May	0.00	0.00	0.25	0.42	0.57	0.70	0.83	0.89	0.81	0.68	0.51	0.28	0.00	6.14
June	0.00	0.00	0.17	0.32	0.42	0.51	0.57	0.55	0.52	0.40	0.28	0.16	0.00	4.02
July	0.00	0.00	0.00	0.21	0.30	0.37	0.41	0.42	0.39	0.30	0.21	0.00	0.00	2.95
August	0.00	0.00	0.15	0.28	0.45	0.56	0.61	0.63	0.55	0.44	0.32	0.17	0.00	4.28
September	0.00	0.00	0.18	0.37	0.53	0.64	0.75	0.79	0.75	0.60	0.41	0.20	0.00	5.17
October	0.00	0.00	0.16	0.38	0.54	0.67	0.74	0.78	0.74	0.62	0.38	0.17	0.00	5.03
November	0.00	0.00	0.17	0.39	0.57	0.73	0.80	0.80	0.76	0.61	0.41	0.18	0.00	5.42
December	0.00	0.00	0.16	0.34	0.55	0.68	0.77	0.79	0.70	0.56	0.41	0.18	0.00	5.17

Table 6.6.4: Percentage Frequency and Cumulative Percentage Frequency Distributions of Daily Global Solar Radiation (at Goa)

Month		0.5	1	1.5	2	2.5	3	3.5	4	4.5	5	5.5	6	6.5	7	7.5	8	8.5	9		
		Class Interval kWh/sq.m																			
		0 - 0.5	0.5 - 1	1 - 1.5	1.5 - 2	2 - 2.5	2.5 - 3	3 - 3.5	3.5 - 4	4 - 4.5	4.5 - 5	5 - 5.5	5.5 - 6	6 - 6.5	6.5 - 7	7 - 7.5	7.5 - 8	8 - 8.5	8.5 - 9	9 - 9.5	9.5 - 10
January	% F							1	1	3	5	14	54	21	2						
February	% F									1	1	2	12	43	37	6					
March	% F										1	2	6	17	37	34	3				
April	% F									1		1	4	13	28	36	16				
May	% F					1	1	1	1		2	3	5	14	23	27	20	3			
June	% F		1	4	4	6	9	5	7	7	7	10	9	9	7	11	5				
July	% F		1	4	6	9	11	9	11	7	7	12	9	9	3	1					
August	% F		2	1	3	5	3	5	11	8	11	12	16	12	7	4					
September	% F			2	2	5	3	5	7	3	7	9	11	18	16	10	1				
October	% F		1	1		1	2	2	4	3	5	12	23	24	21						
November	% F			1	1		1	1	3	4	7	15	26	38	4						
December	% F					1			2	4	13	22	53	5							
January	% CF							1	1	4	9	23	77	98	100						
February	% CF										1	4	15	58	94	100					
March	% CF										1	3	9	26	63	97	100				
April	% CF									1	2	2	6	20	48	84	100				
May	% CF					1	2	3	4	5	7	9	14	27	50	78	97	100			
June	% CF		1	5	9	15	24	29	35	43	49	59	68	77	84	95	100				
July	% CF		1	5	12	21	32	41	51	59	66	77	87	96	99	100					
August	% CF		2	2	5	11	13	19	30	38	49	61	78	90	96	100					
September	% CF			2	4	8	11	16	23	26	34	43	54	73	89	99	100				
October	% CF		1	1	2	3	4	6	11	14	19	31	54	78	100						
November	% CF			1	1	1	2	4	7	11	17	32	58	96	100						
December	% CF					1	1	1	3	7	20	42	95	100							

Table 6.6.5: Percentage Frequency and Cumulative Percentage Frequency Distributions of Daily Global Solar Radiation (at Mangalore)

Class interval		0 - 0.5	0.5 - 1	1 - 1.5	1.5 - 2	2 - 2.5	2.5 - 3	3 - 3.5	3.5 - 4	4 - 4.5	4.5 - 5	5 - 5.5	5.5 - 6	6 - 6.5	6.6 - 7	7 - 7.5	7.5 - 8	8 - 8.5	8.5 - 9	9 - 9.5	9.5 - 10
Month																					
January	% F						1		4	6	10	34	35	10	1						
February	% F									1	4	11	25	36	21	1					
March	% F									1	1	6	13	33	30	16					
April	% F									2	1	7	14	23	31	18	3				
May	% F			1	1	2	2	2	2	4	5	11	12	26	19	11	1				
June	% F	1	4	5	5	10	10	7	9	10	11	10	8	7	3						
July	% F	2	5	10	12	10	12	11	10	8	8	6	3	1	1						
August	% F		2	4	5	7	8	7	11	13	15	14	7	5	2						
September	% F				2	4	2	5	6	11	14	13	16	15	9	2					
October	% F				1	3	5	4	6	12	19	16	17	17	1						
November	% F		1	2			3	1	5	4	12	27	32	13							
December	% F						1		2	7	25	42	22								
January	% CF						1	1	5	11	21	54	89	99	100						
February	% CF									1	6	17	42	77	99	100					
March	% CF									1	2	8	21	54	84	100					
April	% CF									3	4	11	24	47	78	97	100				
May	% CF			1	3	5	7	9	11	15	20	30	43	69	88	99	100				
June	% CF	1	5	10	15	25	34	41	50	60	71	81	89	97	100						
July	% CF	2	7	18	29	39	51	62	72	81	89	95	98	99	100						
August	% CF		2	6	11	17	25	32	43	56	71	85	92	97	100						
September	% CF			1	3	7	9	15	20	32	45	59	75	89	98	100					
October	% CF				1	3	9	13	19	31	49	66	82	99	100						
November	% CF		1	3	3	3	6	7	12	16	28	55	87	100							
December	% CF						1	2	4	11	36	78	100								

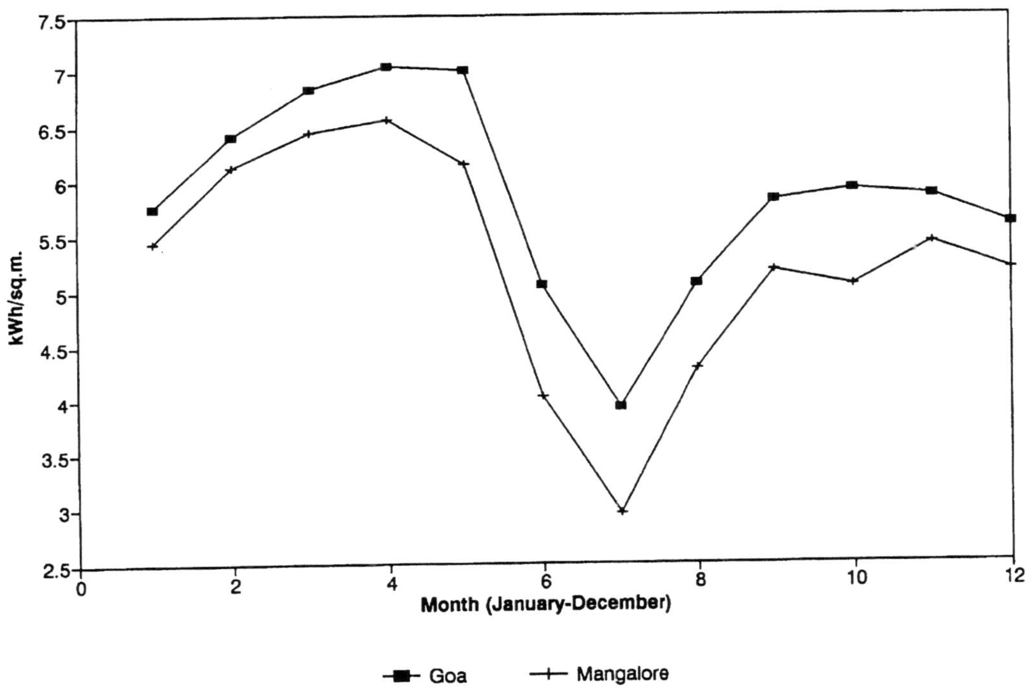

Figure 6.4: Median Values of Hourly, Daily Global Solar Radiation

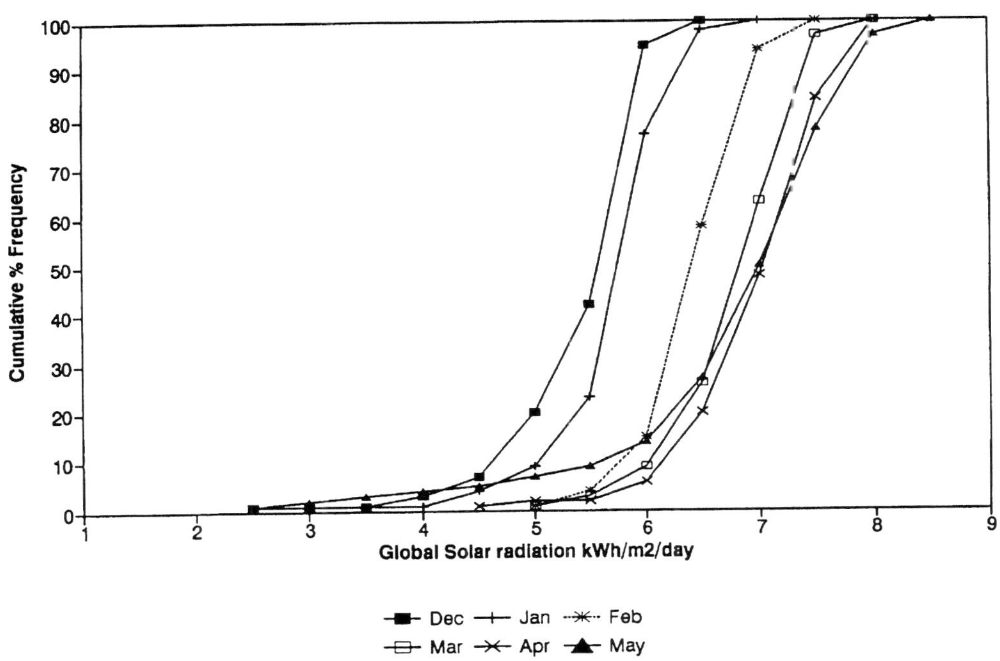

Figure 6.5.1: Cumulative Percentage Frequency Distribution (December-May) at Goa

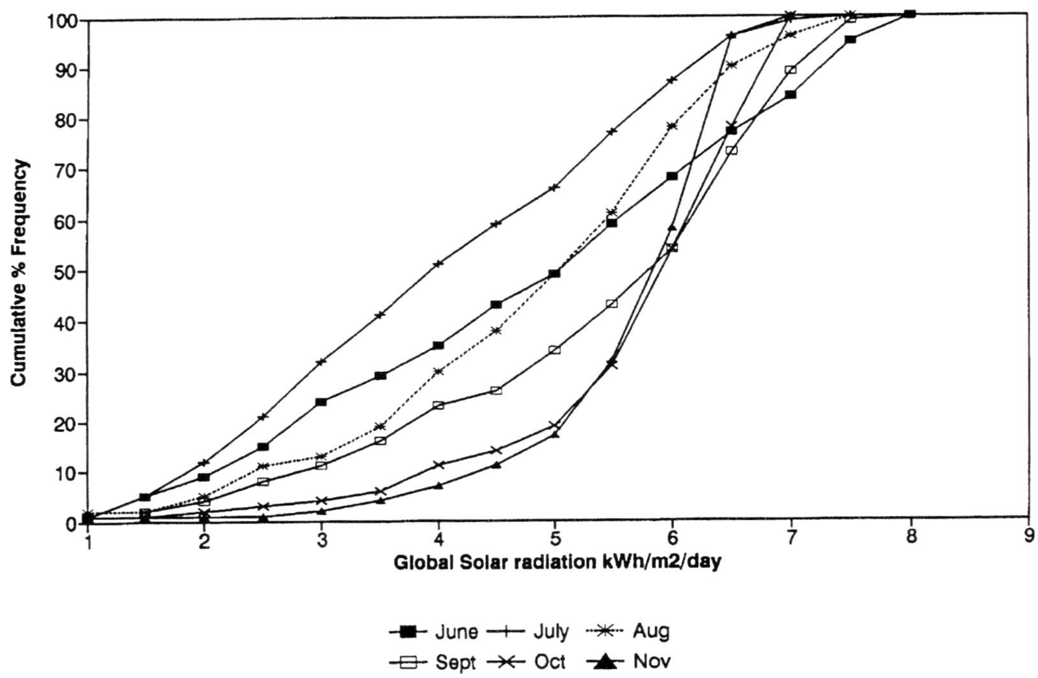

Figure 6.5.2: Cumulative Percentage Frequency Distribution (June-November) at Goa

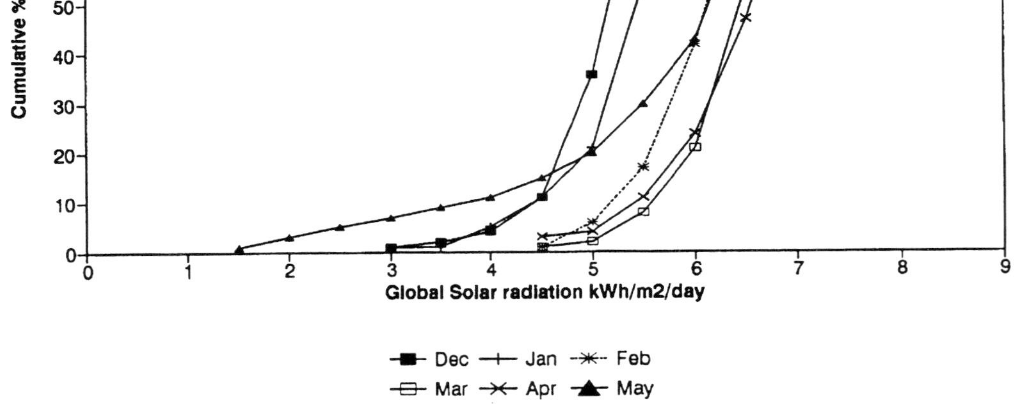

Figure 6.6.1: Cumulative Percentage Frequency Distribution (December-May) at Mangalore

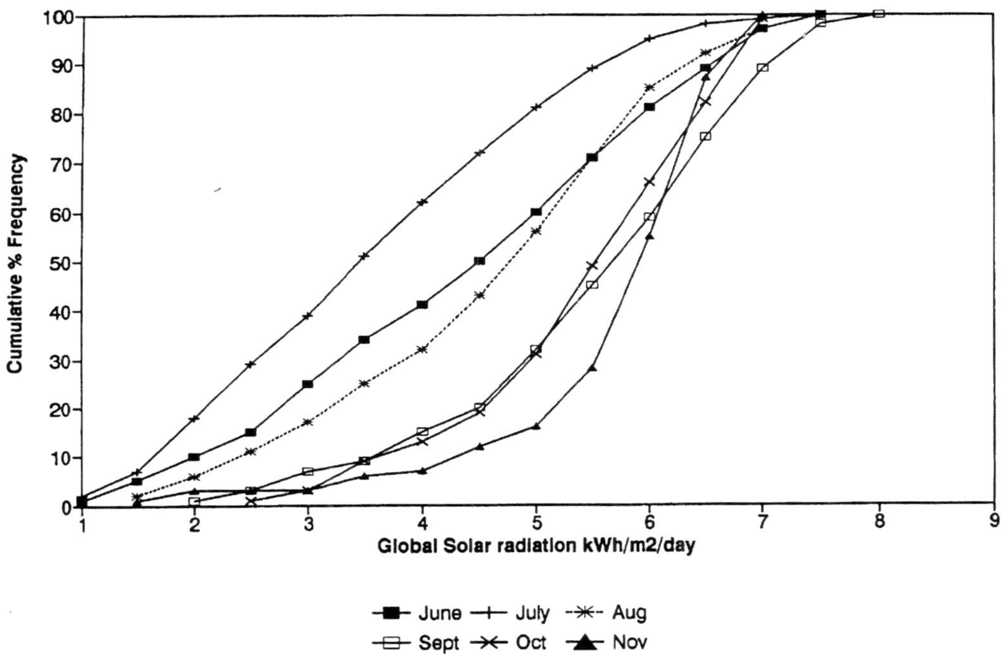

Figure 6.6.2: Cumulative Percentage Frequency Distribution (June-November) at Mangalore

Place:Goa————Karwar———Honnavar————Bhatkal————Mangalore
Distance |---->60<-----|---->74<------|------->40<----|--------->120<-------------|
(in km)

Karwar and Honnavar are the closest to the Goa radiation station. While, Bhatkal is closer to Mangalore. The India Meteorological Department has set up climatological stations at Karwar (14° 47'N latitude, 74° 08' longitude, elevation 4 m, established in 1877), Honnavar (14° 17'N latitude, 74° 27' longitude, elevation 26 m, established in 1935) and Bhatkal [Shirali (14° 05'N latitude, 74° 32' longitude, elevation 45 m, established in 1974)], where observations are made at least once per day, including daily readings of extreme temperatures and amount of precipitation.

The computations of daily sums of global radiations at Karwar, Honnavar and Bhatkal (Shirali) are carried out mainly by using regression relations between solar radiation and sunshine duration. These regression relations are derived from 25 years of data collected at Goa and 20 years at Mangalore. It is found that computed values of solar radiation agree within 5-10% of the observed values.

6.5.0 Estimation of Solar Energy in the Coastal Belt of Uttara Kannada District

Data of hours of sunshine, mean temperature, minimum and maximum daily temperature, relative humidity and total rainfall, all as monthly averages for Karwar (for 1952-89), Honnavar (for 1939-89) and Shirali (for 1974-89) provided by the India Meteorological Department (at Pune) have been analysed and used to estimate solar radiation in the coastal belt of Uttara Kannada. Knowledge of the regression relation between global radiation and sunshine hours at Goa and Mangalore (which agrees with observed values within 2-5%), is interpolated to determine global radiation in the coastal belt. Radiation data of Mangalore for 1963-1983 and Goa for 1963-1988, obtained from the India Meteorological Department and Handbook of solar radiation of the Department of Science and Technology, Government of India (Mani, 1980; Mani and Rangarajan, 1982), show the relation between G (global radiation) and hours of sunshine (equations. 6 and 7) as

$$G'/ETR = a + b\,(n/N') \quad (6)$$
$$G\text{-}G' = G.R.[(0.25\, n/N') + 0.60\,(1\text{-}n/N')] \quad (7.1)$$

In addition, in the following section, the influence of other climatological parameters such as Mean Temperature (Tm), Relative Humidity (RH), Specific Humidity (SH), Rainfall (R') and ratio of Minimum and Maximum Temperature (ψ) either in terms of single parameter or multi parameters on global radiation is investigated. The specific humidity (SH) is used instead of Relative Humidity (RH) to take care of the relatively large variations in relative humidity (Garg and Garg, 1982) and is given by,

$$SH = RH(4.7923 + 0.3647\, Tm + 0.55\, Tm^2 + 0.0003\, Tm^3) \quad (7.2)$$

The different relations obtained considering various climatological parameters are:

$$(G'/ETR) = a + b\,(n/N') \quad (6)$$
$$(G'/ETR) = c_1 + c_2\,(n/N') + c_3\,(Tm) \quad (8)$$
$$(G'/ETR) = d_1 + d_2\,(n/N') + d_3\,(\psi) \quad (9)$$
$$(G'/ETR) = e_1 + e_2\,(n/N') + e_3\,(\psi) + e_4(SH) \quad (10)$$
$$(G'/ETR) = f_1 + f_2\,(n/N') + f_3\,(Tm) + f_4(SH) \quad (11)$$
$$(G'/ETR) = g_1 + g_2\,(n/N') + g_3(Tm) + g_4(RH) \quad (12)$$
$$(G'/ETR) = h_1 + h_2\,(n/N') + h_3\,(Tm) + h_4\,(SH) + h_5\,(R') \quad (13)$$
$$(G'/ETR) = j_1 + j_2\,(n/N') + j_3\,(Tm) + j_4\,(RH) + j_5\,(R') \quad (14)$$
$$(G'/ETR) = k_1 + k_2\,(n/N') + k_3\,(n/N')^2 \quad (15)$$

Table 6.7.1 and Table 6.7.2 list the results of various relations tried for data at Goa and Mangalore. From Table 6.7.1 and Table 6.7.2, it is found that equations 11 and 13 are most appropriate, as they have the least error of y-estimate compared to other equations. Using these relations and climatological parameters of Karwar, Honnavar and Bhatkal, the monthly global radiation received at these places is estimated. Since Karwar and Honnavar are located close to Goa, their values are estimated based on the relation derived at Goa. For Bhatkal, the relation derived for Mangalore (since Bhatkal is closer to Mangalore than to Goa) is used. The global solar radiation computed using equations 11 and 13 for Goa and Mangalore is listed in Table 6.8.1. It is seen that the computed and observed values of G agree within ± 5% for Goa and Mangalore. The observed and computed values are presented in Figure 6.7.1 for Goa and Figure 6.7.2 for Mangalore. Estimated values of GR for Karwar, Honnavar and Bhatkal are listed in Table 6.8.2 using equations 11 and 13. The global radiation measured at Goa and Mangalore and computed GR values using equation 11 for Karwar, Honnavar and Bhatkal (Shirali) are given in Figure 6.8.1. Figure 6.8.2 presents the same for computed GR using equation 13, which clearly shows that GR undergoes the same seasonal changes as Goa or Mangalore (also the magnitude of GR at these sites falls within that of Goa and Mangalore). Karwar has GR in the range 5.7-6.9 kWh/m^2 during January-May and minimum of 4-5 during the monsoon months July-September. At Honnavar, the GR range during January-May is 5.47-6.5 kWh/m^2 and is minimum during monsoon months. Thus, this study in five locations of coastal area of the Western Ghats region demonstrates that good solar energy is available in this region during most months. For only a few days in the monsoon months July-August, cloud cover in this region was thick and solar radiation was of the order 3.5-4.5 kWh/m^2/day.

Table 6.7.1: Regression Constants, Correlation Coefficient Computed for Various Relationships (Goa)

Place	Equation	Reg. Const.	R^2	R	Std. Error
Goa	6	a = 0.260 b = 0.577	0.946	0.972	0.0322
	8	c_1 = 0.523 c_2 = 0.646 c_3 = -0.010	0.962	0.980	0.0295
	9	d_1 = 0.330 d_2 = 0.565 d_3 = -0.082	0.946	0.972	0.033
	10	e_1 =0.763 e_2 =0.507 e_3 =-0.360 e_4 =-0.005	0.992	0.995	0.0170
	11	f_1 =0.389 f_2 =0.519 f_3 =0.005 f_4 =-0.007	0.994	0.996	0.0125
	12	g_1 =1.075 g_2 =0.457 g_3 =-0.0114 g_4 =-0.00553	0.982	0.990	0.0214
	13	h_1 =0.350 h_2 =0.640 h_3 =0.0032 h_4 =-0.0068 h_5 =0.00010	0.995	0.997	0.0117
	14	j_1 =0.789 j_2 =0.624 j_3 =0.0125 j_4 =-0.0028 j_5 =0.00082	0.994	0.996	0.0128
	15	k_1 =0.269 k_2 =0.535 k_3 =0.039	0.9460	0.972	0.033

Table 6.7.2: Regression Constants, Correlation Coefficient Computed for Various Relationships (Mangalore)

Place	Equation	Reg. Const.	R^2	R	Std. Error
Mangalore	6	a = 0.208 b = 0.518	0.973	0.9864	0.020
	8	c_1 =0.251 c_2 =0.527 c_3 =-0.0017	0.977	0.9884	0.020
	9	d_1 =-0.366 d_2 =0.612 d_3 =0.674	.978	0.9889	0.019
	10	e_1 =-0.622 e_2 =0.651 e_3 =0.908 e_4 =0.0014	0.979	0.9897	0.019
	11	f_1 =0.2948 f_2 =0.5692 f_3 =-0.0077 f_4 =0.0028	0.994	0.9969	0.012
	12	g_1 =0.0506 g_2 =0.581 g_3=-0.00024 g_4 = 0.00160	0.995	0.997	0.0121
	13	h_1 =0.397 h_2 =0.132 h_3 =0.0059 h_4 =-0.0014 h_5=-0.00025	0.9943	0.9971	0.0112
	14	j_1 =0.489 j_2 =0.135 j_3 =0.0024 j_4 =-0.005 j_5=-0.00082	0.994	0.996	0.0011
	15	k_1 =0.131 k_2 =0.853 k_3 =-0.303	0.982	0.990	0.017

Table 6.8.1: Actual and Estimated Values of GR and % Error computed for Goa and Mangalore in kWh/sq.m./day

Month	Goa					Mangalore				
	Actual	Estimated		% Error		Actual	Estimated		% Error	
	GR	Eq 11	Eq 13	Eq 11	Eq 13	GR	Eq 11	Eq 13	Eq 11	Eq 13
January	5.68	5.80	5.91	-2.03	-4.12	5.33	5.53	5.38	-3.82	-1.03
February	6.37	6.51	6.47	-2.30	-1.66	6.03	5.97	5.76	0.96	4.51
March	6.74	6.49	6.51	3.62	3.44	6.39	6.22	6 17	2.60	3.35
April	6.90	6.82	6.83	1.10	0.90	6.43	6.38	6 32	0.78	1.65
May	6.72	6.68	6.70	0.57	0.32	5.64	6.09	5 82	-8.08	-3.24
June	4.76	4.91	4.79	-3.18	-0.65	3.88	4.03	4.03	-3.75	-3.85
July	3.97	4.05	4.00	-1.96	-0.89	3.07	3.19	3.14	-4.05	-2.43
August	4.83	4.64	4.83	3.82	-0.05	4.06	3.90	3.75	4.08	7.69
September	5.43	5.22	5.23	3.73	3.60	4.95	4.47	4.66	9.59	5.83
October	5.67	5.98	5.94	-5.55	-4.76	4.94	5.16	5.30	-4.53	-7.30
November	5.55	5.33	5.31	3.97	4.33	5.16	5.40	5.35	-4.61	-3.74
December	5.36	5.52	5.44	-2.96	-1.55	5.07	5.09	5.25	-0.39	-3.54

Table 6.8.2: Estimated Values of GR at Karwar, Honnavar and Shirali using Eq 11 and Eq 13.

Month	Karwar		Honnavar		Shirali	
	Eq 11	Eq 13	Eq 11	Eq 13	Eq 11	Eq 13
January	5.67	5.97	5.47	5.86	5.35	5.74
February	6.27	6.40	6.10	6.32	5.73	5.98
March	6.38	6.58	6.24	6.51	6.04	6.30
April	6.69	6.86	6.39	6.64	6.30	6.53
May	6.48	6.55	6.23	6.38	6.21	6.46
June	5.10	4.54	4.93	4.28	4.50	3.91
July	4.02	3.68	4.17	3.57	3.50	3.41
August	4.84	4.70	4.86	4.59	4.78	4.07
September	5.10	5.32	5.27	5.43	5.18	5.26
October	5.41	5.81	5.35	5.78	5.25	5.63
November	5.13	5.66	5.03	5.61	5.03	5.53
December	5.12	5.65	4.97	5.53	4.79	5.30

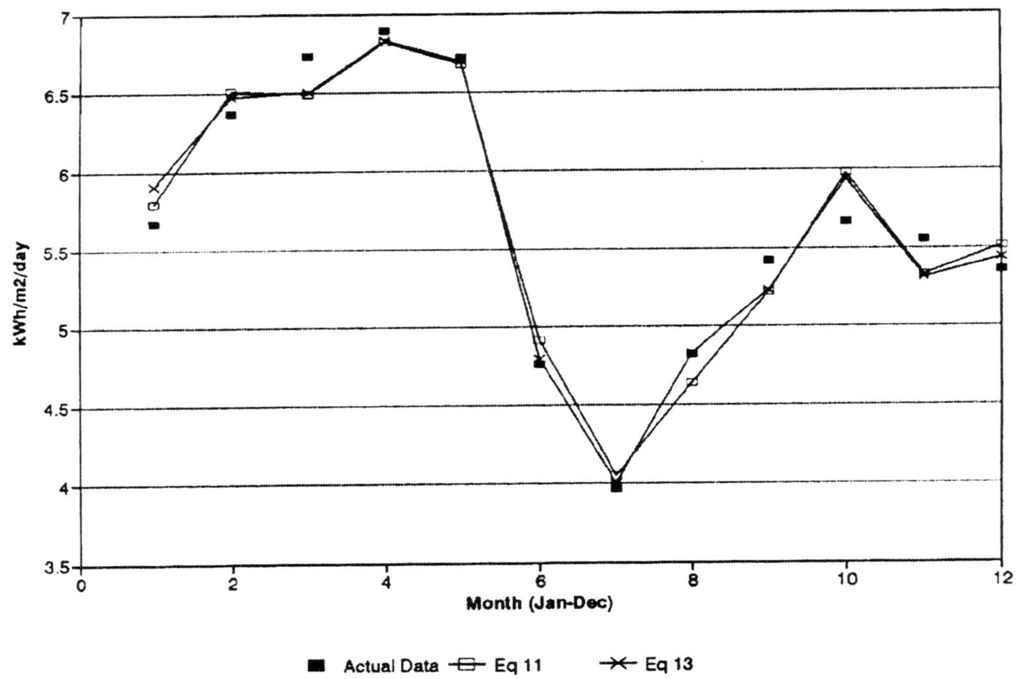

Figure 6.7.1: Global Radiation (Estimated and Actual Values) at Goa

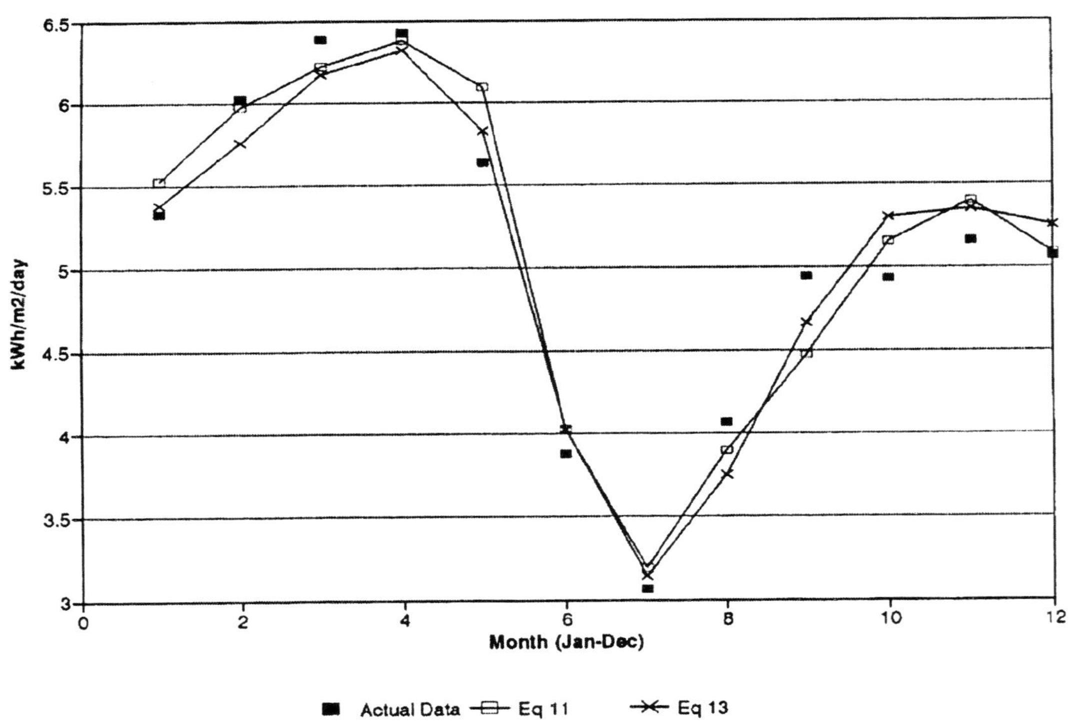

Figure 6.7.2: Global Radiation (Estimated and Actual Values) at Mangalore

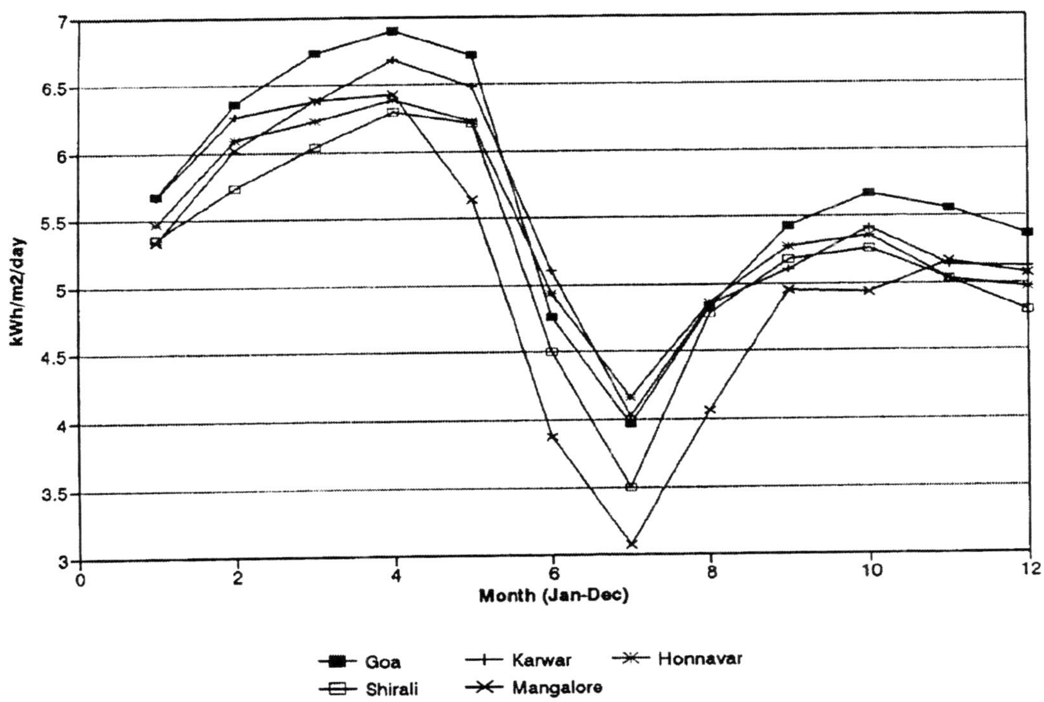

Figure 6.8.1: Global Radiation at Goa, Karwar, Honnavar, Shirali and Mangalore (as per equation.11)

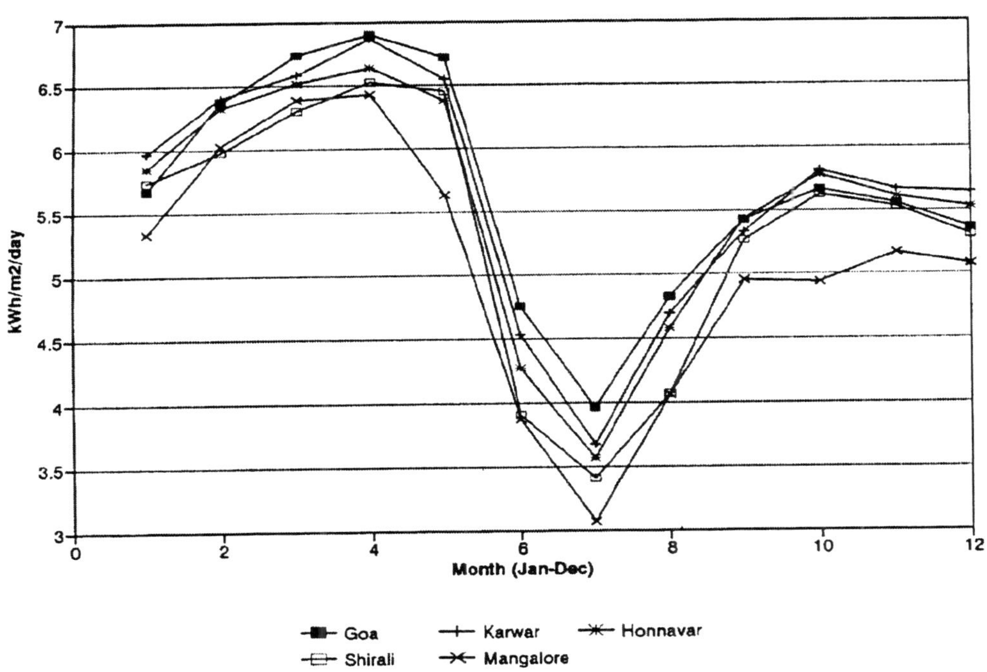

Figure 6.8.2: Global Radiation at Goa, Karwar, Honnavar, Shirali and Mangalore (as per equation.13)

6.5.1 Solar Energy Potential Assessments in the Coastal Belt of Uttara Kannada District

In determining a realistic potential for solar energy converting systems, it is quite evident that one should take into account the availability of solar energy associated with time, such as daily variation due to day and night cycle, seasonal change due to the earth's orbit around the sun, and variation due to local weather conditions. With detailed investigation of solar energy availability and changes at various places in the coastal area, assessment of the potential that could be harnessed to meet the community energy needs for light, heat and power in the immediate future is undertaken. Due to the dilute nature of solar energy flux at the earth's surface, large collecting efforts are required in many applications. This makes it necessary to look at land use patterns in a region and land available for harnessing solar energy. Solar energy systems would be the most interesting option economically, if the peak demand for electricity coincides with the highest insolation level. With integrated system of solar/wood/fossil fuels (backup), peak power requirements during the evening hour could be supplied by other sources. Solar and conventional power plants are, to a large extent, based on the same principles. Numerous components identical to both systems may be integrated in one solar-wood/fossil hybrid. In this way, the total investment is significantly reduced over that required for two single plants. The hybrid system may be optimised for maximum power increase by the solar part of the system by using solar boosters, if the peak load is expected in the afternoon hours. Apart from this, there are options like storage through batteries (not a solution for more than few hours), pumped storage hydro backup (region and location specific), and production of hydrogen through electrolysis of water (rather poor efficiency of 50% or less for the production of hydrogen and reconversion to electricity).

The coastal belt in Uttara Kannada is densely populated, and human influence on forests is quite evident from the barren hill tops (due to deforestation) in this area. Table 6.9.1 lists talukwise barren/un-cultivable/wasteland in the coastal region of Uttara Kannada. This varies from 3.34% of the total geographical area (Bhatkal) to maximum of 8.86% (Ankola). For the purpose of installing solar devices to harness solar energy in this belt, it is assumed that 1- 5% of present wasteland/barren land could be made available. With the assumption that 5% of the present wasteland in each taluk could be conveniently used for harnessing solar energy, the monthly solar energy that could be harnessed either through Photovoltaic or Solar Thermal Power Stations is computed for each taluk. Table 6.9.2 lists the amount of solar energy harnessed by utilising 5% of present wasteland as solar collector area. It is found that about 95.72 million units (mkWh) could be generated annually from the coastal belt (40.90 mkWh annually at Karwar, 21.98 at Ankola, 22.60 at Kumta, 4.176 at Honnavar and 6.045 at Bhatkal). Table 6.9.3 shows that by utilising 1% of available wasteland in the respective taluks of coastal Uttara Kannada, 19.14 million units per year could be harnessed.

Table 6.9.1: Barren / Wasteland Available in Coastal Taluks of Uttara Kannada District

	Karwar	Ankola	Kumta	Honnavar	Bhatkal	U.K. Total Area
Total Geo Area (Hect)	70975	91127	56663	74555	34380	1029100
% to Dist Area	6.90	8.86	5.51	7.24	3.34	
Barren/Uncultivable land (Hect)	4066	2186	2247	423	635	
Barren as % of Taluk's T.Area	5.73	2.40	3.97	0.57	1.85	
Barren/Uncultivable land (sq. m)	406600	218600	224700	42300	63500	
Barren/Uncultivable land (5%) sq.m.	20330	10930	11235	2115	3175	
Barren/Uncultivable land (1%)	4066	2186	2247	423	635	

Annual electricity consumption in Uttara Kannada in various sectors is about 294.049 million units. By utilising 5% of currently available wasteland in the coastal area for harnessing solar energy, at least 32.55% of present demand of electricity could be met by solar option.

6.6.0 TECHNO-ECONOMIC ANALYSES OF SOLAR DEVICES FOR HARNESSING SOLAR ENERGY

Solar Thermal Systems and Photovoltaic systems are the two fundamentally different concepts of solar energy conversion. For solar thermal conversion, a heat carrying medium (usually liquid or a gas) flows continuously through a heat receiver, thus transporting the absorbed heat to the site of thermal use for generation of electricity in a conventional bottoming steam cycle. While photovoltaic systems employ semiconducting compound having characteristics of an energy threshold, contributing directly to the electricity. The conversion efficiency is independent of system capacity and incoming radiation intensity. As a consequence, a modular design from a few watts to the Mega watt scale becomes possible without significant loss in performance even in the case of only moderate insolation. Different solar concepts and possible applications are listed in Table 6.10.1.

6.6.1 Solar Thermal Electric Devices

Parabolic trough, parabolic dish and central receiver are the three main types of solar thermal electric technologies. Among these, the most advanced type of the "concentrator" systems is the parabolic trough (De Laquill, Kearney, Geyer and Diver, 1993). Parabolic troughs track the sun along one axis, concentrating the energy into a receiver tube located at the trough's focal line. Concentration ratios of 10-100 are typically achieved, with operating temperatures of about 400°C. The receiver tube usually has water or oil running through it as the heat transfer medium. This fluid is then piped from each of the parabolic troughs to a central area, where the energy is converted to electricity.

The parabolic dish operates on the same principle as the parabolic trough, but it tracks the sun on two axes, concentrating the energy at the focal point of the dish because it is always pointed at the sun. The parabolic dish's concentration ratios are considerably higher than the trough, of the order of 600-2000, and the operating temperature exceeds 1500°c. The power generating equipment for use with parabolic dishes may be mounted at the focal point of the dish itself, or as with the trough, energy may be collected from a number of separate installations and converted to electricity at a central point. The former option is perhaps the most promising, as the use of the dish technology makes it very well suited to remote or stand-alone applications.

The central receiver is a very promising technology for large-scale grid connected power generation, even though it is at an early stage of development compared to parabolic trough technology. In this case, heliostats and flat tracking mirrors concentrate the sun's energy onto a central receiver tower. Concentration ratios are 300-1500 and systems can operate at temperatures of 500-1500°c. Energy losses from thermal energy transport are minimised as solar energy is being directly transferred by reflection from the heliostats to a single receiver rather than being moved through a transfer medium from several receivers to one central point, as with parabolic troughs. Solar to electric efficiencies for test systems are in the range of 10-15%.

Table 6.9.2: Solar Radiation Potential kWh/month by using 5% of Current Barren Land for Harnessing Solar Energy in Coastal Taluks

Month	Karwar		Ankola		Kumta		Honnavar		Bhatkal	
	Eq 11	Eq 13	Eq 11	Eq 13	Eq 11	Eq 13	Eq 11	Eq 13	Eq 11	Eq 13
January	3571862.6	3763254.2	1920337.3	2023235.1	1973924.1	2079693.1	358347.8	384025.8	527062.3	564895.9
February	3569323.9	3644035.8	1918972.4	1959139.8	1972521.1	2013809.3	361310.3	374489.3	509328.9	531204.6
March	4020100.6	4149169.9	2161323.1	2230714.6	2221634.5	2292962.3	408823.9	427132.8	594301.2	620420.8
April	4081315.4	4186932.4	2194234.0	2251016.8	2255463.8	2313831.1	405490.3	421113.9	599750.2	621787.7
May	4085840.5	4126572.0	2196666.9	2218565.2	2257964.5	2280474.0	408152.9	418125.6	611482.0	635418.9
June	3109938.0	2766944.1	1671993.2	1487589.7	1718649.9	1529100.7	312790.4	271496.3	428625.0	372448.5
July	2534637.3	2318434.8	1362694.8	1246458.0	1400720.6	1281240.3	273089.3	234359.0	344487.5	335624.4
August	3051231.8	2959221.3	1640431.0	1590963.5	1686207.0	1635359.1	318651.3	300674.0	470022.2	401069.5
September	3111272.9	3247377.9	1672710.9	1745884.9	1719387.7	1794603.6	334640.4	344550.1	493157.0	501300.7
October	3406521.5	3664514.4	1831445.2	1970149.6	1882551.4	2025126.4	350772.4	379273.7	517145.4	554110.1
November	3131027.0	3454860.8	1683331.3	1857433.8	1730304.4	1909265.2	319273.4	355993.3	478793.3	526754.2
December	3227546.4	3558419.0	1735222.9	1913109.7	1783644.1	1966494.7	325651.9	362353.5	471425.7	521782.4
Total	40900617.9	41839736.6	21989363.2	22494260.7	22602973.1	23121959.7	4176994.2	4273587.5	6045580.8	6186817.9
Total per year from all Coastal Taluks (in Million Units- mkWh):								95.72		

Table 6.9.3: Solar Radiation Potential kWh/month by using 1% of Current Barren Land for Harnessing Solar Energy in Coastal Taluks

Month	Karwar		Ankola		Kumta		Honnavar		Bhatkal	
	Eq 11	Eq 13	Eq 11	Eq 13	Eq 11	Eq 13	Eq 11	Eq 13	Eq 11	Eq 13
January	714372.5	752650.8	384067.5	404647.0	394784.8	415938.6	71669.6	76805.2	105412.5	112979.2
February	713864.8	728807.2	383794.5	391828.0	394504.2	402761.9	72262.1	74897.9	101865.8	106240.9
March	804020.1	829834.0	432264.6	446142.9	444326.9	458592.5	81764.8	85426.6	118860.2	124084.2
April	816263.1	837386.5	438846.8	450203.4	451092.8	462766.2	81098.1	84222.8	119950.0	124357.5
May	817168.1	825314.4	439333.4	443713.0	451592.9	456094.8	81630.6	83625.1	122296.4	127083.8
June	621987.6	553388.8	334398.6	297517.9	343730.0	305820.1	62558.1	54299.3	85725.0	74489.7
July	506927.5	463687.0	272539.0	249291.6	280144.1	256248.1	54617.9	46871.8	68897.5	67124.9
August	610246.4	591844.3	328086.2	318192.7	337241.4	327071.8	63730.3	60134.8	94004.4	80213.9
September	622254.6	649475.6	334542.2	349177.0	343877.5	358920.7	66928.1	68910.0	98631.4	100260.1
October	681304.3	732902.9	366289.0	394029.9	376510.3	405025.3	70154.5	75854.7	103429.1	110822.0
November	626205.4	690972.2	336666.3	371486.8	346060.9	381853.0	63854.7	71198.7	95758.7	105350.8
December	645509.3	711683.8	347044.6	382621.9	356728.8	393298.9	65130.4	72470.7	94285.1	104356.5
Total	8180123.6	8367947.3	4397872.6	4498852.1	4520594.6	4624391.9	835398.8	854717.5	1209116.2	1237363.6
Total per year from all Coastal Taluks (in Million Units- mkWh):								19.14		

6.6.2 Photovoltaic Devices

Photovoltaic (PV) modules are made from a number of materials and fabricated in a variety of designs. The principle is the same for silicon and other semiconductor devices. When sunlight falls on these materials, it frees an electron from fixed sites. The wavelength of the sunlight absorbed depends on the band gap of the material. The materials are designed so that the electrons cannot return to these sites easily except by flowing through an external circuit, thus generating a current. A typical solar cell consists of a layer of semiconductor material sandwiched between conducting top and bottom layers. PV modules are made up of several interconnecting solar cells, as the individual PV cells do not provide much power. PV modules are generally less than one square meter in size and deliver between 50 and 150 watt of electric power (Thornton and Brown, 1992). The whole set up is encapsulated in clear waterproof form arrays, either tracking or nontracking. The former is complex and more expensive, but they can optimise the system's performance. The efficiency of single crystal silicon modules is about 10-15%, while that of amorphous silicon modules is about 3-5%. However, the manufacturing cost of the former is much lower.

6.6.3 Cost of Electricity Generation from Solar Thermal Electric Technologies

The annualised capital cost (AC) is computed assuming r= 0.10 and lifetime t= 30 years, and is given by

$$AC = \frac{C*r*(1+r)^n}{(1+r)^n-1} \qquad (16)$$

$$\text{Cost of electricity} = \frac{AC + O\&M}{E}$$

where E is the number of units of kWh produced annually (kWh/yr).

The cost is computed from the quoted capital cost, assuming 2% of capital cost as operation and maintenance (O&M) cost, and the life of the plant as 30 years. The levelised cost or annual capital recovery factor is computed assuming a discount rate of 10%. The cost computed for solar thermal devices for various load factor and efficiencies is listed in Table 6.10.2.

Table 6.10.1: Various Solar Devices and Possible Applications

Solar Thermal Devices	Energy Transfer Medium	Applications
Flat Plate Collector	Water/Air, water	Low Temp system η=50-70%: Water Heating, Process Heating, Drying
Parabolic Troughs	Thermal oil, Steam	Mid-Temp system η=15-20%, (with storage): Electricity Generation, Process Heat
Heliostats	Gas, salt/gas, steam	High temp. system, η=10-20%: (with storage), Electricity Generation, Process Heat
Photovoltaic Devices / solar cells / flat plate collector	Electricity	Electricity η=15-20%: Irrigation, Cooling, Drying, Lighting, Telecommunications, Battery Charging, Aerospace.

Table 6.10.2: Cost per kWh at Various Efficiencies (Solar thermal systems)

Size (Megawatt)	Efficiency (%)	Load factor	Cost (Rs./kWh)
30	12	0.45	19.98
30	12	0.65	13.83
30	12	0.85	10.58
30	30	0.45	6.85
30	30	0.65	4.74
30	30	0.85	3.62
30	50	0.45	4.79
30	50	0.65	3.32
30	50	0.85	2.53
30	70	0.45	3.42
30	70	0.65	2.37
30	70	0.85	1.81

6.6.4 Solar Energy for Water Heating Purposes

Solar energy for water heating would decrease fuel wood consumption in the domestic sector. Results of a survey of rural households in Kumta concerning trends in fuel wood consumption in the domestic sector (kg/capita/day) are listed in Table 6.11.1.

Table 6.11.2 and Table 6.11.3 list population by taluk and seasonal requirement of fuel wood and agricultural residues (converted to fuel wood equivalent) for water heating and cooking. It is seen, on average, that the coastal belt of Uttara Kannada requires 222.56 kilo tonnes of fuel wood for water heating and 185.81 kilo tonnes of fuel wood for cooking. A fuel wood survey conducted in the villages of Kumta revealed that about 46.7% of the population take a bath every day, 32.45% twice a day during the agricultural season, 32% once every 2-3 days, 12% once every 4 days and 9.3% once a week.

Table 6.11.1: Per Capita per Day Fuel Consumption

	Type	Summer	Winter	Monsoon
Cooking	fuel wood	0.78	0.90	0.98
Water Heating	agri-residue, fuel wood	0.60	1.20	1.40

Table 6.11.2: Seasonwise Fuel wood Consumption (in tonnes) for Water Heating

Taluk	Population	Summer	Winter	Monsoon	All season
Karwar	136470.00	10235.25	19651.68	22926.96	52813.89
Ankola	103466.00	7759.95	14899.10	17382.29	40041.34
Kumta	107963.00	8097.23	15546.67	18137.78	41781.68
Honnavar	129650.00	9723.75	18669.60	21781.20	50174.55
Bhatkal	97539.00	7315.43	14045.62	16386.55	37747.59
Total for Coastal area		43131.60	82812.67	96614.78	222559.06
Total for Coastal (kilo tonnes):		43.13	82.81	96.61	222.56
30% anticipated replacement in kilo tonnes:					66.77

Table 6.11.3: Fuel wood Consumption (in tonnes) per Season for Cooking

Taluk	Population	Summer	Winter	Monsoon	All season
Karwar	136470	13305.83	14738.76	16048.87	44093.46
Ankola	103466	10087.94	11174.33	12167.60	33429.86
Kumta	107963	10526.39	11660.00	12696.45	34882.85
Honnavar	129650	12640.88	14002.20	15246.84	41889.92
Bhatkal	97539	9510.05	10534.21	11470.59	31514.85
Total for Coastal area		56071.08	62109.50	67630.35	185810.93
Total for Coastal (kilo tonnes):		56.07	62.11	67.63	185.81
5% anticipated replacement in kilo tonnes:					9.29

Solar water heating technology has been well established and functioning satisfactorily in some places in Bangalore city for the past ten years. An integrated solar system with electric immersion heater is satisfactorily supplying the energy needs for all periods and seasons of the year. Only during very cloudy days immersion heaters are used. The cost of such devices is about Rs. 12000 for 200 litres capacity and Rs. 8000 for 100 litres capacity.

In the coastal area, it is seen that the climate is favorable to set up solar water heating devices to meet the requirement of hot water for bathing purposes. If 30% of the population switches over to solar water heating devices, the fuel wood savings would be of the order of 66.77 kilo tonnes per year from five coastal taluks of Uttara Kannada. Lack of knowledge of technology and non-availability of proper service/maintenance backup in this region are the main reasons that energy efficient devices and alternative technology are not finding significant usage in the rural energy scene.

Rural industries like cashew processing industries are major consumers of fuel wood in rural coastal areas of Uttara Kannada. A primary survey conducted in the cashew processing industries has revealed that some industries use fuel wood of the order of 4-8 kg for processing a kilogram of cashews (in boilers, dryers, etc.). Also, these industries are seasonal and operate at a peak load during the pre-monsoon period (February-June). It is possible for them to switch over to solar energy for conditioning, mild roasting and Brooma drying operations (solar energy is also plentiful during this period).

Drying is one of the most important processes in the post harvest handling of agricultural products such as cardamom, pepper, chillies, copra and arecanut. Drying of agricultural products is an energy intensive process. It is estimated that to remove a kilogram of water 2.5 MJ of heat energy is required (Malhotra and Imre, 1990), mainly through fuel wood and other bio residues.

6.7.0 CONCLUSIONS

Goa and Mangalore receive global solar radiation of 5.5-7.5 kWh/m^2/day during November-May period. During June-August and September-October the solar radiation received is 2.5-5.0 and 3.5-6.5 kWh/m^2/day respectively. These analyses show that solar potential at Goa and Mangalore is 5-7 kWh/m^2/day for about nine months a year and 3.5-5 kWh/m^2/day during the monsoon season. Dense cloud cover for about 10-12 days, on average, in July and 5-6 days in August are the lean periods for solar energy in this region. Computed global radiation using equation 6 (linear relationship between global radiation and sunshine hours) and Observed values of global radiation varies within the range of 2-7%.

In order to improve accuracy using the available data, attempts are made to estimate global radiation using parameters such as sunshine duration, temperature, relative humidity, specific humidity and rainfall (data collected for 25 years at Goa and 20 years at Mangalore). Based on the high correlation coefficient and minimum standard error of 'y' estimate, it can be concluded that equations 11 and 13 are appropriate to estimate global solar radiation. It is found that observed and computed values of solar radiation agree within ±5% of the observed values. With this experience, GR is computed at Honnavar, Karwar and Bhatkal based on climatological data of 50 years at Honnavar, 39 years at Karwar, and 15 years at Bhatkal (Shirali) using equations 11 and 13.

Karwar has a GR range of 5.5-6.5 kWh/m^2 during January-May and is in the minimum range of 4-5 during monsoon months July-September. While at Honnavar, GR during January-May is 5.47-6.5 kWh/m^2 and minimum during the monsoon months. Thus, this study of the coastal area of the Western Ghats region demonstrates that good solar energy potential is available in this region during most months of the year.

The amount of solar energy that could be harnessed by utilising 5% of the present wasteland in coastal taluks as solar collection area is found to be of the order of 95.72 million units (mkWh) annually from coastal belt (40.90 mkWh annually at Karwar, 21.98 at Ankola, 22.60 at Kumta, 4.175 at Honnavar and 6.045 mkWh annually at Bhatkal). This constitutes 32.55% of present electricity demand from various sectors of Uttara Kannada.

6.8.0 ACKNOWLEDGEMENTS

I am grateful to Prof. J. Srinivasan and Dr. G.S. Bhat for their suggestions. I thank the India Meteorological Department (IMD), Pune, for providing climatological data. Helpful communication with Dr. C.D. Padmanaban, Dr. G.L. Mokashi, and Dr. M.I. Ali of IMD, is gratefully acknowledged. I thank Dr. Tulsidas, Director, IMD, Bangalore for providing daily surface data of recent years for Karwar, Honnavar, and Shirali meteorological observatories.

6.9.0 REFERENCES

1. Kimball H.H., 1919. *Monthly Weather Review*, 47:769.
2. Angstrom, A., 1924. Solar and terrestrial radiation. *Qr. J. R. Meteor. Soc.*, 50:121.
3. Angstrom, A., 1956. On computation of global radiation from records of sunshine. *Arkiv Geophisk*, 3(23):551.
4. Page, J.K., 1980. Climatic considerations and energy conservation. In: *Interactions of Energy and Climate*. Batch, W., Pankarth, J. and Williams, J. (Eds.), pp. 73-88. Reidel Publishing, Dordrecht.
5. WMO, 1981. Meteorological aspects of the utilisation of solar energy as an energy source. *WMO Technical Note*, No. 172, WMO, Geneva.
6. Page, J.K., 1984. Solar radiation and climatic data. In: *Proc. Solar World Congress*, Perth, Australia, Szokolay, S.V. (Ed.), 4:2066-2082, Pergamon Press, Oxford
7. Tasdemiroglu, E. and Sever, R., 1989. Estimation of solar radiation from bright sunshine hours in Turkey. *Energy*, 14(12):827-830.
8. Page, J.K., 1989. Operational aspects of using meteorology for energy purposes. *Solar and Wind Technology*, 6(4):441-491.

9. Hoyt, D.V., 1977. Percent of possible sunshine and total cloud cover. *Monthly Weather Review*, 105:648-655.
10. Hay, J.E., 1979. Calculations of monthly solar radiation for horizontal and inclined surfaces. *Solar Energy*, 23:301-307.
11. op cit. Ref. 10.
12. Collares-Pereira, M. and Rabl, A., 1979. The average distribution of solar distribution-correlations between diffuse and hemispherical and between daily and hourly insolation values. *Solar Energy*, 22:155-164.
13. Mani, A. (Ed.), 1980. *Handbook of Solar Radiation Data for India*. Allied Publishers Private Limited, New Delhi, India.
14. Mani, A. and Rangarajan, S., 1982. *Solar Radiation Over India*. Allied Publishers Pvt. Ltd., New Delhi, India.
15. Garg, H.P. and Garg, S.N., 1982. Prediction of global solar radiation from bright sunshine and other meteorological parameters, Solar-India. In : *Proceedings of National Solar Energy Convention*. pp. 1004-1007, Allied Publishers, New Delhi, India.
16. Laquill III, P. De., David, K., Michael, G. and Richard, D., 1993. Solar thermal electric technology. In: *Renewable Energy: Source for Fuel and Electricity*. Johansson, T.B., Kelly, H., Reddy, A.K.N., Williams, R.H. (Eds.). Island Press, Washington D.C., U.S.A.
17. Thornton, J. and Brown, L., 1992. The promise and payoff of photovoltaics. *The Electricity Journal*, 5(3):34-41.
18. Malhotra, A.K. and Imre, L. 1990. Role of solar agricultural drying in developing countries. *International Journal of Ambient Energy*, 11(4):205-210.

6.10.0 APPENDIX

6.10.1 Different Radiation Parameters

G: Global solar radiation, is the total quantity of shortwave solar radiant energy emitted by the sun's disc as well as that scattered diffusively by the atmosphere and clouds passing through a unit horizontal area in units of time. It is also referred to as incoming total shortwave radiation.

D: Diffuse solar radiation, is that part of shortwave radiation scattered by the atmosphere reflected diffusely and transmitted by clouds and passing through a unit horizontal area in units of time. This radiation comes from a solid angle of 2π with the exception of the solid angle subtended by the sun's disc.

I_H: Direct solar radiation on a horizontal surface, is the quantity of solar radiant energy emitted from the solid angle subtended by the visible disc of the sun and passing through a unit horizontal area in units of time. This is also called the vertical component of the direct solar radiation. $I_H = G\text{-}D$

I_N: Direct solar radiation at normal incidence, is the quantity of shortwave radiation energy emitted by the solid angle subtended by the visible disc of the sun and passing through a unit area held normal to the solar beam at the earth's surface in units of time. It is given by $I_N = (G\text{-}D)/\sin(h)$, where 'h' is the angle of elevation of the sun.

6.10.2 Mean Daily Radiation and Sunshine parameters

ETR: extra terrestrial radiation on a horizontal surface for a whole day, (kWh/sq.kms.)
n: actual sunshine duration during a day (hours)
N: maximum possible duration of sunshine, in hours and tenths
N': maximum possible sunshine during a day with solar elevation $\geq 5°$, in hours and tenths
G: daily sum of global solar radiation, in kWh/m^2
D: daily sum of diffuse solar radiation in kWh/m^2
I_H: daily sum of direct solar radiation on a horizontal surface in kWh/m^2
G(c): daily sum of global solar radiation on a clear sky under average conditions in kWh/m^2
C.I.: clearness index = G/ETR (dimensionless)

WS: sunset hour angle, degrees; when divided by 7.5 gives the duration of the day (hours)

6.10.3 Clear Sky Noon Radiation Parameters

I: Direct solar radiation for clear sky noon under average conditions kW/m^2
I(H): Direct solar radiation on a horizontal surface under average conditions kW/m^2 or $kWh/m^2/day$
D: Diffuse solar radiation on a horizontal surface under average conditions kW/m^2 or $kWh/m^2/day$
G: Global solar radiation on a horizontal surface under average conditions kW/m^2 or $kWh/m^2/day$
W: total precipitable water vapor in the atmosphere gm/sq.cm
P: barometric pressure at station level (millibar)

6.10.4 Mean Hourly Meteorological Parameters

T: mean hourly temperature, °C
RH: mean hourly relative humidity, percent (U)
W: mean wind speed in km/hr for certain specified synoptic hours ranging from 1 to 6 in a day, km/h
R: mean total rainfall for a day from 0830 I.S.T. to 0830 I.S.T next day, millimeter
T(m): mean temperature (Tm)
SH: specific humidity (SH)
ψ: ratio of minimum and maximum temperature (ψ)
I.S.T.: The Indian Standard Time - is the mean solar time for longitude 82° 30'

Chapter 7

PART I: ENERGY CONSUMPTION PATTERNS IN THE DOMESTIC SECTOR

7.0 SUMMARY

Traditional Cook Stove

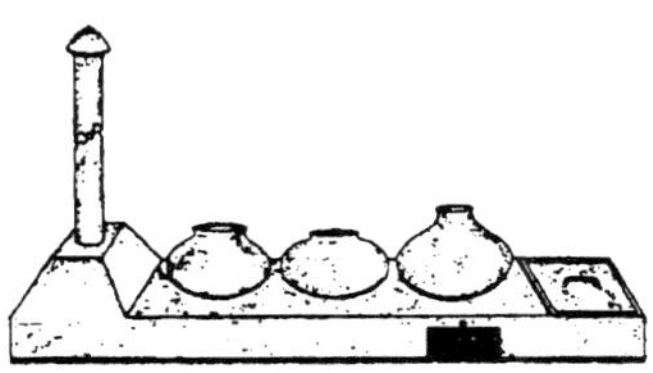

Improved Cook Stove

Energy planning of any region should be based on existing levels of energy consumption. Sectorwise disaggregated information of energy usage is developed for Uttara Kannada district to assist in the regional energy planning exercise.

This chapter provides comparative analyses of village level domestic energy consumption pattern across coastal, interior, hilly and plain zones considering regional and seasonal variations. Cooking, water heating and space heating are the major end use activities. The results based on eighteen months of field research in five taluks of the district reveal that the average energy consumption norm does vary significantly for cooking and water heating in various seasons across the zones.

Among the five taluks, 90 villages (out of 119) and all divisions of Kumta town in Kumta taluk, and 190 - 220 randomly selected households in selected villages of Sirsi, Mundgod, Siddapur and Ankola were studied.

Survey of 1304 households from 90 villages in Kumta taluk shows that most of them still use traditional stoves for cooking (97.92%) and water heating (98.3%). Average consumption (kg per person per day) for cooking ranges from 2.01 ± 1.49 (coastal) to 2.32 ± 2.09 (hilly). Seasonwise fuel wood requirement for cooking in coast and hilly zones ranges from 1.98 and 2.22 (summer) to 2.22 and 2.51 (monsoon) respectively. While, for water heating (for bath and washing) it ranges from 1.17 ± 0.02 (coast) to 1.63 ± 0.05 (hilly). Seasonal variation is evident from the range 1.12 and 1.53 (summer) to 1.22 and 1.73 (monsoon) for coastal and hilly zones respectively.

Analysis of other sources of energy for domestic purposes shows that kerosene is used for cooking and lighting in the coast. Kerosene consumption (litres per person per month) for cooking ranges from 0.05 (hilly) to 0.34 (coast) and for lighting from 0.74 (coast) to 0.98 (hilly). Availability of bioresources in hilly zone is the main reason for less consumption of kerosene for cooking. In hilly zone, electrification of all households has not been possible as they are scattered. Due to erratic supply during all seasons (especially monsoon), electrified households also depend on kerosene for lighting.

Based on fuel consumption norms (regionwise, seasonwise and end usewise), (a) the total fuel wood required (cooking, water heating, space heating, jaggery making and parboiling) works out to 16,68,698.25 tonnes/year, (b) electricity demand excluding irrigation, about 32.65 million kWh/year and (c) kerosene demand for cooking and water heating about 15.86 million litres per year.

Keywords: Energy efficiency, household energy, rural energy, fuel wood, per capita fuel consumption, traditional stoves, improved stoves, sustainable development.

7.1.0 Introduction

Energy is a fundamental and strategic tool even to attain the minimum quality of life. Sustainable development of a region depends critically on the health of renewable resources like soil, water, vegetation, livestock and genetic diversity. The integrated development of all these components is essential for environmentally sound development. The procurement of energy is also responsible in varying degrees for the ongoing deforestation, and loss of vegetation and top soil. While energy availability is a determining factor for agricultural productivity, traditional use of agricultural residues for energy production leads to soil impoverishment. The current inefficient energy use in various sectors is certainly responsible for detrimental impacts on the environment. Hence, sound policy and management decisions must involve three elements: economics, environment and energy, which must be considered in the search for ways to improve current energy supplies. This necessitates promotion of conservation activities among local communities and application of their traditional environmentally sound technologies.

Energy use patterns are closely linked to agro-climatic and socio-economic conditions. Hence, a detailed energy survey was carried out to understand the energy use patterns in various agro-climatic zones and seasons. Energy problems in rural areas are closely linked to soil fertility, landholding, livestock holding, etc. Energy planning of any region should be based on the existing levels of energy consumption. However, existing programs of development lack disaggregated information base. Regional developmental activities have to be based on detailed information from each sector. In this direction, a detailed investigation of all sectors of the district was carried out to develop a disaggregated energy database. In this chapter, household energy consumption patterns in five taluks of this district of Western ghats, based on detailed field work for 18 months, and strategies to minimise the potential negative impacts of energy system on natural and human systems are discussed.

7.2.0 Objectives

The main objectives of this study are to:

(1) determine fuel consumption patterns in various agro-climatic zones of Uttara Kannada district and find out parameters involved in variation and level of consumption,
(2) measure and estimate the daily per capita fuel wood consumption in traditional and improved stoves for cooking and water heating, and
(3) identify different end uses and fuels used in rural and urban areas.

These studies are carried out to identify the basic energy utilising activities and end uses of energy in a region (zone) and suggest methods to meet the basic needs.

7.3.0 Study Area

The study was carried out in Kumta (Longitude 74° 24' to 74° 45' E, Latitude 14° 17' to 14° 35' N), Sirsi (74° 35' to 78° 5', 14° 27' to 14° 48'), Mundgod (74° 52' to 75° 7', 14° 42' to 15° 04') and Ankola (74° 18' to 74° 43', 14° 29' to 14° 48') taluks of Uttara Kannada district (Figure 1.1). Ankola and Kumta represent coastal region, Mundgod plain region and Sirsi hilly region. In terms of bioresource availability, Kumta, Ankola and Mundgod are categorised as bioresource scarce and Sirsi as bioresource surplus regions.

7.4.0 METHODOLOGY

An exploratory survey was carried out in various taluks initially. Socio-economic and energy data was collected from randomly selected samples in Sirsi, Siddapur, Kumta and Ankola taluks. The preliminary results of the survey in households using fuel efficient stoves in Sirsi and Kumta taluks showed fuel wood requirement for cooking to be 1.80 and 1.78 (summer) and 2.25 and 1.98 (monsoon) kgs/person/day respectively. Significant regional variation is noticed only during monsoon, mainly due to incessant rain (in hilly terrain) when people in Sirsi prefer hot food and water. During winter, fuel consumption at Sirsi and Kumta is 1.94 and 1.90 kgs/person/day respectively. Similar variation in fuel requirement is noticed for water heating. During monsoon and winter, hot water is preferred by villagers for bathing and washing, while in summer it is used only for bathing. Hence, fuel consumption is higher during monsoon - winter than summer. It is also noticed that, fuel consumption for water heating during all seasons is higher at Sirsi than Kumta. The results also revealed that the levels of energy use and mix of energy sources depend on climatic and geographic factors. In order to get an insight into energy consumption pattern and extent of variation due to these factors in the domestic sector, detailed energy surveys were carried out in the district, consisting mainly of secondary and primary data collection.

7.4.1 Collection of Secondary Data from Government Agencies

(i) Landholding particulars (agriculture, horticulture, etc.) for each household was collected from Village Accountants' offices (14 offices covering all villages in Kumta, 6 in Sirsi covering 36 villages, 2 in Mundgod covering 12 villages and one in Ankola covering 10 villages. Out of these, Kumta and Ankola represent coastal zone, Sirsi hilly and Mundgod plain zone).

(ii) Data on villagewise demography, and occupational and infrastructural facilities were collected from Tahsildar's office at respective taluk head quarters, District Statistical Office at Karwar and Directorate of Census Operation, Bangalore.

(iii) Data on livestock population, etc., was collected from respective veterinary departments of each taluk.

(iv) Villagewise land use maps were collected from the Survey of Land Settlement and Records Office (Revenue Department).

7.4.2 Collection of Primary Data

The secondary information was analysed to select households for stratified sampling (based on landholdings, community, etc.) for energy survey. Households in each village were categorised into landless, small, medium and large farmers based on the landholdings. Under each category, households were grouped communitywise and samples were selected from each category.

7.4.3 Detailed Investigation of Energy Consumption Pattern

Kumta taluk was considered to study intra-taluk variation. It was divided into 3 zones viz. (a) Villages along the coast - Aghnashini, Kagal, Gokarna, Kodibag, etc., (b) Villages in the interior - Hegde, Divgi, Manaki, etc., and (c) Villages in the hilly area - Yana, Sandolli, Morshe, Santeguli, etc. Ninety two villages and Kumta town were chosen to represent all the three zones with rural and urban population of Kumta taluk. In each village, representative households of all communities and different landholding categories were chosen for fuel consumption survey. Data on seasonal fuel procurement and consumption was also collected. The classification adopted based on landholding is: (i) Landless, (ii) Marginal farmers (0 - 0.5 hect), (iii) Small farmers (0.5 - 1 hect), (iv) Medium farmers (1 - 1.5 hect), (v) Large farmers (1.5 - 2.5 hect) and (vi) Very large farmers (>2.5 hect), keeping in mind the fragmented landholding scenario in the thickly populated coastal taluks of the district.

7.5.0 LITERATURE SURVEY

7.5.1 Energy Studies in India

(i) *ESCI Study*: The first attempt to estimate and forecast total energy consumption and sources of supply was made by a 1965 Energy Survey Committee of India (ESCI). Data on consumption of noncommercial fuels was derived from the 1962 All India Sample Surveys of Rural Households and the 1958 Household Sample Surveys in Bombay, Calcutta and Delhi conducted by the National Council of Applied Economic Research (NCAER, 1959). The study analysed the consumption trends of commercial sources of energy viz., coal, oil and electricity in relation to economic growth and GNP. It has also estimated the consumption of noncommercial sources of energy including firewood, and animal and agricultural wastes. The energy demand till 1980-81 was projected. It also conducted survey of rural energy consumption of traditional and noncommercial energy sources. The committee reported that twelve million tonnes of agricultural wastes was burned each year largely in villages (GOI, 1965).

The surveys estimated average per capita domestic energy consumption to be 0.38 and 0.40 Tonne of Coal Replacement (TCR) in rural and metropolitan areas respectively. Income elasticity of demand for domestic energy was estimated only for city dwellers and was found to be 0.4 for the group with per capita income above Rs. 300 each year. Since energy use estimates for urban areas other than the three metropolitan cities were not available, the committee assumed it to be 0.39 TCR per capita. It also assumed that energy consumption during the preceding decade had increased by 4.5 per cent, equivalent to the income elasticity of energy demand in cities. The past per capita energy use patterns were reconstructed on that basis. This data and rural and urban population estimates were used to arrive at the total domestic energy consumption. To arrive at the noncommercial energy use, commercial energy consumption (for which relatively better data was available) was deducted from the estimated total energy consumption. Within noncommercial sources, estimated contribution of fuel wood, animal dung and agricultural residues was based on the assumption that their relative shares did not change over time. To project the energy demand till 1980-81, the committee assumed that:

- average per capita energy consumption was 0.38 and 0.39 TCR in rural and urban areas respectively,
- per capita energy consumption would rise by 25 and 15 per cent in rural and urban areas respectively. This was based on the assumptions that; (i) the total real per capita consumption may nearly double, (ii) electricity may be more widely available so that around 50 per cent of the urban and 30 per cent of the rural population would belong to the category with income elasticity of energy demand, and (iii) income elasticity of energy demand across the country would be 0.5 per cent. It may, however, be noted that the committee's report is not specific on this point and the reference seems to be to the average income elasticity of energy demand. It may also be noted that the NCAER earlier estimated the income elasticity of commercial energy demand in rural areas at 0.18-0.20 (NCAER, 1960),
- commercial energy consumption would double every ten years, as in the past, and
- use of farm residues would increase in proportion to the expected increase in agricultural output, but the proportionate contribution of fuel wood and dung cake in the remaining noncommercial energy use would stay at the 1963 level.

During 1961-81 the total energy demand was expected to increase by 94 per cent, with the demand for electricity, coal and oil expected to increase by more than 11, 10 and 5 times respectively, while commercial energy as a whole was expected to increase by 6 times. Compared to that, the demand for noncommercial energy, fuel wood and dung cake was expected to increase by only 43, 31 and 17 per cent respectively. A perusal of the data also shows that, within the noncommercial group, fuel wood's share was expected to decline from 65 to 60 per cent whereas, the share of farm residues was expected to increase to 28 from 20 per cent.

The committee with the expectation of a major shift towards commercial energy sources, suggested that growth in noncommercial energy use should be curbed, and measures to achieve that objective included:

(a) enhanced production and subsidised supply of commercial energy in general, and soft coke briquettes in particular and

(b) imposition of octroi duties and other taxes on fuel wood.

The committee did, however, emphasise planned development of noncommercial energy sources.

For instance, it suggested that vigorous measures be adopted to enhance fuel wood production. A specific suggestion was to set up a fuel wood division in the then Ministry of Food and Agriculture. But, such administrative actions could hardly be expected to lead to a developed fuel wood market. Similarly, octroi duties on fuel wood, even if feasible, could not be expected to substantially reduce illicit cutting, inefficient use or even misuse of the material (at the same time, the suggestion to enhance fuel wood output should be welcomed, as enhanced production of soft coke briquettes could possibly not be a financially, socially and environmentally preferred alternative). The committee's demand forecasts were not validated - the observed consumption levels in 1963 were close to the expectations for 1961, and the 1979 observations were substantially below those for 1981. At the same time, animal dung has been an exception as its observed use in 1979 was significantly higher than the estimate for 1981. This may mean that dung continued to be a preferred source of household energy due to its inherent characteristics such as slow burning, relatively assured supply and perception of dung work as an integral part of household chores, even in land owning families in rural India.

(ii) *NCAER Study*: The NCAER also used the results of their 1962 survey to project the energy demand of rural households up to 1976. It was assumed that:

- rural population by the mid 1970s would be around 477 million,
- per capita income during 1962-76 would increase by about 70 per cent,
- rural areas income elasticity of demand for commercial fuels would be 0.18 to 0.20 (NCAER, 1960), and
- besides the income effect, consumer preferences would shift towards commercial fuels by 5 to 7 per cent.

But the council's forecast didn't come true, mainly due to absolute shortages of non commercial fuels. Therefore, the forecast was reassessed with additional assumptions that:

- fuel wood supply from forest and non-forest areas could be around 70 mt (million tonnes) a year as estimated by the Ministry of Food and Agriculture (GOI, 1962),
- fuel wood consumption in urban areas would be negligible,
- use of crop residues as fuels should be restricted to 17 mt to enhance their use for compost making,
- success of the scheme to move milk colonies from cities would limit animal dung use to rural areas,
- soft coke use would be enhanced to meet shortfalls in supply of fuel wood and agricultural residues,
- around 2,02,000 villages would be electrified and 1.77 tWh a year would be available in rural areas,
- effective per capita consumption of lighting energy in rural areas would increase from 1 to 6 kWh, and
- kerosene supply would be 8 to 9 mt, of which about 6 mt would be available to rural households.

(iii) *Fuel Policy Committee*: In 1974, the Fuel Policy Committee (FPC) estimated the household use of commercial and noncommercial fuels (GOI, 1974), using the data generated by NCAER surveys. It was also assumed that the annual per capita energy consumption was 0.38 and 0.40 TCR in rural and urban areas respectively; and that fuel wood, animal dung and agricultural residues contributed 65, 15 and 20 per cent respectively to the total noncommercial energy. With these assumptions, and the estimated rural and urban population, domestic consumption of noncommercial and total energy during 1961-71 was reconstructed.

A comparative analysis of ESIC and FPC studies shows that actual increase in domestic energy use during 1961-71 was estimated at 22 per cent, whereas an earlier committee expected it to be 37 per cent. Similarly, consumption of commercial energy in the domestic sector increased by only 70 per cent against the earlier committee's forecast of nearly 150 per cent.

The FPC also forecast the demand up to 1991. It assumed that:

- per capita energy consumption may not change considerably,
- almost all urban and 70 per cent of rural households would use electricity for lighting,
- about 20 per cent of rural households would use kerosene for lighting,
- nearly 60 and 10 per cent of urban and rural households respectively would use kerosene for cooking,
- up to 1984, shares of fuel wood, dung and agricultural residues in the non commercial energy use would be the same as estimated by ESIC for 1963, and
- absolute consumption of crop residues would stabilise starting 1984, whereas that of fuel wood and animal dung would stay in the same ratio as of 1984 (the ESIC had assumed that consumption of agricultural residues will increase in proportion to farm output).

The committee's forecasts of household energy demand in 1991 are summarised as - electricity 25 tWh (25 MTCR - million tonnes of coal replacement), kerosene 6 mt (49.8 MTCR), soft coke 20 mt (30.0 MTCR), LPG 2 mt (16.6 MTCR), firewood 122 mt (115.9 MTCR), dung cakes 53 mt (21.2 MTCR) and agricultural residues 46 mt (43.7 MTCR).

The 1989 data with respect to household use of kerosene and LPG show that the forecasts for 1991 by FPC may be fairly validated. Similarly, the 1984 electricity data shows that the observed use was comparable to the forecast. Comparable data for soft coke was not available. FPC projections also show that, while commercial and total energy use was anticipated to increase by about 150 and 25 per cent, noncommercial energy was expected to decline by 8-10 per cent due to decline in fuel wood and animal dung use. Such shifts from noncommercial energy sources are welcome to facilitate the formulation and execution of sustainable energy policies. However, an important issue at the same time is whether fuel wood should continue to be treated as noncommercial.

The Royal Commission on Agriculture in 1928 believed that the use of animal dung as fuel could be reduced with assured fuel wood supplies at reasonable prices, that is, prices which the rural households can afford to pay. The reasoning seems to be more relevant at present than it was in 1920s. The relevance of observing nature's rules is better and more widely appreciated than ever before. Those who burn animal dung seem to be aware, but can be better educated of its opportunity cost. Since the society is to benefit by shifts from noncommercial to commercial, non renewable to renewable energy sources and from animal dung to fuel wood, an important step must be to enhance fuel wood supplies beyond a critical minimum (there has been a policy emphasis on enhanced fuel wood production but the programmes and policies do not seem to match).

(iv) *National Commission on Agriculture:* The National Commission on Agriculture (NCA) in 1976 projected the fuel wood demand up to year 2000 (GOI, 1976). The commission accepted the FPC's 1971 estimates of fuel wood consumption of 220 kgs per capita or total 120 mt, but argued that 15-20 per cent of that is first used as timber and then as fuel wood. Thus, the net per capita fuel wood use was estimated at about 194 kgs a year. It was assumed that, starting around 1985, per capita fuel wood consumption would decline at the rate of one per cent per annum and may be about 166 kgs by the turn of the century (the Commission has not stated the basis for the assumption). The demand projections arrived on that basis for fuel wood is 115.5 mt in 1970 and 1975, 128.8 mt in 1980, 141.4 mt in 1985 and 157.5 mt in 2000. The commission did not expect an appreciable shift away from noncommercial fuels. Within that category, however, it suggested deliberate policy interventions to reduce the share of dung cake from 15 to 5 per cent. A large part of the gap caused by that was expected to be covered by enhanced availability of agricultural residues due to an anticipated increase in farm productivity (GOI, 1976).

(v) *Working Group on Energy Policy*: A Working Group on Energy Policy (WGEP) in 1979 also adopted the FPC's methodology to reconstruct domestic energy use during 1954 and 1976. The group adopted a scenario approach to make a reference level and an optimum level forecast of domestic energy demand up to 2000. The Reference Level Forecast (RLF) pertained to a situation without additional policy efforts for demand and supply management. The first step was to estimate the number of households in

1976 which used electricity and kerosene for lighting, and LPG, kerosene, soft coke, and non commercial fuels for cooking. Following that, the group assumed that:

- the annual per capita energy consumption would continue to be 0.38 and 0.4 TCR in rural and urban areas respectively,
- the number of electrified households would increase by 2.14, 3.6, 4.0 and 5.0 million during 1976-83, 1983-88, 1988-93 and 1993-2001 respectively,
- soft coke output would increase from 6.5 mt in 1963 to 9 mt, 14 mt and 30 mt in 1988, 1993 and 2000 respectively,
- the proportion of households using noncommercial fuels would decline to 70 per cent in rural and 10 per cent in urban areas by the turn of the century, which would be mainly due to enhanced kerosene consumption,
- LPG would be a perfect substitute for kerosene and the growth in kerosene demand would be reduced in proportion to the indigenous production of LPG,
- both rural and urban per capita consumption of kerosene and soft coke for cooking would be 240 litres and 1.75 tonnes, and
- the per capita consumption of kerosene and electricity for lighting would be 24 litres and 300 kWh in rural, and 36 litres and 400 kWh in urban areas.

On the other hand, the Optimum Level Forecast (OLF) pertained to a situation where suggested policy interventions had been adopted and assumptions regarding the likely changes in the energy system validated. On the policy side, the group emphasised that the production potential of lands under forests, roadsides and tank bunds be harnessed. This was expected to enhance fuel wood supply beyond forecasted requirements. It was also argued that financial resources required to enhance the soft coke output and its transport throughout the country be better used for improved land management (GOI, 1979).

The group's assumptions regarding the energy use system were that:

- efficiency of household electricity consumption would improve by 2.5, 5 and 10 per cent by 1988, 1993 and 2001 respectively,
- appliances can and would be improved to enhance the consumption efficiency of fuel wood and agricultural residues by at least 2, 5 and 10 per cent by 1988, 1993 and 2001 respectively,
- growth rate of soft coke production would be limited to 6.7 per cent, the rate projected for 1982-1987, so that its output by 2000 would be limited to 16.75 mt rather than 30 mt estimated in the case of RLF and its demand reduced by 1.25 and 7.25 mt (coal equivalent) by 1993 and 2001 respectively,
- by 2000, 8.7 per cent of rural households could depend on kerosene for cooking against 22.3 per cent estimated in the case of RLF, and the demand reduced by 0.91 and 1.24 mt by 1993 and 2001 respectively, and
- quantities of animal dung and agricultural residues used for cooking could stabilise at 1983 level.

The resulting projections by the OLFs and RLFs for 1983 and 1988 are comparable. The 1993 OLFs and RLFs with respect to electricity, LPG, fuel wood, animal dung and agricultural residues are also comparable, but for kerosene and soft coke, OLFs are substantially lower than RLFs. More importantly, the two scenarios for 2001 are comparable only with respect to LPG and electricity, and the OLFs regarding fuel wood, kerosene, and soft coke are substantially lower, while for dung cake and farm residues they are substantially higher than the RLFs. Part of the reason for the divergent expectations may be due to substantially lower OLFs, compared to RLFs, for commercial energy in 1993 and 2001. At the same time, the total noncommercial energy use visualised under the two scenarios is strictly comparable for all the reference years. Such results must raise a number of issues including their value for policy making, the desirability of encouraging or even anticipating fuel wood substitution with animal dung and farm residues in spite of environmental concerns, and the prospects of substantial increase in fuel wood outputs with marginal improvements in forest and non-forest wasteland management.

(vi) *Fuel wood Study Committee:* The Government of India formed a Fuel wood Study Committee in 1981 (GOI, 1982), which agreed with the OLFs of the 1979 Working Group, but made some additional observations and assumptions on land availability and management for fuel wood production. A few of these are:

- Forests have not been managed for fuel wood supplied except for casual supervision to avoid the right holders' "excesses". But sixty million hect. (mha) of forest land could annually give at least 30 mt recorded fuel wood output (against the current 15 mt) a year, that is, output excluding removals by right holders. Another 30 mt is, and can continue to be, obtained from private lands, gardens and trees around houses; and about 25 mt may be available from social forestry in the near future.
- Private forests have been managed to raise industrial wood, but fuel wood is also produced and can be produced where required.
- With identification of appropriate tree species to minimise competition with field crops, training, demonstrations and incentives, farm plans may include fast growing trees to, at least, meet family needs for fuel wood.
- Large tracts of low productivity land without clearly defined ownership or organised management, particularly in arid and semi-arid tracts, primarily support scattered trees, bushes or shrubs. This land provided forage and substantial quantities of fuel wood to the rural areas, adjoining urban areas. Its scientific management is difficult due to undefined ownership and high biotic pressures, but its potential for enhanced fuel wood production is extremely high.

In essence, the committee emphasised an integrated approach to enhance biomass production. Fuel wood plantations at the rate of 1.5 mha a year were considered an absolute necessity. It was estimated that at least 20 mha, comprising wastelands, degraded forests, roadsides, railroad sides and canal banks, could be available for that purpose. It was also suggested that:

- the State agricultural universities should take up fuel wood farming with their research and extension education programmes, and fuel wood should be included in major farming systems,
- schools, colleges, local bodies and voluntary organisations should be encouraged to raise tree nurseries,
- hypothecation of standing tree crops should be permitted to facilitate institutional finance for tree farmers and
- tree growers should be assured remunerative prices and assistance for organised marketing of the produce, and the scope of the Agricultural Prices Commission should be enlarged for that purpose.

Though the emphasis once again was on supply management, it was considered feasible with better use of forest and non-forest wasteland, harnessing the complementarity between industrial wood and fuel wood production and inducing the farmers to grow fuel wood.

(vii) *Advisory Board on Energy*: As an ongoing effort of national energy planning and to continuously review the energy situation in the country in global context, and for proposing future energy plans on an integrated and coordinated basis, the Government of India set up an Advisory Board on Energy (ABE) in March 1983. ABE got a set of studies carried out regarding the long-term perspectives on demand and supply of various energy sources. The projected energy requirements for the year 2004-05 of ABE were also computed based on two alternative growth rates of 4 and 5%, and three alternative population values. The report postulates a minimum level of energy in household sector, in terms of useful energy per person per day, as 620 kcal for cooking, 30 kcal for space heating and 30 kcal for lighting. ABE made the demand for firewood based on an efficiency of 8% for wood stoves. Use of improved wood stoves will drastically reduce the projected demand of firewood (GOI, 1984).

The Advisory Board on Energy in 1985 presented an energy demand supply perspective for 2005. The forecasts for the household sector were based on:

- three alternative population levels; 1,003 million (m) 1,046 m and 1,115 m,
- a rural and urban population mix of 67.2 and 32.8 per cent respectively,
- a minimum per capita per day useful energy requirement of 630 kcal for cooking (computations based on the assumption that a 15 kg LPG cylinder with 1,54,500 kcal and 60 per cent appliance efficiency would serve a five member household for 30 days, that is 618 kcal/day/person, assumed to be uniform throughout the country), and 30 kcal each for space heating and lighting,
- no significant change in relative share of cooking fuels,

- chulah efficiency at eight per cent (the need for improved efficiency was emphasised) and
- per capita energy requirements for lighting as observed for greater Bombay.

The Board, however, made two important observations on the data. It would be undesirable to use 200 to 220 mt of dung cakes a year as fuel as that would deprive it for biogas production and manure. Thus, it was emphasised that dung use for household energy ought to be stabilised at 75-80 mt a year, that is, at the 1979 level. This was considered possible with enhanced supply of farm residues, which was considered feasible and desirable with an expected rise in farm outputs. It was also suggested that large-scale commercial and industrial markets for farm residues should not be encouraged [NCAER, on the contrary, has emphasised the enhanced use of agricultural residues for compost making (NCAER, Section 2)]. Alternatively, it was estimated that an additional 56.4 mt fuel wood would stabilise animal dung use at 75-80 mt. This meant that 375 mt fuel wood would be annually required by 2005. It was argued that this is feasible because the productivity of forest land could be raised by at least 300 per cent, and at least 50 per cent of India's forest and non-forest wasteland, estimated at 60 mha, could be used for fuel wood plantations (GOI, 1985). However, it was feared that fuel wood supplied in Uttar Pradesh, Punjab, Haryana, Tamil Nadu and Kerala may fall short of requirements. Hence, fuel efficiency measures, including briquetting of farm residues, enhanced soft coke production and their supply to deficit States, were recommended.

As the Indian economy grew at much lower rates during the final phase of IVth and beginning of Vth Five Year Plans, FPC projections were found to be generally on the higher side. In order to review the energy demand projections afresh, the Planning Commission, Government of India constituted the 'Working Group on Energy Policy' (WEP) in 1977 with a view to carry out a "comprehensive review of the present energy situation in the light of recent developments, both within the country and outside, develop a perspective to next five to fifteen years and recommend appropriate policy measures for optimal utilization of available resources including non-conventional sources of energy." WEP submitted its final report in 1979 (WEP, 1979). As per the report, the total primary energy input during 1978-79 was estimated to be 8.87×10^{18} joules of which noncommercial sources provided 4.42×10^{18} joules. Based on certain GDP, sectoral and population growth projections, WEP had forecast the demand for both commercial and noncommercial sources of energy till the year 2000-01. Though both the trend and forecast show gradual decrease in the relative share of noncommercial energy sources, it is certain that these sources will continue to dominate the rural energy sector for many decades to come.

(viii) *Perspective Planning Division, Planning Department, Government of Karnataka* (GOK, 1981): In Karnataka, study of domestic energy consumption patterns was carried out by the Perspective Planning Division, Planning Department, Government of Karnataka (1981) based on 32nd round of National Sample Survey (1977-78). 6745 (Rural 4247, Urban 2498) households covering all the 19 districts of the State were surveyed, which revealed that the per capita monthly expenditure on energy is a major expenditure both in rural and urban households. In rural households, it varied from Rs. 43.33 to Rs. 74.91, with a mean of Rs. 58.69. The firewood consumption varied from 13.69 to 73.55 kgs per month (mean value 28.38 kg) and kerosene used for lighting varied from 0.33 to 0.77 litres per month (mean value 0.54 litres).

In recent years, the Government of India has shown some interest in the promotion and utilisation of new and renewable sources of energy. It has set up a high powered Commission of Additional Sources of Energy (CASE) and Department of Non conventional Energy Sources (DNES) in 1981. While CASE formulates the policies and programmes, DNES is responsible for their implementation. By the end of March 1993, DNES had constructed 0.17 million biogas plants (family size) against targeted 0.15 million, saving 0.65 million tonnes of wood per year. Large size community and institutional biogas plants based on animal dung, industrial effluent and sewage are being constructed. 1009 such plants were operational as on March, 1993. The total number of family size biogas plants in India as on December 1996 was 2.15 million. The Department has also launched the National Improved Chula (Woodstove) Programme in 1983. The total number of improved woodstoves in the country by March 1998 was 2,56,97,410. Their efficiencies are found to vary from 20 to 35 per cent. There are more than 12,517 domestic and 6142 industrial solar water heating systems of different sizes operating in the country. More than 71 solar kilns, 61 solar crop dryers and 10,195 distillation systems are at various stages of completion. This does not include nearly 2,88,028 solar cookers in use. More than 1772 solar photovoltaic pumps have been installed in rural areas providing water for drinking and micro irrigation. 30,569 solar photovoltaic street lights, 42,845 domestic lighting systems and 88,920 solar lanterns have been provided in more than 450 villages, where electricity would perhaps never reach. Wind farms of a total capacity of 968 MW were installed in the coastal areas of Gujarat, Tamil Nadu, Maharashtra and Orissa. DNES has also drawn up an ambitious programme to set up 5000 MW of wind turbine capacity by the year 2000. The concept of harnessing locally available renewable energy sources in an integrated manner for supplementing energy supply options through Rural Renewable Energy Systems has been put into practice in about 392 villages of 13 States (DNES, 1994).

7.5.2 Studies by Individual Researchers

As a first step to understand rural energy problems, ASTRA (Centre for Application of Science and Technology for Rural Areas) of IISc conducted a detailed survey in six villages in Kunigal taluk (Tumkur district) in Karnataka during 1975-76 (ASTRA, 1976) based on observations, discussions, measurements and checks. Some of the findings are: (a) firewood is a dominant energy source (81.6%) used mainly for household activities, (b) cooking is a major activity consuming human and firewood energy, and efficiencies of chulahs are in the range of 5.08%, (c) human energy in hours/day/household (especially women and children) was inefficiently used in firewood gathering (2.6), cooking (3.68), carrying food to farms (1.82), fetching water (1.53), taking cattle for grazing (5.54), etc., with the share of domestic burden between men, women and children being 24, 20 and 20% respectively, (d) kerosene consumption for lighting is about 4.3 litres/non-electrified house, 78% of the houses being non-electrified, and (e) industrial consumption is very small.

Roger Revelle (1976) estimated the total energy utilized in rural India for the year 1971. As per his study and analysis, only 10 per cent (1.2×10^{14} kcal per year) of the total energy (11.42×10^{14} kcal per year) in rural areas was derived from the commercial sources, mainly contributed by kerosene, diesel, chemical fertilizers and electricity from hydro sources. The remaining 90 per cent is derived from traditional sources of energy viz. human and animal labour, firewood, crop residues and animal wastes. Revelle concludes that rural people in India are tied to poverty and misery mainly because they use too little energy and that too inefficiently, and secure all required energy by their own physical effort. A transformation of rural India, he suggested, can be brought about by increasing the quantity and improving the technology of energy use.

Gerald A. Leach (1988) reviews the major features of residential energy use that have bearing on demand and supply options, and also examines the profound changes that have occurred over the past few years in attitudes to wood fuel problems and their implications for energy policies and planning. Data from 15 country assessments show that households account for 30-35% of total energy use, compared to 25-30% for industrialised countries. The highest proportions are found in poorer countries where households exclusively depend on biomass fuels.

Reddy and Krishna Prasad (1977) in their critical analysis of the energy scene in India showed with facts and figures, the highly skewed nature of the energy consumption patterns existing in the country. Their estimates on per capita energy consumption of rural population, relying mainly on noncommercial energy sources, lies somewhere between 4.7 to 7 kWh/day, which is much below the satisfactory minimum of 36 kWh/day. In their opinion, the technologies appropriate for growth and development, biased in favour of rural India, may well turn out to involve small-scale decentralized energy systems, viz., biogas, solar and wind, with supplementary power coming, if necessary, from grid electricity. The authors also give a brief but exhaustive list of technological alternatives to harness new and renewable sources of energy like biogas, solar and wind, and also methods to improve utilisation of animate sources of energy. The suggestions, if implemented at National level, will not only improve the energy supplies needed but also enhance the quality of life and rural environment.

The energy use patterns of eight rural communities in the semi-arid agro-climatic zones were analysed by Bowonder et al. (1985). Irrigation is found to be the most significant factor influencing energy consumption and demand. There is a progressive trend towards monetization of fuel wood, animal wastes and agricultural residues in these communities. There is emphasis on the need to generate disaggregate information on rural energy consumption patterns, though it involves collection of enormous data covering all households, income levels and landholding patterns for effective energy planning.

Energy use patterns and resource assessment of village Islamnagar, Bhopal district, Madhya Pradesh was done systematically by Maheshwari et al. (1981). This study shows that out of a total energy consumption of 2.46 GJ/capita/year, domestic activities consumed 2.08 GJ, out of which cooking (mainly fuel wood) accounted for 2.03 and lighting 0.05 GJ.

Ravindranath and Chanakya (1986) give a detailed picture of energy consumption in a village ecosystem. Their study in Ungra village, Karnataka shows that biomass fuels account for 81% of the village energy budget, while commercial energy accounts for 11%. The per capita consumption of energy is 13.1 GJ per year.

Ramakumar and Hughes (1981) and Ramakumar (1983) give a conceptual model of an Integrated Renewable Energy System for a village, where available biomass sources are converted to biogas to supplement the output of a sub-system designed to integrate solar radiation and wind resources.

The obvious feature of residential energy use in developing countries is the great variation in energy use and the mix of fuels. This diversity is apparent from National and regional averages. Attempt to quantify average consumption, based on nearly 350 household surveys and rough estimates for 88 countries by the United Nations Food and Agriculture Organisation (FAO) shows (De Montalembert and Clement, 1983) higher consumption for cold mountain areas (reflecting space heating needs) and warmer regions of sub-Saharan Africa and Latin America compared to North Africa, the Middle East and Asia (reflecting more abundant biomass resources). The lower proportions of biomass use in Latin America reflect greater rural incomes, infrastructure development and access to modern fuels. Elsewhere, biomass use is typically more than 90-95% of total residential consumption, the remainder being kerosene (and occasionally electricity) for lighting. The heavy dependence on crop residues and animal wastes in Asia reflects low forest cover and farm tree resources. This study reveals much greater variation at local level ranging from 2.4 to 59.2 GJ (150-3700 kgs fuel wood equivalent). In Nepal, survey of 25 villages (Bajracharya, 1981) shows per capita consumption ranging from 0.1 to 2.55 m^3. A study of 74 Indian villages (Leach, 1987; Aiyasamy, et al., 1983; Bowonder, et al., 1983) involving more than 5200 households, found the village average biomass fuel use to be 4-6 GJ per capita annually in 30% and 2-8 GJ in 65% of the villages, but the complete range was 1.5-20.5 GJ.

7.5.3 Fuel Consumption Pattern in Karnataka

State level energy consumption was studied before taking up district/taluk level studies in order to know the trend in levels and types of energy consumption at macro level. Sourcewise energy consumption in Karnataka reveals that 53.20% of the total energy is met by noncommercial sources of energy like firewood (43.60%), cow dung cake (1.40%) and agrowastes (8.20%). While commercial energy like coal (5.80%), oil (11.60%), kerosene (2.60%), LPG (0.70%) and electricity (26.10%) constitute 46.80%. A significant part of these noncommercial energy sources cater the heating (domestic) needs of the rural population (about 70 - 80%), followed by village industries. Demographic, land use and energy consumption patterns in Kumta taluk are discussed in the following sections.

7.6.0 Data Analyses

Per Capita Fuel Consumption (PCFC) is computed to

(i) determine fuel consumption pattern in various agro-climatic zones,
(ii) find out various parameters involved in variation and level of consumption, and
(iii) estimate the daily per capita consumption of fuel wood in traditional and improved stoves for cooking and water heating.

7.6.1 Computation of PCFC (Per Capita Fuel Consumption)

PCFC = FC/p, where FC = fuel consumed in kgs/day and p = number of adult equivalents, for whom food was cooked.

Computation of PCFC considering the following:

(i) More than one type of fuel is used for cooking and water heating in any household. The quantity of fuel consumed is determined by subtracting the weight of the remaining fuel from its initial weight. The daily consumption of different fuels is calculated separately from the fuel weights of the consecutive days.

(ii) These daily consumption values in each household are converted to their equivalent dry weights using the measured moisture content values, which are then converted into equivalent value using the net calorific value of each type of residue. These are added to get the daily energy consumption for each household.

(iii) The daily energy consumption of each household is further converted to per adult energy consumption using the adult equivalent of the number of people, which is computed assuming conversion factors listed below:

Standard adult equivalents used in analysis;

Family Size	Standard Adult Equivalent
men 18-59 yrs.	1
women 18-59 yrs.	0.8
men >59 yrs.	0.8
women >59 yrs.	0.8
boys 5-18 yrs.	0.5
girls 5-18 yrs.	0.5
kids 1-5 yrs.	0.35
child <1 yr.	0.25

(iv) For each family, daily per adult energy consumption is computed seasonwise. The average value, standard deviation, maximum and minimum values are computed for annual and seasonal consumption.

(v) Data is grouped based on household income, landholding category, community and village separately. All the above parameters were computed for these groups.

7.7.0 RESULTS AND DISCUSSION

7.7.1 Area and Population

Detailed investigation of energy consumption is carried out in Kumta taluk , which has 14 mandals [A Mandal represents a cluster of villages. For administrative purposes, like land revenue collection, etc., each taluk is divided into Mandals (rural area) and Towns (urban area)]. Out of 14 mandals, 10 are in zone-1 (coastal), 2 in zone-2 (interior) and 2 in zone-3 (hilly). Population in these mandals for the last four decades is listed in Table 7.1. It shows that mandals located in zone-1 have an overall increase in population of 78.84% during the period 1951-91, followed by 95.524% in zone-2 and 132% in zone-3. This is shown pictorially in Figure 7.1.

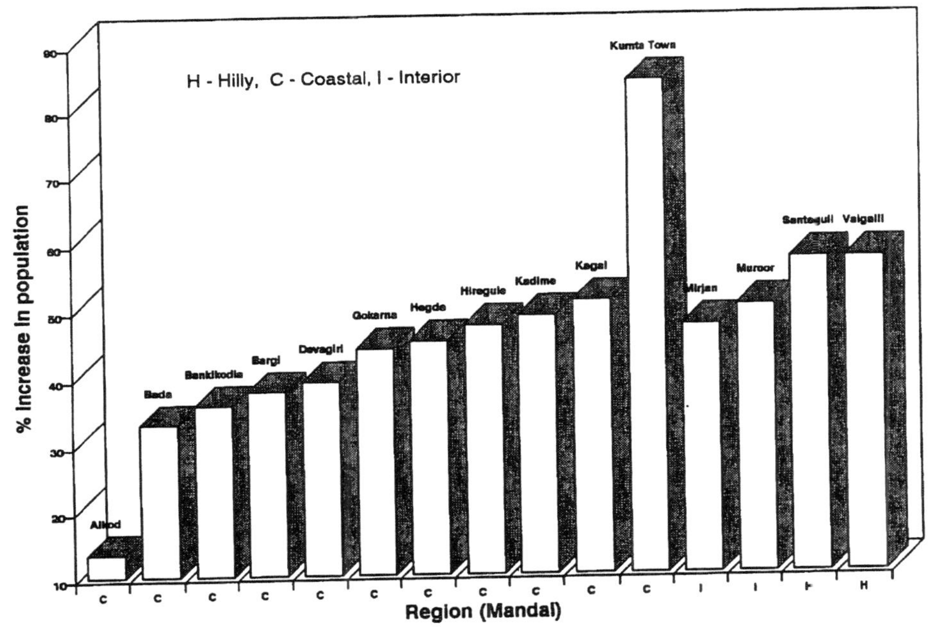

Figure 7.1: Percentage Population Increase - Mandalwise and Regionwise (1951-1991)

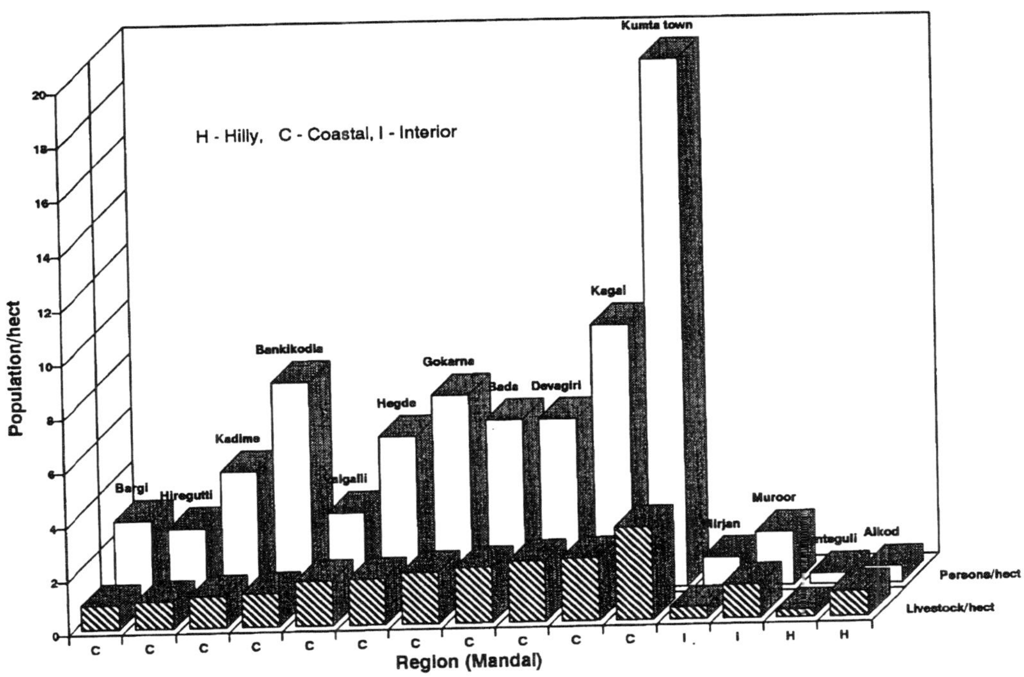

Figure 7.2: Human and Livestock Density Regionwise

Table 7.1: Mandalwise Population and its Percentage Increase during Four Decades

Zone	Mandal	Total Population					% increase in population				
		1951	1961	1971	1981	1991	1951-61	1961-71	1971-81	1981-91	1951-91
1	Bargi	4860	4800	4614	5054	5611	-1.23	-3.88	9.54	11.02	13.38
1	Hiregutt	3599	3862	4243	4978	5369	7.31	9.87	17.32	7.85	32.97
1	Kadime	3927	3992	4397	5042	6105	1.66	10.15	14.67	21.08	35.68
1	Bankikodla	3478	3887	4374	4871	5587	11.76	12.53	11.36	14.70	37.75
1	Valgalli	8800	9175	9547	11085	12253	4.26	4.05	16.11	10.54	39.24
1	Hegde	5833	6569	7397	9081	10412	12.62	12.60	22.77	14.66	43.98
1	Gokarna	4790	5439	6135	7614	8704	13.55	12.80	24.11	14.32	44.97
1	Bada	3805	4976	5847	6425	7235	30.78	17.50	9.89	12.61	47.41
1	Devagiri	4556	4936	6033	7473	8872	8.34	22.22	23.87	18.72	48.65
1	Kagal	2533	2898	3565	4180	5146	14.41	23.02	17.25	23.11	50.78
1	Kumta town	12600	16223	19112	23385	29833	28.75	17.81	22.36	27.57	83.89
2	Mirjan	3113	4353	5038	5450	5902	39.83	15.74	8.18	8.29	47.26
2	Muroor	4058	4942	5973	7258	8119	21.78	20.86	21.51	11.86	50.02
3	Santeguli	4354	5327	6538	8358	10109	22.35	22.73	27.84	20.95	56.93
3	Alkod	3671	4833	6069	7071	8539	31.65	25.57	16.51	20.76	57.01
	Total	61377	86212	98882	117325	137796					
1	Coast	58781	66757	75264	89188	105127	13.569	12.743	18.500	17.871	78.845
2	Interior	7171	9295	11011	12708	14021	29.619	18.462	15.412	10.332	95.524
3	Hilly	8025	10160	12607	15429	18648	26.604	24.085	22.384	20.863	132.374

Table 7.2: Mandalwise Number of Persons/hectare, Livestock/hectare and Percentage Forest Cover

Zone	Mandal	Persons / hectare	Total livestock		Total livestock/hect			Total forest	% forest cover
			Cattle	Buffalo	Cattle	Buffalo	Livestock		
1	Bargi	2.77	1347	319	0.73	0.17	0.90	599.80	32.32
1	Hiregutt	2.50	1818	370	0.85	0.17	1.02	631.56	29.40
1	Kadime	4.52	1227	219	0.99	0.18	1.17	243.20	19.61
1	Bankikodla	7.78	774	213	0.99	0.27	1.26	12.75	1.62
1	Valgalli	2.94	3711	388	1.51	0.16	1.67	1440.13	58.50
1	Hegde	5.71	2750	345	1.51	0.19	1.70	641.26	35.17
1	Gokarna	7.21	2757	450	1.62	0.26	1.89	479.51	28.20
1	Bada	6.25	2451	411	1.76	0.29	2.05	239.22	17.17
1	Devagiri	6.25	3116	491	1.95	0.31	2.26	335.75	21.03
1	Kagal	9.71	1613	498	1.77	0.55	2.31	223.18	24.43
1	Kumta town	19.50	4850	406	3.17	0.27	3.44		
2	Mirjan	1.02	3020	441	0.38	0.06	0.43	6175.07	77.43
2	Muroor	1.96	3251	498	1.08	0.17	1.24	1963.85	65.12
3	Santeguli	0.38	4208	130	0.28	0.01	0.29	13272.3	87.99
3	Alkod	0.58	3494	509	0.83	0.12	0.95	12930.01	88.13
	Total		35537	5282				39187.59	

Zone	Zone Type	Persons / hectare	Total Livestock		Livestock/hectare			Total forest	% forest cover
			Cattle	Buffalo	Cattle	Buffalo	Livestock		
1	Coastal	5.31	20900	3729	1.39	0.25	1.63	3761.47	24.93
2	Interior	1.35	8171	1155	0.61	0.09	0.70	10179.26	76.44
3	Hilly	0.35	6466	398	0.23	0.01	0.24	25246.86	89.34

Table 7.3.1: Mandalwise and Communitywise Landholding Particulars (in hectares) for Kumta Taluk

Zone	Mandal	PAT	HB	GOU	NAI	GSB	TRA	GOVT	PT	MUK	FM	CH	OT2	MU	TOTAL
1	Kagal	138.27	29.48	77.46	110.34	11.70	23.84	0.20	29.78	1.73	27.95	5.90	2.78	6.10	465.52
1	Bargi	418.27	18.89	25.85	163.77	12.48	8.74	0.61	6.45	2.66	83.83	2.86	0.92	8.56	753.89
1	Kumta Town	116.35	116.78	103.42	165.58	96.38	67.87	74.59	7.43	5.14	11.62	20.87	15.75	14.43	816.20
1	Kadime	129.29	171.83	241.58	203.54	23.18	44.05	22.88	10.21	15.12	19.99	8.47	13.20	0.47	903.79
1	Valgalli	190.11	328.05	150.25	119.56	39.44	30.64	4.52	22.64	39.55	2.18	8.62	8.16	0.28	944.00
1	Bankikodla	118.65	59.95	488.44	176.89	17.77	25.28	11.52	3.80	1.86	10.33	20.39	19.71	15.79	970.36
1	Bada	244.78	339.27	7.14	178.90	29.93	30.80	33.20	40.48	55.86	36.49	2.94	46.91	3.79	1050.50
1	Gokarna	162.17	215.01	469.35	95.28	37.25	34.40	19.94	3.65	13.45	4.31	0.74	10.35	1.77	1067.67
1	Devagiri	464.23	209.45	25.77	212.57	91.36	40.87	28.83	29.69	35.30	15.09	52.33	6.23	1.29	1213.00
1	Hiregutti	227.65	442.22	176.31	371.88	97.53	40.74	18.24	8.55	7.44	8.13	0.00	46.55	4.24	1449.49
1	Hegde	616.94	206.01	104.97	286.05	142.39	26.44	33.30	16.58	69.60	23.79	52.99	5.21	1.24	1585.52
2	Muroor	52.14	460.93	256.73	81.12	11.23	19.46	1.17	13.88	11.76	0.00	5.56	4.38	0.00	918.36
2	Mirjan	410.14	167.89	30.09	195.58	43.93	13.15	0.05	64.66	21.05	17.45	3.15	3.45	1.50	972.07
3	Santeguli	375.10	326.12	320.62	297.14	18.01	19.07	5.67	23.65	17.71	2.32	78.80	7.49	54.60	1546.31
3	Alkod	184.88	512.97	501.19	218.17	33.47	69.82	144.35	91.72	69.88	63.72	50.17	10.74	15.44	1966.51
	Total	3848.96	3604.85	2979.17	2876.37	706.08	495.17	399.05	373.16	368.08	327.18	313.79	201.83	129.50	16623.19

Table 7.3.2: Mandalwise and Communitywise Landholding Particulars (in Percentage) for Kumta Taluk

Zone	Mandal	PAT	BR	GOU	NAI	GSB	TRA	GOVT	PT	MUK	FM	CH	OT2	MU	TOTAL
1	Kagal	29.70	6.33	16.64	23.70	2.51	5.12	0.04	6.40	0.37	6.00	1.27	0.60	1.31	100.00
1	Bargi	55.48	2.51	3.43	21.72	1.66	1.16	0.08	0.86	0.35	11.12	0.38	0.12	1.14	100.00
1	Kumta Town	14.25	14.31	12.67	20.29	11.81	8.32	9.14	0.91	0.63	1.42	2.56	1.93	1.77	100.00
1	Kadime	14.30	19.01	26.73	22.52	2.56	4.87	2.53	1.13	1.67	2.21	0.94	1.46	0.05	100.00
1	Valgalli	20.14	34.75	15.92	12.67	4.18	3.25	0.48	2.40	4.19	0.23	0.91	0.86	0.03	100.00
1	Bankikodla	12.23	6.18	50.34	18.23	1.83	2.61	1.19	0.39	0.19	1.06	2.10	2.03	1.63	100.00
1	Bada	23.30	32.30	0.68	17.03	2.85	2.93	3.16	3.85	5.32	3.47	0.28	4.47	0.36	100.00
1	Gokarna	15.19	20.14	43.96	8.92	3.49	3.22	1.87	0.34	1.26	0.40	0.07	0.97	0.17	100.00
1	Devagiri	38.27	17.27	2.12	17.52	7.53	3.37	2.38	2.45	2.91	1.24	4.31	0.51	0.11	100.00
1	Hiregutti	15.71	30.51	12.16	25.66	6.73	2.81	1.26	0.59	0.51	0.56	0.00	3.21	0.29	100.00
1	Hegde	38.91	12.99	6.62	18.04	8.98	1.67	2.10	1.05	4.39	1.50	3.34	0.33	0.08	100.00
2	Muroor	5.68	50.19	27.96	8.83	1.22	2.12	0.13	1.51	1.28	0.00	0.61	0.48	0.00	100.00
2	Mirjan	42.19	17.27	3.10	20.12	4.52	1.35	0.01	6.65	2.17	1.79	0.32	0.35	0.15	100.00
3	Santeguli	24.26	21.09	20.73	19.22	1.16	1.23	0.37	1.53	1.15	0.15	5.10	0.48	3.53	100.00
3	Alkod	9.40	26.09	25.49	11.09	1.70	3.55	7.34	4.66	3.55	3.24	2.55	0.55	0.78	100.00
	Total	23.15	21.69	17.92	17.30	4.25	2.98	2.40	2.24	2.21	1.97	1.89	1.21	0.78	100.00

PAT	Patgar	NAI	Naik	TRA	Traders	FM	Fishermen
HB	Havyak Brahmin	GSB	Gouda Saraswath Brahmin	GOVT	Government	CH	Christian
GOU	Gouda	MUK	Mukri	PT	Palm Tappers	OT	Others
		MU	Muslim				

Table 7.4.1: Mandalwise and Categorywise Land Distribution (Frequency Distribution) Kumta Taluk

Zone	Mandal	0 Hectare	0-0.5	0.5-0.75	0.75-1.25	1.25-1.75	1.75-2	2-4	4-6	6-8	>8	Total
1	Kagal	0	616	163	52	70	27	19	6	3	0	956
1	Bargi	5	354	324	196	65	48	27	1	0	3	1023
1	Devagiri	3	589	268	147	58	29	37	2	1	1	1135
1	Valgalli	13	655	265	146	58	52	50	15	6	9	1269
1	Bankikodla	5	586	276	179	105	54	59	9	1	4	1278
1	Hiregutt	3	604	233	166	111	67	131	48	18	25	1406
1	Kadime	3	916	317	138	66	43	37	10	2	4	1536
1	Hegde	6	1170	254	112	74	57	102	26	4	4	1809
1	Bada	12	1073	329	178	124	47	72	7	4	3	1849
1	Gokarna	11	1223	280	122	74	42	94	22	10	3	1881
1	Kumta town	10	2255	252	85	40	19	45	6	3	8	2723
2	Muroor	3	491	256	164	74	68	85	13	4	0	1158
2	Mirjan	19	445	314	179	84	50	72	6	0	4	1173
3	Santeguli	1	163	177	200	164	122	201	36	9	3	1076
3	Alkod	1	418	248	207	147	85	155	31	10	10	1312

Table 7.4.2: Mandalwise and Categorywise (Percentage) Land Distribution in Kumta Taluk

Zone	Mandal	0 Hectare	0-0.5	0.5-0.75	0.75-1.25	1.25-1.75	1.75-2	2-4	4-6	6-8	>8	or 1.75-8
1	Bargi	0.49	34.60	31.67	19.16	6.35	4.69	2.64	0.10	0.00	0.29	7.72
1	Hiregutt	0.21	42.96	16.57	11.81	7.89	4.77	9.32	3.41	1.28	1.78	20.55
1	Bankikodla	0.39	45.85	21.60	14.01	8.22	4.23	4.62	0.70	0.08	0.31	9.94
1	Valgalli	1.02	51.62	20.88	11.51	4.57	4.10	3.94	1.18	0.47	0.71	10.40
1	Devagiri	0.26	51.89	23.61	12.95	5.11	2.56	3.26	0.18	0.09	0.09	6.17
1	Bada	0.65	58.03	17.79	9.63	6.71	2.54	3.89	0.38	0.22	0.16	7.19
1	Kadime	0.20	59.64	20.64	8.98	4.30	2.80	2.41	0.65	0.13	0.26	6.25
1	Kagal	0.00	64.44	17.05	5.44	7.32	2.82	1.99	0.63	0.31	0.00	5.75
1	Hegde	0.33	64.68	14.04	6.19	4.09	3.15	5.64	1.44	0.22	0.22	10.67
1	Gokarna	0.58	65.02	14.89	6.49	3.93	2.23	5.00	1.17	0.53	0.16	9.09
1	Kumta town	0.37	82.81	9.25	3.12	1.47	0.70	1.65	0.22	0.11	0.29	2.97
2	Mirjan	1.62	37.94	26.77	15.26	7.16	4.26	6.14	0.51	0.00	0.34	11.25
2	Muroor	0.26	42.40	22.11	14.16	6.39	5.87	7.34	1.12	0.35	0.00	14.68
3	Santeguli	0.09	15.15	16.45	18.59	15.24	11.34	18.68	3.35	0.84	0.28	34.48
3	Alkod	0.08	31.86	18.90	15.78	11.20	6.48	11.81	2.36	0.76	0.76	22.18
		0.44	53.55	18.33	10.52	6.09	3.75	5.49	1.10	0.35	0.38	11.07

Mandalwise number of persons and livestock per hectare have also been computed and listed in Table 7.2, which shows that mandals located in zone-1 are densely populated with 2.50 (Hiregutti mandal) to 19.50 (Kumta town) persons per hectare (overall average is 5.31). This is followed by zone-2 with 1.02 (Mirjan) to 1.96 (Muroor) persons/hect (overall average is 1.35). Mandals located in zone-3 are least populated having 0.38 (Santeguli) to 0.58 (Alkod) persons per hectare (overall average is 0.35). In zone-1, livestock density ranges from 0.90 (Bargi) to 3.44 (Kumta town). Overall average in zones 1 to 3 is 1.63, 0.70 and 0.24 respectively. Figure 7.2 gives mandalwise population per hectare. The percentage forest cover is 24.93% (zone-1), 76.44% (zone-2) and 89.34% (zone-3). These analyses clearly show that zone-1 is densely populated with humans and livestock. The human pressure on environment in the coastal zone is reflected by less forest cover.

7.7.2. Land use Pattern

Villagewise landholding particulars for each household, collected from respective Village Accountant's office, was computerised. Table 7.3.1 lists aggregated mandalwise and communitywise landholding (in hectares) information, while Table 7.3.2 gives relative percentages. This reveals that Patgars with a share of 23.15% are a dominant agricultural community, followed by Brahmins (21.69%), Goudas (17.92%), Naiks (17.30%), etc.

Land distribution among various categories of farmers is given in Table 7.4.1 and the corresponding percentages in distribution are listed in Table 7.4.2. It is seen that the number of households having 0-0.5 hectare, in most of the mandals in zone-1, is more than 30%. It ranges from 34.60% (in Bargi) to 82.81% (Kumta town), while in zone-3 it ranges from 15.15% (Santeguli) to 31.86% (Alkod). This fragmented landholding scenario in the coastal zone can be attributed to dense population.

7.7.3. Energy Demand Assessment in Kumta taluk

An attempt was made to compute the desirable level of energy consumption in the domestic sector of Kumta taluk. This will then be used, with demographic and agro-climatic features of the district, to estimate the energy demand for various purposes.

There are various ways to calculate the useful domestic energy demand for cooking purposes. It can be based on either energy surveys or certain predetermined energy norms recommended by various agencies, most of which were based on cooking energy requirements in urban areas, and hence are ad hoc. In this context, it is necessary to find a rational and reasonably accurate method of computing domestic energy requirement taking into account seasonal and geographic factors. In this direction, energy survey was carried out in 90 villages as discussed earlier in methodology. Questionnaire used consisted mainly of the following information:

Household:	number of members (male, female and children), educational background, income and landholding particulars.
Devices:	types of commercial and noncommercial devices, purpose and duration.
Livestock:	number and type, seasonal fodder and feed requirement, grazing particulars, and dung yield (seasonal) and usage.
Fuel wood:	mode of collection, quantity per season, types, place of collection, seasonal fuel wood consumption (by recall in most households and actual measurement in at least 5 households representing each landholding category) for cooking, water heating and other purposes like space heating, etc.
Agriculture:	animate and inanimate sources of energy requirement for various operations, type of crop, productivity, etc.
Household industry:	fuel requirements in jaggery making, arecanut (red variety) preparation, etc.
Efficient devices:	Perception about energy efficient devices and new sources of energy.

Data from 1304 households of 90 villages (out of 119) were collected during 18 months of field work covering 88% villages in zone-1, 68% in zone-2 and 70% in zone-3. The role of climatic and geographic factors on levels of energy consumption is explored in the following section.

7.7.4 Seasonal and Regional Variation in Energy Consumption

Fuel wood is collected mainly from nearby forests in zone-3, while in zones 1 and 2 it is purchased from Government distribution outlets with deficit collected from forests. It is mainly used for cooking, while horticulture residues (coconut husk, leaves, etc.) and twigs are also used for water heating along with fuel wood in this region. Based on calorific values, all residues are converted to equivalent fuel wood value for computing per capita fuel wood consumption.

Out of the data collected from 1304 households (stratified samples), 57.05% accounted for households in zone-1 (744), 23.08% for zone-2 and 19.87% for zone-3. Seasonwise and regionwise fuel consumption is computed for cooking (Table 5.1) and bioenergy requirement for water heating (Table 5.2). Kerosene, LPG and biogas (for cooking) and kerosene and electricity (for lighting) are listed in Table 7.5.3.

Table 7.5.1 illustrates seasonal and regional variation in fuel wood consumption. Average consumption ranges from 2.01$\pm$1.49 (zone-1) to 2.32$\pm$2.09 (zone-3) kgs/person/day. Seasonwise fuel wood requirement for zones 1 and 3 ranges from 1.98 and 2.22 (summer) to 2.11 and 2.51 (monsoon) kgs/person/day.

Table 7.5.2 highlights the variation in fuel wood requirement for water heating (for bath and washing), which ranges from 1.17$\pm$0.02 (zone-1) to 1.63$\pm$0.05 (zone-3) kgs/person/day. In zones 1 and 3, seasonal variation ranges from 1.12 and 1.53 (summer) to 1.22 and 1.73 (monsoon) kgs/person/day.

Table 7.5.3 gives overall picture of other sources of energy for domestic purposes. In zone-1 kerosene is used for both cooking and lighting. Kerosene for cooking ranges from 0.05 (zone-3) to 0.34 (zone-1) ltrs/person/month. Less consumption of kerosene in zone-3 compared to zone-1 is due to the availability of bioresources in sufficient quantity, while zone-1 is dependent on commercial sources of energy due to scarcity of bio fuels.

Kerosene for lighting (Table 7.5.4) ranges from 0.75 (zone-1) to 0.99 (zone-3) ltrs/person/month. Zone-3 is dependent more on kerosene for lighting due to non-availability of electricity as the households are spatially distributed, affecting electrification. This is apart from the erratic electricity supply during all seasons (more in monsoon). Electricity consumption for domestic purposes (irrigation is not included) ranges from 1.55 (zone-3) to 2.50 (zone-1) kWh/person/month.

Table 7.5.1: Seasonwise and Regionwise Cooking Fuel wood Requirement in Kumta Taluk

Region	No. of samples	Cooking Fuel wood kgs/person/day							
		Summer		Winter		Monsoon		Average	
		Avg	Sd	Avg	Sd	Avg	Sd	Avg	Sd
Coastal	744	1.98	1.40	1.95	1.34	2.11	1 73	2.01	1.49
Interior	301	2.02	1.34	2.22	1.38	2.32	1.59	2.19	1.44
Hilly	259	2.22	1.56	2.23	1.94	2.51	2.77	2.32	2.09
Total	1304								
Avg		2.07		2.13		2.31		2.17	

Table 7.5.2: Seasonwise and Regionwise Water Heating Fuel Wood Requirement in Kumta Taluk

Region	No. of samples	Water Heating Fuel wood kgs/person/day							
		Summer		Winter		Monsoon		Average	
		Avg	Sd	Avg	Sd	Avg	Sd	Avg	Sd
Coastal	744	1.12	0.50	1.18	0.55	1.22	0.54	1.17	0.02
Interior	301	1.23	0.66	1.35	0.79	1.46	0.89	1.35	0.09
Hilly	259	1.53	0.63	1.64	0.73	1.73	0.75	1.63	0.05
Total	1304								
Avg		1.29		1.39		1.47		1.38	

Table 7.5.3: Kerosene, Biogas and LPG Requirement for Cooking in Domestic sector

Region	No. of samples	Cooking Energy					
		Kerosene (cooking) ltrs/capita/month		Biogas m3/person/day		LPG kgs/person/month	
		Avg	Sd	Avg	Sd	Avg	Sd
Coastal	744	0.34	0.79	0.23	0.76	0.02	0.10
Interior	301	0.21	0.69	0.26	0.79	0.01	0.05
Hilly	259	0.05	0.19	0.49	1.18	0.00	0.00
Total	1304						
Avg		0.20		0.33		0.01	

Table 7.5.4: Kerosene and Electricity Requirement for Lighting in Domestic Sector

Region	No. of samples	Lighting Energy										
		Electricity kWh/person/ month		Electricity Cost Rs/capita/ month		Kerosene ltrs/person/month						
						Summer		Winter		Monsoon		Average
		Avg	Sd	Avg	Sd	Avg	Sd	Avg	Sd	Avg	Sd	Avg
Coastal	744	2.50	2.62	2.81	4.10	0.69	0.46	0.77	0.47	0.78	0.49	0.75
Interior	301	2.61	3.08	2.66	3.95	0.80	0.45	0.84	0.44	0.84	0.47	0.83
Hilly	259	1.55	2.98	1.93	4.60	0.97	0.60	1.01	0.55	0.98	0.53	0.99
Total	1304											
Avg		2.22		2.47		0.82		0.87		0.87		0.85

7.7.5 Seasonwise and Villagewise Consumption of Energy

The ninety surveyed villages were further grouped into 30 categories (based on location within 10-20 sq.kms diameter - to distinguish the villages closer to sea and hilly tracts). Table 7.6.1 lists villagewise and seasonwise fuel wood consumption for cooking, ranging from 1.61 (Hegde, Masur, etc.) to 2.89 (Antravalli, Bellangi, etc.) kgs/person/day.

Table 7.6.2 lists fuel wood consumption for water heating, which ranges from 0.88 (Kojjalli, Harnir, etc.) to 2.09 (Hegle, Nadumuskeri, etc.) kgs/person/day.

Table 7.6.3 gives villagewise per capita consumption of kerosene, LPG and biogas for cooking (details of number of households dependent on kerosene, biogas, etc., would be discussed in the next section). Average kerosene and biogas consumption is about 0.15 ltrs/person/month and 0.271 m^3/person/day respectively. Kerosene for lighting ranges from 0.40 (Divgi, Manaki-coastal villages) to 1.31 (Sandolli, Yana, Jalghar-remote hilly villages) ltrs/person/month. Electricity consumption shows that villages such as Yana, Sandolli and Medine are still non-electrified.

7.7.6 Communitywise Energy Consumption

In order to see the role of culture and tradition in fuel wood consumption, data was grouped based on community. The data reveals distinct differences in diet and cooking habits. Intercommunity variation in cooking fuel consumption, evident from Table 7.7.1, ranges from 1.62 (Fishermen, Bhandari community) to 2.62 (Achari community) kgs/person/day. Table 7.7.2 lists fuel wood consumption for water heating, which ranges from 0.89 (Madival, Fishermen) to 1.59 (Havyak Brahmin) and 1.62 (Marathi community) kgs/person/day.

Communitywise usage of kerosene, biogas and LPG (for cooking) and kerosene and electricity (for lighting) is listed in Table 7.7.3. It is seen that communities such as Havyak Brahmins (HB) and Gouda Saraswath Brahmins (GSB) have switched over to biogas, kerosene and LPG stoves for cooking. The shift from traditional to improved cookstoves (in certain households kerosene stoves and/or biogas along with improved cookstoves) depends on educational background and economic soundness of the family. Economically and socially backward communities, such as, Mukris and Kumbhi Marathis still prefer traditional stoves (fuel is collected at zero cost). While economically and educationally sound communities have switched over to biogas, LPG, etc. In order to see the role of economic background, number of persons per household and type of devices in consumption pattern, further analyses were carried out.

Table 7.6.1: Villagewise and Seasonwise Cooking Fuel wood Consumption

Sl.No.	Village	No. of samples	Cooking Fuel wood kgs/person/day							
			Summer		Winter		Monsoon		Average	
			Avg	Sd	Avg	Sd	Avg	Sd	Avg	Sd
1	Aghnashini, Bargi, Paduvani	66	1.92	0.96	1.93	0.98	1.97	0.97	1.94	0.97
2	Aigalkurve, Savalkurve, Kelaginastala	32	1.67	0.43	1.76	0.53	1.99	0.61	1.81	0.53
3	Alvekode, Toppalgutta	68	1.63	0.88	1.64	0.89	1.90	1.06	1.72	0.94
4	Antravalli, Bellangi, Bandivala	21	2.73	1.45	2.78	1.48	3.04	1.61	2.89	1.51
5	Bada, Gudeangadi, Manikatta	64	1.50	0.83	1.67	1.34	1.66	1.34	1.61	1.17
6	Divgi, Siragunj, Manaki, Madakibail	33	1.94	1.07	1.91	1.07	2.06	1.07	1.97	1.07
7	Gokarna, Bankikodla, Bavikodla	163	2.23	1.29	2.24	1.36	2.33	1.65	2.27	1.43
8	Harita, Kadambale, Anegundi	23	1.75	1.00	1.81	1.00	2.06	1.00	1.87	1.00
9	Hebbail, Katgal, Yedtare	29	1.82	0.79	1.98	0.86	2.31	0.91	2.04	0.85
10	Hegde, Masur, Lukkeri	86	1.52	1.06	1.50	1.07	1.82	1.28	1.61	1.14
11	Hegle, Nadumaskeri, Harumaskeri	14	2.15	0.52	2.17	0.52	2.17	0.54	2.16	0.53
12	Holangadde, Halkar, Chitragi	52	1.47	2.18	1.49	1.10	2.00	2.26	1.65	1.85
13	Holegadde, Devgiri, Horbag, Harneer	60	2.52	2.59	2.43	2.65	2.88	3.87	2.61	3.04
14	Kagal, Hubbangeri	52	1.95	0.88	1.94	0.88	1.94	0.89	1.94	0.88
15	Kalve, Santeguli, Bastikeri, Basolli,	70	1.94	2.29	2.52	3.03	2.91	4.45	2.46	3.26
16	Koojalli, Kundgani, Harnir	29	1.79	0.53	1.78	0.58	1.77	0.57	1.78	0.56

Table 7.6.1: Villagewise and Seasonwise Cooking Fuel wood Consumption (Cont'd)

Sl.No.	Village	No. of samples	Cooking Fuel wood kgs/person/day							
			Summer		Winter		Monsoon		Average	
			Avg	Sd	Avg	Sd	Avg	Sd	Avg	Sd
17	Malavalli, Santhur, Bandivala	18	2.02	0.46	2.07	0.39	2.07	0.39	2.05	0.41
18	Medine, Santgal, Kavaladi	20	2.35	0.76	2.42	0.77	2.53	0.81	2.43	0.78
19	Mirjan, Nilkod, Kalkod, Tannirhonda	52	2.28	0.69	2.25	0.77	2.26	0.74	2.26	0.73
20	Morse, Moodagi, Mudanalli	17	2.04	0.67	2.02	0.66	2.17	0.62	2.08	0.65
21	Muroor, Kallabe, Karkimakki	91	2.46	1.92	2.40	1.82	2.47	2.23	2.44	1.99
22	Nagur, Kodkani, Masurkurve	34	1.87	0.93	1.91	0.91	2.20	1.17	1.99	1.00
23	Sandolli, Hiregutti, Ennemadi	10	1.56	0.82	1.70	0.85	1.97	1.03	1.74	0.90
24	Shirali, Kadime, Torke, Hoskeri	32	2.44	1.94	2.63	2.31	3.18	3.48	2.75	2.58
25	Talgod, Hervatte, Baggon	61	2.52	1.24	2.57	1.23	2.61	1.25	2.57	1.24
26	Unchalli, Urkeri, Gonehole, Honehalli	23	2.90	1.12	2.83	1.15	2.93	1.40	2.85	1.22
27	Valgalli, Harodi, Horbag	33	1.54	0.84	1.64	0.89	1.81	1.02	1.66	0.92
28	Yana, Uppinapattana, Jalghar	39	1.97	1.17	1.91	1.19	2.08	1.27	1.99	1.21
29	Yelwalli, Nilkod, Yeswanimule	12	1.82	0.33	1.82	0.33	1.82	0.33	1.82	0.33
	Total	1304								
	Avg		2.01		2.06		2.24		2.10	
	Sd		0.38		0.37		0.41		0.37	
	Max		2.90		2.83		3.18		2.89	
	Min		1.47		1.49		1.66		1.61	

Table 7.6.2: Villagewise and Seasonwise Water Heating Fuel wood Consumption

Sl.No.	Village	No. of samples	Water Heating Fuel wood kgs/person/day							
			Summer		Winter		Monsoon		Average	
			Avg	Sd	Avg	Sd	Avg	Sd	Avg	Sd
1	Aghnashini, Bargi, Paduvani	66	1.21	0.49	1.19	0.49	1.21	0.51	1.20	0.01
2	Aigalkurve, Savalkurve, Kelaginastala	32	1.02	0.34	1.13	0.38	1.21	0.37	1.12	0.02
3	Alvekode, Toppalgutta	68	0.99	0.49	1.02	0.48	1.13	0.52	1.05	0.02
4	Antravalli, Bellangi, Bandivala	21	1.60	1.28	1.90	1.58	2.07	1.85	1.86	0.23
5	Bada, Gudeangadi, Manikatta	64	1.13	0.39	1.21	0.66	1.21	0.65	1.18	0.12
6	Divgi, Siragunj, Manaki, Madakibail	33	0.94	0.46	1.04	0.49	1.43	0.86	1.14	0.18
7	Gokarna, Bankikodla, Bavikodla	163	1.20	0.44	1.24	0.47	1.23	0.46	1.22	0.01
8	Harita, Kadambale, Anegundi	23	1.16	0.00	1.23	1.00	1.46	0.00	1.28	0.47
9	Hebbail, Katgal, Yedtare	29	1.38	0.79	1.57	0.88	1.85	0.96	1.60	0.07
10	Hegde, Masur, Lukkeri	86	0.96	0.48	1.01	0.52	1.26	0.60	1.08	0.05
11	Hegle, Nadumaskeri, Harumaskeri	14	1.87	0.54	2.20	0.64	2.20	0.62	2.09	0.04
12	Holangadde, Halkar, Chitragi	52	0.93	0.36	0.97	0.36	0.98	0.35	0.96	0.00
13	Holegadde, Devgiri, Horbag, Harneer	60	1.19	0.53	1.27	0.57	1.27	0.56	1.24	0.02
14	Kagal, Hubbangeri	52	1.03	0.39	1.04	0.38	1.02	0.37	1.03	0.01
15	Kalve, Santeguli, Bastikeri, Basolli,	70	1.79	0.50	1.82	0.55	1.79	0.60	1.80	0.04
16	Koojalli, Kundgani, Harnir	29	0.87	0.23	0.88	0.23	0.88	0.22	0.88	0.00

Table 7.6.2: Villagewise and Seasonwise Water Heating Fuel Wood Consumption (Cont'd)

Sl.No.	Village	No. of samples	Water Heating Fuel wood kgs/person/day							
			Summer		Winter		Monsoon		Average	
			Avg	Sd	Avg	Sd	Avg	Sd	Avg	Sd
17	Malavalli, Santhur, Bandivala	18	1.59	0.52	1.88	0.61	1.90	0.57	1.79	0.04
18	Medine, Santgal, Kavaladi	20	1.90	0.49	2.06	0.64	2.25	0.76	2.07	0.11
19	Mirjan, Nilkod, Kalkod, Tannirhonda	52	0.99	0.28	0.98	0.28	0.99	0.28	0.99	0.00
20	Morse, Moodagi, Mudanalli	17	1.47	0.52	1.55	0.57	1.61	0.61	1.54	0.04
21	Muroor, Kallabe, Karkimakki	91	1.39	0.55	1.47	0.54	1.62	0.62	1.49	0.04
22	Nagur, Kodkani, Masurkurve	34	1.26	0.86	1.47	0.95	1.55	0.92	1.43	0.04
23	Sandolli, Hiregutti, Ennemadi	10	1.46	0.42	1.60	0.49	1.76	0.60	1.61	0.07
24	Shirali, Kadime, Torke, Hoskeri	32	1.28	0.46	1.32	0.48	1.34	0.48	1.31	0.01
25	Talgod, Hervatte, Baggon	61	1.37	0.71	1.49	0.84	1.48	0.80	1.45	0.05
26	Unchalli, Urkeri, Gonehole, Honehalli	23	1.55	0.65	2.03	0.86	1.78	0.76	1.79	0.09
27	Valgalli, Harodi, Horbag	33	1.08	0.51	1.17	0.41	1.32	0.52	1.19	0.05
28	Yana, Uppinapattana, Jalghar	39	1.44	0.78	1.59	0.99	1.69	0.99	1.57	0.10
29	Yelwalli, Nilkod, Yeswanimule	12	1.23	0.29	1.23	0.27	1.24	0.25	1.23	0.02
	Total	1304								
	Avg		1.29		1.40		1.47		1.39	
	Sd		0.28		0.36		0.36		0.33	
	Max		1.90		2.20		2.25		2.09	
	Min		0.87		0.88		0.88		0.88	

Table 7.6.3: Villagewise Kerosene, Biogas and LPG Consumption for Cooking and Kerosene and Electricity for Lighting

Village	Samples N	Cooking Energy						Lighting Energy									
		Kerosene ltrs/person /month		Biogas m3/person /month		LPG kgs/person /month		Electricity kWh/person /month		Electricity Cost Rs/person/month		Kerosene ltrs/person/month					
												Summer		Winter		Monsoon	
		Avg	Sd	Avg	Sd	Avg	Sd	Avg	Sd	Avg	Sd	Avg	Sd	Avg	Sd	Avg	Sd
Aghnashini, Bargi, Paduvani	66	0.15	0.79	0.22	0.49	0.01	0.05	1.99	1.45	3.01	3.10	0.67	0.29	0.75	0.34	0.65	0.28
Aigalkurve, Savalkurve, Kelaginastala	32	0.13	0.40	0.00	0.00	0.00	0.00	2.21	1.85	2.03	2.31	0.94	0.39	0.90	0.37	0.88	0.36
Alvekode, Toppalgutta	68	0.24	1.12	0.25	0.47	0.04	0.16	3.43	3.21	3.73	4.75	0.60	0.49	0.61	0.48	0.77	0.57
Antravalli, Bellangi, Bandivala	21	0.13	0.31	0.00	0.27	0.03	0.15	2.41	2.94	2.05	3.34	0.79	0.56	0.79	0.56	0.72	0.61
Bada, Gudeangadi, Manikatta	64	0.30	0.85	0.38	0.75	0.01	0.06	2.99	2.32	2.34	1.57	0.69	0.31	0.77	0.32	0.65	0.29
Divgi, Siragunj, Manaki, Madakibail	33	0.18	0.41	0.27	0.39	0.00	0.00	1.78	2.08	2.43	2.82	0.46	0.41	0.47	0.42	0.47	0.49
Gokarna, Bankikodla, Bavikodla	163	0.54	0.96	0.10	0.38	0.02	0.09	1.85	2.21	2.16	2.60	0.70	0.42	0.86	0.46	0.86	0.45
Harita, Kadambale, Anegundi	23	0.00	0.00	0.35	0.77	0.00	0.00	1.44	2.00	2.03	3.00	0.94	0.00	0.94	0.00	0.95	0.00
Hebbail, Katgal, Yedtare	29	0.00	0.00	0.00	0.00	0.00	0.00	0.29	0.91	0.43	1.83	0.94	0.42	1.07	0.41	1.09	0.41
Hegde, Masur, Lukkeri	86	0.29	0.77	0.35	0.87	0.05	0.15	3.29	3.39	4.55	8.08	0.38	0.58	0.40	0.54	0.64	0.76
Hegle, Nadumaskeri, Harumaskeri	14	0.00	0.00	0.00	0.00	0.00	0.00	1.14	1.06	1.80	2.75	0.74	0.24	1.01	0.35	0.74	0.24
Holangadde, Halkar, Chitragi	52	0.16	0.41	0.28	0.64	0.01	0.04	2.55	1.53	2.26	1.80	0.64	0.22	0.67	0.26	0.67	0.24
Holegadde, Devgiri, Horbag, Harneer	60	0.43	0.85	0.54	0.97	0.04	0.19	4.19	4.38	4.15	4.99	0.84	0.44	0.75	0.45	0.90	0.49
Kagal, Hubbangeri	52	0.47	0.90	0.23	0.47	0.00	0.00	2.50	1.98	3.26	2.98	0.59	0.23	0.78	0.27	0.61	0.21
Kalve, Santeguli, Bastikeri, Basolli,	70	0.17	0.33	0.80	1.59	0.00	0.00	3.26	3.84	3.51	3.69	0.95	0.58	0.95	0.58	0.95	0.58
Koojalli, Kundgani, Harnir	29	0.37	0.95	0.29	0.50	0.00	0.00	1.92	1.82	1.82	1.84	0.82	0.19	0.82	0.19	0.82	0.19

Table 7.6.3: Villagewise Kerosene, Biogas and LPG Consumption for Cooking and Kerosene and Electricity for Lighting (Cont'd)

Village	Samples N	Cooking Energy						Lighting Energy									
		Kerosene ltrs/person /month		Biogas m3/person /month		LPG kgs/person /month		Electricity kWh/person /month		Electricity Cost Rs/person/month		Kerosene ltrs/person/month					
												Summer		Winter		Monsoon	
		Avg	Sd	Avg	Sd	Avg	Sd	Avg	Sd	Avg	Sd	Avg	Sd	Avg	Sd	Avg	Sd
Malavalli, Santhur, Bandivala	18	0.00	0.00	0.00	0.00	0.00	0.00	0.59	1.28	0.39	0.98	0.86	0.30	1.20	0.65	0.86	0.30
Medine, Santgal, Kavaladi	20	0.00	0.00	0.00	0.00	0.00	0.00	0.00	0.00	0.00	0.00	0.83	0.36	0.77	0.34	0.81	0.34
Mirjan, Nilkod, Kalkod, Tannirhonda	52	0.52	1.01	0.25	0.25	0.02	0.08	2.35	1.56	2.31	1.68	0.62	0.17	0.82	0.23	0.82	0.23
Morse, Moodagi, Mudanalli	17	0.00	0.00	0.00	0.00	0.00	0.00	0.35	1.01	0.51	1.44	1.15	0.61	1.15	0.61	1.12	0.59
Muroor, Kallabe, Karkimakki	91	0.06	0.28	0.67	6.21	0.00	0.00	3.96	3.97	3.92	5.54	1.05	0.53	1.03	0.53	1.09	0.54
Nagur, Kodkani, Masurkurve	34	0.03	0.18	0.10	0.49	0.00	0.00	1.47	2.46	2.15	4.33	0.86	0.35	0.81	0.26	0.84	0.33
Sandolli, Hiregutti, Ennemadi	10	0.00	0.00	0.43	0.89	0.00	0.00	0.00	0.00	0.00	0.00	1.28	0.39	1.31	0.38	1.31	0.38
Shirali, Kadime, Torke, Hoskeri	32	0.00	0.00	0.78	1.37	0.00	0.00	3.25	4.05	5.52	9.89	0.48	0.21	0.60	0.25	0.48	0.21
Talgod, Hervatte, Baggon	61	0.00	0.00	0.06	0.36	0.00	0.00	1.51	1.62	1.24	1.85	0.99	0.63	0.99	0.63	1.00	0.62
Unchalli, Urkeri, Gonehole, Honehalli	23	0.00	0.00	0.00	0.00	0.00	0.00	0.27	0.80	0.08	0.38	1.00	0.44	1.25	0.55	1.00	0.44
Valgalli, Harodi, Horbag	33	0.04	0.19	0.67	0.97	0.00	0.00	2.89	1.76	3.30	2.67	0.68	0.35	0.91	0.36	0.92	0.34
Yana, Uppinapattana, Jalghar	39	0.00	0.00	0.84	1.34	0.00	0.00	0.19	0.84	0.00	0.00	1.40	0.85	1.30	0.56	1.31	0.58
Yelwalli, Nilkod, Yeswanimule	12	0.00	0.00	0.00	0.00	0.00	0.00	0.98	1.78	0.92	1.52	0.61	0.22	0.77	0.17	0.58	0.15
Total	1304																
Avg		0.15		0.27		0.01		1.90		2.13		0.81		0.88		0.85	
Sd		0.17		0.27		0.01		1.21		1.46		0.23		0.22		0.21	
Max		0.54		0.84		0.05		4.19		5.52		1.40		1.31		1.31	
Min		0.00		0.00		0.00		0.00		0.00		0.38		0.40		0.47	

Table 7.7.1: Communitywise and Seasonwise Cooking Fuel Wood Consumption

Sl.No.	Community	No. of samples	Cooking Fuel wood kgs/person/day							
			Summer		Winter		Monsoon		Average	
			Avg	Sd	Avg	Sd	Avg	Sd	Avg	Sd
1	Achari	22	2.50	2.11	2.51	2.53	2.86	3.66	2.62	2.77
2	Bhandari	27	1.55	0.61	1.66	0.73	1.68	0.74	1.63	0.69
3	Christian	12	1.67	1.00	1.68	1.09	2.04	1.07	1.80	1.05
4	Deshbhandari	29	2.12	0.68	2.19	0.68	2.30	0.66	2.20	0.67
5	Ambiga	25	1.62	0.46	1.60	0.43	1.64	0.43	1.62	0.44
6	Gavadi	13	1.97	0.75	1.96	0.78	2.24	0.96	2.06	0.83
7	Gouda	300	2.24	0.91	2.27	0.91	2.33	0.96	2.28	0.93
8	GSB	69	1.93	1.71	1.94	1.87	1.98	2.23	1.95	1.94
9	Gunaga	4	1.86	0.06	1.98	0.21	1.86	0.06	1.90	0.11
10	HB	241	2.08	2.31	2.13	2.51	2.59	3.59	2.27	2.80
11	Madival	20	1.72	0.59	1.77	0.58	1.84	0.64	1.78	0.60
12	Marathi	43	2.07	0.92	2.21	0.93	2.27	1.02	2.18	0.96
13	Mukri	21	1.75	0.57	1.71	0.68	1.88	0.63	1.78	0.63
14	Muslim	22	1.73	0.32	1.90	0.32	1.95	0.52	1.86	0.39
15	Naik	223	1.93	0.94	1.96	1.01	2.03	0.99	1.97	0.98
16	Others	39	1.78	0.94	1.91	0.91	2.00	0.90	1.90	0.92
17	Patgar	173	2.09	1.36	1.98	1.06	2.14	1.11	2.07	1.18
18	Shet	21	2.02	2.18	2.15	2.60	2.68	3.81	2.28	2.86
	Total	1304								
	Avg		1.92		1.97		2.13		2.01	
	Sd		0.23		0.23		0.32		0.25	
	Max		2.50		2.51		2.86		2.62	
	Min		1.55		1.60		1.64		1.62	

Table 7.7.2: Communitywise and Seasonwise Water Heating Fuel Wood Consumption

Sl.No.	Community	No. of samples	Water Heating Fuel wood kgs/person/day							
			Summer		Winter		Monsoon		Average	
			Avg	Sd	Avg	Sd	Avg	Sd	Avg	Sd
1	Achari	22	1.24	0.41	1.28	0.41	1.34	0.44	1.29	0.01
2	Bhandari	27	0.94	0.30	1.02	0.30	1.03	0.31	1.00	0.00
3	Christian	12	1.04	0.53	1.39	0.63	1.46	0.62	1.30	0.04
4	Deshbhandari	29	1.12	0.41	1.11	0.46	1.22	0.61	1.15	0.08
5	Ambiga	25	0.86	0.38	0.91	0.43	0.97	0.41	0.91	0.02
6	Gavadi	13	1.06	0.26	1.05	0.30	1.26	0.42	1.12	0.07
7	Gouda	300	1.31	0.58	1.44	0.71	1.47	0.71	1.41	0.06
8	GSB	69	1.19	0.40	1.22	0.38	1.26	0.34	1.22	0.02
9	Gunaga	4	0.87	0.08	0.91	0.10	0.93	0.12	0.90	0.02
10	HB	241	1.48	0.71	1.60	0.85	1.70	0.91	1.59	0.08
11	Madival	20	0.83	0.31	0.88	0.31	0.96	0.40	0.89	0.04
12	Marathi	43	1.53	0.65	1.61	0.76	1.71	0.86	1.62	0.09
13	Mukri	21	0.97	0.33	1.04	0.40	1.19	0.40	1.07	0.03
14	Muslim	22	1.27	0.33	1.25	0.34	1.29	0.32	1.27	0.01
15	Naik	223	1.15	0.57	1.19	0.59	1.26	0.64	1.20	0.03
16	Others	39	1.14	0.32	1.29	0.42	1.32	0.42	1.25	0.05
17	Patgar	173	1.03	0.54	1.08	0.55	1.15	0.58	1.09	0.02
18	Shet	21	1.17	0.38	1.14	0.41	1.16	0.39	1.16	0.01
	Total	1304								
	Avg		1.12		1.19		1.26		1.19	
	Sd		0.19		0.21		0.22		0.20	
	Max		1.53		1.61		1.71		1.62	
	Min		0.83		0.88		0.93		0.89	

Table 7.7.3: Communitywise Kerosene, LPG and Biogas - for Cooking, and Kerosene and Electricity for Lighting

Community	Cooking Energy						Lighting Energy									
	Kerosene ltrs/person/month		Biogas m3/person/daymonth		LPG kgs/person/month		Electricity kWh/person/month		Electricity Rs/capita/month		Kerosene ltrs/person/month					
											Summer		Winter		Monsoon	
	Avg	Sd	Avg	Sd	Avg	Sd	Avg	Sd	Avg	Sd	Avg	Sd	Avg	Sd	Avg	Sd
Achari	0.10	0.24	0.161	0.549	0.00	0.00	3.61	3.61	4.46	6.37	0.91	0.38	0.95	0.41	0.90	0.37
Bhandari	0.37	0.79	0.212	0.608	0.00	0.00	2.33	1.80	2.28	1.98	0.78	0.40	0.86	0.44	0.82	0.41
Christian	0.23	1.03	0.190	0.628	0.10	0.25	5.24	5.35	7.57	7.21	0.48	0.42	0.50	0.43	0.50	0.43
Deshbhandari	0.18	0.44	0.184	0.252	0.00	0.00	1.53	2.11	1.57	2.28	0.92	0.41	0.96	0.32	0.95	0.44
Ambiga	0.05	0.16	0.000	0.000	0.00	0.00	0.72	0.93	0.78	1.31	0.52	0.20	0.58	0.21	0.50	0.22
Gavadi	0.18	0.45	0.095	0.330	0.00	0.00	2.52	2.17	2.27	1.95	0.87	0.67	0.91	0.66	0.99	0.69
Gouda	0.01	0.07	0.012	0.155	0.00	0.00	0.86	1.29	0.94	1.86	0.78	0.44	0.86	0.47	0.82	0.44
GSB	0.55	0.97	0.683	1.670	0.05	0.14	3.50	2.48	4.68	4.28	0.53	0.29	0.67	0.42	0.67	0.36
Gunaga	0.00	0.00	0.000	0.000	0.00	0.00	2.54	0.36	2.71	1.03	0.71	0.10	0.89	0.10	0.89	0.10
HB	0.51	0.91	0.637	0.576	0.04	0.15	5.05	3.91	5.86	6.80	0.92	0.71	0.93	0.63	0.95	0.66
Madival	0.15	0.42	0.000	0.000	0.00	0.00	2.37	1.31	1.72	1.62	0.43	0.36	0.51	0.40	0.56	0.34
Marathi	0.00	0.00	0.523	0.446	0.00	0.00	0.03	0.18	0.04	0.27	1.06	0.54	1.10	0.51	1.08	0.56
Mukri	0.00	0.00	0.000	0.000	0.00	0.00	0.76	0.95	0.18	0.55	0.70	0.33	0.75	0.33	0.85	0.42
Muslim	0.41	0.94	0.000	0.000	0.00	0.00	1.43	0.82	2.65	1.34	0.60	0.17	0.64	0.15	0.63	0.16
Naik	0.29	0.71	0.097	0.376	0.00	0.04	2.04	2.03	2.11	2.43	0.72	0.36	0.81	0.41	0.82	0.44
Others	0.60	0.93	0.235	0.705	0.02	0.08	2.31	2.65	1.91	3.46	0.84	0.50	0.89	0.50	0.96	0.46
Patgar	0.03	0.20	0.103	0.316	0.00	0.00	1.84	1.67	1.72	1.95	0.71	0.47	0.75	0.46	0.79	0.50
Shet	0.17	1.39	0.141	0.291	0.00	0.00	2.90	1.38	3.88	2.23	0.67	0.24	0.71	0.30	0.73	0.30
Avg	0.21		0.18		0.01		2.31		2.63		0.73		0.79		0.80	
Sd	0.19		0.21		0.03		1.37		1.93		0.17		0.16		0.17	
Max	0.60		0.68		0.10		5.24		7.57		1.06		1.10		1.08	
Min	0.00		0.00		0.00		0.03		0.04		0.43		0.50		0.50	

7.7.7 Energy Consumption Based on Landholding of Households

Households are grouped based on landholdings, such as landless, and those holding 0-0.5 hect, 0.5-0.75 hect, 0.75-1.25 hect, 1.25-1.75 hect, 1.75-2 hect and >2 hect. In the sample households, landless accounted for 2.68%, while families owning 0-0.5 hect accounted for 24.76%, 0.5-0.75 hect 32.13%, 0.75-1.25 hect 17.48%, 1.25-1.75 hect 6.53%, 1.75-2 hect 5.23% and households owning greater than 2 hectares 11.19%. Daily energy consumption for cooking, water heating and lighting was computed for each category. Seasonwise and categorywise fuel wood consumption for cooking, listed in Table 7.8.1, shows variation from 1.83 (landless category) to 2.31 (for landholding > 2 hect category) kgs/person/day.

Table 7.8.1: Fuel wood Consumption for Cooking (kgs/person/day) in Various Landholding Categories

Landholding hectare	No. of samples	Cooking Fuel wood kgs/person/day							
		Summer		Winter		Monsoon		Average	
		Avg	Sd	Avg	Sd	Avg	Sd	Avg	Sd
Landless	35	1.78	0.90	1.73	0.94	1.98	1.11	1.83	0.98
0 - 0.5	323	1.97	1.08	1.95	1.11	2.06	1.34	1.99	1.18
0.5 - 0.75	419	2.05	1.47	2.03	1.30	2.14	1.71	2.07	1.49
0.75 - 1.25	228	2.13	1.25	2.08	1.18	2.21	1.29	2.14	1.24
1.25 - 1.75	85	2.10	1.63	2.11	1.75	2.27	2.12	2.16	1.83
1.75 - 2	68	1.92	1.58	2.16	1.84	2.51	2.67	2.20	2.03
> 2	146	1.94	1.95	2.30	2.54	2.70	3.57	2.31	2.69
Total	1304								
Avg		1.98		2.05		2.27		2.10	

Fuel consumption for water heating, listed in Table 7.8.2, ranges from 1.23 (landless and 0.5-0.75 hect category) to 1.60 (for very large farmers) kgs/person/day. Kerosene, biogas and LPG (for cooking) and electricity and kerosene (for lighting) are listed in Table 7.8.3. Most of the households in all these categories still use firewood for cooking and kerosene stove for boiling milk, tea preparation, etc. Out of 1304 households only 18 use biogas for cooking, out of which 6 households belong to >2 hect and 4 to 1.75-2 hect (this category also uses kerosene stove and fuel wood stoves on special occasions). Eight biogas units were evenly distributed (2 units each) in the remaining categories. Biogas for cooking ranges from 0.276 (landless category) to 0.775 (>2 hectares category) m^3/person/day.

Electricity consumption seemed to have increased with an increase in landholdings. It ranges from 1.99 (landless), 2.14 (0-0.5 hect) to 3.59 (>2 hect) kWh/person/month. Kerosene usage for lighting has decreased with an increase in landholdings. It varies from 0.87 (landless), 0.85 (0-0.5 hect) to 0.75 (> 2 hect) ltrs/person/month.

7.7.8 Energy Consumption Analyses Based on Household Income

Based on income, surveyed households are categorised into low (< Rs. 3600 per year), middle (Rs. 4000-10000 per year) and high (> Rs. 10000 per year) income categories. Pooled data shows 34.5% households belonging to low, 49% middle and the balance 16.5% to high income categories.

Table 7.9.1 lists seasonwise fuel wood consumption for cooking purposes in various income categories. Most of the households are still dependent on fuel wood for cooking which ranges from 2.02 (high income) to 2.09 (low income) kgs/person/day.

Table 7.8.2: Fuel wood Consumption for Water Heating (kgs/person/day) in Various Landholding Categories

Landholding hectare	No. of samples	Water Heating Fuel wood kgs/person/day							
		Summer		Winter		Monsoon		Average	
		Avg	Sd	Avg	Sd	Avg	Sd	Avg	Sd
Landless	35	1.09	0.65	1.22	0.65	1.38	0.77	1.23	0.69
0 - 0.5	323	1.19	0.51	1.28	0.60	1.32	0.58	1.26	0.56
0.5 - 0.75	419	1.17	0.51	1.23	0.57	1.30	0.58	1.23	0.55
0.75 - 1.25	228	1.18	0.63	1.23	0.65	1.33	0.73	1.25	0.67
1.25 - 1.75	85	1.26	0.79	1.38	0.94	1.50	1.09	1.38	0.94
1.75 - 2	68	1.39	0.65	1.43	0.76	1.55	0.83	1.46	0.75
> 2	146	1.52	0.73	1.63	0.80	1.64	0.79	1.60	0.77
Total	1304								
Avg		1.26		1.34		1.43		1.34	0.13

Table 7.8.3: Kerosene, Biogas and LPG Usage for Cooking, and Kerosene and Electricity for Lighting in Various Landholding Categories

Land-holding hectare	No. of samples	Cooking Energy						Lighting Energy										
		Kerosene ltrs/person/ month		Biogas m3/person/ day		LPG kgs/person/ month		Electricity kWh/ person/ month		Electricity Rs/ capita/ month		Kerosene ltrs/person/month						
												Summer		Winter		Monsoon		Average
		Avg	Sd	Avg	Sd	Avg	Sd	Avg	Sd	Avg	Sd	Avg	Sd	Avg	Sd	Avg	Sd	Avg
Landless	35	0.28	0.76	0.276	0.389	0.00	0.00	1.99	3.15	1.18	2.57	0.79	0.74	0.82	0.73	1.00	0.76	0.87
0 - 0.5	323	0.35	0.88	0.295	0.393	0.01	0.10	2.14	2.83	2.33	3.69	0.82	0.45	0.87	0.47	0.88	0.50	0.86
0.5 - 0.75	419	0.19	0.54	0.342	0.718	0.01	0.08	2.18	2.38	2.26	2.72	0.77	0.44	0.86	0.45	0.83	0.45	0.82
0.75 - 1.25	228	0.23	0.67	0.356	0.753	0.00	0.04	2.37	2.75	2.57	4.57	0.72	0.45	0.80	0.43	0.80	0.44	0.77
1.25 - 1.75	85	0.23	0.64	0.393	1.460	0.01	0.08	2.00	2.57	2.17	3.74	0.78	0.62	0.84	0.62	0.85	0.60	0.82
1.75 - 2	68	0.29	0.76	0.536	1.000	0.01	0.07	2.01	2.57	2.86	3.88	0.71	0.37	0.77	0.42	0.79	0.47	0.76
> 2	146	0.23	0.67	0.775	1.410	0.02	0.10	3.59	3.82	4.73	7.06	0.76	0.69	0.75	0.56	0.75	0.57	0.75
Total	1304																	
Avg		0.26		0.42		0.01		2.33		2.59		0.76		0.82		0.84		0.81

Table 7.9.2 provides seasonal fuel wood requirement for water heating, which ranges from 1.26 (low income) to 1.48 (high income) kgs/person/day.

Table 7.9.1: Cooking Fuel wood Consumption in Various Household Income Categories

Income Rs.	No. of samples	Cooking Fuel wood kgs/person/day							
		Summer		Winter		Monsoon		Average	
		Avg	Sd	Avg	Sd	Avg	Sd	Avg	Sd
< 3600	215	2.05	0.96	2.05	0.97	2.17	1.01	2.09	0.98
4000 - 10000	639	2.03	1.36	2.03	1.26	2.12	1.53	2.06	1.38
10500 - 500000	450	1.97	2.13	1.99	2.57	2.11	3.61	2.02	2.77
Total	1304								
Avg		2.02		2.02		2.13		2.06	

Table 7.9.2: Fuel wood Consumption for Water Heating in Various Household Income Categories

Income Rs.	No. of samples	Water Heating Fuel wood kgs/person/day							
		Summer		Winter		Monsoon		Average	
		Avg	Sd	Avg	Sd	Avg	Sd	Avg	Sd
< 3600	215	1.16	0.57	1.27	0.68	1.34	0.74	1.26	0.66
4000 - 10000	639	1.22	0.54	1.28	0.62	1.34	0.63	1.28	0.60
10500 - 500000	450	1.40	0.72	1.48	0.77	1.56	0.78	1.48	0.76
Total	1304								
Avg		1.26		1.34		1.41		1.34	

Table 7.9.3 is an attempt to see the role of income on type of fuel chosen for cooking and water heating in a household. Out of 18 biogas plants in the sample of 1304 households, 2 were in low, 9 in middle followed by 7 in high income categories. Biogas consumption ranges from 0.20 (low income) to 0.59 (high income) m^3/person/day. Households using biogas for cooking also use kerosene to supplement their cooking fuel requirement. Kerosene used for cooking seems to increase with income (Table 7.9.4), with 0.16 (low income), 0.29 (middle income) and 0.33 (high income) ltrs/person/month.

Table 7.9.3: Kerosene, Biogas and LPG for Cooking in Various Household Income Categories

Income Rs.	No. of samples	Cooking Energy					
		Kerosene ltrs/person/month		Biogas m3/person/day		LPG kgs/person/month	
		Avg	Sd	Avg	Sd	Avg	Sd
<3600	215	0.16	0.67	0.200	0.892	0.00	0.00
4000 - 10000	639	0.29	0.73	0.453	0.600	0.02	0.09
10500 - 500000	450	0.33	0.64	0.596	0.359	0.03	0.12
Total	1304						
Avg		0.26		0.42		0.02	

Electricity consumption (Table 7.9.4) also shows similar trend, ranging from 1.34 (low income), 2.21 (middle income) to 4.79 (high income) kWh/person/day. Usage of electric gadgets, such as refrigerators, grinders, mixies, etc., is the main reason for higher consumption (excluding irrigation requirements).

Table 7.9.4: Kerosene and Electricity for Lighting in Various Household Income Categories

Income Rs.	No. of samples	Lighting Energy										
		Electricity kWh/person /month		Electricity Rs/capita /month		Kerosene ltrs/person/month						
						Summer		Winter		Monsoon		Average
		Avg	Sd	Avg	Sd	Avg	Sd	Avg	Sd	Avg	Sd	Avg
<3600	215	1.34	1.86	1.23	2.22	0.83	0.51	0.89	0.48	0.89	0.49	0.87
4000 - 10000	639	2.21	2.48	2.48	3.75	0.73	0.44	0.81	0.45	0.80	0.47	0.78
10500 - 500000	450	4.79	3.85	5.84	6.29	0.76	0.61	0.79	0.61	0.81	0.61	0.79
Total	1304											
Avg		2.78		3.18		0.77		0.83		0.83		0.81

7.7.9 Number of Persons per Household

Out of 1304 households, 12.36% belonged to families with adult equivalent less than 3 per household, 55.59% with 3.10-6 per household, 24.76% with 6.05-9 per household, 4.91% with 9.05-12 per household and the balance 2.38% has >12 adult equivalent per household.

Data shows a reduced average per capita fuel consumption with increase in adult equivalents per household. Table 7.10.1 shows that fuel wood requirement per person per day for cooking ranges from 2.35 kgs (< 3 adults) to 1.42 kgs (> 12 adults). Similarly Table 7.10.2 illustrates that fuel wood requirement for water heating ranges from 1.99 (< 3 adults), 1.33 (3.1 - 6 adults) to 0.73 (>12 adults) kgs/person/day.

Table 7.10.1: Cooking Fuel wood Consumption in Various Categories of Number of Adult Equivalent/Household

Number of persons per household	No. of samples	Cooking Fuel wood kgs/person/day							
		Summer		Winter		Monsoon		Average	
		Avg	Sd	Avg	Sd	Avg	Sd	Avg	Sd
0.80 - 3.00	161	2.25	2.66	2.30	2.86	2.51	4.02	2.35	3.18
3.10 - 6.00	725	2.10	1.20	2.11	1.25	2.24	1.54	2.15	1.33
6.05 - 9	323	1.59	0.72	1.67	0.79	1.75	0.80	1.67	0.77
9.05 - 12	64	1.57	1.59	1.59	0.72	1.70	0.97	1.62	1.09
>12	31	1.40	0.65	1.40	0.68	1.46	0.74	1.42	0.69
Total	1304								
Avg		1.78		1.81		1.93		1.84	

Table 7.10.2: Water Heating Fuel wood Consumption in Various Categories of Number of Adult Equivalent/Household

Number of persons per household	No. of samples	Water Heating Fuel wood kgs/person/day							
		Summer		Winter		Monsoon		Average	
		Avg	Sd	Avg	Sd	Avg	Sd	Avg	Sd
0.80 - 3.00	161	1.82	0.83	2.06	0.96	2.10	1.00	1.99	0.07
3.10 - 6.00	725	1.25	0.49	1.33	0.56	1.42	0.60	1.33	0.04
6.05 - 9	323	1.00	0.44	1.01	0.43	1.07	0.49	1.03	0.03
9.05 - 12	64	0.91	0.38	0.92	0.36	0.96	0.38	0.93	0.01
>12	31	0.70	0.33	0.72	0.33	0.76	0.34	0.73	0.00
Total	1304								
Avg		1.14		1.21		1.26		1.20	

Regression analyses of per capita fuel consumption (PCFC) in kgs/person/day and number of persons per household (NOP) show that PCFC is linearly correlated with NOP, that is,

$$PCFC = 2.341 - 0.122\ (NOP)$$

with a correlation co-efficient of 0.36 and standard error of Y estimate = 0.923.

This interesting reduction in fuel wood consumption with an increase in number of persons per household is due to the coefficiencies of cooking and water heating that result from increased scales of the same. Therefore, proper design of the stoves and size of vessels being used in relation to the number of persons per household are essential parameters in bringing down fuel wood consumption.

7.7.10 Effect of Educational Level of Household Members on Energy Consumption

Qualitative data of educational level of members who are directly involved in domestic activity is quantified as 0 (illiterate), 1 (primary education), 2 (up to 10th standard) and 3 (college education). It is noticed that educated users are more receptive to adopting energy efficient devices and are also efficient in utilising energy even in traditional devices.

Regression analyses carried out between the variables PCFC and level of education of members (EDU) show that these two are linearly correlated as

$$PCFC = 3.21 - 1.021\ (EDU)$$

with correlation co-efficient 0.845 and standard error of Y estimate = 0.506 ($p < 0.001$).

This is mainly due to awareness associated with level of education. This result clearly demonstrates that increase in literacy level among rural population would directly benefit in the form of less energy consumption.

7.7.11 Energy Consumption Based on Type of Devices –Traditional vs Energy Efficient Devices (ASTRA stoves)

Only 2.07% (27 households) and 1.68% (22 households) of the 1304 surveyed households use energy efficient stoves for cooking and water heating respectively. Per capita fuel consumption for cooking, listed in Table 7.11.1, shows that it reduces from 2.14 (traditional stoves) to 0.97 (ASTRA stoves) kgs/person/day. Hence, there is a reduction of 54.67% in daily consumption of fuel wood for cooking. Table 7.11.2 lists fuel consumption for water heating, which shows a saving of 22.30% by switching over to ASTRA stoves (1.01 kgs/person/day) from traditional stoves (1.30 kgs/person/day).

Table 7.11.1: Cooking Fuel wood Consumption based on Type of Devices

Cooking devices	No. of samples	Cooking Fuel wood kgs/person/day							
		Summer		Winter		Monsoon		Average	
		Avg	Sd	Avg	Sd	Avg	Sd	Avg	Sd
Astra	27	0.89	0.68	0.91	0.60	1.11	0.79	0.97	0.69
Traditional	1277	2.06	1.42	2.09	1.49	2.26	1.97	2.14	1.63
Total	1304								
Avg		1.47		1.50		1.68		1.55	

Table 7.11.2: Fuel wood Consumption for Water Heating Based on Type of Devices

Bath stoves	No. of samples	Water Heating Fuel wood kgs/person/day							
		Summer		Winter		Monsoon		Average	
		Avg	Sd	Avg	Sd	Avg	Sd	Avg	Sd
Astra	22	0.89	0.43	1.02	0.50	1.12	0.66	1.01	0.53
Traditional	1282	1.23	0.59	1.28	0.68	1.38	0.70	1.30	0.66
Total	1304								
Avg		1.06		1.15		1.25		1.15	

7.8.0 Energy Consumption for Other Purposes in the Domestic Sector

In the following section, energy requirements for other purposes such as parboiling, jaggery making, etc., in the domestic sector is discussed.

7.8.1 Fuel Consumption for Space Heating

It is seen that 65 out of 744 households in zone-1 and 80 out of 301 households in zone-2 use fuel for space heating during the season at a rate of 0.15 kgs/person/day. While, 80% of households (207 out of 259) in zone-3, due to incessant rain, use fuel wood for space heating and drying of cloth at the rate of 1.11 kgs/person/day.

7.8.2 Fuel Consumption for Areca Boiling (Red Variety Preparation)

Specific fuel consumption (SFC), computed based on data from 12 households, ranges from 1.97 to 2.03 (This means fuel wood required for a quintal of areca processing is 1.97 to 2.03 quintals).

7.8.3 Jaggery Manufacture

Fuel wood required in traditional stoves to convert 300 litres of sugar-cane juice into about 40 litres of viscous jaggery is 195-220 kgs, while in ASTRA stove (jaggery stove designed by ASTRA) it is 160-175 kgs showing 25-30% reduction in fuel wood consumption. SFC computed based on data from 8 households (2 in zone-2 and 6 in zone-3) ranges from 4.88 to 5.52.

7.8.4 Parboiling

Boiled rice is used in most of the households in Kumta taluk. 68% of the surveyed households use boiled rice regularly (mainly in zones 1 and 2). Fuel type used for this purpose ranges from rice husk, fuel wood to coconut residues. SFC computed ranges from 0.70 - 0.89, which means, for manufacturing 1 quintal of rice, fuel wood equivalent required is about 0.7 to 0.89 quintals.

7.9.0 Energy Consumption Pattern in Kumta Town

Kumta town municipality has divided the town into 5 blocks. Approximately 260 households were selected from each block, covering all sections of the society.

Data of 1307 surveyed households was categorised based on annual income. 280 households belong to low income category (< Rs. 3600 per year), 214 to middle class category (Rs. 3600-10,000), 800 to Rs. 10,000-1,00,000 category and 13 households to more than Rs. 1,00,000 income category.

Out of 280 households in low income category, 266 (95%) completely depend on fuel wood for cooking, 13 use kerosene stoves for boiling milk, preparation of tea, etc., and 1 household (landless) uses LPG in addition to fuel wood.

In middle income category of 214 households, 199 (92.2%) completely depend on fuel wood, 3 on kerosene stoves, while the remaining 12 use both kerosene and fuel wood for cooking.

In Rs. 10,000 to 1,00,000 income category, about 62% (496 out of 800 households) use only fuel wood, 18% (144) use kerosene and fuel wood, 16% (128) use biogas, fuel wood and kerosene, and 4% (32) use LPG, fuel wood and kerosene for cooking. In more than Rs. 1,00,000 income category of 13 households, 9 (69.23%) use only fuel wood while the remaining 4 (30.77%) use both LPG and fuel wood.

Fuel consumption for cooking and water heating in these categories is listed in Table 7.12.1. Fuel wood consumption for cooking ranges from 1.59 (Rs. 3600-10,000 category) to 2.03 (> Rs. 1,00,000 category) kgs/person/day, while for water heating it ranges from 0.82 (> Rs. 1,00,000 category) to 1.23 (low income category) kgs/person/day. It is worth noting that middle income category depends mainly on fuel wood supplied by Government Fuel Depot (40 to 55%) and Private Suppliers, while low income category prefers to collect it from forests (in the vicinity of 5 to 10 kms). This is mainly done by women and children. Normally, people from many households collectively go to the forest (Mirjan, Kattalekan, Halkar, etc.) at about 5 a.m. and return at about 1 p.m., each individual carrying an average of 20 to 26 kgs (head load) of fuel wood.

Kerosene and electricity usage for lighting in Kumta town is listed in Table 7.12.2. It is seen that kerosene usage is more or less the same in the first three income categories, while in annual income > Rs. 1,00,000 category, consumption is very less. Electricity consumption analyses show the dependence of higher income categories for lighting compared to first two categories. It ranges from 3.13 (Rs. 3600-10000), 5.96 (Rs. 10,000-1,00,000) to 14.47 (> Rs. 1,00,000 category) kWh/person/month.

Table 7.12.3 lists the results of the analyses based on the number of persons per household. Fuel wood consumption for cooking ranges from 2.14 (<3 adults) to 0.92 (> 12 adults) kgs/person/day, while for water heating it ranges from 1.31 to 0.88 kgs/person/day. This confirms the earlier result (Kumta rural area), that per capita fuel consumption reduces with increase in number of persons per household.

Table 7.12.4 lists communitywise fuel consumption for cooking and water heating. This illustrates the intra-community variation in fuel consumption in kgs/person/day. Havyak Brahmins (2.08 - cooking, 1.27 - water heating), Muslims (2.16 - cooking, 1.50 - water heating) and Gunagas (2.02 - cooking, 2.04 - water heating) seem to be energy intensive communities in Kumta town. Uppar community has minimum per capita consumption (1.17 - cooking, 0.78 - water heating).

Table 7.12.1: Energy Consumption for Cooking and Water Heating based on Household Income in Kumta Town

Income Rs.	No. of samples	Cooking Energy								Bath stove kgs/person/day	
		Fuel wood kgs/person/day		Kerosene ltrs/person/month		Biogas m3/person/day		LPG kgs/person/month			
		Avg	Sd	Avg	Sd	Avg	Sd	Avg	Sd	Avg	Sd
<3600	280	1.80	1.09	0.17	1.48	0.00	0.00	0.04	0.02	1.23	1.65
3600 - 10000	214	1.59	0.92	0.24	1.54	0.00	0.00	0.00	0.00	1.17	0.85
10000 - 100000	800	1.85	0.92	0.18	1.91	0.11	2.23	0.10	0.03	1.04	1.15
> 100000	13	2.03	0.50	0.00	0.00	0.00	0.00	0.23	0.08	0.82	1.03
Total	1307										
Avg		1.82		0.15		0.03		0.09		1.07	

Table 7.12.2: Energy Consumption for Lighting based on Household Income in Kumta Town

Income Rs.	No. of samples	Space heating fuel wood kgs/person/season		Lighting: Kerosene ltrs/person/month		Electricity kWh/person/month		Electricity Cost Rs/capita/month	
		Avg	Sd	Avg	Sd	Avg	Sd	Avg	Sd
<3600	280	0.00	0.00	0.34	0.23	3.58	4.98	6.59	11.69
3600 - 10000	214	0.00	0.00	0.45	0.68	3.13	3.37	4.70	6.74
10000 - 100000	800	0.00	0.00	0.34	0.56	5.96	6.76	11.23	15.64
> 100000	13	0.00	0.00	0.02	0.10	14.47	8.89	42.89	30.48
Total	1307								
Avg		0.00		0.29		6.79		16.35	

Table 7.12.3: Fuel Consumption for Cooking and Water Heating based on Number of Persons/Household

No. of persons/ household	No. of samples	Cooking Energy: Fuel wood kgs/person/day		Kerosene ltrs/person/month		Biogas m3/person/ day		LPG kgs/person/month		Bath stove kgs/person/day	
		Avg	Sd	Avg	Sd	Avg	Sd	Avg	Sd	Avg	Sd
0.8 - 3.0	368	2.14	0.87	0.21	2.49	0.00	0.00	0.00	0.04	1.31	1.38
3.10 - 6	697	1.79	1.00	0.19	1.49	0.13	2.39	0.12	0.02	1.10	1.27
6.05 - 9	186	1.30	0.73	0.24	0.77	0.00	0.00	0.15	0.03	1.07	0.66
9 - 12	41	1.08	0.59	0.22	1.10	0.00	0.00	0.00	0.00	0.92	0.73
>12	15	0.92	0.86	0.35	0.60	0.00	0.00	0.00	0.00	0.88	0.91
Total	1307										
Avg		1.45		0.24		0.03		0.05		1.06	

Table 7.12.4: Communitywise Fuel Consumption for Cooking and Water Heating

Community	No. of samples	Cooking Energy: Fuel wood kgs/person/day		Kerosene ltrs/person/month		Biogas m3/person/day		LPG kgs/person/month		Bath stove kgs/person/day	
		Avg	Sd	Avg	Sd	Avg	Sd	Avg	Sd	Avg	Sd
Uppar	27	1.17	0.64	0.05	0.07	0.00	0.00	0.00	0.00	0.78	0.54
Shetty	24	1.64	0.96	0.18	0.80	0.00	0.00	0.00	0.00	0.90	0.60
Patgar	110	1.99	1.10	0.17	0.63	0.00	0.00	0.00	0.00	1.12	1.57
Others	91	1.53	0.68	0.24	2.38	0.00	0.00	0.00	0.00	0.89	1.45
Naik	240	1.61	0.94	0.21	1.66	0.00	0.00	0.01	0.05	0.96	0.92
Muslim	73	2.16	1.20	0.24	1.98	0.00	0.00	0.00	0.00	1.50	1.43
Havyak Brahmin	190	2.08	1.17	0.32	2.16	0.23	0.31	0.00	0.00	1.27	1.55
Gunaga	9	2.02	1.01	0.31	3.70	0.00	0.00	0.00	0.00	2.04	2.08
GS Brahmin	162	1.85	0.76	0.19	2.10	0.00	0.00	0.00	0.00	1.17	1.27
Gouda	251	1.80	0.86	0.10	1.30	0.00	0.00	0.00	0.03	1.09	1.06
Deshbandari	4	1.58	0.49	0.14	2.23	0.00	0.00	0.00	0.00	1.03	0.33
Christian	52	1.54	0.77	0.23	1.00	0.90	6.41	0.01	0.05	1.02	0.81
Bhandari	31	1.47	0.66	0.21	2.27	0.00	0.00	0.00	0.00	0.96	0.69
Ambiga	19	1.62	0.95	0.05	0.19	0.00	0.00	0.00	0.00	1.19	1.08
Achari	24	1.88	0.74	0.15	1.48	0.00	0.00	0.00	0.00	0.99	0.75
Total	1307										
Avg		1.73		0.19		0.08		0.00		1.13	

Communitywise monthly electricity consumption (Table. 7.12.5) shows that communities like Havyak Brahmins (HB), Shettys (SH), Gouda Saraswath Brahmins (GSB), etc., are some of the electricity intensive communities in Kumta town with 7.09, 6.81 and 6.07 kWh/person respectively.

7.10.0 FUEL WOOD CONSUMPTION PATTERN IN MUNDGOD TALUK

Mundgod taluk with plain terrain differs distinctly from coastal and hilly taluks. In order to see the level of consumption, analysis of data collected from 190 randomly selected households was carried out. Energy consumption for domestic purposes (cooking, water heating, space heating and lighting), based on annual income and number of persons per household, was carried out and is listed in Tables 13.1 and 13.2 and Tables 14.1 and 14.2 respectively.

Table 7.13.1 shows that average fuel requirement for cooking is about 1.65 kgs/person/day. It ranges from 1.37 kg (middle income) to 1.84 kg (for income >Rs. 10,500/year). While average per capita fuel wood consumption for water heating is 1.31 kg.

Table 7.13.2 lists per capita monthly electricity and kerosene consumption for lighting, which shows increased electricity and decreased kerosene consumption with increase in household income.

Table 7.12.5: Communitywise Monthly per Capita Electricity Consumption

Community	No. of samples	Electricity kWh/person/month		Electricity Rs/capita/month	
		Avg	Sd	Avg	Sd
Uppar	27	1.06	1.96	1.26	2.72
Shetty	24	6.81	4.07	12.35	14.13
Patgar	110	5.70	6.26	5.57	18.90
Others	91	3.42	3.25	6.60	8.85
Naik	240	4.61	5.02	6.87	10.32
Muslim	73	3.45	4.39	8.02	14.56
HB	190	7.09	7.65	12.79	17.79
Gunaga	9	2.46	1.74	2.77	1.90
GSB	162	6.07	8.36	11.26	16.57
Gouda	251	4.80	6.40	8.99	15.39
Deshbhandari	4	1.36	1.36	2.90	2.91
Christian	52	5.57	4.15	10.33	13.00
Bhandari	31	4.30	3.81	5.74	5.41
Ambiga	19	2.02	3.33	1.52	2.97
Achari	24	4.91	3.78	7.24	4.95
Total	1307				
Avg		4.24		6.95	

Table 7.13.1: Energy Conumption for Cooking and Water Heating in Mundgod

Income Rs.	No. of samples	Cooking Energy: Fuel wood kgs/person/day		Kerosene ltrs/person/month		Biogas m3/person/day		LPG kgs/person/month		Bath stove kgs/person/day	
		Avg	Sd	Avg	Sd	Avg	Sd	Avg	Sd	Avg	Sd
< 3600	112	1.74	1.22	0.35	0.60	0.43	4.55	0.01	0.08	1.29	1.28
4000-10000	40	1.37	1.76	0.48	0.11	0.00	0.00	0.00	0.00	1.27	1.46
> 10500	38	1.84	1.86	0.59	0.56	0.00	0.00	0.00	0.00	1.36	1.49
Total	190										
Avg		1.65		0.47		0.14		0.00		1.31	

Table 7.13.2: Energy Consumption for Space Heating and Lighting (based on Household Income) in Mundgod Taluk

Income Rs.	No. of samples	Space heating Fuel wood kgs/person/season		Lighting: Kerosene ltrs/person/month		Electricity kWh/person/month		Electricity Cost Rs/capita/month	
		Avg	Sd	Avg	Sd	Avg	Sd	Avg	Sd
< 3600	112	0.00	0.00	0.54	0.32	3.68	3.97	11.72	22.22
4000-10000	40	0.00	0.00	0.52	0.22	1.30	2.21	1.77	2.98
> 10500	38	0.00	0.00	0.34	0.18	6.01	5.91	10.16	18.01
Total	190								
Avg		0.00		0.47		3.66		7.88	

Table 7.14.1: Energy Consumption for Cooking and Water Heating in Mundgod (based on Number of Persons/Household)

No. of persons/ household	No. of samples	Cooking Energy									
		Fuel wood kgs/person/day		Kerosene ltrs/person/month		Biogas m3/person/day		LPG kgs/person/month		Bath stove kgs/person/day	
		Avg	Sd	Avg	Sd	Avg	Sd	Avg	Sd	Avg	Sd
0.8 - 3.0	62	1.92	1.07	0.69	0.18	0.00	0.00	0.01	0.11	1.41	1.65
3.0 - 6.0	107	1.69	1.22	0.78	0.60	0.45	4.63	0.00	0.00	1.59	1.23
6.0 - 9.0	18	1.82	1.61	0.42	0.57	0.00	0.00	0.00	0.00	1.25	1.13
9.0 - 12	3	1.24	1.23	0.00	0.00	0.00	0.00	0.00	0.00	1.22	1.19
Total	190										
Avg		1.67		0.47		0.11		0.00		1.37	

Table 7.14.1 confirms the earlier conclusion (Kumta taluk), that per capita fuel wood consumption decreases with increase in number of persons per household. Fuel wood requirement for cooking ranges from 1.92 (0.80-3.0 person/household) to 1.24 (9.0-12 person/household), while for water heating it ranges from 1.41 to 1.22 kgs/person/day.

Table 7.14.2 listing energy requirement for lighting shows the variation from 5.38 (0.8-3.0 person/household) to 1.28 (9.0-12 person/household) kWh/person/month.

Table 7.14.2: Energy Consumption for Space Heating and Lighting in Mundgod (based on Income)

No. of persons/ household	No. of samples	Space heating Fuel wood kgs/person/season		Lighting					
				Kerosene ltrs/person/month		Electricity kWh/person/month		Electricity Cost Rs/capita/month	
		Avg	Sd	Avg	Sd	Avg	Sd	Avg	Sd
0.8 - 3.0	62	0.00	0.00	0.56	0.34	5.38	6.47	13.20	29.26
3.0 - 6.0	107	0.00	0.00	0.34	0.12	3.10	2.94	8.13	12.75
6.0 - 9.0	18	0.00	0.00	0.45	0.22	3.13	1.64	5.37	5.22
9.0 - 12	3	0.00	0.00	0.32	0.10	1.28	0.93	3.47	0.46
Total	190								
Avg		0.00		0.42		3.22		7.54	

7.11.0 Fuel wood Consumption for Cooking and Water Heating in Sirsi taluk

The average per capita fuel consumption in 202 sample households located in villages of Sonda, Mogadde, Bhairumbhe, Neernalli and Kalave for cooking is 2.22 (avg)±0.92(Sd), while for water heating it is 1.72±0.68 kgs/person/day.

End use efficiency experiment carried out in households located in Sirsimakki-Mundgesara catchment shows consumption in traditional stoves as 1.92±1.02 and improved stoves as 1.1±0.78 kgs/person/day, showing a saving of about 42% in fuel consumption in improved cookstoves (ASTRA design).

The fuel requirement for water heating is mainly met by residues of coconut and areca plantations, and partly by fuel wood. The average fuel consumption in traditional stoves is 1.68±0.80 and improved bath stoves 1.36±0.63 kgs/person/day, showing a saving of 19-24% in improved design (ASTRA).

7.12.0 Regionwise Comparison

Table 7.15 is a compilation of surveys carried out in various taluks (regions) of Uttara Kannada district which shows the variation in fuel wood consumption from 1.65 (Mundgod) to 2.32 (Kumta hilly) for cooking in kgs/person/day. Figure 7.3 depicts variation in fuel wood consumption due to geographic factors. It illustrates that households located in hilly regions consume 40.6% more fuel wood than those in the plains.

Figure 7.4 illustrates the regionwise variation in fuel consumption for water heating, which ranges from 1.07 (Kumta town) to 1.72 (Sirsi-hilly region) kgs/person/day. In most of the surveyed households, 65 to 75% of the fuel requirement is met by bioresidues, such as, coconut husk, coconut leaves, areca residues, etc. Based on these calculations, the overall average fuel wood consumption for cooking is about 2.02 and water heating about 1.42 kgs/person/day.

Table 7.15: Regionwise Fuel Requirement for Cooking and Water Heating (Based on Average Values of Per Capita Fuel Consumption)

	Region	Cooking Avg kgs/person/day	Water Heating Avg kgs/person/day
1	Mundgod	1.65	1.31
2	Kumta Town	1.82	1.07
3	Sirsi (Semi-Urban)	1.92	1.68
4	Kumta Coast	2.01	1.17
5	Kumta Interior	2.19	1.35
6	Sirsi Hilly	2.22	1.72
7	Kumta Hilly	2.32	1.63
	Avg	2.02	1.42

These surveys have revealed that the levels of energy use and mix of energy sources depend on the following factors.

(1) Geographic and climatic factors (space heating, drying needs, altitude, seasonal variation in cooking energy requirement),
(2) Culture and traditions (diet, cooking habits, etc.),
(3) Household income (urban),
(4) Household ownership of assets such as land, cattle, etc. (rural),
(5) Household size (number of persons per household),
(6) Educational level of household members (educated women are found to be more receptive to alternatives such as fuel efficient stoves, biogas, etc.), and
(7) End use technologies (thermal efficiency of cooking and water heating devices).

7.13.0 Total Demand for Fuel in Uttara Kannada District

Based on fuel consumption norms determined for various regions in these surveys, talukwise fuel demand for cooking and water heating is computed and listed in Table 7.16.1. Currently, about 60% (approx.) fuel for water heating is met by agricultural residues. With this assumption, fuel wood required for domestic cooking and water heating is 12,02,615.64 tonnes and agriculture residues about 3,67,861.44 tonnes (fuel wood equivalent). Thus, total fuel wood required for the domestic sector is about 15,70,477.07 tonnes.

Due to incessant rain in this region, fuel wood (normally logs) is used to keep the place warm and dry clothes. This requirement in the coastal and hilly regions is about 0.15 and 1.11 kgs/person/day respectively, amounting to 53,772.5 tonnes during monsoon.

Table 7.16.2 lists talukwise kerosene demand for cooking and lighting, and electricity for domestic purpose (excludes irrigation). Kerosene demand of the district for lighting and cooking is about 12.73 and 3.32 million litres respectively. LPG required is about 226.40 tonnes (or 15613 cylinders) per year. Electricity demand in the domestic sector (lighting and All Electric homes) is about 32.65 million kWh per year. Table 7.16.3 lists activitywise total annual requirement of fuel wood. Total fuel wood equivalent required for the domestic sector is about 15,70,477.07 tonnes and for space heating about 53,772.5 tonnes.

Uttara Kannada district produces about 1,49,698 tonnes of sugar-cane per year (1992-93 and 1993-94 production), out of which jaggery is prepared for domestic purposes. The fuel required for this is about 41,316.65 tonnes.

Areca is one of the major horticultural crops in Uttara Kannada district. Boiling of Areca (to manufacture red variety for day to day use) is another fuel intensive operation carried out here, requiring about 1399.88 tonnes of fuel wood per year.

Boiled rice is commonly used in coastal taluks. Fuel wood and rice husks are used for parboiling. The quantity of fuel wood required is about 1732.15 tonnes. Thus, the total fuel wood demand in the domestic sector of Uttara Kannada district (Table 16.3) is about 16,68,698 tonnes.

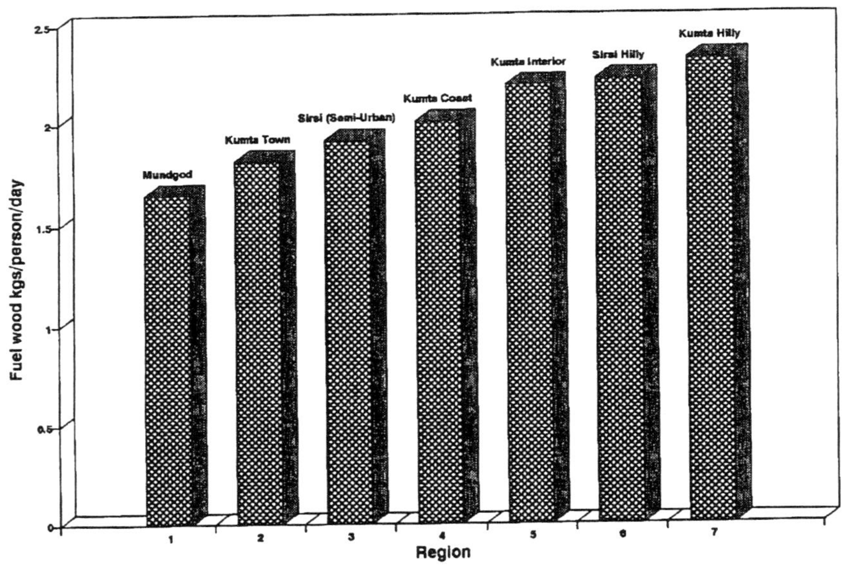

Figure 7.3: Fuel wood Consumption (for Cooking) Regionwise

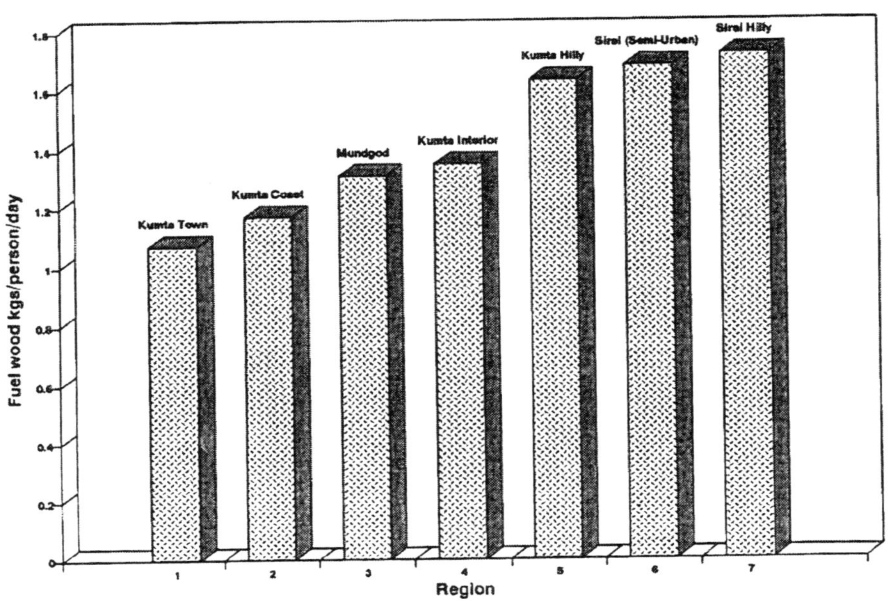

Figure 7.4: Fuel wood Consumption (for Water Heating) Regionwise

Table 7.16.1: Talukwise Fuel wood Consumption for Cooking and Water Heating

Sl.No.	Taluk	Area (in hectare)	Population 1991	FW demand for Cooking (tonne)	FW demand for Bathing (tonne)	FW (cook + 100% bath) Total	Fuel wood Total cook+40%bath	Agricultural residues	Space Heating
1	Ankola	91870.00	91310	67100.68	38993.94	106094.61	82698.25	23396.36	1232.68
2	Bhatkal	35731.96	129017	94810.29	55096.71	149907.00	116848.98	33058.03	1741.73
3	Haliyal	86489.33	94363	75314.26	46497.37	121811.62	93913.20	27898.42	1273.90
4	Honnavar	77546.13	145842	107174.42	62281.83	169456.25	132087.15	37369.10	1968.86
5	Karwar	73210.00	142845	104972.03	61001.96	165973.99	129372.81	36601.17	1928.40
6	Kumta	58486.81	134144	98577.95	57286.20	155864.15	121492.43	34371.72	1810.94
7	Mundgod	66890.00	77939	46938.76	37266.53	84205.30	61845.38	22359.92	1052.17
8	Siddapur	85930.00	94202	79770.25	56045.48	135815.73	102188.45	33627.29	9410.78
9	Sirsi	133342.62	152935	129505.36	90988.68	220494.04	165900.83	54593.21	15278.21
10	Supa	189514.23	99519	84272.69	59208.83	143481.52	107956.22	35525.30	9941.95
11	Yellapur	130110.00	81410	68937.99	48434.88	117372.87	88311.94	29060.93	8132.86
	District	1029121.08	1243526	957374.68	613102.39	1570477.07	1202615.64	367861.44	53772.48
	Computation from district average			986447.7249	627877.0028	1614324.73	1237598.53		

Table 7.16.2: Talukwise Energy Consumption for Cooking and Lighting

Sl.No.	Taluk	Area (in hectares)	Population 1991	Kerosene Cooking x 1000 ltrs	LPG Cooking x 1000 kgs	Kerosene Lighting x 1000 ltrs	Electricity Lighting x 1000 kWh	Electricity Lighting x 1000 Rs.
1	Ankola	91870.00	91310	372.54	21.91	818.14	2739.30	3078.97
2	Bhatkal	35731.96	129017	526.39	30.96	1155.99	3870.51	4350.45
3	Haliyal	86489.33	94363	237.79	11.32	936.08	2955.45	3012.07
4	Honnavar	77546.13	145842	595.04	35.00	1306.74	4375.26	4917.79
5	Karwar	73210.00	142845	582.81	34.28	1279.89	4285.35	4816.73
6	Kumta	58486.81	134144	547.31	32.19	1201.93	4024.32	4523.34
7	Mundgod	66890.00	77939	196.41	9.35	773.15	2441.05	2487.81
8	Siddapur	85930.00	94202	56.52	11.30	1115.35	1752.16	2181.72
9	Sirsi	133342.62	152935	91.76	18.35	1810.75	2844.59	3541.97
10	Supa	189514.23	99519	59.71	11.94	1178.30	1851.05	2304.86
11	Yellapur	130110.00	81410	48.85	9.77	963.89	1514.23	1885.46
	District	1029121.08	1243526	3315.13	226.40	12540.23	32653.27	37101.18
				2984.46	149.22	12733.71	33127.53	36808.37

Table 7.16.3: Total Requirement of Fuel wood for Uttara Kannada District

Activity	Fuel wood requirement
Cooking and Water heating	1570477.07
Space heating	53772.48
Jaggery manufacture	41316.65
Areca processing	1399.88
Parboiling	1732.15
Total (tonnes)	1668698.23

7.14 CONCLUSIONS

The average consumption of fuel wood for cooking and water heating varies considerably among agro-climatic zones (in terms of quantity of fuel used). Significant seasonal variation in consumption of fuel is noticed in all zones during monsoon among various communities. The amount of cooking fuel used per person does not vary substantially with household income in both rural and urban areas. Higher income groups are also dependent on fuel wood significantly.

In rural areas, scarce income combined with freely available biomass fuels lead people to continue to depend on biomass fuel. When firewood is scarce, residents depend on crop residues and dung. Poor distribution of fuels, such as kerosene in rural areas and LPG in towns, is an impeding factor in switching over to modern fuels.

Fuel wood consumption analysis illustrates seasonal and regional variation. Average consumption ranges from 2.01 ± 1.49 (coastal) to 2.32 ± 2.09 (hilly) kgs per person per day. Seasonwise fuel wood requirement for coastal and hilly zones ranges from 1.98 and 2.22 (summer) to 2.11 and 2.51 (monsoon) kgs/person/day respectively.

Data also reveals variation in fuel wood requirement for water heating (for bath and washing), which ranges from 1.17 ± 0.02 (coast) to 1.63 ± 0.05 (hilly) kgs per person per day. Seasonal variation is also evident for coastal and hilly zones from 1.12 and 1.53 (summer) to 1.22 and 1.73 (monsoon) kgs/person/day respectively.

The overall picture of other sources of energy for domestic purposes shows that kerosene is used for both cooking and lighting in the coastal zone. Kerosene for cooking ranges from 0.05 (hilly) to 0.34 (coast) ltrs/person/month. Survey results reveal that the availability of bioresources in the hilly region is the main reason for less consumption of kerosene for cooking compared to the coastal region.

Kerosene for lighting ranges from 0.74 (coast) to 0.98 (hilly) ltrs/person/month. Hilly zone is more dependent on kerosene for lighting due to non-availability of electricity as households are spatially distributed, hindering electrification. This is apart from the erratic electricity supply during all seasons (more in monsoon).

2.68% of the sample accounted for landless category, while families owning 0-0.5, 0.5-0.75, 0.75-1.25, 1.25-1.75, 1.75-2 and > 2 hect of land accounted for 24.76, 32.13, 17.48, 6.53, 5.23 and 11.19% of the sample respectively. The daily energy consumption computed for each category shows variation from 1.83 (landless) to 2.31 (>2) for cooking and 1.23 (landless) to 1.60 (>2) kgs/person/day for water heating. Most of the households in all these categories still use firewood for cooking and kerosene stove for boiling milk, tea preparation, etc.

Out of 1304 households only 18 use biogas for cooking supplemented by kerosene stove (6 belong to >2 hect, 4 to 1.75-2 hectare and 8 evenly distributed, 2 units each in the remaining categories). Fuel wood stoves are used only on special occasions. Biogas for cooking ranges from 0.276 (landless) to 0.775 (> 2 hect) m^3/person/day.

Electricity consumption increased with increase in landholding, ranging from 1.99 (landless), 2.14 (0-0.5 hect) to 3.59 (>2 hect) kWh/person/day. Kerosene usage for lighting decreased with increase in landholding, varying from 0.87 (landless), 0.85 (0-0.5 hect) to 0.75 (> 2 hect) ltrs/person/month.

Based on fuel consumption norms (regionwise, seasonwise and end usewise) obtained from detailed household surveys, it is estimated that (a) fuel wood required in the domestic sector (for cooking, water heating, space heating, jaggery making and parboiling) of Uttara Kannada district works out to 16,68,698.25 tonnes/year, (b) electricity demand in the domestic sector (excluding irrigation) about 32.65 million kWh/year and (c) kerosene demand for cooking and lighting about 15.86 million litres per year.

Fuel efficiency studies reveal that there is scope for saving 27.45 to 42% fuel wood by switching over to improved stoves. Specific consumption of biomass fuels used in the domestic sector in most of the villages of Kumta taluk is quite high. Considering the low efficiency of devices at present, there is scope for improvement. In the sample of 1304 households only 22 use improved stoves. Lack of knowledge, proper training and service backup facilities are the main contributing factors for not adopting efficient devices. By switching over to energy efficient devices, such as ASTRA stoves, the saving of fuel wood in the domestic sector would be 4,50,548 to 7,00,853 tonnes per year.

7.15.0 ACKNOWLEDGEMENTS

I thank Mr. V.G. Bhat for helping us in end use efficiency experiments carried out in Sirsimakki microcatchment area, Mr. C.M. Shastri for extending his support in carrying out socio-economic survey in 220 households in Sirsi taluk, and Mr. Kumar, Mr. Wilson Alexander and Mr. Ajit D. Nagarakatte for their assistance in collecting field data in Mundgod taluk. We thank Village Accountants in Kumta and Sirsi taluks for providing landholding data. Officials at Block Development Offices (Kumta, Sirsi and Siddapur), Zilla Panchayat (Karwar) and Village Panchayats (Kumta, Sirsi) provided information pertaining to renewable energy programme in the respective regions. We thank officials at Veterinary Department (Kumta, Sirsi), Forest Department and Agriculture Department for providing statistical information required for this work. This research is funded by grants from the Ministry of Environment and Forests, Government of India and from Indian Institute of Science.

7.16.0 REFERENCES

1. NCAER, 1959. *Domestic Fuels in India,* National Council of Applied Economic Research. Asia Publishing House, New Delhi, India.
2. GOI, 1965. *Report of the Energy Survey of India.* Government of India, New Delhi, India.
3. NCAER, 1960. *Demand for Energy in India: 1960-1975.* National Council of Applied Economic Research, New Delhi, India.
4. op.cit. Ref. 3.
5. GOI, 1962. *Timber Trends and Prospects in India: 1960-1975.* Ministry of Food and Agriculture, Government of India, New Delhi, India.
6. GOI, 1974. *Report of the Fuel Policy Committee.* Government of India, New Delhi, India.
7. GOI, 1976. *Report of the National Commission on Agriculture, Part IX, Forestry.* Ministry of Agriculture and Irrigation, Government of India, New Delhi, India.
8. op.cit. Ref. 7.
9. GOI, 1979. *Report of the Working Group on Energy Policy.* Planning Commission, Government of India, New Delhi, India.
10. GOI, 1982. *Report of the Fuel wood Study Committee.* Planning Commission, Government of India, New Delhi, India.
11. GOI, 1984. *A Perspective on Energy Demand for Energy in India up to 2004-05.* Advisory Board on Energy, Government of India, New Delhi, India.
12. GOI, 1985. *Towards a Perspective on Energy Demand and Supply in India in 2004-05.* Advisory Board on Energy, Government of India, New Delhi, India.
13. WEP, 1979. *Report of Working Group on Energy Policy.* Government of India, New Delhi, India.
14. GOK, 1981. *An Analyses of 1977-79 National Sample Survey on Households Energy Consumption (32nd Round) - Karnataka.* Study Report, Perspective Planning Division, Planning Department, Government of Karnataka, Bangalore, India.
15. DNES, 1994. *Urja Bharati,* Special Issue on Rural Energy. Department of Non-Conventional Energy, Government of India, India.
16. ASTRA, 1976. Rural Energy Consumption Patterns: A Field Study Report, Indian Institute of Science, Bangalore.
17. Revelle, R., 1976. Energy use in rural India. *Science,* 192:969-974.
18. Leach, G.A., 1988. Residential energy in the third world. *Annual Rev. Energy,* 13:47-65.
19. Reddy, A.K.N and Prasad, K.K., 1977. Technological alternatives and the Indian energy crisis. *Economic and Political Weekly,* XII (33-34) : 1465-1502.
20. Bowonder, B., Rao, N.P., Dasgupta, B. and Prasad, S.S.R., 1985. Energy use in eight rural communities in India. *World Development,* 13(12):1263-1286.
21. Maheshwari, R.C., Srivastava, P.K. and Bohra, C.P., 1981. *Energy Consumption and Resource Assessment of Islamnagar in the Bhopal district,* Central Institute of Agricultural Engineering, Bhopal, India.

22. Ravindranath, N.H., and Chanakya, H.N., 1986. Biomass based energy system for a South Indian village, *Biomass*, 9: 215-233.
23. Ramakumar, R. and Hughes, W.L., 1981. Renewable energy sources and rural development in developing countries. *Trans. IEEE on Education*, E-24(3):242-251.
24. Ramakumar R., 1983. Renewable energy sources and developing countries. *Trans. IEEE on Power Apparatus and Systems*, PAS 102(2):502-510.
25. Montalembert, M.R.De and Clement, J., 1983. *Fuel wood supplies in the developing countries*. pp. 125. FAO, Rome.
26. Bajracharya, D., 1981. *Implications of Fuel and Food Needs for Deforestation*. Ph.D. Thesis, University of Sussex, England.
27. Leach, G., 1987. *Household Energy in South Asia*. pp. 111. Elsevier Applied Science, London and New York.
28. Aiyasamy, P.K., 1983. *Social Forestry Project in Tamilnadu: Survey Report No. 2.*, Tamilnadu Agriculture University, Coimbatore, India.
29. Bowonder, B., Shankar, R.K. and Prasad, S.S., 1983. *Energy Implications of Social Forestry*, Admin. Staff College of India, Hyderabad, India. Proj. P-068.

PART II: END USE EFFICIENCIES IN THE DOMESTIC SECTOR OF UTTARA KANNADA DISTRICT

7.17.0 SUMMARY

End use efficiency experiments carried out in some households of Masur village (based on 4 days experiment) of Kumta taluk, show that there is scope for saving 27 to 42% of energy by switching over to improved devices, designed to maximise combustion rate (of fuel, ensuring the presence of sufficient oxygen), radiant heat transfer (from fire to vessels, keeping them as close to fire as possible), convection (to pass maximum hot gas over vessels, reducing drafts), conduction (heat is concentrated near the vessels by using insulating material for the stove,) and user satisfaction (with user friendly design). Thermal performance of improved stoves designed at the Indian Institute of Science, Bangalore (ASTRA - Application of Science and Technology to Rural Areas) have also been studied in the field. Irrespective of type of fuel, community, etc., improved stoves show significant saving in fuel. By switching over to energy efficient devices 4,50,548 to 7,00,853 tonnes of fuel wood can be saved per year in the district.

Keywords: End use efficiency, Traditional Cookstoves (TC's), Improved Cookstoves (IC's), education, domestic energy, adult equivalent.

7.18.0 INTRODUCTION

The integrated development of renewable resources such as soil, water, vegetation, livestock and genetic diversity is essential for environmentally sound development. This necessitates promotion of conservation activities among local communities and application of their traditional environmentally sound technologies. Primary energy demand can be reduced substantially, while maintaining a given level of energy services, through higher end use, transport and conversion efficiencies. Surveys conducted in villages of Kumta, Sirsi, Siddapur, Mundgod and Ankola taluks show that bioresources represent about 82-88% of the energy supply. This extensive use of biofuels has resulted in serious deforestation contributing to detrimental impact on the environment. As the quantity and quality of forest and water resources decline, women and children spend more time in fuel and water collection. The severity of energy crisis is quite evident in densely populated taluks viz., Kumta, Ankola, etc., along the coastal belt. Most households still use three stone fire or home-made mud stoves without proper combustion and heat transfer facilities. With deforestation, evident from barren hill tops, women and children have to walk 5 to 10 kms every day to fetch fuel wood. Raising the efficiency of household fuel usage increases the available energy for basic needs in poor households, reducing women's workload in fuel collection and its use, releasing time for more productive work. Hence, improving efficiencies of Traditional Cookstoves (TC's) has attracted a lot of attention in energy programmes. As a response to these multiple problems, the Ministry of Non-conventional Energy Resources, Government of India, initiated the national programme on Improved Cookstoves (IC's) with the following objectives: (i) conserve and optimise the use of fuel wood in rural areas, arresting the rate of resource degradation and eliminating the drudgery of women (removal of smoke in the kitchen, improvement in hygiene, etc.), (ii) provide employment to rural youth in construction and dissemination of stoves, and (iii) accelerate research and development activities to improve the models and propagate new devices.

In this section, household energy consumption patterns in five taluks of this district of Western Ghats based on detailed field work for 18 months and strategies to minimise the potential negative impacts of energy system on natural and human systems are discussed.

7.19.0 OBJECTIVES

(1) Measure and estimate the daily per capita fuel wood consumption in traditional and improved stoves for cooking and water heating.
(2) Determine the extent of use of IC's.
(3) Determine the factors influencing acceptability of improved stoves.
(4) Study the extent of fuel saved using alternate devices.

7.20.0 METHODOLOGY

Computation of Per Capita Fuel Consumption (PCFC) was done for households using traditional and improved stoves, considering the daily and seasonal variation in fuel wood consumption, number of persons (adult equivalents) per household, household income, landholding, community and village separately. A few households were selected from Kumta and Sirsi taluks to conduct detailed cooking and water heating experiments (based on actual measurements of daily consumption - input/output measurements) in order to see the role of end use technologies on variation in consumption patterns. These studies were carried out in the following hamlets.

- Sankuli, Masur, Halasmavu, Kalbavi and Hoskeri of Masur village in Kumta taluk, and
- Sirsimakki, Vadgeri, Bellikeri and Mundgesara-Abrimane in Sirsimakki-Mundgesara microcatchment, located in the Aghnashini basin of the Ghats, east of the crestline.

In order to assess the quantity and type of fuel and cooking devices used, a questionnaire based survey of the stratified random sample of 118 households from 6 hamlets of Masur village and 23 households from Sirsimakki-Mundgesara catchment was conducted. Data pertaining to (a) type of stoves, (b) type of fuel - agricultural residues, etc., (c) mode of fuel collection / procurement, (d) cooking time and (e) cooking energy requirement, was collected.

In order to measure fuel consumption (per day per household), the actual quantity of fuel required for a day's cooking was weighed beforehand (W1) and the leftover fuel was weighed (W2) next day using single pan balance of 50 gram accuracy. The difference in W1 and W2 would give the fuel consumption value. This experiment was conducted in sample households for 4 consecutive days to account for changes in quantity and type of food cooked.

Stratified random sampling technique, based on landholding categories, was adopted to give an insight into the inter-relationship among households and the use of non-conventional sources of energy (fuel wood, agricultural residues, biogas, etc.). Data collection was done with emphasis on:

- consistent measurement of consumption of different energy sources,
- accounting for different kinds of fuels - twigs, branches, logs, etc.,
- proper classification as fuel, fodder and building materials according to end use,
- time spent collecting fuels by human labour and
- efficiency measurement of different end use devices.

Similar experiment was conducted in all 82 households in Sirsimakki-Mundgesara catchment and 22 households surrounding it, to find out fuel consumption for water heating.

7.20.1 Types of Stoves

(i) *Conventional Wood Burning Stoves (Traditional Stoves)*:
Size of stoves differed depending on the number of persons per household. These stone and mud stoves are without proper air inlet and outlet systems. The three stone stove used in landless labour class households is the least efficient. Fuel efficiency studies revealed a low efficiency of 11% in these stoves.

(ii) *Fuel Efficient Stoves (Improved Stoves)*:
Fuel efficient stoves, popularly known as ASTRA stoves designed at Indian Institute of Science, is characterised by complete fuel combustion with as little excess air as practicable to generate the

highest temperature of flue gases (ASTRA, 1983). This is achieved by combustion of fuel over a grate in an enclosed fuel box with ports of suitable size for entry of air. The grate helps in the entry of primary air below the fuel bed to burn the char and separate the ash. Air required for burning the volatile matter released as a consequence of fuel heating, called the secondary air, enters through a port at a level slightly above the grate. After generating highest temperature of the flue gases, heat gets transferred by the mechanism of conduction, convection and radiation for useful work, that is to the vessels. To reduce the transfer of heat to the walls of the stove, an inner lining of mud mortar with rice husk is provided, which gives a higher thermal insulation. Chimney of suitable height and diameter helps to create draught and disperse smoke away from the cooking zone. The efficiency of these stoves are in the range of 32-41%.

7.20.2 Diffusion of Stoves among the Population

The exploratory survey revealed the per capita fuel wood consumption in this locality to be quite high in the range of 2.5 to 4.25 kgs per day, of which 1.26 kgs was used for water heating and 1.92 kgs for cooking. This requirement is met by minor forests, leaf forests (*soppinabettas*) and partly by areca and coconut residues. In order to reduce fuel wood consumption and hence pressure on forests, dissemination of fuel efficient stoves for cooking and water heating was taken up by CES (Centre for Ecological Sciences, I.I.Sc, Bangalore) under an ecodevelopment programme which was done after consultations with the villagers and local NGO's (Non Governmental Organisations). Few youths were trained to build stoves taking into account the kind of vessel, etc., suitable for local requirements. This has also generated local employment in a limited sense as some of the skilled masons have taken up construction of fuel efficient stoves as a profession.

7.21.0 Study Area

The study was carried out in the following villages:

(i) *Masur village*: Located at 14°28' N lat 74°23' E long, 4 kms from Kumta town, it is a cluster of 9 hamlets on an island in the estuary of river Aghnashini. This island is on a hillock, at an altitude of 75 m, surrounded by brackish water on all sides. The soil is lateritic, acidic and reddish brown in colour. Mean annual precipitation is 3200 m, mainly from May to November (Kamath, 1985). The total area is 358.3 ha with minor forest for 103.0 ha, private grasslands 23.4 ha, upland paddy cultivation 22.3 ha, coconut groves 2.2 ha, freshwater paddy cultivation 55.3 ha, areca 2.0 ha, brackish water paddy field 87.6 ha and river course 62.5 ha (District Statistical Officer, 1991). This region has a population density of 949 per sq.km with 15 endogamous communities, the major ones being Patgars, Halakki Goudas, Havyak Brahmins, Ambigas and Madivals. The average landholding per family is 0.5 ha, and there are about 86 landless families.

(ii) *Sirsimakki Mundgesara*: Study was conducted in the hamlets of Sirsimakki, Vadgeri, Bellikeri, Mundgesara-Abrimane in Sirsimakki-Mundgesara microcatchment, located in the Aghnashini basin of the Ghats, east of the crestline, 5 km from Sirsi town bound on either side by Sirsi-Kumta and Sirsi-Siddapur roads. It is at 14° 35' N lat. and 74° 48' E long. with an altitude ranging from 545 to 618 m spread over an area of 386.5 ha of which 324.5 ha comprises hills. The whole terrain is greatly undulating with 26 peaks, with a mean annual precipitation of 2600-3000 mm mainly between June and October. Eighty two households (with 661 people) are clustered in these hamlets with 316 males and 345 females. There are 105 children in the age group of 0 to 6 years. The major communities are 1) Havyak Brahmins (65 households - mostly arecanut land holders), 2) Goudas (8 households) and 3) Others (Naik - 1, Devadiga - 3, Adidravida - 2, Shet - 1, Uppar - 1 and Poojari - 1).

7.22.0 Results and Discussion

7.22.1 Efficiency Measurements of End use Devices

In order to take into account efficiency of end use devices (for cooking and water heating) and day to day fluctuation in fuel consumption, detailed efficiency experiments were carried out in few households located in Masur village. It is noticed that considerable quantity of fuel wood could be saved by using improved design. In this regard the factors responsible for the adoption of energy efficient devices were studied.

7.22.2 Efficiency Measurement of Water Heating Stoves

Efficiency of water heating stoves in bathrooms is done by water heating tests. Heat gained by 50 litres of water, heated from 27 to 44°C, is measured by burning weighed quantity of wood, found to be 0.64 to 1.15 kgs. The thermal efficiency of fuel efficient stoves is in the range 23.86±5.21 (average efficiency: 23.86%, standard deviation: 5.21), while for traditional stoves it is 7.89±3.22 (average efficiency: 7.89%, Sd:3.22%). The results of fuel efficient stoves are listed in Table 7.17.

Table 7.17: Thermal Efficiency and Quantity of Wood Required (in kgs) for Heating 50 Litres of Water in Water Heating Devices

Sampled Household number	Time (min.)	Burning rate of wood kg/hr	Heat utilized kcal	Thermal efficiency	Quantity of wood required
1	27.69	2.50	921.0	19.34	1.15
2	13.20	5.09	1609.0	19.96	1.12
3	29.25	2.35	1656.0	19.54	1.14
4	16.05	3.12	1430.0	26.77	0.83
5	26.23	2.35	1458.0	21.77	1 02
6	15.76	2.59	1725.0	32.79	0.68
7	16.49	2.89	1135.0	28.06	0.79
8	27.04	2.64	1100.0	18.77	1.19
9	28.09	2.28	907.5	20.90	1.07
10	13.44	3.24	1771.0	30.80	0.72
11	12.25	3.16	2845.0	34.60	0.64
12	15.50	3.65	1919.0	23.66	0.94
13	28.00	2.66	1040.0	22.88	1.02
14	28.00	3.47	1140.0	17.35	1.21
15	35.00	2.91	1131.0	20.69	1.09
Avg		2.99		23.86	
Sd		0.69		5.21	

7.22.3 Fuel Consumption Pattern in Cookstoves

In order to assess the quantity and type of fuel, and the types of cooking devices used, a questionnaire survey, based on actual measurements, was conducted in the stratified random sample of 118 households in 6 hamlets of Masur village. The actual quantity of fuel required for a day's cooking was weighed beforehand and the fuel leftover was weighed the next day using single pan balance of 50 gram accuracy.

This experiment was conducted for 4 consecutive days (thus collected three days cooking data) in order to account for day to day change in type of food, quantity, etc. The cooking fuel need in improved cookstoves (sample size 20) is 0.835±0.225 (average value: 0.835, standard deviation: 0.225) and traditional stoves (sample size 98) is 1.151±0.65 (avg. value: 1.151, standard deviation: 0.65) kgs/person/day. This means that there is 27.45% saving in fuel wood consumption by switching over to fuel efficient cookstoves.

7.22.4 User Reaction to the Improved Cooking Stoves

In order to assess the user's perception of energy efficient improved stoves, the survey and subsequent analyses reveal that:

(a) Two-pan IC's were preferred by 65.2% households comprising of Patgar, Naik, Halakki Gouda and Ambiga communities, while 34.8% preferred Three-pan stoves (majority are Havyak Brahmins, Gouda Saraswat Brahmins and Deshbhandaris).

(b) 31.25% of households complained about space problem in switching over to IC's.

(c) 46.87% preferred traditional 3 stone stove despite lack of energy efficiency compared to the needs of space heating, protection from insects provided by smoke and flexibility for using a variety of fuels (type and size) in different seasons.

(d) Out of 92 stoves constructed by various agencies, the share of Block Development Office is 73.91% (68 numbers) and CES 26.09% (24 numbers). It is found that 92% have modified the design to suit their vessels and cooking needs.

(e) 8-10% households having thatched roofs expressed their fear of fire hazard due to the presence of chimney in the IC's. During its usage, unburned particles often settle on the inside of the chimney due to difficulty in maintenance and poor durability. These particles are highly inflammable and often pose fire hazard.

7.22.5 Comparison of Fuel Consumption in IC's and TC's

The average daily consumption of fuel wood in IC's is 0.835, while in traditional it is 1.151 kgs/person. A 27.45% reduction in fuel wood consumption is seen by switching over to IC's. For statistically comparing sets of measurements for IC's and TC's, t - test was performed. "t" value [Zar, 1984] is computed by,

$$t = \frac{X_{TC} - X_{IC}}{\left[\frac{S_{TC}^2}{n_{TC}} + \frac{S_{IC}^2}{n_{IC}}\right]^{\frac{1}{2}}}$$

Where X_{IC}, S_{IC}, n_{IC} are mean, standard deviation and number of households for improved cookstoves, while X_{TC}, S_{TC}, n_{TC} represent the same for traditional cookstoves. The computed t value is compared to values in student t-table to determine if the mean of one group is significantly greater than the mean of other.

The t-table is used by comparing the calculated t-value to the numbers in the table at the appropriate degrees of freedom. It can be said that the mean from one group of tests is greater than the mean from the other, at a certain level of significance, if the computed t value is greater than the number in the table at that level.

The t value computed for data from 20 households using IC's and 98 households using TC's is 3.79. There are 116 degrees of freedom (since 118 experiments were conducted) and two parameters (the mean for each group). Based on entries in the t-table at 116 degrees of freedom, the t-value is found to be greater than the number in the table at the level of significance of 0.5%. Thus, this test further confirms that households using IC's have a higher probability of saving fuel compared to those using TC's.

Confidence limits for the difference between means: The 95% confidence interval for difference in means is computed by

$$X_{tc}-X_{ic} \pm t_{0.5\,(2),v}\, S_{Xtc\,-Xic}$$
$$= (1.15-0.835) + (1.981)*(0.0829) = 0.4792$$
$$= (1.15-0.835) - (1.981)*(0.0829) = 0.1507$$

At 95% confidence interval, the difference in fuel consumption in households using traditional stoves and improved stoves is not lower than 0.1507 (or the probability of this difference being less than 0.1507 is 0.025).

7.22.6 Communitywise Fuel wood Consumption

In order to look at the variation in consumption pattern, if any, data is further analysed based on community, number of persons in each house and type of fuel used. Table 7.18 lists communitywise fuel wood consumption. Only four communities use IC's in these villages.

(i) The average fuel consumption in Patgar community using IC's is 0.818 (standard deviation 0.258) and TC's is 1.066 (standard deviation 0.528) kgs/person/day, showing a saving of 23.18% in fuel consumption. The t-value computed with data of 14 Patgar households using IC's and 52 using TC's is 2.46. From t-table for 64 degrees of freedom, the computed t-value is greater than the number in the table at the level of significance of 1% but less than 0.5%. The change in average values for IC's and TC's and lower mean for IC's illustrates the scope for fuel saving.

(ii) The average fuel wood consumption among Halakki Goudas using TC's is 1.137 and IC's is 0.907 kgs. The t-value computed for data collected from 3 households using IC's and 15 households using TC's is 1.53. For degrees of freedom of 16, the computed t-value is greater than t-table value at the level of significance of 10%, but less at 5%.

This series of statistical tests for various communities further illustrates the scope for saving fuel wood using IC's. Communitywise fuel wood saving by switching over to IC's is 20.25% for Halakki Goudas, 23.18% for Patgars and 37.52% for Brahmins.

7.22.7 Number of Persons per Household

A scatter plot of per unit fuel consumption versus number of persons per household shows declining trend for both households using TC's and IC's with increase in number of persons per household. A regression analysis of 3 days cooking data from 98 households using TC's and 20 households using IC's is carried out. Based on correlation co-efficient and low percentage error, the linear relationship among the variables PCFC (per capita fuel consumption) and NOP (number of persons per household) per day is found to be the best relationship.

The relationship for households using IC's is

$$(PCFC)_{IC} = 0.9368-0.012\ (NOP)$$

with R^2 = 0.442, R = 0.665, % error of Y estimate = 13.42 and n = 60 (20 household data for 3 days). Plot of PCFC versus NOP revealed scatter, which is removed by filtering.

With partial removal of scatter (removing scatter points beyond range [avg±Sd]),

$$(PCFC)_{IC} = 0.8416-0.0106\ (NOP)$$

with R^2 = 0.542, R = 0.735 and % error of Y estimate = 9.42.

For TC's,

$$(PCFC)_{TC} = 1.615 - 0.065\ (NOP)$$

with R^2 = 0.516, R = 0.72, n = 296 (98 household data for 3 days) and % error of Y estimate = 8.68.

Table 7.18: Fuel wood Consumption using TCs and ICs (Communitywise)

(a) Improved cook stoves (ICs)			
Community type	No. of households	Fuel consumption in kgs/person/day	
		Avg	Sd
GOU	3	0.907	0.145
HB	1	0.846	
Naik	2	0.877	0.017
Patgar	14	0.818	0.258
Total	20		

(b) Traditional cook stoves (TCs)			
Community type	No. of households	Fuel consumption in kgs/person/day	
		Avg	Sd
Ambiga	3	1.061	0.407
HB	9	1.355	0.523
Deshbhandari	2	1.231	0.053
GOU	15	1.137	0.483
Gunaga	4	0.955	1.346
Kodeya	2	1.453	0.216
Madival	2	1.085	0.207
Mukri	1	0.950	
Naik	5	0.908	0.358
GSB	3	1.377	0.358
Patgar	52	1.066	0.528
Total	98		

With partial removal of the scatter

$$(PCFC)_{TC} = 1.590\text{-}0.0560\ (NOP)$$

with $R^2 = 0.594$, $R = 0.77$ and % error of Y estimate = 6.67.

These results reveal that per capita fuel consumption decreases both in TC's and IC's, while cooking for large number of persons. This decrease is due to the coefficiencies of cooking and water heating that result from increasing their scales. Hence, proper design of stoves and size of vessels taking into consideration number of persons in a household are the essential factors bringing down fuel wood consumption. Table 7.19 lists fuel consumption in IC's and TC's based on number of persons per household. It is seen that for most of the categories, fuel consumption in IC's is lower than TC's. For example, for a household of 4 persons, fuel consumption in TC is 1.095 and IC is 0.967 kgs/person/day. This study is carried out to understand the impact of large and small families on fuel wood consumption.

Table 7.19.0: Fuel Consumption (in kgs/person/day) Based on Number of Persons per Household

(a) Improved cook stoves (ICs)			
No. of persons per household	No. of households	Fuel consumption in kgs/person/day	
		Avg	Sd
15	1	0.473	
14	1	0.757	
12	2	0.852	0.233
9	4	0.84	0.079
8	2	0.775	0.084
7	4	0.727	0.18
6	2	0.948	0.101
5	3	1.122	0.083
4	1	0.967	
	20		

(b) Traditional cook stoves (TCs)			
No. of persons per household	No. of households	Fuel consumption in kgs/person/day	
		Avg	Sd
30	1	0.437	
20	1	0.977	
17	1	0.515	
16	1	0.837	
15	1	0.415	
14	1	0.507	
13	1	0.692	
12	5	0.858	0.139
11	2	0.550	0.182
10	3	1.150	0.420
9	7	0.931	0.232
8	6	0.822	0.204
7	13	1.104	0.514
6	20	1.083	0.292
5	15	1.191	0.453
4	7	1.095	0.406
3	7	2.175	1.500
2	5	2.076	0.623
1	1	1.850	
	98		

7.22.8 Type of Fuel Used

Table 7.20 lists type of fuel consumption in IC's and TC's, where IC's use fuel better than TC's. The difference in level of fuel consumption is evident in households using species like *Terminalia paniculata* (TC: 1.093, IC: 0.765) and *Memecylon umbellatum* (TC: 1.231, IC: 0.786). It is evident from these results, that fuel consumption is significantly lower in IC's compared to TC's, irrespective of various parameters such as (a) community, (b) number of persons per household and (c) type of fuel used.

7.23.0 CONCLUSIONS

Fuel efficiency studies reveal that there is scope for saving 27.45 to 42% fuel wood by switching over to improved stoves. Specific consumption of biomass fuels, used in the domestic sector in most of the villages of Kumta taluk, is quite high. Considering the low efficiency of devices at present, there is scope for improvement. In the sample of 1304 households, only 22 use improved stoves. Lack of knowledge, proper training and service backup facilities are the main contributing factors for not adopting efficient devices. By switching over to energy efficient devices, such as ASTRA stoves, the saving of fuel wood in the domestic sector would be 4,50,548 to 7,00,853 tonnes per year.

7.24.0 REFERENCES

1. ASTRA, 1983. *Proceedings of the Astra Seminar.* Indian Institute of Science, Bangalore, India.
2. Kamath, S.U. (Ed.), 1985. *Karnataka State Gazetteer: Uttara Kannada district.* Government of Karnataka, Bangalore, India.
3. District Statistical Officer, 1991. *Statistical Abstract of Uttara Kannada District*, Government of Karnataka, Bangalore.
4. Zar, J.H., 1984. *Biostatistical Analysis.* Princeton Hall, Inc. Englewood Cliffs, N.J., USA.

Table 7.20: Fuel Consumption (Based on Type of Fuel Wood) in kgs/person/day

(a) Improved cook Stoves (ICs)

Species	No. of households	Fuel consumption in kgs/person/day	
		Avg	Sd
Acacia auriculiformis	1	0.690	
Cinnamomum zeylanicum	1	0.846	
Artocarpus integrifolia	1	0.847	
Memecylon umbellatum	4	0.786	0.119
Terminalia paniculata	5	0.765	0.200
Eugenia jambolana	3	0.896	0.336
Terminalia sps.	2	1.021	0.264
Aporosa lindleyana	3	1.139	0.225
	20		

(b) Traditional cook stoves (TCs)

Species	No. of households	Fuel consumption in kgs/person/day	
		Avg	Sd
Acacia auriculiformis	2	0.517	0.101
Spondias mangifera	1	1.355	
Areca catechu	1	1.661	
Caryota urens	6	0.930	0.232
Cocus nucifera	3	1.715	0.630
Cinnamomum zeylanicum	1	0.920	
Anacardium occidentale	3	0.942	0.261
Memecylon umbellatum	12	1.231	0.764
Terminalia paniculata	23	1.093	0.471
Pterocarpus marsupium	1	1.384	
Terminalia sps.	13	0.985	0.271
Xylia xylocarpa	1	0.988	
Dillenia pentagyna	1	0.904	
Careya arborea	1	2.550	
Mappia oblonga	1	0.608	
Careya arborea	1	0.722	0.227
Mangifera indica	9	1.170	0.389
Eugenia jambolana	14	1.503	1.120
Ochrocarpos longifolius	2	0.709	0.076
Cocus nucifera	2	0.892	0.165

ANNEXURE 7.1: SCHEDULE FOR ENERGY SURVEYS

Indian Institute of Science, Energy Research Group, Centre for Ecological Sciences,
Bangalore 560012
Tel: 91-080-3600985/3092506, fax: 91-080-3601428/3600085/3600683 (Attn: CES/TVR)
Schedule for Research on "Integrated Regional Energy Planning": HH and AGRI

Date **Investigator**
Village: **Gram Panchayat :** **Taluk:** **District:**
Respondent Relationship with Family Head:
Educational Qualification of Respondent .... Age .... Sex ....
Name of Head of Household ....

Ia. **Housing:**

Type			No of		Facility in Each Room			Other
Roof	Wall	Floor	Rooms	Electrified	Bulb	Type	Wattage	Details

Ib. **Family Particulars:**

				Occupation		Relationship
Members' Names**	Age	Sex	Education	Primary	Sec	To the Head

** includes all members normally eating at home (including labours, servants, etc - if any whose food is cooked daily)

IIa. **Festivals Celebrated:**

No. of Festivals in a year:

Type of Extra Energy used: (fuel wood / kerosene / other (specify))

Quantity used:

No of Persons During Festival & Other Special Occasion:

IIb. **Ownership of Assets and Utilisation:**

a1. Commercial Energy Consuming Assets

(If possible efficiency of items 1-6 is to be computed)

Sl. No.	Items	Type	Number	Size	Fuel Type	Hours Of Use	Remarks
1	Chula(fuel wood)						
1.	Stove (fuel wood)*						
2	Sigri						
3	Kerosene Stove						
4	Biogas Stove						
5	Biogas Mantle						
6	LPG Stove						
7	Bulbs-tube						
8	Bulbs-incandescent						
9	Radio						
10	Tape Recorder						
11	T.V						
12	Fans						
13	Mixer						
14	Fodder Cutter						
15	Flour Mill						
16	Tractor						
17	Thresher						
18	Pumpsets						
19	Moped/scooter						
20	Truck						
21	Jeep						
22	Others Specify						
23							
24							

* Traditional /Fuel efficient stove

IIc. **Non-Commercial Energy Consuming Assets:**

Sl. No.	Items	Number	Hours Of Use	Remarks
1	Manual-fodder Cutter			
2	-Thresher			
3	-Grinder			
4	-Sewing machine			
5	Cycle			
6	Bullock Cart			
7	Horse Cart			
8	Rickshaw			

IIIa. **Livestock details:** (NOTE: Gr=grazing, gm=grams, Agri=agriculture)

Type: Da: Draught Animals MA: Mulching Animals Y: Young Calves, M: Mature
HY: Hybrid L: Local

Sl.No.	Livestock Type	HY/L	Number	Stall fed	Taken out for Grazing
1.	Bullock				
2.	Buffalo				
3.	Cattle				
4.	Sheep				
5.	Goat				
6.	Pigs				
7.	Others (Specify)				

IIIb. **Fodder Consumption and Dung Collection:**

SUMMER

8. Qty. of fodder fed/animal
   a. Green fodder (gm)
   b. Dry fodder (gm)
   c. Grains (gm)
9. Dung collected/animal/day
   a. on a normal day
   b. when taken out for Gr.
   c. when taken out to fields
10. No. of agriculture days
11. If cattle is taken out for grazing, how frequently?
   a. per day (hours)
   b. per week (days)
12. Average milk produced/animal/day

WINTER

8. Qty.of fodder fed/animal
   a. Green fodder (gm)
   b. Dry fodder (gm)
   c. Grains (gm)
9. Dung collected/animal/day
   a. on a normal day
   b. when taken out for Gr.
   c. when taken out to fields
10. No. of agriculture days

11. If cattle is taken out for grazing, how frequently?
   a. per day (hours)
   b. per week (days)
12. Average Milk produced/animal/day

MONSOON

8. Qty.of fodder fed/animal
    a. Green fodder (gm)
    b. Dry fodder (gm)
    c. Grains (gm)
9. Dung collected/animal/day
    a. on a normal day
    b. when taken out for Gr.
    c. when taken out to fields
10. No of agriculture days
11. If cattle is taken out for grazing, how frequently?
    a. per day (hours)
    b. per week (days)
12. Average Milk produced/animal/day

IV. **Preparation & Utilisation of Dungcakes:**

Particulars	Summer March to June	Monsoon July to Sept	Winter Oct to Feb
1. Preparation of Dung cakes			
a. No. of cakes made/day			
b. approx. weight of wet dung/dung cake			
c. approx. weight of dry dung/dung cake			
2. Utilization of dung cake			
a. No. of dung cake/month			
i. put in for storage			
ii.taken out of storage			
b. No. of dung cakes/month			
i. sold			
ii. distributed free			
iii.purchased			
iv. obtained free			
c. No./qty. of dung cake/day			
i. used as fuel			
ii. in sigri			
iii.kept aside for manure (kg)			
iv. fed into biogas plant (kg)			

V. **Collection of Fuels and Source of Collection:**

Particulars	Summer	Winter	Monsoon
1. **COWDUNG**			
a. qty collected			
b. place of collection			
c. number of persons			
d. time spent on collection per day (hours) per month (days)			
e. storage available YES/NO			
2. **CROP RESIDUE**			
a. qty collected (kg/qtl/tonne/other unit)			
b. place of collection			
c. number of persons			
d. time spent on collection			
i. per day (hours)			
ii.per month (days)			
e. storage available YES/NO			
f. time spent on transport			
g. months in which			
i. scarcity is felt			
ii.not available			
3. **FIREWOOD/TWIGS**			
a. qty collected (kg/qtl/tonne/other units)			
b. place of collection			
i. own farms / nearby forests /roadside			
ii. others			
c. How far is it from your home			
d. practice/means of collection			
i. do you cut the tree			
ii.if yes how often			
1. once in a season			
2. once in a year			
3. cut branches/twigs			
e. time spent on collection			
i. in felling the tree			
ii. transporting it home			
iii.chopping wood/br/twigs			
iv. how often wood is chopped in a week / in a month			
f. number of persons deployed			
g. any other relevant information			

VI. **Domestic Energy Consumption:**

		Cooking			Water Heating		Lighting	
Type of Fuel	Units	Qty.	Device	Qty.	Qty	Device.	Qty.	Device
SUMMER								
a. Dung								
b. crop residue								
type..........								
c. firewood								
type..........								
d. twigs & branches								
e. kerosene								
f. biogas								
g. electricity								
h. others								
(please specify)								
WINTER								
a. Dung								
b. crop residue								
type								
c. firewood								
type								
d. twigs & branches								
e. kerosene								
f. biogas								
g. electricity								
h. others								
(please specify)								
MONSOON								
a. Dung								
b. crop residue								
type..........								
c. firewood								
type..........								
d. twigs & branches								
e. kerosene								
f. biogas								
g. electricity								
h. others								
(please specify)								

VIa. **Domestic Energy Consumption:**

Type of Fuel	Units	Space Heating Qty.	Space Heating Device	Fodder Preparation Qty.	Fodder Preparation Device	Fed to Biogas Plant Qty.	Fed to Biogas Plant Device
SUMMER							
a. Dung							
b. crop residue type..........							
c. firewood type..........							
d. twigs & branches							
e. kerosene							
f. biogas							
g. electricity							
h. others (please specify)							
WINTER							
a. Dung							
b. crop residue type..........							
c. firewood type..........							
d. twigs & branches							
e. kerosene							
f. biogas							
g. electricity							
h. others (please specify)							
MONSOON							
a. Dung							
b. crop residue type..........							
c. firewood type..........							
d. twigs & branches							
e. kerosene							
f. biogas							
g. electricity							
h. others (please specify)							

VIb. **Domestic Energy Consumption (Household Industry):**

Type of Fuel	Units	Household Industry Qty.	Household Industry Device	Other Qty.	Other Device
a. Dung					
b. crop residue type..........					
c. firewood type..........					
d. twigs & branches					
e. kerosene					
f. biogas					
g. electricity					
h. others (please specify)					
i. Type of raw materials					
ii. qty. raw materials in kgs/tonnes					
iii.brought from own lands					
iv. end products and its use					

VIc. **House is Electrified** Yes / No

Electricity Consumption (as per Electrical Bill)

	During 1994 Units	During 1994 Rs.	During 1995 Units	During 1995 Rs.	During 1996 Units	During 1996 Rs.
Jan						
Feb						
March						
April						
May						
June						
July						
Aug						
Sept						
Oct						
Nov						
Dec						

VII. **Land Particulars** (in Hectare)

Total Land Scattered Joint

Irrigated Non irrigated Barren Land

Pastures Wasteland No. of Trees Owned

Type of Trees:

VIII. **Agricultural Energy Consumption:**

Agricultural Energetics for One Cycle of Crop Production

Note: details to be given for each activity under corresponding month

	SEASON I (rabi)		SEASON II (kharif)	
PARTICULARS	CROP1	CROP 2	CROP1	CROP 2
1. name of crop (s)				
2. area (hectares)				
3. month of sowing				
4. harvesting month				
5. **TILLING**				
a. no. of preparatory tilling				
i. if bullocks used				
ii.if tractor used				
iii.H.P.of tractor				
b. area covered (hectares)				
i. by bullocks				
ii.by tractor				
c. time required/tilling/hectare				
i. using bullocks				
1. hours/day				
2. no. of days				
ii. using tractors				
1. hours/day				
2. no. of days				
d. no. of persons employed				
i. using bullocks				
1. family members				
2. hired labourers				
ii. using tractors				
1. family workers				
2. hired labourers				
6. **SOWING**				
a. area sown (hectares)				
i. by bullocks				
ii. by tractors				
iii.H.P.of tractors				
b. time required per hectare				
i. using bullocks				
1. hours/day				
2. no. of days				
ii. using tractors				
1. hours/day				
2. no of days				
d. no of persons employed				
i. using bullocks				
1. family members				
2. hired labourers				
ii.using tractors				
1. family workers				
2. hired labourers				
7. **IRRIGATION**				
a. area rainfed				
b. area irrigated				

c. no. of irrigation using
 i. diesel pumpset
 ii. using electric pumpset
 iii.canals/others
d. H.P. of machine used
 i. diesel
 ii. electric
e. time spent/irrigation/hectare
 i. using diesel pump (hours/hectare)
 ii.using electric pump (hours/hectare)
 iii.canal/well/others
f. no of human hours per hectare
 i. family workers
 ii. hired labourers
g. no. of bullock hours/hectare
 i. family
 ii.hired

8. **INTERCULTURE /TRANSPLANTATION**
a. area covered
b. time required (hours/day)
 i. using mechanical method
  1. hours/day
  2. no of days
 ii. by spraying /dusting
  1. hours/day
  2. no of days
 iii.no of persons employed
  1. family workers
  2. hired workers

9. **MANURE APPLICATION**
a. area covered
b. no of times applied in season
c. qty. applied/hectare
d. time spent/hectare
  1. hours/day
  2. no of days
e. no of persons employed
 i. family workers
 ii.hired workers

10. **FERTILIZER APPLICATION**
a. area covered
 i. urea
 ii. DAP.
 iii.other
b. no. of times applied in the season
 i. urea
 ii. DAP.
 iii.other
c. qty. applied/hectare
 i. urea
 ii.DAP.
 iii.other
d. time spent/hectare

i. urea
1. hours/day
2. no of days
ii. DAP
1. hours/day
2. no of days
e. no of persons employed
i. family workers
ii.hired workers

11. **HARVESTING**
a. area harvested (hectares)
i. by manual labour
ii. by tractors
iii.H.P.of tractors
b. time required per hectare
i. using manual labour
1. hours/day
2. no of days
ii. using tractors
1. hours/day
2. no of days
c. no. of persons employed
i. using manual labour
1. family members
2. hired labourers
ii. using tractors
1. family workers
2. hired labourers

12. **THRESHING & WINNOWING**
a. qty. threshed (qtls)
i. using bullocks
ii. using thresher
iii.H.P. of thresher
b. time required
i. using bullocks
1. hours/day
2. no. of days
ii.using thresher
1. hours/day
2. no of days
c. no. of persons employed
i. family workers
ii.hired workers

IX. **Crop By-products and its Use:**

Crop Byproducts	Total Production	Self Consumption	Quantity Stored for Seeds Fuel Fodder	Give Away In form of Wages	Sold Qty. Rs.
1.**Ragi**					
a. grains					
b. stalk					
c. straw					
d.other					

2. **Paddy**
   a .grains
   b. stalk
   c. straw
   d. other
3. **Jowar**

4. **Wheat**

5. **Gram**

X. **Water Requirement:**

Season	Source	Qty/day Of Water Source	Distance	Hours Spent	No.of Persons	Storage At Home	Purpose

XIa: **Biogas plant Information:**

Capacity of the plant:
Year of Installation
Input source
Quantity/day
Is Gas is sufficient for domestic purpose
Problems encountered with biogas plant
Suggestion/ solution to the problem
If Biogas plant is not installed, reasons:

XIb. **Energy Used in Transport:**

Trips	------Purpose -------				
	Agriculture	**Domestic**	**Business**	**Services**	**Others**
a. no of trips/day					
during summer					
winter					
monsoon					
b. no of persons/trips					
c. mode of travel					
d. place & distance travelled					
e. farm output carried					
-output -qty					
f. inputs brought in					
-details -qty					
g. other details					

XII. **Family Income:**

Agriculture	Rs.	Horticulture	Rs. ....
Business	Rs.	Services	Rs. ....
Other	Rs.		

XIIIa. **Any Other Information Relevant to this Research:**

XIIIb. **Suggestions Regarding Energy Sources, Scarcity, New Technology & Expectation by Villagers from Researchers/Scientists:**

This data is authentic and collected personally by interviewing villagers.

Date **Investigator** **Signature**

Address for Communication: T.V.Ramachandra,
Energy Research Group [CES], Centre for Ecological Sciences,
Indian Institute of Science, Bangalore 560 012
Tel: 91-080-3600985/309 2506,
E mail: cestvr@ces.iisc.ernet.in or cestvr@hamsadvani.serc.iisc.ernet.in

Chapter 8

ENERGETICS IN PADDY CULTIVATION

8.0 SUMMARY

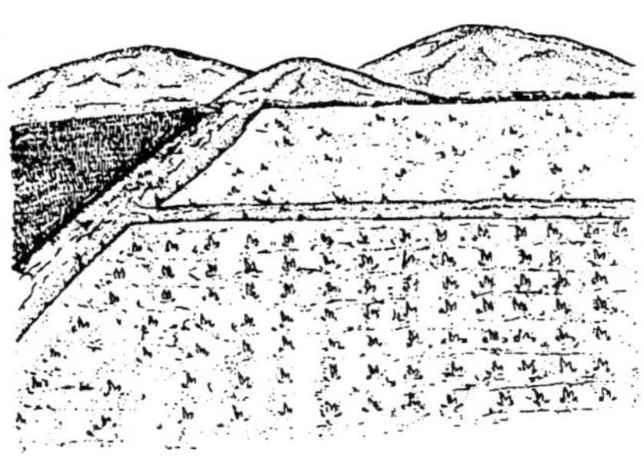

Agriculture requires three major resources, land, water and energy. Land being a fixed resource, agricultural productivity could be linked directly to the availability of water (rain or irrigation) and energy inputs. Land preparation, irrigation, harvesting, threshing and transportation are the main tasks dependant on mechanical energy apart from manure input, seed, etc. Since rainfed paddy is the major crop, agriculture in Uttara Kannada is less dependent on irrigation. However, there is scope to grow a second crop through irrigation. Water and associated average daily energy requirements depend on the area irrigated, type of crop, sources of water, total period of irrigation and irrigation efficiency. Farm Yard Manure (FYM) is a major component of input energy in paddy cultivation. However, the level of input changes from zone to zone (hilly, interior and coastal) and across various landholdings. The energetics in rainfed paddy cultivation is discussed in this chapter. Detailed analyses of energy input in various categories of farmers, based on stratified random sampling, show that marginal farmers get higher yield compared to others. Energy input in this category in the form of FYM is almost double compared to that applied by large farmers (>2 hectare). This greater usage of FYM by marginal farmers is attributed to higher dung availability (livestock per hectare in marginal farmers' category is almost twice that of medium farmers or four times that of large farmers).

Keywords: Agriculture, Paddy, Energy Inputs, Energetics, Farm Yard Manure, Irrigation, Productivity.

8.1.0 INTRODUCTION

Agriculture sector represents both consumers and producers of various forms of energy. Valuable byproducts of agriculture, like crop residues, are used as animal feed, cooking fuel and as raw material in industries. Animals provide draught power and their residues in turn are used as manure and fuel. The level and pattern of energy use in agriculture, as well as its contribution to energy supplies depends on a variety of agronomic and socio-economic factors. In India, agriculture not only provides food for all, but also employment to 70% of the population, generates 40% of the National income and consumes about 10% of the commercial forms of energy.

Crop cultivation requires application of both animate (bullock and human power) and inanimate (tractors, tillers, etc.) forms of energy at different stages. Nutrients are provided through FYM, chemical fertiliser or both. Pesticides are required to check or prevent pest attack. Irrigation is done either manually

(manually and animal operated) or through diesel/electric pumpsets (to lift ground water). Besides this, energy is required for processing the output and transporting it to the consumer centers.

Available data compiled by government machineries is inadequate to show a detailed energy consumption pattern in agriculture. Aggregate information of human labourers and number of pumpsets does conceal the wide disparity in regional energy consumption. Also, these aggregates exclude energy input into fertiliser production. This required a detailed survey to get information on various inputs, their quantity and time of application.

Energy analyses in agriculture include computation of energy content in inputs that go into crop production and comparison of the same with energy content in the output. For instance, energy input includes energy content of seeds, irrigation, human and animal power, etc. Studies indicate that output/input ratio under Indian conditions is in the range 1.5-2.7 (Pimentel and Pimentel, 1979).

During field research in Kumta taluk, farmers expressed that yields have been declining and higher fertiliser application was needed every year to maintain the yield. This was true with at least 60% of the farmers who have switched over to inorganic fertilisers. Analyses of data on yield show that, despite rising input levels, yields have been declining or were stagnant in the last decade. Yield declines are strongly associated with the duration that intensive production has been practiced in each taluk. Similar trends have been reported in other regions where intensive agriculture has been practiced. This indicates land degradation as the main factor in reducing productivity.

Earlier study of Karnataka's agricultural sector (Subramanian and Ramachandra, 1991) reveals a stagnant food production for the last 15 years in spite of the increase in irrigation facility, pesticide and fertiliser input. This implies that investments made in this sector, during sixth and seventh plan periods, have not yielded any net benefit through an increase in per capita yields, either to the farmers or to the economy.

There has been rising concern over intensive agriculture through chemical fertilisers, as it may not be sustainable and might damage the environment or other productive sectors, such as fisheries (through water pollution), etc. Decisive conclusions here are possible only after detailed studies relating to various energy inputs and corresponding yields at village and taluk levels are undertaken. In view of these, detailed investigations were carried out in 90 villages of Kumta taluk to identify the factors responsible for variation in yield.

8.2.0 Objectives

The objectives of the study are to

(i) measure the quantity of energy inputs in paddy cultivation,
(ii) find out its type and share in productivity,
(iii) analyse the regional variation in the amount and type of energy used,
(iv) examine the relationship between landholding and energy consumption, and
(v) identify factors responsible for difference in levels of energy consumption.

8.3.0 Study Area

Uttara Kannada district in the mid-western part of Karnataka state is selected for this study. It lies 74° 9' to 75° 10' east longitude and 13° 55' to 15° 31' north latitude extending over an area of 10,291 sq.kms, 5.37% of the total area of the state, with a population above 1.2 million. It is a region of gentle undulating hills, rising steeply from a narrow coastal strip bordering the Arabian sea to a plateau at an altitude of 500 m with occasional hills rising above 600 to 860 m.

This district with 11 taluks can be broadly categorised into three distinct regions - coast lands (Karwar, Ankola, Kumta, Honnavar and Bhatkal taluks), mostly forested Sahyadrian interior (Supa, Yellapur, Sirsi and Siddapur taluks) and the eastern margin where the table land begins (Haliyal, Yellapur and Mundgod taluks). Climatic conditions ranged from arid to humid due to physiographic conditions ranging from plains, mountains to coast.

8.4.0 MATERIAL AND METHODS

Secondary data on production, yield and area was collected from the agriculture department and from annual reports from the office of Principal Agricultural Officer, Karwar.

8.4.1 Household Energy Consumption for Agricultural Activities

Detailed survey was carried out in 90 villages spread over coast, interior and hilly zones of Kumta taluk, covering all categories of landholdings. Questionnaire on power sources, such as human, animal, prime movers and machinery in various farm operations, and organic (FYM) and inorganic (fertilisers, pesticides, etc.) inputs, etc., was used to collect data from 1304 households by stratified random sampling during 18 months of field research. Out of these, 1068 households contained complete information on agriculture, which was analysed to calculate the operationwise and sourcewise energy consumption (regionwise and based on landholding) on a unit area basis.

8.5.0 COLLECTION OF DATA

Questionnaire data was collected periodically by a combination of recall and actual measurements (in few farms – at least one from each category in each village).

Operationwise energy flow patterns were studied to find out energy consumption. The energy data pertaining to ploughing, sowing, transplantation/broadcasting, irrigation, weeding, harvesting and threshing were obtained from sample households. Since pesticide's proportion to total energy is insignificant, it has not been included in the energy computation. These values are converted to common energy units - Giga Joules (GJ). The energy conversion figures have been obtained from various literatures (Makhijani and Poole, A., 1975; Revelle, 1976; Avadani et al, 1988; Pimentel and Pimentel, 1979) as listed in annexure I.

8.6.0 LITERATURE REVIEW

Joy Tivy (1990) discusses the reciprocal interactions between agricultural or agriculturally associated organisms and their physical habitat. This study illustrates varying intensities of rice production, in terms of type and amount of direct and indirect energy subsidies, in Japan, Borneo and California. In Borneo, methods of cultivation are simple and yield is dependent on the inherent fertility of the soil. Rice yield is about 7,318,080 kcal/hectare and energy efficiency is of the order of 7.08%. In Japan, rice cultivation is more labour intensive as a higher proportion of the direct energy subsidy is accounted for by mechanisation and indirect subsidies are twice that of the direct subsidy. Rice yield is 17,598,240 kcal/hectare and energy efficiency is 2.45%. The level of inputs is highest in California with 22,360,800 kcal/hectare yield and energy efficiency of 1.55%. This study shows that energy output is the function of the intensity of agriculture. There is an inverse relationship between energy efficiency and productivity.

Pimentel et al. (1973) analyzed the changes in US maize production over a quarter of a century and showed that yields have increased by 138%, achieved at the cost of a 310% increase in fossil fuel consumption, while the energy ratio decreased from 3.72 to 2.8. The energy ratio, first applied by Black (1971), is the ratio of the calorific value of food to the energy required to produce it.

Pimentel D. and Pimentel M. (1979) provide comparative analyses in energy input for rice production. Energy inputs for a hectare of rice total 1.0 x 10^6 kcal, with about 2/3 expended for manpower and the balance for seeds. The yield in Borneo is about 2 tonne/hect or about 7.3 x 10^6 kcal food energy. The output/input ratio is 7:1, denoting a relatively high return for the investment. USA uses high energy inputs (14.4 x 10^6 kcal), particularly fossil fuel. The average yield per hectare is 6.16 tonnes (22.4 x 10^6 kcal). The energy output/input ratio is 1.6:1. Rice production in Japan is still labour intensive requiring about 1730 hours (8.03 x 10^6 kcal) of manpower per hectare. The average yield per hectare is about 4.8 tonnes (17.6 x 10^6 kcal) resulting in an output/input ratio of 2.45:1, compared to USA's 1.6:1, reflecting more efficiency in production.

Kamal Rijal, et al. (1991) examine the total energy requirements and outputs of subsistence agriculture in rural Nepal with special emphasis on animate energy inputs. They analyse the availability of animate energy, usage pattern and various issues of energy demand at micro levels and their impact at macro level. The output/input ratio computed for maize, paddy and wheat is relatively higher here (2.4-7.5) compared to highly mechanised agriculture of developed countries (1.5-3.5). They conclude that energy output/input ratio declines as the level of mechanisation increases.

Sathyajith Mathew, et al. (1993) analyse energy flow patterns in rainfed paddy cultivation under three puddling treatments - bullock drawn plough, power tiller and tractor. Their study reveals that (a) energy consumption per hectare for treatment with bullocks, tractor and tiller was 14.2, 14.2 and 15.0 GJ respectively, (b) output/input ratio for tractor, bullock and tiller treatment was 7.63, 6.58 and 5.4 respectively, (c) fertilisers and chemicals constitute a major portion of total energy input, and (d) the share of human power is maximum in planting and harvesting operations.

Carlos A. Bonetto and Arturo J. Carcano (1995) compare rice productivity in three kinds of treatment - control without fertiliser application, fertiliser application with urea (applied in two split dressings equivalent to 20 kgs per hectare each) and application of *Azolla caroliniana* (incorporated at planting at a rate equivalent to 5 tonne dry weight per hectare) - and conclude based on the grain yield of 7.3: 7.8: 8.5 tonne/hectare, that Azolla treated yield is significantly greater than the other two.

8.7.0 Results and Discussion

8.7.1 Land Utilisation in Uttara Kannada

Forest and cropped area at the end of 1993-94 was about 80.96% (8296 sq.kms.) and 10.94% (1121 sq.kms.) respectively.

8.7.2 Crops in Uttara Kannada District

Paddy is the major crop in all taluks of the district which record a marginal increase in the area under paddy (3.15%) for the last twenty years. Haliyal (168.54 sq.kms) and Mundgod (136.34 sq.kms) taluks in the plains lead in the area under paddy.

8.7.3 Production of Various Crops

The introduction of high yielding varieties (H.Y.V) of paddy in early 70s increased production from 2.26 (in 1973-74) to 3.03 million tonnes (in 1980-81). In spite of 3.15% increase in total cultivated area, production declined by mid 80s coming down to 1.75 million tonne (in 1992-93). The yield declined from 3.3 to 1.9 tonne/hectare despite the continued use of H.Y.V, greater application of inorganic fertilisers (26.66% increase) and pesticides. H.Y.V. yield changed from 2.5 to 2.2, while improved variety changed from 3.7 to 1.6 tonne/hectare. This fluctuation in production and yield along with area, rainfall and fertiliser (inorganic) is depicted in Figure 8.1. Commonly cited reasons for the decline are untimely or failure of rain, pests and decline in organic inputs.

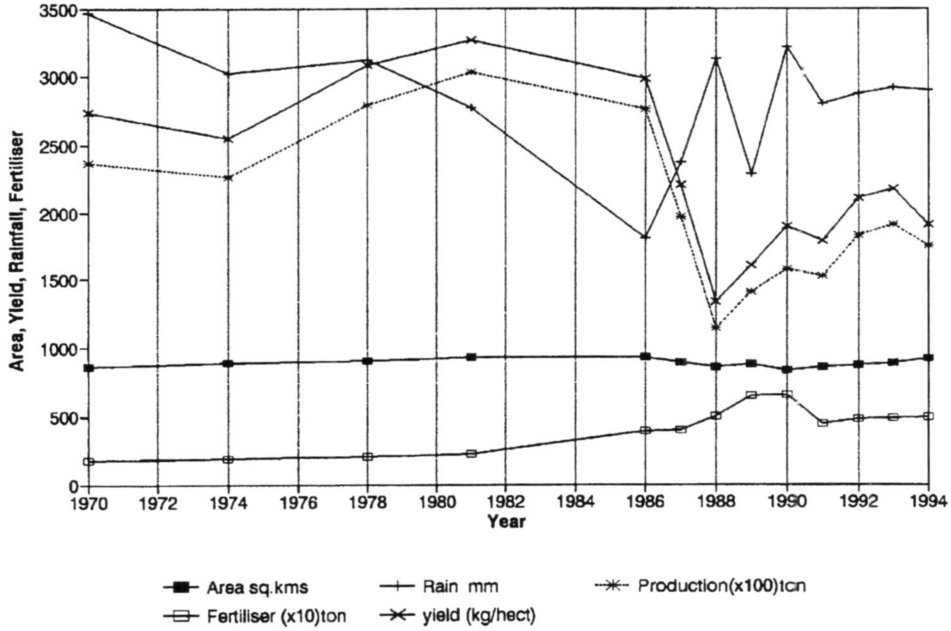

Figure 8.1: Area, Production, Yield - Paddy (Temporal Changes in Uttara Kannada)

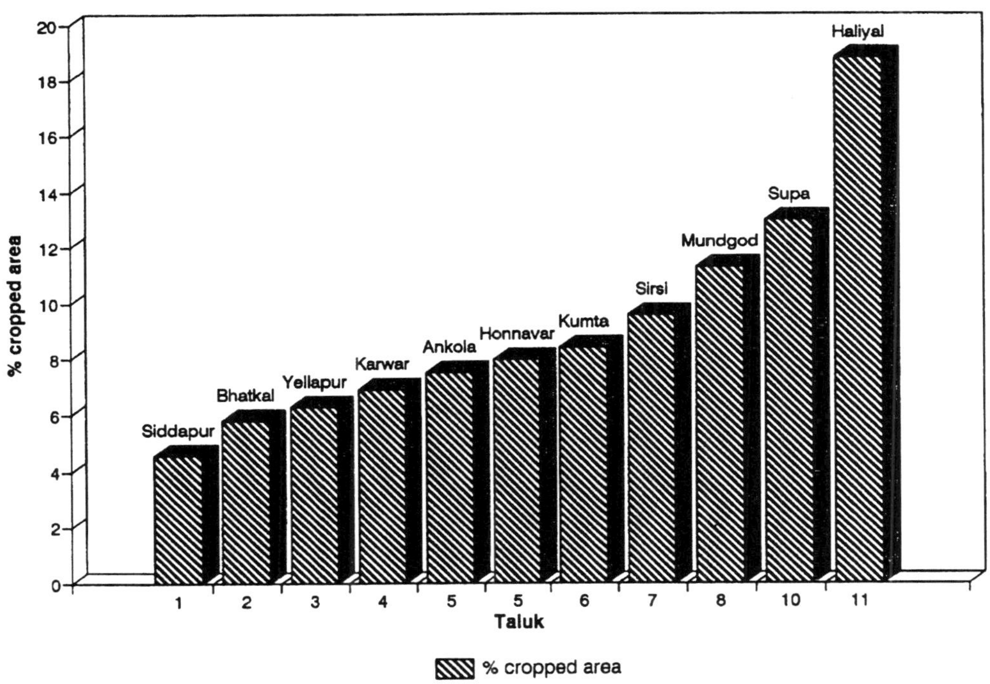

Figure 8.2: Talukwise Cropped Area (%)

Figure 8.2, shows Haliyal (18.76%) and Siddapur (4.56%) with highest and lowest percentage area under crops among 11 taluks.

Talukwise rainfall data (in mm) of last twenty years shows that, coastal taluks - Bhatkal (3994), Honnavar (3637), Ankola (3543), Kumta (3377) and Karwar (3214) get higher rainfall compared to hilly taluks - Siddapur (2982), Supa (2534), Yellapur (2438) and Sirsi (2379) or plain taluks - Mundgod (1203) and Haliyal (1179).

In order to see the effect of inorganic fertiliser input on yield over a period, regression analyses (both linear and non linear) was carried out with 13 years data (Paddy Yield kg/hectare, fertiliser tonnes). The relationship is found to be linear, of the form

Yield (kg/hectare) = 3465 - 2.92 (fertiliser-tonnes),
with R = correlation coeff. = 0.79.

This shows that yield declined at the rate of 2.92 with every additional fertiliser input, which may be due to excess or blind usage of fertiliser (based on advertisement in mass media) without considering soil condition (saline or otherwise) and its nutrient contents.

Most of the households practice rainfed paddy cultivation while, only few use irrigation (for the second crop March-April). In the sample of 1068 households in Kumta taluk, 12 grow two crops in a year.

In order to see the effect of annual changes in rainfall and fertiliser input on yield, multiple regression of the following variables was carried out: cultivated area (hectare), annual rainfall (mm), inorganic fertiliser (tonnes) and yield (kg/hectare). The probable relationships are

Yield = -9401.17+0.132 (area), with R = 0.66.
Yield = -13753.3 + 0.1665 (area)+0.4615 (rain), with R = 0.72.
Yield = -2039.2+0.0632 (area)-0.09 (rain)-0.2448 (fertiliser), with R = 0.86.

Most of the farming systems here use high quantity of organic manure. The torrential south-west monsoon depletes the soil of nutrients due to rapid soil erosion. This requires large quantities of leaf and organic manure. The leaf manure helps in providing soil cover and hence moisture in the soil. Decay of these materials enriches soil with nutrients. In the hilly taluks, due to reasonably good forest cover and dung availability (at present), substantial quantities of FYM and green manure are available. While in coastal taluks, dense human population and less forest cover deprive the soil of required nutrients.

The stagnation and declining yield in late eighties imply that investments made through fertilisers, irrigation and related areas have not yielded net benefit either to farmers or the economy through increase in per capita yields. In view of this, to get insight into energy requirements in agriculture and factors contributing to variation in yield, a detailed survey of 1304 households spread over 90 villages covering hilly, interior and coastal zones of Kumta taluk was undertaken. Out of these, reliable data on paddy cultivation was obtained from 1068 households. It is seen that population density varies from 0.35 (hilly), 1.35 (interior) to 5.31 (coastal) persons/hectare and forest cover changes from 25% (coastal), 76% (interior) to 89% (hilly). Livestock density varies from 0.24 (hilly), 0.70 (interior) to 1.63 (coastal) animals/hectare. There is a positive linear correlation between livestock and human density. Human and animal pressures on forests for fuel, fodder and green manure have resulted in declining forest cover in the coast.

8.7.4 Energy Use Pattern in Agriculture

Tables 8.1, 8.2 and 8.3 list operationwise (tilling, threshing, etc.) energy requirements for various landholdings in the coastal, hilly and interior zones.

In the coast, total energy input in GJ per hectare varies from 33.22 (<0.4 hect), 34.65 (0.8-1.2 hect), 26.14 (1.2-1.6 hect) to 19.72 (>2 hect). Output in GJ per hectare also varies from 56.62 (<0.4 hect), 39.99 (0.8-1.2 hect) to 38.14 (>2 hect). Output-input ratio varies from 1.13 (1.2-1.6 hect) to 1.93 (>2 hect).

Similar trends are noticed in interior and hilly zones. In hilly zone input energy varies from 55.95 (<0.4 hect) to 17.50 (>2 hect) and corresponding output varies from 58.36 to 30.64. In interior zone input energy varies from 29.18 (< 0.4 hect) to 19.82 (>2 hect) while, output varies from 32.87 (0.4 hect) to 28.08 (>2 hect).

It is evident from these Tables that paddy cultivation mainly depends on organic manure (FYM), which constitutes 70 to 74% of total input. Yield in all the three zones is dependent on the inherent fertility of soil and successful maintenance of nutrient cycle. The productivity in all zones is dependent on the level of inputs, which is the intensity of agriculture as organic inputs, etc. As the field sizes are small and fragmented, there are problems in cultivating with the help of tractors, tillers, etc. The design of machineries to substitute manual tasks under such conditions, is complex and expensive which makes mechanisation of agriculture inappropriate.

It is to be noted that the new inputs typical of many phases of development in paddy cultivation are divisible. New seeds and fertiliser can be bought in any quantity according to farmers' inclination and financial position. However, indiscriminate usage of fertiliser has resulted in significant reduction in yield over a period of time. Some farmers who switched over to inorganic fertiliser in the early eighties, have returned to organic farming due to declined yield and water pollution problems. Due to large gap between the average farmer's yield and maximum attainable yield, total fertiliser use during 1993-94 came down in Kumta, Ankola, Honnavar, Karwar and Siddapur taluks. Organisational improvements and careful operation make important contributions to raise output. Tables 8.1, 8.2 and 8.3 demonstrate that skilled and experienced small farmers are in just as good a position as large farmers to raise the productivity of the land

Table 8.1: Energy Requirements (GJ/Hectare) for Various Operations in Paddy Cultivation - Coastal Zone

Operation	Category (Hect)	0 - 0.4	0.4 - 0.8	0.8 - 1.2	1.2 - 1.6	1.6 - 2	> 2	All Categories	
	En input	Avg	Avg	Avg	Avg	Avg	Avg	Avg	Sd
	N	248	112	40	23	12	25	460	
	Land (hect)	0.31	0.66	1.09	1.53	1.94	3.01	0.69	0.69
Tilling	BH	1.57	3.81	5.46	9.07	10.03	9.25	3.45	3.55
	HH	0.13	0.28	0.34	0.57	0.55	0.63	0.24	0.32
Sowing	S	0.97	1.52	2.51	3.62	3.97	4.01	1.58	1.20
	BH	0.36	0.55	1.04	1.17	1.20	1.01	0.56	0.79
	HH	0.02	0.03	0.06	0.07	0.07	0.06	0.03	0.04
Irrigation	HH	0.04	0.09	0.12	0.09	0.13	0.11	0.07	0.06
	BH	0.00	0.00	0.00	0.00	0.00	0.00	0.00	0.00
Weeding	HH	0.06	0.10	0.12	0.11	0.16	0.13	0.09	0.05
Transplant	BH	0.00	0.00	0.00	0.00	0.00	0.01	0.00	0.00
	HH	0.13	0.30	0.61	0.89	1.51	2.89	0.40	0.89
Manure	FYM	6.15	14.95	24.26	19.68	16.47	31.62	13.59	20.90
Fertiliser	DAP.	0.03	0.05	0.00	0.00	0.00	0.00	0.03	0.47
	Other	0.00	0.00	0.00	0.00	0.00	0.00	0.00	0.00
Harvest	HH	0.14	0.25	0.40	0.77	0.91	1.63	0.30	0.41
	BH	0.00	0.00	0.00	0.00	0.00	0.00	0.00	0.00
Threshing	HH	0.07	0.14	0.26	0.25	0.42	0.48	0.14	0.14
	BH	0.00	0.00	0.00	0.00	0.00	0.00	0.00	0.00
Total Input	Total GJ	10.45	23.73	37.88	40.08	40.21	59.27	22.20	22.51
Output	GJ	17.81	29.01	43.72	45.57	77.37	114.61	30.85	29.21
Input/hect	GJ/hect	33.22	35.94	34.65	26.14	20.70	19.72	32.14	
Output/hect	GJ/hect	56.62	43.94	39.99	29.72	39.83	38.14	44.65	
O/I	Output/Input	1.70	1.22	1.15	1.13	1.92	1.93	1.39	1.30

Table 8.2: Energy Requirements (GJ/Hectare) for Various Operations in Paddy Cultivation – Hilly Zone

	Category (Hect)	0 - 0.4	0.4 - 0.8	0.8 - 1.2	1.2 - 1.6	1.6 - 2	> 2	All Categories	
Operation	En input	Avg	Avg	Avg	Avg	Avg	Avg	Avg	Sd
	N	160	101	30	29	8	13	341	
	Land (hect)	0.30	0.73	1.17	1.56	2.01	3.55	1.01	1.09
Tilling	BH	1.89	4.56	7.41	9.39	11.00	8.36	4.92	4.39
	HH	0.10	0.26	0.40	0.49	0.60	0.46	0.27	0.24
Sowing	S	1.23	1.99	2.96	2.73	4.68	3.04	2.10	1.53
	BH	0.18	0.43	0.65	0.79	1.01	1.02	0.47	0.46
	HH	0.02	0.03	0.05	0.05	0.05	0.08	0.03	0.04
Irrigation	HH	0.04	0.08	0.13	0.11	0.25	0.11	0.09	0.15
	BH	0.00	0.00	0.00	0.00	0.00	0.00	0.00	0.00
Weeding	HH	0.06	0.10	0.13	0.13	0.19	0.15	0.10	0.10
Transplant	BH	0.00	0.00	0.00	0.00	0.00	0.01	0.00	0.00
	HH	0.32	0.54	0.92	0.98	1.62	3.56	0.87	1.46
Manure	FYM	11.87	16.46	13.99	22.05	31.14	34.31	21.11	14.83
Fertiliser	DAP	0.00	0.00	0.00	0.00	0.00	0.00	0.00	0.00
	Other	0.00	0.00	0.00	0.00	0.00	0.00	0.00	0.00
Harvest	HH	0.11	0.27	0.48	0.39	0.86	1.71	0.40	0.57
	BH	0.00	0.00	0.00	0.00	0.00	0.00	0.00	0.00
Threshing	HH	0.06	0.13	0.18	0.18	0.29	0.52	0.15	0.17
	BH	0.00	0.00	0.00	0.00	0.00	0.00	0.00	0.00
Total Input	Total GJ	16.62	26.64	30.21	41.16	56.66	62.09	33.10	12.19
Output	GJ	17.33	33.01	48.00	49.08	67.00	108.70	38.40	35.53
Input/hect	GJ/hect	55.95	36.55	25.72	26.37	28.20	17.50	32.90	
Output/hect	GJ/hect	58.36	45.29	40.87	31.44	33.35	30.64	38.17	
O/I	Output/Input	1.04	1.24	1.59	1.19	1.18	1.75	1.16	2.91

N - No of Samples BH - Bullock Hour S - Seed HH - Human Hour F - Fertiliser FYM - Farm Yard Manure

Table 8.3: Energy Requirements (GJ/hectare) for Various Operations in Paddy Cultivation – Interior Zone

	Category (Hect)	0 - 0.4	0.4 - 0.8	0.8 - 1.2	1.2 - 1.6	1.6 - 2	> 2	All Zones	
Operation	En input	Avg	Avg	Avg	Avg	Avg	Avg	Avg	Sd
	N	90	100	27	23	13	14	267	
	Land (hect)	0.38	0.54	1.10	1.55	1.95	3.45	0.69	0.68
Tilling	BH	2.30	1.38	5.03	6.27	9.58	9.61	2.79	3.09
	HH	1.20	1.33	0.70	0.87	0.67	0.67	0.44	1.40
Sowing	S	1.41	0.82	2.63	2.75	4.23	4.10	1.63	1.42
	BH	0.22	0.19	0.43	0.40	1.07	0.62	0.20	0.33
	HH	0.02	0.01	0.05	0.05	0.06	0.10	0.03	0.03
Irrigation	HH	0.05	0.05	0.08	0.08	0.10	0.06	0.06	0.05
	BH	0.00	0.00	0.00	0.00	0.00	0.00	0.00	0.00
Weeding	HH	0.06	0.06	0.10	0.10	0.23	0.19	0.08	0.07
Transplant	BH	0.00	0.00	0.00	0.00	0.00	0.01	0.00	0.00
	HH	0.42	0.33	0.52	0.90	1.20	3.73	0.41	0.95
Manure	FYM	12.78	45.52	29.82	30.45	36.12	38.35	16.20	10.12
Fertiliser	DAP	0.00	0.00	0.00	0.00	0.00	0.00	0.00	0.00
	Other	0.00	0.00	0.00	0.00	0.00	0.00	0.00	0.00
Harvest	HH	0.15	0.14	0.36	0.63	1.00	1.32	0.28	0.34
	BH	0.00	0.00	0.00	0.00	0.00	0.00	0.00	0.00
Threshing	HH	0.09	0.07	0.19	0.22	0.38	0.55	0.12	0.15
	BH	0.00	0.00	0.00	0.00	0.00	0.00	0.00	0.00
Total Input	Total GJ	11.09	10.16	42.53	46.55	59.65	68.37	24.94	19.83
Output	GJ	12.49	8.81	56.50	78.60	82.98	96.86	34.42	40.11
Input/hect	GJ/hect	29.18	18.81	38.62	30.04	30.59	19.82	36.36	29.00
Output/hect	GJ/hect	32.87	16.32	51.31	50.73	42.55	28.08	50.17	58.66
O/I	Output/Input	1.13	0.87	1.32	1.69	1.39	1.41	1.38	2.02

N - No of Samples BH - Bullock Hour S - Seed HH - Human Hour F - Fertiliser FYM - Farm Yard Manure

8.7.5 Regionwise Variation in Energy Consumption Pattern in Agriculture

Table 8.4 lists regionwise energy consumption (GJ/hect) in paddy cultivation. Energy input from all sources (such as FYM, etc.) is in the range 35.47 (interior), 37.24 (coastal) to 43.21 (hilly), the major component being organic (FYM). Hilly zone with 89.78% forest cover and dung yield (kgs/animal/day) of 5.8 ± 1.8 is rich in organic matter compared to coast (forest cover 27.66%, dung yield 3.26) and interior (forest cover 78.14%, dung yield 3.94). The average paddy yield is about 1.73 tonne/hect. Regionwise variation shown in Figure 8.3 depicts output more or less same in all zones, while input energy is higher in hilly zone.

Table 8.4: Regionwise Variations in Different Energy Inputs and Outputs (GJ) for Paddy Cultivation

Zone	N		BH	HH	S	FYM	Fert	Total Input	Total Output	Paddy Out/In	Rice Out/in	Paddy Yield (0.1 tonne)
All	1068	avg	6.62	1.46	3.05	27.42	0.03	38.57	50.36	1.31	0.49	17.89
		sd	4.51	1.68	2.59	98.07	0.68	39.91	27.40	2.31	0.87	9.76
Coastal	460	avg	6.56	1.46	2.98	26.20	0.04	37.24	51.76	1.39	0.52	18.40
		sd	4.38	1.35	2.32	89.07	0.82	36.12	30.10	0.34	0.75	10.70
Hilly	341	avg	6.61	1.43	3.16	32.01	0.00	43.21	50.14	1.16	0.43	17.83
		sd	4.83	0.99	2.59	119.70	0.00	36.12	29.11	0.24	0.76	10.35
Interior	267	avg	6.72	1.51	3.02	24.18	0.05	35.47	49.05	1.38	0.52	17.40
		sd	4.31	2.43	2.88	82.28	0.85	33.43	21.30	0.26	1.09	7.63

In order to assess the effect of landholding on energy input, data is categorised based on landholding. Table 8.5 shows that marginal farmers (< 0.4 hectare) get maximum yield of 2.1 tonne/hectare compared to other categories. Figure 8.4 shows marginal farmers with maximum input of 46.09 compared to 23.46 GJ/hect (> 2 hect category). This significant variation in energy input and yield across zones and landholdings necessitated detailed analyses to see changes in energy consumption within a zone across various landholdings. Table 8.6 lists categorywise and zonewise energy consumption patterns. In all landholding categories (except 1.2-1.6 hectare) energy input is higher in the hilly zone. Energy intensive agriculture is practiced by marginal and small farmers (0.4 - 0.8 hect) due to higher availability of organic matter (in the form of dung as livestock in these categories range from 6.9 to 7.9 per hect). Similarly, energy output per hectare is higher in these categories. It appears marginal and small farmers have better output per hectare due to better management. Categorywise and zonewise energy input and output are shown in Figures 8.5 and 8.6, while Figure 8.7 depicts the inverse relationship between productivity and output/input ratio. Yield per hectare plotted in Figure 8.8, corroborates the earlier findings on small and marginal farmers.

Table 8.5: Energy Consumption Pattern in Paddy Cultivation Among Various Categories of Farmers (Landholding)

LH	N		BH	HH	S	FYM	Fert	Total Input	Total Output	Paddy Out/In	Rice out/in	Paddy Yield (0.1 tonne)
All	1068	avg	6.62	1.46	3.05	27.42	0.03	38.58	50.55	1.31	0.49	17.96
		sd	4.51	1.68	2.59	98.12	0.68	98.65	27.79	2.35	0.89	9.89
0 - 0.4	498	avg	6.97	1.71	3.94	33.45	0.03	46.09	59.08	1.28	0.48	21.01
		sd	5.17	2.15	3.30	121.50	0.73	122.10	30.65	0.25	0.82	10.90
0.4 - 0.8	313	avg	6.49	1.35	2.50	21.80	0.05	32.19	45.99	1.43	0.54	16.31
		sd	4.23	1.14	1.47	60.65	0.87	61.04	18.23	2.74	1.03	6.54
0.8 - 2.0	205	avg	6.54	1.15	2.20	24.03	0.00	33.93	39.93	1.18	0.45	14.20
		sd	3.31	1.01	0.94	91.78	0.00	91.83	27.35	2.17	0.82	9.73
> 2.0	52	avg	4.42	0.99	1.18	16.87	0.00	23.46	38.05	1.62	0.61	13.53
		sd	2.16	0.43	0.93	17.46	0.00	18.28	21.99	1.89	0.72	7.82

N - No of Samples BH - Bullock Hour S - Seed HH - Human Hour F - Fertiliser FYM - Farm Yard Manure
LH Landholding Category

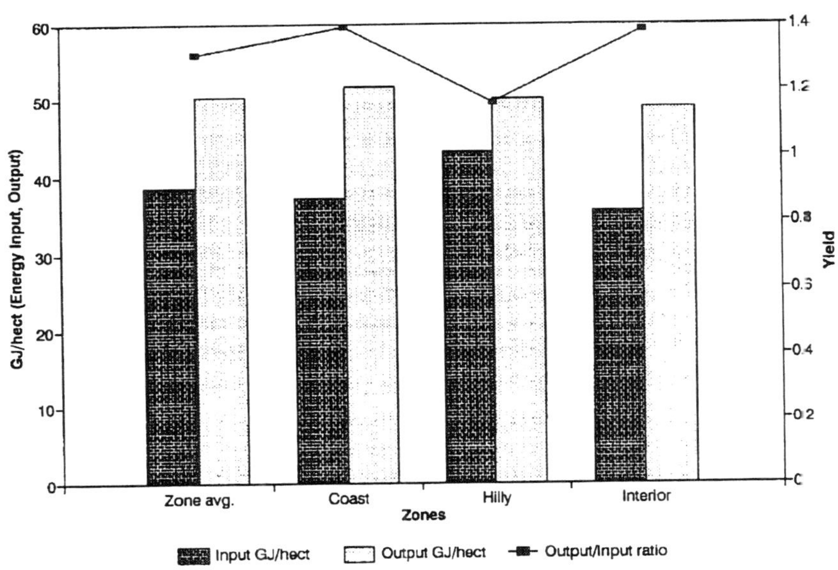

Figure 8.3: Energetics in Paddy Cultivation - Zonewise, Kumta Taluk

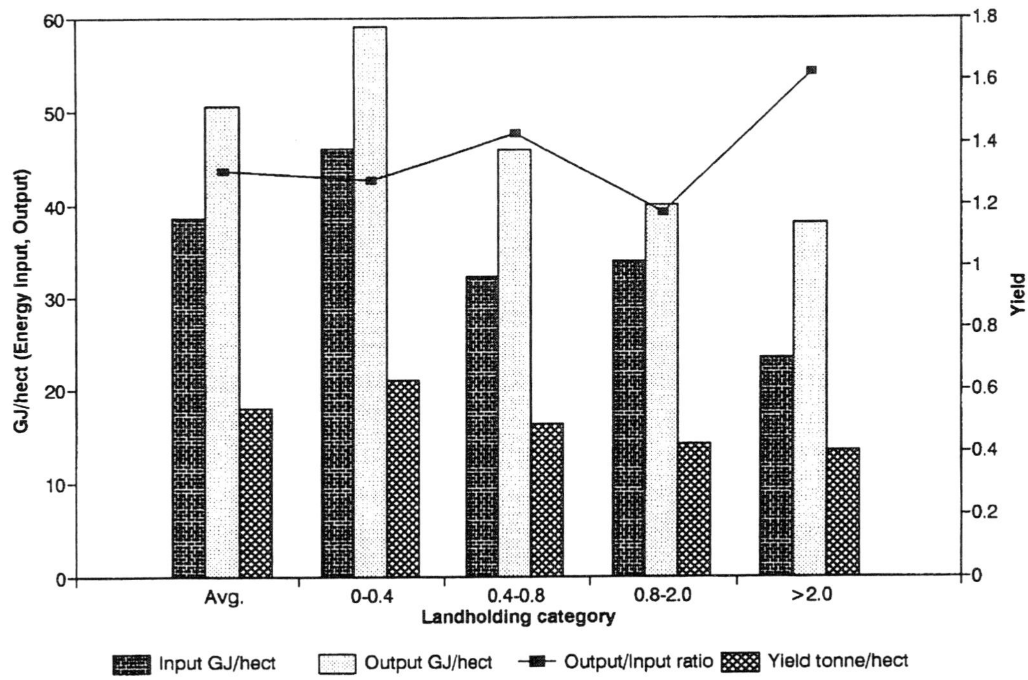

Figure 8.4: Energetics in Paddy Cultivation - Landholdingwise, Kumta Taluk

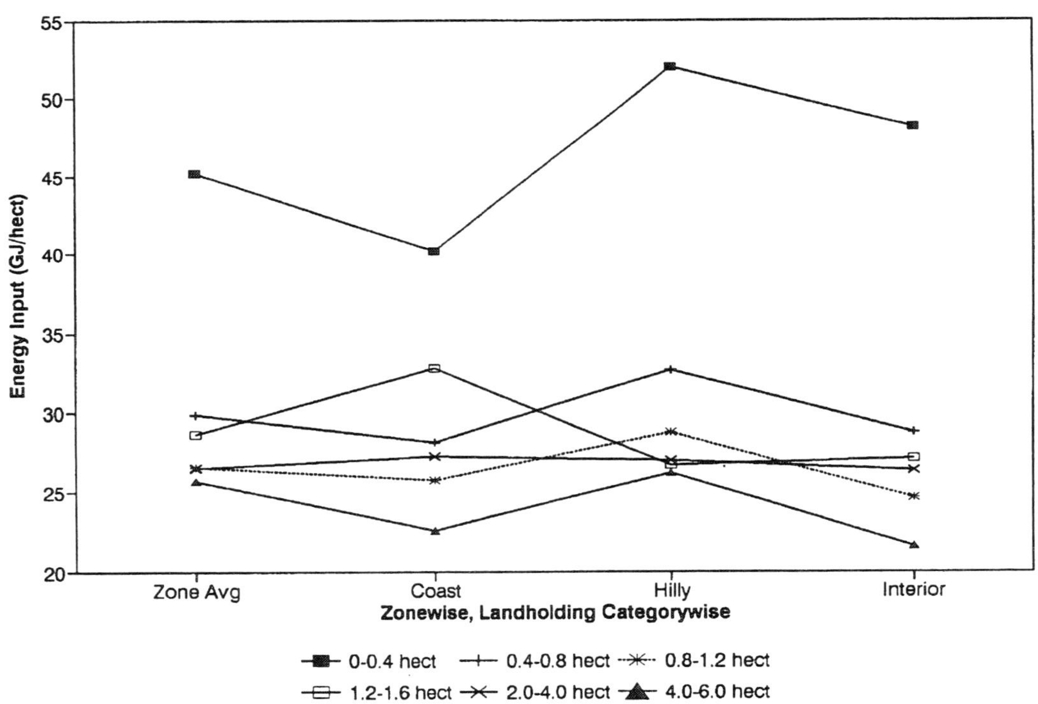

Figure 8.5: Energy Input (GJ/hect) in Paddy Cultivation

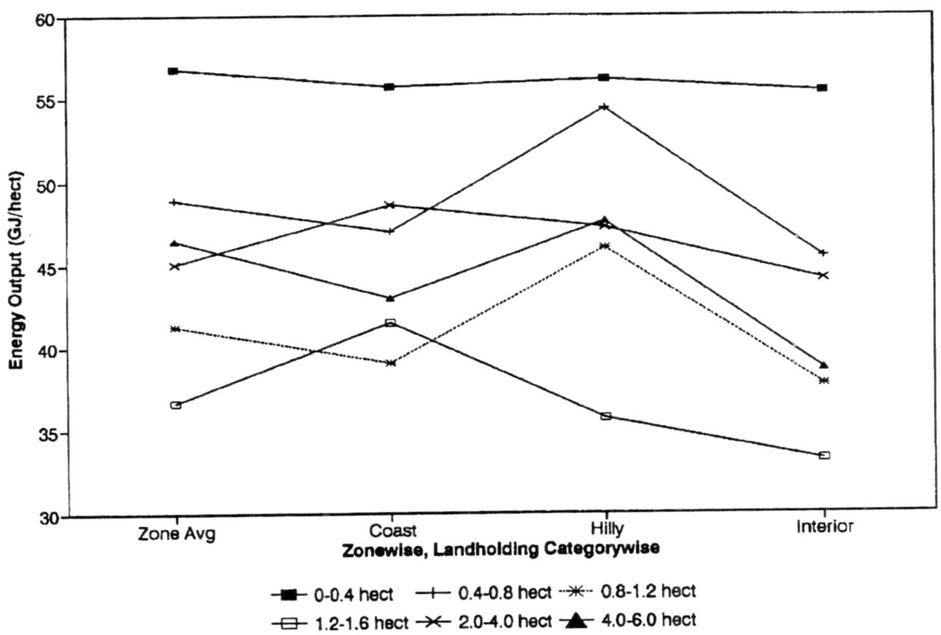

Figure 8.6: Energy Output (GJ/hect) in Paddy Cultivation

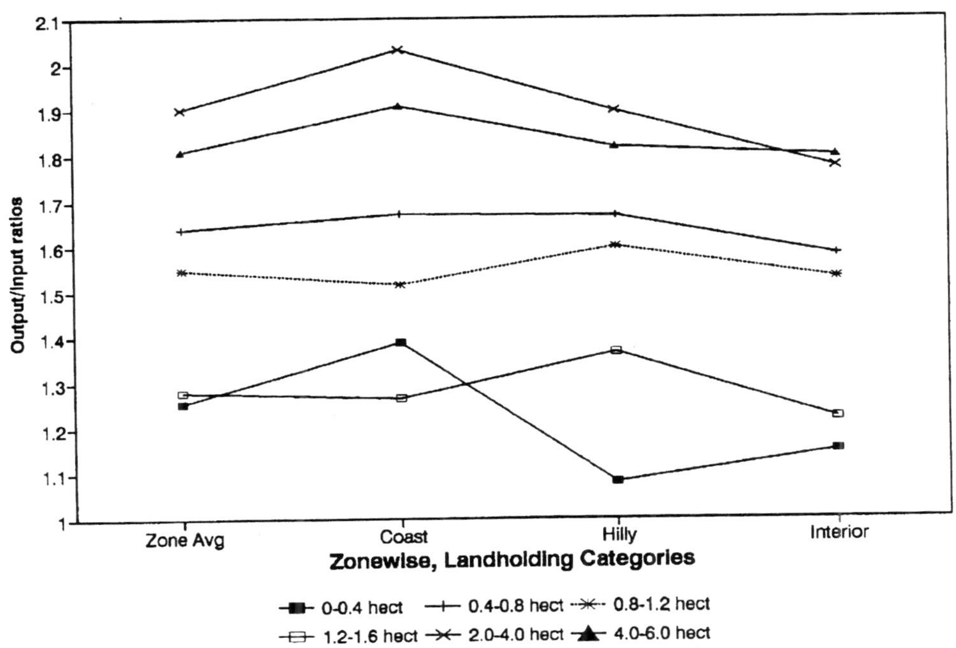

Figure 8.7: Output/Input Ratios for Various Categories of Landholding

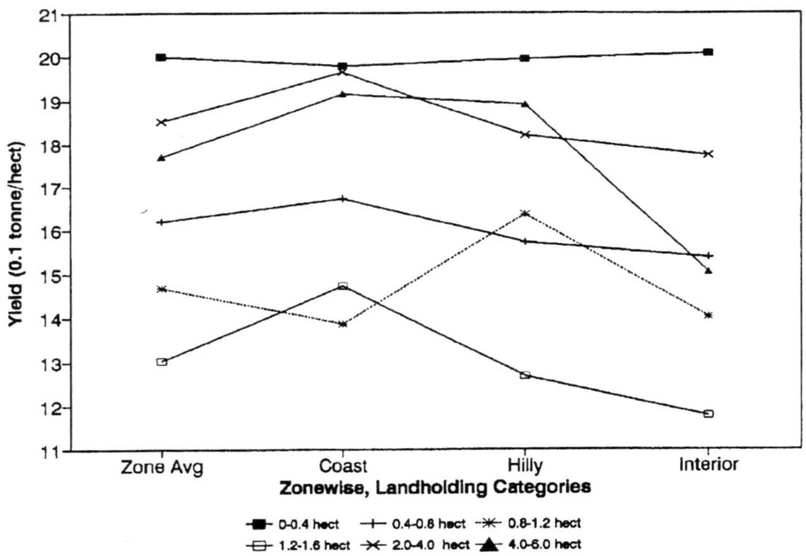

Figure 8.8: Paddy Yield (0.1 tonne/hect) – Zonewise

8.7.6 Factors Affecting the Level of Energy Consumption in Agriculture

Earlier discussions show that bullock power (BH), human power (HH), seed (S), FYM and inorganic fertilisers (F) are the major components in energy input. In order to see the variation in yield due to the consumption levels of these factors, regression analyses were carried out and the results are given in Table 8.7. This analysis is carried out first with all data points, and later by partial removal of scatter (i.e. removal of data points beyond (avg$\pm$sd)).

To investigate the factors responsible and their contributions for variations in yield, step-wise regression analysis is carried out zonewise (Table 8.8) and landholdingwise (Table 8.9) by adding the variables one by one. Standardised regression coefficients {(X coefficient)(Sd of X-variables)/(Sd of Y-variables)} have also been computed, which help in identifying the importance of each variable in the least square formula. These values for coastal zone, Land (0.418), BH (0.107), HH (0.0184), S (0.176), FYM (0.119) and F (0.00085), show that FYM, S, L and BH contribute significantly to the variation of Y. Similar coefficients computed for hilly zone, Land (0.167), BH (0.004), HH (0.114), S (0.123) and FYM (0.0945), show that HH contribute to Y. In all these cases, variable land also contributes significantly (depending on its nutrient status) in the variation of paddy yield. This trend continues even with normalised variables, that is Xi =(Xi-Xavg)/Xsd and Yi= (Yi-Yavg)/Ysd.

Based on the experience in the field and discussions with progressive farmers, the results are interpreted as:

(1) Variation in yield due to variable land is attributed to salinity and varying nutrient content of soil across the zones.

(2) High dung availability due to higher livestock density and good management practices have resulted in higher yield in the coast. While interior zone has scarcity of green manure and dung availability, FYM contribution in hilly zone is significant.

(3) Usage of high yield paddy, improved paddy or mixture of both may contribute towards yield. High yield variety demands more soil nutrients than local varieties. But, dwindling resource base in the form of green manure, etc., has contributed to further decline in the yield. Changes in physical and chemical properties of soil, such as, changing quality and quantity of organic matter and tilling of land, also play a role in declining yields. Apart from this, erratic and untimely rainfall, salinity and waterlogging have also contributed to variation in yield.

Table 8.6: Energy Consumption Pattern in Various Categories of Landholding within a Region

Zone	N	BH	HH	S	FYM	Fert	Total Input	Total Output	Paddy Out/In	Paddy yield (0.1 tonne)
0-0.4 hect										
All	498	6.96	1.71	3.91	32.64	0.03	45.24	56.76	1.25	20.00
Coastal	248	6.67	1.70	3.82	27.84	0.09	40.13	55.55	1.39	19.79
Hilly	160	7.12	1.58	4.06	39.22	0.00	51.94	56.07	1.08	19.94
Interior	90	6.90	2.05	3.55	35.74	0.00	48.23	55.35	1.15	20.07
0.4-0.8 hect										
All	313	6.51	1.35	2.49	21.82	0.05	29.87	48.95	1.64	16.19
Coastal	112	6.55	1.36	2.45	17.80	0.00	28.15	47.02	1.67	16.73
Hilly	101	6.40	1.42	2.58	29.66	0.00	32.64	54.37	1.67	15.72
Interior	100	6.57	1.28	2.46	18.35	0.15	28.81	45.49	1.58	15.37
0.8-1.2 hect										
All	97	6.58	1.23	2.38	16.40	0.00	26.60	41.27	1.55	14.67
Coastal	40	6.93	1.11	2.43	15.30	0.00	25.77	39.02	1.52	13.87
Hilly	30	5.72	1.15	2.37	14.36	0.00	28.79	46.01	1.60	16.36
Interior	27	7.04	1.50	2.33	20.08	0.00	24.66	37.71	1.53	14.02
1.2-1.6 hect										
All	75	6.49	1.06	1.93	19.20	0.00	28.69	35.67	1.28	13.04
Coastal	23	6.76	1.03	1.95	23.00	0.00	32.74	41.48	1.27	14.75
Hilly	29	7.02	1.26	1.82	16.61	0.00	26.72	35.68	1.37	12.69
Interior	23	5.56	0.85	2.05	18.66	0.00	27.12	33.16	1.22	11.78
1.6-2 hect										
All	33	6.59	1.12	2.29	20.87	0.00	30.88	45.06	1.46	12.35
Coastal	12	6.62	1.31	1.87	21.82	0.00	31.62	48.63	1.54	15.42
Hilly	8	6.09	1.07	2.53	22.91	0.00	32.60	47.26	1.45	11.17
Interior	13	6.87	0.98	2.53	21.25	0.00	31.63	44.12	1.40	10.48
2-2.4 hect										
All	25	5.55	1.26	1.49	24.77	0.00	33.07	39.49	1.75	14.04
Coastal	11	5.61	1.20	1.31	21.82	0.00	29.95	30.19	1.43	10.73
Hilly	4	6.69	1.49	1.26	26.52	0.00	35.97	51.72	1.48	18.39
Interior	10	5.03	1.24	1.78	27.30	0.00	35.35	44.82	2.20	15.94
2.4-4.0 hect										
All	11	4.56	0.85	1.58	19.50	0.00	26.49	47.74	1.90	18.52
Coastal	5	5.09	0.96	1.59	19.62	0.00	27.25	55.25	2.03	19.65
Hilly	4	3.77	0.67	1.76	20.78	0.00	26.99	51.15	1.90	18.19
Interior	2	4.83	0.91	1.61	19.03	0.00	26.38	46.75	1.77	17.72

Table 8.6: Energy Consumption Pattern In Various Categories of Landholding within a Region (Cont'd)

Zone	N	BH	HH	S	FYM	Fert	Total Input	Total Output	Paddy Out/In	Paddy yield
4.0-6.0 hect										
All	16	2.55	1.41	1.51	20.19	0.03	25.68	46.47	1.81	17.69
Coastal	9	2.45	0.65	1.42	18.01	0.00	22.53	42.95	1.91	19.15
Hilly	5	2.78	0.72	1.29	21.36	0.05	26.20	47.62	1.82	18.90
Interior	2	2.44	0.56	1.33	17.20	0.03	21.56	38.70	1.80	15.02

N - No of Samples BH - Bullock Hour S - Seed HH - Human Hour F - Fertiliser FYM - Farm Yard Manure
LH Landholding Category

Table 8.7: Regression Analyses of Yield and Various Energy Input Parameters

Zone	N	Dependent variable (y) (tonne/hect)	Independent variable/s (x)	R^2	Std.Error of Y est.	Probable relationship
Coastal	397	Yield	l (hect)	0.209	4.246	y = 0.8233+(-0.0939)l
			bh (GJ/hect)	0.086	4.336	y = 0.6891+(0.0516)bh
			hh (GJ/hect)	0.101	4.320	y = 0.6977+(0.197)hh
			s (GJ/hect)	0.257	4.196	y = 0.6017+(.2931)s
			fym (GJ/hect)	0.386	4.005	y = 0.6955+(0.01146)fym
			input (GJ/hect)	0.398	3.983	y = 0.6728+(0.0118) input
Hilly	342	Yield	l (hect)	0.231	4.098	y = 0.8106+(-0.1160)l
			bh (GJ/hect)	0.056	4.195	y = 0.6895+(0.0293)bh
			hh (GJ/hect)	0.172	4.139	y = 0.6171+(.4480)hh
			s (GJ/hect)	0.203	4.114	y = 0.6179+(0.2010)s
			fym (GJ/hect)	0.116	4.173	y = 0.7084+(0.002470)fym
			input (GJ/hect)	0.123	4.170	y = 0.7030+(0.00271) input
Interior	329	Yield	l (hect)	0.377	3.093	y = 0.8463+(-0.17822)l
			bh (GJ/hect)	0.161	3.295	y = 0.6282+(0.0755)bh
			hh (GJ/hect)	0.105	3.320	y = 0.6895+(0.914)hh
			s (GJ/hect)	0.321	3.162	y = 0.5986+(0.2261)s
			fym (GJ/hect)	0.111	3.338	y = 0.7101+(0.00024)fym
			input (GJ/hect)	0.107	3.337	y = 0.7063+(0.00074) input

Table 8.8: Regionwise Analyses of Factors Affecting the Yield in Paddy Cultivation

Zone	N	Dependent variable (y) (tonne/hect)	Independent variable/s (x)	R^2	Std.Error of Y est.	Probable relationship
Coast	397	Yield	l	0.21	4.25	y = 0.823+(-0.0938)l
			l, bh	0.22	4.24	y = 0.784+(-0.08892)l+(0.0345)bh
			l, bh, hh	0.22	4.24	y = 0.755+(-0.08645)l+(0.0296)bh+(0.121)hh
			l, bh, hh, s	0.29	4.17	y = 0.641+(-0.05682)l+(0.0223)bh+(0.069)hh+(0.2297)s
			l, bh, hh, s, fym	0.47	3.85	y = 0.594+(-0.0543)l+(0.032)bh+(0.064)hh+(0.2027)s+(0.0124)fym
			l, bh, hh, s, fym, fert	0.47	3.85	y = 0.594+(-0.054)l+(0.0321)bh+(0.0642)hh+(0.2025)s+(0.01235)fym+(0.0542)fert
Hilly	342	Yield	l	0.23	4.09	y = 0.811+(-0.116)l
			l, bh	0.23	4.09	y = 0.794+(-0.1132)l+(-0.0146)bh
			l, bh, hh	0.27	4.06	y = 0.716+(-0.1042)l+(0.00247)bh+(0.3542)hh
			l, bh, hh, s	0.29	4.04	y = 0.651+(-0.0863)l+(0.00492)bh+(0.291)hh+(0.1235)s
			l, bh, hh, s, fym	0.31	4.02	y = 0.646+(-0.08398)l+(-0.00247)bh+(0.2939)hh+(0.121)s+(0.00197)fym
			l, bh, hh, s, fym, fert	0.31	4.02	y = 0.646+(-0.0834)l+(-0.00247)bh+(0.299)hh+(0.121)s+(0.0019)fym+(0.0443)fert
Interior	328	Yield	l	0.39	2.85	y = 0.836+(-0.1729)l
			l, bh	0.42	2.82	y = 0.761+(-0.166)l+(0.06175)bh
			l, bh, hh	0.42	2.82	y = 0.746+(-0.1605)l+(0.05928)bh+(0.0568)hh
			l, bh, hh, s	0.47	2.75	y = 0.703+(-0.143)l+(0.02717)bh+(0.03211)hh+(0.1435)s
			l, bh, hh, s, fym	0.47	2.76	y = 0.703+(-0.143)l+(0.02471)bh+(0.0321)hh+(0.14326)s+(-0.0000247)fym
			l, bh, hh, s, fym, fert	0.47	2.76	y = 0.703+(-0.1433)l+(0.02717)bh+(0.03211)hh+(0.143)s+(-0.2E-04)fym+(0.04445)fert

l - Land (hect) s - Seed (GJ) hh - Human hour (GJ) fert - Fertiliser (GJ) bh - Bullock Hour (GJ)
fym - Farm yard manure N - No of samples R - Correlation Co-efficient

Table 8.9: Regression Analyses (Landholdingwise) of the Factors Affecting the Yield

LH hect	N	Dependent variable (y) (tonne/hect)	Independent variable/s (x)	R^2	Std Error of Y est	Probable relationship
0-0.4	491	Yield	l	0.23	3.24	y = 1.047+(-.7435)l
			l, bh	0.28	3.20	y = 0.948+(-.65)l+(0.06175)bh
			l, bh, hh	0.28	3.20	y = 0.932+(-.639)l+(0.059)bh+(0.05187)hh
			l, bh, hh, s	0.30	3.18	y = 0.843+(-.4794)l+(0.0518)bh+(0.0396)hh+(0.079)s
			l, bh, hh, s, fym	0.31	3.18	y = 0.840+(-.476)l+(0.0518)bh+(0.0395)hh +(0.0815)s+(0.000494)fym
			l, bh, hh, s, fym, fert	0.31	3.18	y = 0.841+(-.49)l+(.052)bh+(.039)hh+(0.082)s +(0.0005)fym+(0.047)fert
0.4-0.8	312	Yield	l	0.15	2.47	y = 0.867+(-.309)l
			l, bh	0.15	2.48	y = 0.869+(-.309)l+(-0.002)bh
			l, bh, hh	0.17	2.47	y = 0.850+(-.306)l+(-0.005)bh+(0.091)hh
			l, bh, hh, s	0.18	2.47	y = 0.865+(-.294)l+(-0.005)bh+(0.117)hh+(0.072)s
			l, bh, hh, s, fym	0.19	2.47	y = 0.858+(-.289)l+(-0.007)bh+(0.116)hh+(0.069)s +(0.001)fym
			l, bh, hh, s, fym, fert	0.19	2.48	y = 0.853+(-.282)l+(-0.007)bh+(0.116)hh+(0.069)s +(0.001)fym+(0.047)fert
0.8-1.2	96	Yield	l	0.01	2.66	y = 0.571+(0.02)l
			l, bh	0.01	2.68	y = 0.574+(0.02)l+(-0.002)bh
			l, bh, hh	0.06	2.69	y = 0.552+(0.027)l+(-0.002)bh+(0.072)hh
			l, bh, hh, s	0.24	2.63	y = 0.516+(-0.064)l+(-0.002)bh+(0.042)hh+(0.368)s
			l, bh, hh, s, fym	0.29	2.60	y = 0.465+(-0.027)l+(-0.017)bh+(0.054)hh+(.343)s +(0.012)fym
			l, bh, hh, s, fym, fert	0.29	2.60	y = 0.465+(-0.027)l+(-0.017)bh+(0.054)hh+(.343)s +(0.012)fym+(0.045)fert
1.2-1.6	75	Yield	l	0.03	2.66	y = 0.664+(-0.089)l
			l, bh	0.21	2.62	y = 0.670+(-0.02)l+(-0.106)bh
			l, bh, hh	0.23	2.63	y = 0.721+(-0.064)l+(-0.114)bh+(0.158)hh
			l, bh, hh, s	0.27	2.62	y = 0.476+(0.042)l+(-0.119)bh+(0.156)hh+(0.277)s
			l, bh, hh, s, fym	0.27	2.63	y = 0.437+(0.057)l+(-0.111)bh+(0.143)hh+(.254)s +(0.005)fym
			l, bh, hh, s, fym, fert	0.27	2.63	y = 0.437+(0.057)l+(-0.111)bh+(0.143)hh+(.254)s +(0.005)fym+(0.045)fert
1.6-2.0	33	Yield	l	0.15	1.36	y = 0.725+(-0.158)l
			l, bh	0.18	1.38	y = 0.812+(-0.18)l+(-0.037)bh
			l, bh, hh	0.18	1.40	y = 0.813+(-0.18)l+(-0.037)bh+(0.096)hh
			l, bh, hh, s	0.24	1.41	y = 0.539+(0.077)l+(-0.042)bh+(-0.015)hh+(0.175)s
			l, bh, hh, s, fym	0.24	1.43	y = 0.554+(-0.084)l+(-0.042)bh+(-0.022)hh+(0.178)s +(0.001)fym
			l, bh, hh, s, fym, fert	0.24	1.43	y = 0.554+(-0.084)l+(-0.042)bh+ (-0.022)hh+(0.178)s+(0.001)fym+(0.044)fert

Table 8.9: Regression Analyses (Landholdingwise) of the Factors Affecting the Yield (Cont'd)

LH hect	N	Dependent variable (y) (tonne/hect)	Independent variable/s (x)	R^2	Std Error of Y est	Probable relationship
2.0-2.4	25	Yield	l	0.29	3.57	y = 0.731+(-0.222)l
			l, bh	0.29	3.65	y = 0.711+(-0.225)l+(0.012)bh
			l, bh, hh	0.29	3.73	y = 0.745+(-0.215)l+(0.072)bh+(-.309)hh
			l, bh, hh, s	0.30	3.82	y = 0.740+(0.222)l+(0.059)bh+(-0.17)hh+(-0.1146)s
			l, bh, hh, s, fym	0.44	3.68	y = 0.786+(0.232)l+(.375)bh+(-1.49)hh+(.319)s+(0.047)fym
			l, bh, hh, s, fym, fert	0.44	3.68	y = 0.786+(0.232)l+(.375)bh+(-1.49)hh+(.319)s +(0.047)fym+(0.039)fert
2.4-2.8	27	Yield	l	0.41	2.89	y = 0.935+(-0.086)l
			l, bh	0.41	2.95	y = 0.921+(-0.084)l+(0.01)bh
			l, bh, hh	0.50	2.86	y = 0.888+(-0.054)l+(.48)bh+(-3.038)hh
			l, bh, hh, s	0.50	2.93	y = 0.847+(-0.049)l+(.472)bh+(-3.03)hh+(0.114)s
			l, bh, hh, s, fym	0.77	2.20	y = 0.456+(-0.015)l+(-0.165)bh+(-.408)hh+(1.03)s +(0.083)fym
			l, bh, hh, s, fym, fert	0.77	2.20	y = 0.456+(-0.015)l+(-0.165)bh+(-.408)hh+(1.03)s +(0.083)fym+(0.038)fert
All	1059	Yield	l	0.28	3.05	y = 0.788+(-0.104)l
			l, bh	0.29	3.04	y = 0.764+(-0.101)l+(0.02)bh
			l, bh, hh	0.30	3.04	y = 0.741+(-0.096)l+(0.02)bh+(0.094)hh
			l, bh, hh, s	0.34	2.99	y = 0.675+(-0.082)l+(0.012)bh+(0.064)hh+(0.136)s
			l, bh, hh, s, fym	0.34	2.99	y = 0.671+(-0.079)l+(0.012)bh+(0.054)hh+(0.133)s +(0.001)fym
			l, bh, hh, s, fym, fert	0.34	2.99	y = 0.670+(-0.079)l+(0.012)bh+(0.064)hh+(0.133)s +(0.001)fym+(0.057)fert

l - Land (hect)
bh - Bullock hour (GJ)
hh - Human hour (GJ)
LH - Landholding category
s - Seed (GJ)
fym - Farm yard manure
fert - Fertiliser (GJ)
N - No of samples
R - Correlation Co-efficient
Dependent variable (y) yield tonne/hect

8.8.0 CONCLUSIONS

(1) Paddy yield showed decline from 3.3 (1980-81) to 1.9 (1992-93) tonne/hect in spite of continued use of H.Y. varieties, and greater application of inorganic fertiliser (25.6% increase) and pesticides. H.Y. variety yield (tonne/hect) changes from 2.5 (1984-85) to 2.2 (1992-93), while improved variety changes from 3.7 (1984-85) to 1.6 (1992-93).

Population density varies from 0.35 (hilly), 1.35 (interior) to 5.31 (coastal) persons/hectare, while forest cover changes from 25% (coastal), 76% (interior) to 89% (hilly). Livestock density varies from 0.24 (hilly), 0.70 (interior) to 1.63 (coastal) animals/hectare. There is a positive linear correlation between livestock and human density.

(2) Energy input from all sources (such as FYM, etc.) is in the range 35.47 (interior), 37.24 (coast) to 43.21 (hilly), the major component being organic (FYM). Hilly zone with 89.78% forest cover and dung yield (kgs/animal/day) of 5.82±1.8 is rich in organic matter compared to coast (forest

cover 27.66%, dung yield 3.26) and interior (forest cover 78.14%, dung yield 3.94). The average paddy yield is about 1.73 tonne/hect.

In the coast, total energy input in GJ per hectare varies from 33.22 (<0.4 hect), 34.65 (0.8-1.2 hect), 26.14 (1.2-1.6 hect) to 19.72 (>2 hect). Output in GJ per hectare also varies from 56.62 (<0.4 hect), 39.99 (0.8-1.2 hect) to 38.14 (>2 hect). Output-input ratio varies from 1.13 (1.2-1.6 hect) to 1.93 (>2 hect)

Similar trends are noticed in interior and hilly zones. In hilly zone input energy varies from 55.95 (<0.4 hect) to 17.50 (>2 hect) and corresponding output varies from 58.36 to 30.64. In interior zone input energy varies from 29.18 (< 0.4 hect) to 19.82 (>2 hect), while output varies from 32.87 (0.4 hect) to 28.08 (>2 hect). Marginal and small farmers have better output per hectare due to better management.

(3) It is evident that paddy cultivation mainly depends on organic manure (FYM) which constitutes 70 to 74% of the total input. Yield in all the three zones is dependent on the inherent fertility of soil and successful maintenance of nutrient cycle. The productivity in all zones is dependent on the levels of input, that is the intensity of agriculture in the form of organic inputs, etc. As the field sizes are small and fragmented there are problems in cultivating with the help of tractors, tillers, etc. The design of machineries to substitute manual tasks under such conditions is complex and expensive, which makes mechanisation of agriculture inappropriate.

New seeds and fertiliser can be bought in any quantity according to farmers' inclination and financial position. However, indiscriminate usage of fertiliser has resulted in significant reduction in yield over a period. Some farmers who switched over to inorganic fertiliser in the early eighties, have returned to organic farming, due to declined yield and water pollution problems. This may be due to indiscriminate usage of fertiliser without considering soil condition.

This study shows that, it is not necessary always to increase the energy inputs in agriculture to get higher production. Initial increments in yield due to energy inputs in the form of fertiliser, etc., have dwindled in the later years. In view of these, it is necessary to practice environmentally sound management practices for sustainable agriculture without affecting other components of the ecosystem.

8.9.0 Acknowledgements

My sincere thanks to Mr. Srikanth V. Gunaga and Mr. R.B. Harikantra for their help in collection of seasonal data related to agricultural operations. Thanks are due to the officials of the Agriculture Department at Karwar, Sirsi, Kumta and Bangalore for providing statistical information and useful discussions. This research is supported with grants from the Ministry of Environment and Forests, Government of India and Indian Institute of Science.

8.10.0 References

1 Pimentel, D. and Pimentel, M., 1979. *Food Energy and Society*. Edward Arnold Ltd., London.

2. Subramanian, D.K. and Ramachandra T.V., 1991. Some aspects of agriculture and irrigation in Karnataka. In: *Karnataka State of Environment Report V*. Cecil J. Saldanha (Ed.), Chapter XI, pp. 95-116. Centre for Taxonomic Studies, Bangalore, India.

3. Makhijani, A. and Poole, A., 1975. *Energy and Agriculture in the Third World*. Ballinger Publishing Co., Cambridge.

4. Revelle, R., 1976. Energy use in rural India. *Science*, 192: 969-975.

5. Avadani, M.H.V., Kumar, V. and Krishna, G., 1988. Energy optimisation in the application of fertiliser. In: *Food Energy Nexus and Ecosystems*. Moulik, T.R. (Ed.), pp. 412-429, Oxford and IBH Publishing Co., New Delhi, India.

6. op. cit Ref.1

7 Tivy, J., 1990. *Agricultural Ecology*. Longman Scientific and Technical, John Wiley & Sons, Inc., New York.

8. Pimentel, D., Hurd, L.E., Belloti, A.C., Forester, M.J., Oka, I.N., Sholes, O.D. and Whitman, R.J., 1973. Food production and the energy crisis. *Science,* 102:443-449.
9. Black, J.N., 1971. Energy relations in crop production: A preliminary survey. *Proc. Assoc. Appl. Biol.,* 67:272-278.
10. op.cit Ref 1
11. Kamal Rijal, Bansal N.K. and Grover, P.D., 1991. Energy in subsistence agriculture: A case study of Nepal. *International Journal of Energy Research*, 15:109-122.
12. Sathyajith Mathew, Geetha Susan Philip, Jayachandran, S., Jayashree, G.S., Shobhalatha, P.K. and Suresh Kumar, P., 1993. Energy flow pattern in paddy cultivation with special reference to puddling. *Energy Environment Monitor*, 9(2):81-83.
13. Carlos A. Bonetto and Arturo J. Carcano, 1995. Azolla utilisation, rice yield and nitrogen balance in Argentina. *International Journal of Ecology and Environmental Sciences*, 21: 37-44.

8.11.0 ANNEXURE I.

Item	Unit	Energy Equivalent (GJ/unit)
Human Hour	hr	0.0010442
Draught Animal	hr	0.019168
Farm Yard Manure	kg	0.047990
Paddy Straw	kg	0.013300
Paddy Husk	kg	0.014170
Rice	kg	0.015195
DAP-Fertiliser	kg	0.065800
Complex	kg	0.030340
Urea	kg	0.079500
Pesticide	kg	0.101325

Chapter 9

ENERGY UTILISATION IN INDUSTRIAL SECTOR

9.0 SUMMARY

Energy is a vital component in rural industries like agro-processing, brick and tile manufacture, etc. In rural areas, population pressure and low agricultural productivity are some of the factors forcing people into marginal and ecologically fragile lands. Forests are being logged to meet the energy demand of the region in addition to clearing them for agriculture to meet the food demand. Soil erosion, siltation and floods are the consequences of the destruction of forests. Low productivity of agricultural lands and lack of alternate employment opportunities in rural areas have resulted in large-scale migration of rural people to urban areas in search of jobs. In this context, industries in rural areas could help in arresting this migration and improve the economic status of the local people. For the sustainable development of a region, industrialisation has to be supplemented with integrated energy planning. This necessitates the search for alternate sources of energy, that are renewable and which could be harnessed in a decentralised way, and improvement of energy efficiency of end use devices. Numerous opportunities exist for developing industrial processes with greater efficiencies. Energy auditing carried out in sample rural industries has revealed wide disparity in energy consumption to produce same quantity of similar products. About 50% of sample units have Percentage Production Capacity Utilisation (PPCU) less than 50% and higher Specific Energy Consumption (SEC) and Energy Intensity (EI). The wide variation in EI in cashew industries, from 4.49 to 8.66 kgs of fuel wood/kg of kernel, reveals the scope for energy conservation in the order of 30 to 48%. It has been noticed that optimal utilisation of installed capacity (better PPCU) would help in energy conservation. The energy analyses carried out reveal that the relationship between SEC and Pr is power law with an exponent less than one. This indicates improvement of SEC with increase in production which in turn implies the optimal utilisation of installed production capacity. This chapter focuses on energy use pattern in rural industries in Uttara Kannada district, by collecting secondary data from various Government agencies and conducting a primary survey of sample industries in Kumta taluk.

Keywords: Specific Energy Consumption (SEC), Energy Efficiency, Rural Industries, Percentage Production Capacity Utilisation (PPCU), Energy Intensity (EI).

9.1.0 INTRODUCTION

Rural energy supply is closely knit with agriculture and forest production, with an overall crisis of agricultural productivity and environmental degradation linked to modernisation and changes in land use.

Rural community depends mainly on non-commercial energy for domestic and industrial needs. If new energy strategies are to be successful, rural energy needs have to be placed squarely in the context of integrated approaches to rural development, basic needs and general questions of rural poverty, land and other scarce resources. With the emphasis on rural development by the government, rural industries are gaining importance. This chapter attempts to focus on the various ecological parameters, such as, energy conservation and raw material availability, etc., involved in the successful functioning of rural industries. Rural industries are important for the stability and growth of a regional economy. They play a significant role in averting the migration of rural people to urban areas/cities, which in turn results in new slums and hence leads to environmental/ecological problems. They process local raw materials and rely on local resources like firewood and agricultural residues for their heat energy needs.

Wood is the traditional domestic fuel in rural areas. With the depletion of forest resources, the financial viability of many industries is threatened due to increase in fuel wood prices and lack of technical and financial assistance to switch over to alternate fuels. Although conventional fuels like oil, etc., are available in many rural areas, their cost per unit of useful energy is usually much higher than those of traditional sources. With the use of conventional fuels, industries become more dependent on distribution networks which are liable to interruption, and it is not desirable to use high quality energy like electricity for low quality work like heating, etc. This unreliability in supply and the high cost of conventional sources of energy, demand for viable alternatives to meet the demands of rural industries locally. The supply of energy from wood and residues is more flexible and reliable. The increasing demand for wood and residues, both for energy and other domestic, agricultural and industrial purposes, put further pressure on traditional sources, often with adverse effects on other end-users, stability and productivity of the rural environment. This in turn demands for increasing the biomass stocks by raising energy plantations to meet the growing fuel demand of the population.

Rural industries provide majority of off-farm employment and income in rural areas. In Uttara Kannada, there are 1283 villages within 11 taluks. The population of the district is 12,43,526 with a density of 120.84 persons/sq.km and 76.51% of the total population being in rural areas. Deforestation and loss of soil fertility has resulted in serious reduction in food production by traditional agriculture. Increase in population, leading to fragmentation in landholding, has resulted in further degradation of agricultural land. Fuel shortages, usually tied with environmental deterioration, have also affected food production. Rural industries are playing a significant role in this regard by providing employment to poor peasants, especially women. This division of labour has influenced women's ability to produce food and provide the required nutrition for their families.

9.1.1 Karnataka's Energy Scene

Karnataka does not have any coal deposits. It gets its coal from outside. Initially, the electrical energy for the state was purely hydro and with the commissioning of Raichur thermal power station, it gets electrical energy from coal also. The other major source of commercial energy - oil - being unavailable, the main source of commercial energy for the state comes from hydroelectric plants. These plants have large reservoirs to store rain-water throughout the year. The dams being built in forest areas submerge sizable tracts of prime forest. It is shown that a comparable quantity of energy can be obtained from forest biomass instead of water from the submerged area. An ideal solution will be to go in for a set of peaking power plants with minimum storage which utilize rain-water during monsoon and can be supplemented by firewood based thermal power plant.

Karnataka state depends both on commercial and noncommercial forms of energy (Subramanian, D.K., 1985). Noncommercial energy constitutes 53.2%, provided mainly by sources like firewood, agricultural residues, charcoal and cow dung. While, commercial energy's share is 46.8%, mainly by electricity, oil, coal, etc. Table 9.1 lists the sourcewise consumption of energy during the year 1990-91.

But commercial energy sources like coal, oil and electricity meet only a small part of the energy demand of the state. Major energy share comes from firewood. Table 9.1 gives the consumption pattern for Karnataka for 1990-91. Electricity represents 55.67% of commercial energy for 1990-91, showing an increase in the per cent share. Firewood consumption is around 7.44 million tons of oil equivalent. Agro wastes are also used for energy purposes. The share of industries in energy is around 44%. The state generated 12,430 million units of electrical energy in 1990-91 and 12,887 million units during 1991-92.

The share of industries in electricity is 4,482 million units in 1990-91 and 4,316 million units in 1991-92. It is seen that the share of irrigation pumpsets is gradually increasing. 8.82% of pumpsets energized so far is already guzzling 36.26% of total electrical energy. This is followed by industries with a share of 34.34%, domestic lighting & AEH with a share of 15.35%, LT industries with a share of 9.90%, commercial lighting 2.23%, public lighting 1.24% and others 0.68%.

Present study mainly concentrates on energy use pattern in rural industries in Karnataka by collecting secondary data from various government agencies and conducting a primary survey of samples at Kumta taluk in Uttara Kannada district.

Table 9.1: Energy in Karnataka during 1990-91 (in millions of tonnes of oil equivalent – mtoe)

Source	Demand (mtoe)	% share
Coal	1.005	5.81
Kerosene	0.445	2.57
Oil (HSD, LDO, etc.)	2.014	11.64
LPG	0.130	0.75
Electricity	4.510	26.06
Commercial Energy Total	8.104	46.83
Agricultural Residues	1.510	8.73
Firewood	7.440	43.00
Biogas, cow dung, etc.	0.250	1.44
Noncommercial Energy Total	9.200	53.17
Total energy during 1990-91	17.304	100.00

9.1.2 Industries in Karnataka

Industries are playing a pivotal role in the development of the state. The secondary sector in the state accounts for nearly a quarter of the state income in real terms of the total output in the economy (Government of Karnataka, 1995). The average annual growth rate of industrial production over the period from 1981-82 to 1990-91 was about 6.4% as against 7.9% in the country as reflected in the index of industrial production.

9.1.3 Small-scale Industries: An Overview

The basic frame work for the industrial policy of the government was first spelled out in the Industrial Policy Resolution of 1956, with emphasis on the development of small-scale industries. The small, medium and large industries have been assigned a mutually complementary role with a view to facilitate an integrated and harmonious growth of the industrial sector as a whole with the objective of economic growth with social justice.

The policy measures, announced in 1991 by the government, provided further impetus to the growth of small sector. The primary objective of the 1991 policy measure is to impart more vitality and growth impetus to the sector to enable it to contribute fully to the economy particularly in terms of growth of output and employment. As per the policy, the investment in fixed assets in plant and machinery in small-scale industries, whether held on ownership, lease or hire purchase terms, does not exceed Rs. 6 millions.

The spectrum of industries ranges from unorganised traditional sectors, modern small-scale sectors to large and medium scale industries. The traditional sector and small-scale sector provide maximum employment (41.34 million in 1989-90) and constitute an important component of the economy. In terms of value added, they are estimated to contribute 50% in the manufacturing sector. The uniform distribution and growth in this sector besides resulting in preponderance of self-employment and under dispersal of industrial and economic activities, ensures maximum utilisation of both human and material resources. The economics of production is important in small-scale industries. Energy consumption plays a key and dominant role in the production economics of these industries. The small-scale industries sector plays an important role in the industrial economy of the state. It contributes substantially to the industrial production and employment. These industries are dispersed all over the state, however, there is disparity in regional distribution. The growth of small-scale industries in last ten years is phenomenal. Table 9.2 illustrates the growth of small-scale industries in Karnataka since 1969-70. The growth in the last decade is phenomenal as the number of industries has increased from 37,148 to 1,29,915 with percentage annual growth rate of 15.01% and 13.58% growth in manpower employed. Table 9.3 gives districtwise cumulative information of SSI's from 1982-83 to 1991-92. While, Table 9.4 gives percentage yearly increase in number of industries, etc. From Table 9.3 it is evident that Bangalore district with 27,691 industries leads all other district. Districtwise increase in number of industries in last ten years is given in Table 9.5.1. Belgaum has annual average growth of 18.61%, while for Bangalore it is 9.31%. Districtwise information regarding number of industries, investment and manpower employed in SSI sector at the end of financial year 1991-92 is listed in Table 9.5.2. The percentage of units varies from a low of 1.15% (for Kolar district) to a high value of 21.97% (for Bangalore district). Investment per job computed for each district reveals a variation from Rs.9539.20 per person for Dharwad to Rs.22652.09 per person for Dakshina Kannada, Rs.20,571.43 for Bidar and Rs.20,097.12 for Chikmagalur. This indicates that some industries are capital intensive.

Workers in manufacturing and household industries are listed in the tenth column of Table 9.5.2. The share of manpower in SSI to the total manpower in the industrial sector shows variation of 16.67% for Chitradurga, 18.27% for Dakshina Kannada, 88.34% for Dharwad, 93.74% for Chikmagalur and 97.08% for Raichur. Lower values for Chitradurga and Dakshina Kannada is mainly due to the presence of large number of household industries in these districts. Most of these units are located in rural areas.

Table 9.2: Growth of Small-Scale Industries in Karnataka

	Year	No of SSI Units (Registered)	Investment (Rs.in lakhs)	Manpower Employed	SSI Units (cum)	Investment (Rs.in lakhs) (cum)	Manpower (cum)	Investment (Rs./Job)	% Growth (yearwise) Units	Investment	Manpower
1	1969-70	3890	3456.70	47960	3890	3456.70	47960	7207.46			
2	1970-71	1908	2279.43	44295	5798	5736.13	92255	6217.69	49.05	65.94	92.36
3	1971-72	2372	1309.39	21343	8170	7045.52	113598	6202.15	40.91	22.83	23.13
4	1972-73	2272	1350.10	22490	10442	8395.62	136088	6169.26	27.81	19.16	19.80
5	1973-74	3043	1638.23	21814	13485	10033.85	157902	6354.48	29.14	19.51	16.03
6	1974-75	1907	3991.27	56043	15392	14025.12	213945	6555.48	14.14	39.78	35.49
7	1975-76	1562	1641.36	12783	16954	15666.48	226728	6909.81	10.15	11.70	5.97
8	1976-77	1420	1482.51	15406	18374	17148.99	242134	7082.44	8.38	9.46	6.79
9	1977-78	1621	1517.11	24750	19995	18666.10	266884	6994.09	8.82	8.85	10.22
10	1978-79	1975	1451.46	16957	21970	20117.56	283841	7087.62	9.88	7.78	6.35
11	1979-80	2910	3255.01	34376	24880	23372.57	318217	7344.85	13.25	16.18	12.11
12	1980-81	2776	3041.83	26164	27656	26414.40	344381	7670.11	11.16	13.01	8.22
13	1981-82	3396	4955.16	41375	31052	31369.56	385756	8131.97	12.28	18.76	12.01
14	1982-83	6096	6255.29	46420	37148	37624.85	432176	8705.91	19.63	19.94	12.03
15	1983-84	7479	6396.51	44282	44627	44021.36	476458	9239.29	20.13	17.00	10.25
16	1984-85	11962	6324.31	55849	56589	50345.67	532307	9458.01	26.80	14.37	11.72
17	1985-86	11634	3787.09	60796	68223	54132.76	593103	9127.04	20.56	7.52	11.42
18	1986-87	11179	9408.47	56883	79402	63541.23	649986	9775.78	16.39	17.38	9.59
19	1987-88	10530	10165.19	52498	89932	73706.42	702484	10492.26	13.26	16.00	8.08
20	1988-89	9811	10532.77	50448	99743	84239.19	752932	11188.15	10.91	14.29	7.18
21	1989-90	9700	11247.02	51521	109443	95486.21	804453	11869.71	9.72	13.35	6.84
22	1990-91	9884	11843.16	53568	119327	107329.37	858021	12508.94	9.03	12.40	6.66
23	1991-92	10588	18623.11	61903	129915	125952.48	919924	13691.62	8.87	17.35	7.21
TOTAL		129915	125952.48	919924					16.26	17.50	14.76*

Note: 1 million = 10 lakhs

Table 9.3: Number of Units, Investment and Manpower Details in Small-Scale Industries (Districtwise – Cumulative)

Year ---->	1982-83			1983-84			1984-85			1985-86		
District	Units Cum	Investment Cum (lakhs)	Persons Cum	Units Cum	Investment Cum (lakhs)	Persons Cum	Units Cum	Investment Cum (lakhs)	Persons Cum	Units Cum	Investment Cum (lakhs)	Persons Cum
Bangalore	11368	14381.78	156691	13138	15589.10	165515	13127	15248.40	149436	15254	16942.11	158560
Belgaum	2010	1856.93	16082	3438	2319.77	20715	4102	2490.98	22693	5283	3329.12	27628
Bellary	1530	788.16	8133	2041	1121.29	11271	2135	1222.58	11531	2634	1395.70	14156
Bidar	625	400.79	5790	989	703.05	7920	1077	707.17	8036	1330	917.22	9364
Bijapur	1282	604.15	12979	1693	799.25	14973	1890	919.56	15573	2321	1052.57	17579
Chickmagalur	598	380.12	4503	980	636.59	6065	917	492.67	5182	1134	618.78	5932
Chitradurga	1638	979.72	11417	1867	1086.13	12221	2024	1293.00	12278	2460	1582.85	14668
Dakshina Kannada	2641	2027.17	23989	3984	2444.58	29618	3708	3602.86	29445	4471	4987.61	35304
Dharwad	2444	4094.75	64714	3178	4702.68	69050	4094	5092.78	72204	5280	5698.78	77730
Gulbarga	850	642.62	7830	1232	775.54	9014	1385	794.71	9100	1793	877.45	10529
Hassan	824	550.90	6133	1085	685.31	7476	1155	739.67	7203	1422	1036.12	9274
Kodagu	460	959.61	10936	600	986.80	11487	643	962.27	11299	772	1022.28	11674
Kolar	1816	1124.05	11672	2170	1298.88	13779	2246	1256.32	15482	2614	1769.74	17947
Mandya	1087	624.54	6275	1392	759.36	7580	1293	807.94	7387	1545	952.36	9595
Mysore	2871	2619.16	23153	4314	3140.62	30863	5075	3361.04	33320	6241	3919.89	41007
Raichur	855	1767.17	19670	1179	1968.01	21106	1372	2088.51	21966	1800	2664.52	24550
Shimoga	1904	2022.83	13779	2543	2428.31	15830	2796	2721.31	15832	3379	3216.48	18703
Tumkur	1492	1279.13	17737	2243	1900.44	21479	2591	2389.19	23466	3340	2939.23	27116
Uttara Kannada	798	537.23	9214	989	619.41	10584	1083	692.74	11010	1274	767.98	11923
Total	37093	37640.81	430697	49055	43965.12	486546	52713	46883.70	482443	64347	55690.79	543239

Table 9.3: Number of Units, Investment and Manpower Details in Small-Scale Industries (Districtwise – Cumulative) (Cont'd)

Year --- >	1986-87			1987-88			1988-89		
District	Units Cum	Investment Cum (lakhs)	Persons Cum	Units Cum	Investment Cum (lakhs)	Persons Cum	Units Cum	Investment Cum (lakhs)	Persons Cum
Bangalore	17185	19021.37	167351	18872	21106.14	177084	20520	23851.00	188173
Belgaum	6487	4252.23	32022	7610	5150.28	36564	8445	6057.54	40190
Bellary	3137	1543.34	16801	3639	1745.10	18889	4085	1965.11	20848
Bidar	1596	1065.92	10777	1856	1426.66	12111	2101	1771.19	13047
Bijapur	2707	1237.00	19894	3121	1394.67	21507	3516	1652.16	23345
Chickmagalur	1357	722.90	7542	1560	819.21	8771	1741	962.18	9490
Chitradurga	2872	2007.70	17375	3276	2515.74	19356	3658	3000.97	21279
Dakshina Kannada	5745	5935.33	40685	6488	7502.01	45476	7175	8996.06	49523
Dharwad	6109	7097.61	84202	6944	7696.27	88735	7844	8309.61	93316
Gulbarga	2226	1198.69	13201	2658	1558.29	15695	3118	2015.43	18025
Hassan	1673	1210.40	10125	1930	1404.03	11157	2167	1527.04	12102
Kodagu	872	1149.05	12696	982	1199.39	13058	1084	1246.53	13419
Kolar	3003	2123.51	20339	3327	2505.42	22445	3676	2903.63	24676
Mandya	1799	1106.78	10948	2008	1267.09	11972	2208	1425.29	12885
Mysore	7203	4498.07	45912	8331	5199.00	50848	9129	5774.33	54930
Raichur	2236	3002.03	26507	2695	3397.33	28521	3108	3722.88	30245
Shimoga	3892	3573.76	20964	4402	4004.76	22808	4906	4417.87	24681
Tumkur	3942	3485.98	29999	4671	4410.03	33889	5523	5089.36	38403
Uttara Kannada	1485	847.59	12787	1686	943.03	13739	1863	1059.04	14493
Total	75526	65079.26	600127	86056	75244.45	652625	95867	85777.22	703070

Table 9.3: Number of Units, Investment and Manpower Details in Small-Scale Industries (Districtwise – Cumulative) (Cont'd)

Year	1989-90			1990-91			1991-92		
District	Units Cum	Investment Cum (lakhs)	Persons Cum	Units Cum	Investment Cum (lakhs)	Persons Cum	Units Cum	Investment Cum (lakhs)	Persons Cum
Bangalore	22402	26783.42	202480	24303	30583.86	219098	27691	37102.85	242868
Belgaum	9291	7065.21	43859	10185	7997.54	47422	11080	9172.65	51078
Bellary	4526	2092.18	22557	4946	2229.09	24034	5257	2589.05	25522
Bidar	2346	2839.02	14575	2593	3242.00	16571	2826	3674.88	17864
Bijapur	3905	1952.46	25099	4330	2276.53	27066	4688	2814.06	28967
Chickmagalur	1918	1128.22	10215	2118	1302.33	10891	2293	1581.79	11555
Chitradurga	4037	3550.66	23390	4429	4087.58	25367	4860	5516.86	27451
Dakshina Kannada	7821	10138.97	52913	8399	11658.68	55875	8955	13458.51	59414
Dharwad	8750	8822.88	97656	9680	9329.64	101537	10506	10134.64	106242
Gulbarga	3499	2311.77	19822	3932	2679.36	21888	4179	3064.89	23470
Hassan	2411	1648.01	12992	2634	1834.21	13970	2791	1997.74	14778
Kodagu	1189	1282.15	13795	1344	1335.25	14271	1444	1441.89	14649
Kolar	4030	3306.00	27039	4379	3850.74	29047	4709	4740.34	31262
Mandya	2422	1770.87	14008	2670	1952.39	15173	2851	2275.39	16154
Mysore	9829	6199.24	58535	10609	6757.93	62720	11374	7550.83	67326
Raichur	3509	4302.50	32021	3920	4557.01	33479	4287	5144.87	35191
Shimoga	5402	4711.19	26359	5929	5202.65	28641	6410	5769.07	30711
Tumkur	6219	5882.27	41767	6792	6564.26	44589	7437	7784.95	48202
Uttara Kannada	2061	1237.22	15512	2259	1426.35	16523	2401	1675.25	17361
Total	105567	97024.24	754594	115451	108867.40	808162	126039	127490.51	870065

Table 9.4: Percentage Increase in Industrial Units in Karnataka (Districtwise)

Year ---->	% Variation 82-83 to 83-84			% Variation 83-84 to 84-85			% Variation 84-85 to 85-86		
District	Units	Investment	Persons	Units	Investment	Persons	Units	Investment	Persons
Bangalore	15.57	8.39	5.63	-0.08	-2.19	-9.71	16.20	11.11	6.11
Belgaum	71.04	24.93	28.81	19.31	7.38	9.55	28.79	33.65	21.75
Bellary	33.40	42.27	38.58	4.61	9.03	2.31	23.37	14.16	22.76
Bidar	58.24	75.42	36.79	8.90	0.59	1.46	23.49	29.70	16.53
Bijapur	32.06	32.29	15.36	11.64	15.05	4.01	22.80	14.46	12.88
Chickmagalur	63.88	67.47	34.69	-6.43	-22.61	-14.56	23.66	25.60	14.47
Chitradurga	13.98	10.86	7.04	8.41	19.05	0.47	21.54	22.42	19.47
Dakshina Kannada	50.85	20.59	23.46	-6.93	47.38	-0.58	20.58	38.43	19.90
Dharwad	50.85	20.59	23.46	-6.93	47.38	-0.58	28.97	11.90	7.65
Gulbarga	44.94	20.68	15.12	12.42	2.47	0.95	29.46	10.41	15.70
Hassan	31.67	24.40	21.90	6.45	7.93	-3.65	23.12	40.08	28.75
Kodagu	30.43	2.83	5.04	7.17	-2.49	-1.64	20.06	6.24	3.32
Kolar	19.49	15.55	18.05	3.50	-3.28	12.36	16.38	40.87	15.92
Mandya	28.06	21.59	20.80	-7.11	6.40	-2.55	19.49	17.88	29.89
Mysore	50.26	19.91	33.30	17.64	7.02	7.96	22.98	16.63	23.07
Raichur	37.89	11.37	7.30	16.37	6.12	4.07	31.20	27.58	11.76
Shimoga	33.56	20.05	14.88	9.95	12.07	0.01	20.85	18.20	18.13
Tumkur	50.34	48.57	21.10	15.51	25.72	9.25	28.91	23.02	15.55
Uttara Kannada	23.93	15.30	14.87	9.50	11.84	4.02	17.64	10.86	8.29
Total	32.25	16.80	12.97	7.46	6.64	-0.84	22.07	18.78	12.60

Table 9.4: Percentage Increase in Industrial Units in Karnataka (Districtwise) (Cont'd)

Year --- >	% Variation 85-86 to 86-87			% Variation 86-87 to 87-88			% Variation 87-88 to 88-89		
District	Units	Investment	Persons	Units	Investment	Persons	Units	Investment	Persons
Bangalore	12.66	12.27	5.54	9.82	10.96	5.82	8.73	13.01	6.26
Belgaum	22.79	27.73	15.90	17.31	21.12	14.18	10.97	17.62	9.92
Bellary	19.10	10.58	18.68	16.00	13.07	12.43	12.26	12.61	10.37
Bidar	20.00	16.21	15.09	16.29	33.84	12.38	13.20	24.15	7.73
Bijapur	16.63	17.52	13.17	15.29	12.75	8.11	12.66	18.46	8.55
Chickmagalur	19.66	16.83	27.14	14.96	13.32	16.30	11.60	17.45	8.20
Chitradurga	16.75	26.84	18.46	14.07	25.30	11.40	11.66	19.29	9.93
Dakshina Kannada	28.49	19.00	15.24	12.93	26.40	11.78	10.59	19.92	8.90
Dharwad	15.70	24.55	8.33	13.67	8.43	5.38	12.96	7.97	5.16
Gulbarga	24.15	36.61	25.38	19.41	30.00	18.89	17.31	29.34	14.85
Hassan	17.65	16.82	9.18	15.36	16.00	10.19	12.28	8.76	8.47
Kodagu	12.95	12.40	8.75	12.61	4.38	2.85	10.39	3.93	2.76
Kolar	14.88	19.99	13.33	10.79	17.98	10.35	10.49	15.89	9.94
Mandya	16.44	16.21	14.10	11.62	14.48	9.35	9.96	12.49	7.63
Mysore	15.41	14.75	11.96	15.66	15.58	10.75	9.58	11.07	8.03
Raichur	24.22	12.67	7.97	20.53	13.17	7.60	15.32	9.58	6.04
Shimoga	15.18	11.11	12.09	13.10	12.06	8.80	11.45	11.06	8.21
Tumkur	18.02	18.60	10.63	18.49	26.51	12.97	18.24	15.40	13.32
Uttara Kannada	16.56	10.37	7.25	13.54	11.26	7.45	10.50	12.30	5.49
Total	17.37	16.86	10.47	13.94	15.62	8.75	11.40	14.00	7.73

Table 9.4: Percentage Increase in Industrial Units in Karnataka (Districtwise) (Cont'd)

Year	% Variation 88-89 to 89-90			% Variation 89-90 to 90-91			% Variation 90-91 to 91-92		
District	Units	Investment	Persons	Units	Investment	Persons	Units	Investment	Persons
Bangalore	9.17	12.29	7.60	8.49	14.19	8.21	13.94	21.32	10.85
Belgaum	10.02	16.63	9.13	9.62	13.20	8.12	8.79	14.69	7.71
Bellary	10.80	6.47	8.20	9.28	6.54	6.55	6.29	16.15	6.19
Bidar	11.66	60.29	11.71	10.53	14.19	13.69	8.99	13.35	7.80
Bijapur	11.06	18.18	7.51	10.88	16.60	7.84	8.27	23.61	7.02
Chickmagalur	10.17	17.26	7.64	10.43	15.43	6.62	8.26	21.46	6.10
Chitradurga	10.36	18.32	9.92	9.71	15.12	8.45	9.73	34.97	8.22
Dakshina Kannada	9.00	12.70	6.85	7.39	14.99	5.60	6.62	15.44	6.33
Dharwad	11.55	6.18	4.65	10.63	5.74	3.97	8.53	8.63	4.63
Gulbarga	12.22	14.70	9.97	12.37	15.90	10.42	6.28	14.39	7.23
Hassan	11.26	7.92	7.35	9.25	11.30	7.53	5.96	8.92	5.78
Kodagu	9.69	2.86	2.80	13.04	4.14	3.45	7.44	7.99	2.65
Kolar	9.63	13.86	9.58	8.66	16.48	7.43	7.54	23.10	7.63
Mandya	9.69	24.25	8.72	10.24	10.25	8.32	6.78	16.54	6.47
Mysore	7.67	7.36	6.56	7.94	9.01	7.15	7.21	11.73	7.34
Raichur	12.90	15.57	5.87	11.71	5.92	4.55	9.36	12.90	5.11
Shimoga	10.11	5.92	6.80	9.76	10.43	8.66	8.11	10.89	7.23
Tumkur	12.60	15.58	8.76	9.21	11.59	6.76	9.50	18.60	8.10
Uttara Kannada	10.63	16.82	7.03	9.61	15.29	6.52	6.29	17.45	5.07
Total	10.12	13.11	7.33	9.36	12.21	7.10	9.17	17.11	7.66

Table 9.5.1: Districtwise Data for the years 1982-83 and 1991-92

Year --->	1982-83			1991-92			% (Variation) growth			Percent average annual growth		
District	Units Cum	Investment Cum (lakhs)	Persons Cum	Units Cum	Investment Cum(Lakhs)	Persons Cum	Units	Investment	Persons	Units	Investment	Persons
Bangalore	11368	14381.78	156691	27691	37102.85	242868	143.59	157.99	55.00	9.31	9.94	4.48
Belgaum	2010	1856.93	16082	11080	9172.65	51078	451.24	393.97	217.61	18.61	17.32	12.25
Bellary	1530	788.16	8133	5257	2589.05	25522	243.59	228.49	213.81	13.14	12.63	12.12
Bidar	625	400.79	5790	2826	3674.88	17864	352.16	816.91	208.53	16.29	24.81	11.93
Bijapur	1282	604.15	12979	4688	2814.06	28967	265.68	365.79	123.18	13.84	16.63	8.36
Chickmagalur	1638	979.72	11417	4860	5516.86	27451	196.70	463.11	140.44	11.49	18.87	9.17
Chitradurga	598	380.12	4503	2293	1581.79	11555	283.44	316.13	156.61	14.39	15.32	9.88
Dakshina Kannada	2641	2027.17	23989	8955	13458.51	59414	239.08	563.91	147.67	12.99	20.84	9.49
Dharwad	2444	4094.75	64714	10506	10134.64	106242	329.87	147.50	64.17	15.70	9.49	5.08
Gulbarga	850	642.62	7830	4179	3064.89	23470	391.65	376.94	199.74	17.26	16.91	11.60
Hassan	824	550.9	6133	2791	1997.74	14778	238.71	262.63	140.96	12.98	13.75	9.19
Kodagu	460	959.61	10936	1444	1441.89	14649	213.91	50.26	33.95	12.12	4.16	2.97
Kolar	1816	1124.05	11672	4709	4740.34	31262	159.31	321.72	167.84	10.00	15.48	10.35
Mandya	1087	624.54	6275	2851	2275.39	16154	162.28	264.33	157.43	10.12	13.80	9.92
Mysore	2871	2619.16	23153	11374	7550.83	67326	296.17	188.29	190.79	14.76	11.17	11.26
Raichur	855	1767.17	19670	4287	5144.87	35191	401.40	191.14	78.91	17.49	11.28	5.99
Shimoga	1904	2022.83	13779	6410	5769.07	30711	236.66	185.20	122.88	12.91	11.05	8.34
Tumkur	1492	1279.13	17737	7437	7784.95	48202	398.46	508.61	171.76	17.43	19.79	10.51
Uttara Kannada	798	537.23	9214	2401	1675.25	17361	200.88	211.83	88.42	11.64	12.04	6.54
Total	37093	37640.81	430697	126039	127490.51	870065	239.79	238.70	102.01	13.01	12.97	7.28
Average	1952.26	1981.10	22668.26	6633.63	6710.03	45792.89	273.94	316.57	141.04	13.81	14.49	8.92
Maximum	11368.00	14381.78	156691	27691.00	37102.85	242868	451.24	816.91	217.61	18.61	24.81	12.25
Minimum	460.00	380.12	4503.00	1444.00	1441.89	11555.00	143.59	50.26	33.95	9.31	4.16	2.97
Standard Deviation	2325.76	3061.92	34132.97	5794.99	7858.93	51721.10	88.62	174.78	54.25	2.68	4.65	2.66

Table 9.5.2: Districtwise Data for the Year 1991-92

Year --->	1991-92				1991-92									
District	Units Cum	Investment Cum (lakhs)	Persons Cum	% SSI	% Investment	% Persons	Investment per job	% of population employed in SSI	Area sq.kms	Population	Workers in Maf. & HHI	No of pers/sq.km	% SSI/ total mp	% SSI/ pop. total
Bangalore	27691	37102.85	242868	21.97	29.10	27.91	15276.96	3.73	8213.21	6512356	513755	792.91	47.27	3.73
Belgaum	11080	9172.65	51078	8.79	7.19	5.87	17958.12	1.43	13530.92	3583606	128966	264.85	39.61	1.43
Bellary	5257	2589.05	25522	4.17	2.03	2.93	10144.39	1.35	9884.87	1890092	44872	191.21	56.88	1.35
Bidar	2826	3674.88	17864	2.24	2.88	2.05	20571.43	1.42	5494.8	1255799	20551	228.54	86.93	1.42
Bijapur	4688	2814.06	28967	3.72	2.21	3.33	9714.71	0.99	17165.1	2927990	86950	170.58	33.31	0.99
Chickmagalur	4860	5516.86	27451	3.86	4.33	3.16	20097.12	2.70	7239.69	1017283	29283	140.51	93.74	2.70
Chitradurga	2293	1581.79	11555	1.82	1.24	1.33	13689.23	0.53	10836.08	2180443	69328	201.22	16.67	0.53
Dakshina Kannada	8955	13458.51	59414	7.10	10.56	6.83	22652.09	2.21	8509.03	2694264	325131	316.64	18.27	2.21
Dharwad	10506	10134.64	106242	8.34	7.95	12.21	9539.20	3.03	13778.61	3503150	120270	254.25	88.34	3.03
Gulbarga	4179	3064.89	23470	3.32	2.40	2.70	13058.76	0.91	16215.43	2582169	50233	159.24	46.72	0.91
Hassan	2791	1997.74	14778	2.21	1.57	1.70	13518.34	0.94	6895.22	1569684	18009	227.65	82.06	0.94
Kodagu	4709	4740.34	31262	3.74	3.72	3.59	15163.27	6.40	4108.6	488455	32705	118.89	95.59	6.40
Kolar	1444	1441.89	14649	1.15	1.13	1.68	9842.92	0.66	8268.49	2216889	59567	268.11	24.59	0.66
Mandya	2851	2275.39	16154	2.26	1.78	1.86	14085.61	0.98	5005.48	1644374	28508	328.51	56.66	0.98
Mysore	11374	7550.83	67326	9.02	5.92	7.74	11215.33	2.13	12107.05	3165018	107787	261.42	62.46	2.13
Raichur	4287	5144.87	35191	3.40	4.04	4.04	14619.85	1.52	14049.02	2309887	36249	164.42	97.08	1.52
Shimoga	6410	5769.07	30711	5.09	4.53	3.53	18785.03	1.61	10579.46	1909663	49199	180.51	62.42	1.61
Tumkur	7437	7784.95	48202	5.90	6.11	5.54	16150.68	2.09	10633.26	2305819	62632	216.85	76.96	2.09
Uttara Kannada	2101	1675.75	17361	1.90	1.31	2.00	9649.50	1.42	10335.07	1220260	33108	118.07	52.44	1.42
Total	126039	127490.51	870065	100.00	100.00	100.00	14652.99	1.93	192849.39	44977201	1817103	233.22	47.88	1.93
Average	27691	37102.85	242868	21.97	29.10	27.91	22652.09	6.40	17165.10	6512356	513755	792.91	97.08	6.40
Maximum	1444	1441.89	11555	1.15	1.13	1.33	9539.20	0.53	4108.60	488455	18009	118.07	16.67	0.53
Minimum	6633.63	6710.03	45792.89	5.26	5.26	5.26	14512.24	1.90	10149.97	2367221.11	95637.00	242.34	59.90	1.90
Standard Deviation	5794.99	7858.93	51721.10	4.60	6.16	5.94	3941.02	1.34	3585.84	1269462.66	119489.90	142.55	25.66	1.34

Table 9.6.1: Dispersal of Units

	Rural		Urban		Metropolitan		
District	No of Units	Units	%	Units	Percent	Units	Percent
Bangalore (Urban)	6051	305	5.04	939	15.52	4807	79.44
Belgaum	2755	1002	36.37	1753	63.63	0	0.00
Bellary	2027	625	30.83	1402	69.17	0	0.00
Bijapur	1712	875	51.11	837	48.89	0	0.00
Bidar	972	459	47.22	513	52.78	0	0.00
Chikmagalur	1002	448	44.71	554	55.29	0	0.00
Chitradurga	2139	685	32.02	1454	67.98	0	0.00
Kodagu	305	187	61.31	118	38.69	0	0.00
Dharwad	4814	2401	49.88	2413	50.12	0	0.00
Gulbarga	1419	500	35.24	919	64.76	0	0.00
Hassan	1164	394	33.85	770	66.15	0	0.00
Uttara Kannada	795	513	64.53	282	35.47	0	0.00
Dakshina Kannada	3867	2066	53.43	1801	46.57	0	0.00
Mandya	955	550	57.59	405	42.41	0	0.00
Kolar	1315	424	32.24	891	67.76	0	0.00
Mysore	2211	709	32.07	1502	67.93	0	0.00
Raichur	1594	745	46.74	849	53.26	0	0.00
Shimoga	2928	1425	48.67	1503	51.33	0	0.00
Tumkur	1803	833	46.20	970	53.80	0	0.00
Bangalore(Rural)	697	321	46.05	376	53.95	0	0.00
Total	40525	15467	38.17	20251	49.97	4807	11.86

Table 9.6.2: Dispersal of Units by Nature of Activities

	Rural		Urban		Metropolitan		Total	
	units	%	Units	%	Units	%	Units	%
Manufacturing/Assembling	5368	31.47	8867	51.98	2824	16.55	17059	42.10
Processing	5880	54.85	4317	40.27	523	4.88	10720	26.45
Job work only	789	21.98	1925	53.64	875	24.38	3589	8.86
Repairing & servicing	1714	33.57	3189	62.46	203	3.98	5106	12.60
Combined	1716	42.36	1953	48.21	382	9.43	4051	10.00
Total	15467	38.17	20251	49.97	4807	11.86	40525	100.00
Energy consumption by location (Lakh Rs.)								
Electricity,firewood,charcoal, etc.	1650	34.64	2253	47.30	860	18.06	4763	

Table 9.7: Sectorwise Details of Number of Units, Manpower and Production

NIC Code	Description	Units	%	Manpower	%	Production Lakh Rs.	%
20 & 21	Food products	10221	25.22	47363	19.41	71615	28.30
22	Beverages, tobacco	550	1.36	6812	2.79	6086	2.41
23	Cotton textiles	52	0.13	659	0.27	641	0.25
24	Wool, silk & synthetic textiles	28	0.07	197	0.08	101	0.04
25	Jute hemp & mesta tex	9	0.02	47	0.02	28	0.01
26	Hosiery & garments	1750	4.32	10203	4.18	5824	2.30
27	Wood products	3779	9.33	17354	7.11	12826	5.07
28	Paper products	2623	6.47	14541	5.96	28972	11.45
29	Leather products	1381	3.41	4926	2.02	3800	1.50
30	Rubber & plastic	1570	3.87	11576	4.74	13829	5.47
31	Chemicals	1587	3.92	17086	7.00	18598	7.35
32	Non-metallic mineral	1422	3.51	20280	8.31	8103	3.20
33	Basic metal products	546	1.35	8961	3.67	14326	5.66
34	Metal products	4173	10.30	24920	10.21	31349	12.39
35	Machinery parts	2915	7.19	21114	8.65	12348	4.88
36	Electric Machinery	699	1.72	9408	3.86	13278	5.25
37	Transport equipment	808	1.99	5988	2.45	4285	1.69
38	Miscellaneous manufacture	394	0.97	3222	1.32	2048	0.81
97	Repair services	5041	12.44	15392	6.31	2833	1.12
99	Services not classified elsewhere	87	0.21	375	0.15	286	0.11
OT	Other services	890	2.20	3615	1.48	1838	0.73
	Total	40525	100.00	244039	100.00	253014	100.00

9.1.4 Dispersal of small-scale units

The second census of registered small-scale units (Development Commissioner, 1990) conducted by the Directorate of Industries and Commerce of 40,525 industries reveals that 38.14% of total industries are located in rural areas, while 49.97% are in urban areas, and 11.86% units on the whole in the metropolitan city of Bangalore as indicated in Table 9.6.1. A look at the employment generated in these industries reveals that 40,525 units have provided employment to 2,44,039 persons. The food products sector (National Industry Code: NIC 20 & 21) leads with 19.4% of total employment. This is followed by machinery and parts units (8.65%) and non-metallic mineral products 8.33%. Table 9.6.2 gives information regarding dispersal of units according to nature of activities (major activities such as repairing, servicing, etc.).

The total production of 40,525 units surveyed was to the tune of Rs. 25,268.7 million. Out of this, 28.32% is by food products, metal products 12.39% and paper products 11.46%. This is illustrated in Table. 9.7. Food products sector constitutes major component in SSI sector and most of these units are located in rural areas. As rural industries play a major role in the rural energy scene, further study is carried out on industries located in rural areas.

9.2.0 Case Study: Energy Studies in Cashew Processing Industries

Cashew processing industries constitute a major component of Uttara Kannada's industrial sector. Therefore, these industries are taken up for further studies to explore their role in rural economy and the role of traditional fuels in their survival. Cashew processing industries, located mainly in rural areas, are labour as well as energy intensive, depending mainly on fuel wood and other agriculture residues for their heat energy requirements. They employ mainly women and hence have helped the village women to attain economic independence and self-sufficiency. These industries are located mainly in coastal belts of Karnataka due to availability of raw material, namely, cashew-nuts in the region. Coastal areas are densely populated and relatively biomass scarce. The non-availability of fuel wood and agricultural residues has often caused serious problems.

9.3.0 Objectives

The objectives were to:

(i) explore the relationship between energy consumed and production in rural industries based on the data collected from them,
(ii) identify the pattern, type and efficiency of energy use in rural industries, particularly, in their more energy intensive processes like heating,
(iii) study ecological perspectives of rural industries, and
(iv) identify technical measures for improving energy efficiency.

9.4.0 Methodology

This study was carried out in five stages:

(i) Initial information regarding cashew industries in Karnataka was collected from government agencies like the Department of Industries, Directorate of Economics and Statistics, District Industries Centre, etc.

(ii) A questionnaire based survey, covering firms/units in the cashew processing industrial sector was conducted in Kumta taluk of Uttara Kannada district to determine levels and patterns of energy consumption and the technologies in use.

(iii) Brief (one to two days) energy audits, and management and employees interviews were conducted in representative firms to check or cross verify the data collected from the secondary sources and identify opportunities for energy efficiency improvements.

(iv) Computation of Specific Energy Consumption (SEC) and Energy Intensity (EI) to find out the level of disparity among firms.

(v) Establishment of quantitative relationship among various parameters, such as, Energy (En), Production (Pr), Specific Energy Consumption (SEC) and Percentage (installed capacity) Production Capacity Utilisation (PPCU).

9.5.0 LITERATURE REVIEW

Henderson, P.D. (1975) gives an overall picture of energy scene in India, with a good deal of background and supporting information - technical, statistical and historical. In the fourth chapter, the author gives a review of the situation and development in three main energy industries - coal mining, petroleum and electric power.

Jin-ping Huang (1993) examines the effect of structural change and improvements in energy intensities of individual industries in aggregate industrial energy intensity changes in China during 1980-88. To decompose the energy intensity effect and structural effect, the Divisia Index was used. A major finding of this analysis is that the main contribution to changed aggregate energy intensity was from the changes in energy intensities of individual industries during the period. Structure of production (structural effect) was found to contribute little to the aggregate energy changes.

Parikh, J.K., et al. (1991) quantify per capita, aggregate consumption levels, environmental consequences of population growth and urbanisation in selected developed and developing countries. They also examine the resource needs for overall economic development, the consequences of continual poverty and the trade-offs and priorities among local, regional and global environmental problems. Some policy options to promote sustainable development in an equitable way at both the national level of developing countries and the global level are given in this report.

Schurr, S.H. and Eliasberg, V.F. (1962) investigated the long-term relationship between energy consumption and real G.N.P. (Gross National Product) in the United States. They conclude that from about 1880 to 1915 there was a steady rise in energy requirements per unit of G.N.P. From 1915 through 1955, the use of energy per unit of G.N.P. declined steadily. The total period of 75 years witnessed a 55 per cent rise in the amount of energy used per unit of G.N.P. Authors attribute this to (i) increased efficiency of energy production, resulting in dramatic increase in the useful energy output obtained from the raw energy input, (ii) generation of electricity has experienced especially significant technology improvements that have resulted in greater efficiency, and (iii) changes in the structural composition of G.N.P., which partially explain the decline in relative energy output.

Schurr, S.H. (1985) discusses rapid multi factor productivity growth and strong improvements in energy efficiency which have occurred simultaneously over a long period of time in the USA. After 1973, there is substantial gains in energy efficiency while productivity growth languished. Schurr argues that the key to achieve both energy conservation and multi factor productivity growth is to use energy in conjunction with technologies that achieve more efficient means of combining all production inputs, instead of emphasising only on technologies in which energy conservation alone is the overriding objective.

Peter E. de Janosi and Leslie E Grayson (1972) discuss the role of energy utilisation and economic growth in a quantitative way where both cross-section and time-series approaches are utilised. They conclude that energy consumption tends to be more responsive to economic growth in less developed countries than in advanced countries. However, they also point out that this relationship between energy consumption and economic growth activity is affected by various other factors. Multivariate tests suggest that the industrial structure of the economy and the composition of energy consumption are especially significant additional variables.

Jonathan E. Sinton and Mark D. Levine (1994) analysed three different sets of energy consumption and output value data using a Laspeyres Index method to determine the relative role of structural shift and real intensity change in China's industrial sector between 1980 and 1990. The authors found that real intensity change (economic intensity of industrial subsectors, composed of physical intensity change and other non-structural factors) accounted for most of the large apparent drops in industrial energy intensity in the 1980's. Comparison of economic and physical energy intensity indicators suggests that physical intensity is not the only major cause of real intensity change.

Ang B.W. (1991) examines energy coefficient from statistical viewpoint - the basic properties, its relationship with energy-output ratio, and the effects of measuring it on a per capita basis On a yearly basis the coefficient tends to be unstable. Several methods of deriving the trend, which provides an insight into the underlying energy-output relationship, are presented. The effectiveness of these methods is examined by applying them to the time series data covering 1960-86.

Ulf Hansen (1990) discusses energy conservation as a three-pronged policy. The pattern of industrial development has been moving away from basic materials and heavy industry towards product refinement and service industry. The trend is accompanied by a reduction in specific energy input. He identifies improved energy efficiency as the major contributor to delink energy consumption from economic growth in Germany. The largest increase in energy efficiency is found in industrial sectors. The time element and the pace of technological developments are major determinants in energy conservation. Energy policies have emphasised market forces flanked by substantial investment in R and D for new energy technologies.

Ramachandra, T.V. and Subramanian, D.K. (1993) based on primary survey of grain mills spread over Karnataka state carried out detailed energy analyses. This study shows that there is scope for saving energy of the order of 23.12% in the food grain sector and 18.67% in paddy dehusking sector by structural adjustments. Energy saved in this sector is around 55 million units per year.

Mark Ross (1992) explains the potential for energy efficiency in manufacturing sectors. With new technology and increasing affluence, the composition of production is changing, such that, energy and material consumption is growing more slowly than economy. He elaborates how through new technologies and appropriate public policies, the energy intensities of all production processes also continue to decline.

Siddayao, C.M., et al. (1987) present the results of a study of the demand elasticities for energy and non-energy inputs in food processing and textile industries of Bangladesh, Philippines and Thailand. Estimates of substitution elasticities in production were derived by fitting data to a translog cost function for the period 1970-80 for Bangladesh, 1970-78 for Philippines, and 1974-77 for Thailand. Results are compared with other developing and industrialised countries, and substitutiveness among inputs was found to be greater in the manufacturing sector of the developing countries than those of industrialised countries. The generally high elasticities between labour and energy have implications for relative pricing and use in developing countries.

Sergio Alvarado and Jose Iribarne (1990) carried out energy analyses in the production of copper, wood pulp and steel, which account for 56% of total manufacturing sectors in Chile. Specific energy requirement was found to be 25.6, 23.7 and 24.2 MJ/kg respectively. This study shows that there exists a large potential for improvement in energy efficiency in all these three cases.

Wen S. Chern and William E. James (1985) examine the historical relationship between energy demand and economic growth, to analyse the factors affecting energy demand, evaluate the improvement in energy productivity, present scenarios of future energy demand and discuss implications for energy conservation. This study also focusses on eight industries of the manufacturing sector in Philippines. Analyses of the fuel and total energy productivities reveal that there is scope for efficiency improvement in a food processing sector. The fuel weighted productivity calculated provides better measure of energy efficiency.

Rabadi, N.J., et al. (1991) quantify and assess the potential of electricity load management within the industrial sector in Jordan. They consider different load management options that would result in reducing the maximum demand on electricity by the industrial sector. The potential for reducing the peak load of the Jordan electric power system by load management within the industrial sector (medium industrial plants) is estimated to be 22.2 MW.

Douglas Gardner (1993) discusses both the magnitude and pattern of energy use in the Ontario industrial sector during 1962-84 period. Divisia Index approach is used to disaggregate changes in the aggregate energy intensity of the industrial sector into structural and intensity components. Structural components measure changes due to shifts in the composition of industrial sector output. The intensity component measure changes due to energy intensity shifts of individual industries. Changes were noticed before pre oil crisis (1962-73) and post oil crisis (1973-84), as well as, differences in the trends of fossil fuel and electricity use. Structural changes were found to be of little importance in the 1962-73 period, but of substantial importance in the 1973-84 period, in reducing aggregate energy intensity in the industrial sector. While changes in the intensity of individual industries were found to be significant in 1962-73 period, the significance waned in the later period.

Fakhre Mahmud and Salim Chishti (1990) examine the extent of interfuel substitution and substitution between energy and non-energy inputs in the large-scale manufacturing sector of Pakistan by estimating input demand for various energy components and hence an aggregate Divisia Index. Later, this index was used as an instrument to estimate aggregate input demand for capital, labour and energy along with their price and substitution elasticities. The results show that energy and labour are substitutes, while energy and capital are complements.

Wolde-Ghiorgies, W. (1991) discusses energy utilisation patterns in three factories involved in cement production, textile manufacturing and food processing in Ethiopia. Based on specific energy consumption analyses, he showed the areas of apparent energy losses and corresponding conservation possibilities.

Susan McFaddin and Maclyn Clouse (1993) developed a new growth model which enables the evaluation of the interdependence of financial objectives, operational performance and sustainable growth in the energy industry. This model includes both the effect of inflation and growth produced from equity issues. This model is statistically compared with both actual growth rates and Higgin's model.

Reitler, W., et al. (1987) analysed the trends in electricity and fuel consumption in industries and changes in these factors due to changes in production quantity, production structure and specific consumption.

Anandalingam, G. (1985) examines the most energy intensive industries and methods by which fuel efficiency could be improved. It is shown that investing in energy efficiency is more economical Btu per Btu than investing in the enhancement of domestic energy resources. He also assessed the changes in the economics of conservation of private firms when there are government incentives and government policies that can overcome economic and non-economic disincentives for investing in energy conservation.

Parikh, J.K. (1990) through statistical analyses based on data from the Annual Survey of Industries attempts to answer how the manufacturing industries moved to conserve energy, and how significant the energy costs are with respect to value of the output, other material costs, invested capital or value added. At the all-industries level, there is 14% reduction over 1973 values in energy costs per output, both deflated by their respective prices (1970=100). Energy intensive industries, such as, basic metals, alloys and textiles seem to have changed due to managerial options (soft options) judging from price elasticities, but output elasticity close to unity. While, long-term changes have been made by judging from the overall elasticities of 0.73 for all industries.

Ramachandra, T.V. and Subramanian D.K. (1991) carried out detailed energy analyses of small-scale industries in Karnataka. They show wide disparity in Specific Energy Consumption within and across various sectors. Out of a sample of 153 industries, about 60 were found to have Percentage Production Capacity Utilisation less than 50%. Substantial saving in energy is possible by better utilisation of installed capacity.

Mark D. Levine, et al. (1991) explore efficiency improvement as an energy strategy for developing countries. They describe evolving energy patterns in these countries and outline the role that energy efficiency can play in the near and mid range future. They also summarise the main impacts of these patterns and the issues that is to be confronted in improving energy efficiency in many countries.

Hafiz A. Pasha, et al. (1989) quantify the economic cost of power outages in the industrial sector to allow for long run adjustments by firms to outages like investment in generators and renegotiation of labour contracts with flexible timing provisions, and by incorporating multiplier effect on other sectors. Variation in anticipated and unanticipated outage costs per kWh by type of industry, firm size and location are highlighted by analysing data collected from a sample of 843 industrial units. They show that, while outage costs in the industrial sector are not as high as claimed by some of the industrial associations in the country, they are still large enough (about 1.8% of GDP) to warrant the expanded programme of investment in energy generation of the medium run and the pursuit of a loss minimising load management strategy and pricing policy in the short run.

Richard B. Howarath, et al. (1991) examine the trends in manufacturing energy use in eight OECD countries, decomposing the changes that occurred between 1973 and 1987 into the effects of changes in aggregate manufacturing activity, industry structure and energy intensities measured at the industry group level. While manufacturing, production grew in every country except the UK, the growth rate varied from nation to nation. Structural changes led to modest reductions in energy use in most countries, although in Norway it led to substantial growth in energy consumption. The reduction in energy intensities is uniform across all nations, ranging from 20% (Norway) to 36% (Japan) over the period of analysis

Nand, S. (1993) presents energy consumption scenarios in Indian fertiliser industry. He outlines the potential and approach for energy saving in various plants while highlighting some of the measures implemented in Ammonia and Urea plants.

Ang B.W. (1987) analyses the nature of the correlation between energy consumption and output based on cross sectional data for 100 countries. The bias in the correlation is due to the use of nominal income and the convention adopted in measuring noncommercial energy consumption. It is found that, when real income is used and noncommercial energy consumption is measured in commercial energy replacement value, the cross sectional energy elasticity increases rapidly from about unity for the low income developing countries to between 1.6 and 1.8 for the high income developing countries, and declines marginally thereafter in the industrial country income range. The results show a certain amount of disparity in energy-output ratio across countries, with higher income countries consuming disproportionately more energy per unit of real output.

Ramachandra T.V. and Subramanian D.K. (1997) showed that about 27.72% energy could be saved in industrial sector, which means about 1541 million units of electricity could be saved (as per electricity statistics, industrial sector in Karnataka has consumed about 5560 million units during 1991-92). This saved electricity is equivalent to electricity output of 300 MW installed capacity of generating stations. Environmental problems associated with mega projects, the increasing demand of states/country and resource depletion make it increasingly imperative that the planners divert their attention towards improving efficiency in all sectors. This illustrates that, there is ample scope to conserve energy in the industrial sector.

Ramachandra T.V. and Subramanian D.K. (1992) study of energy efficiencies of end use devices in an electro metallurgical industry has shown that the efficiency of welding sets is about 14%, furnace 10.4%, diesel generators 36.5% and electro plating process 36.5%. This shows possible saving in industrial sector by switching over from conventional motor generator welding sets to transformer-rectifier welding sets with micro processor based numeric controller. Motors and drives consume a large percentage of the electrical energy used in the industrial sector. High efficiency motors (85.5 - 95% efficiency), currently available in the market, are highly competitive compared to conventional motors, considering the cost of motor losses.

Carlsmith, R.S., et al. (1990) give information on advanced processes in the steel industry. These are mostly major process changes that could revolutionise the iron and steel sector. The plassmelt method involves smelting of partially reduced iron powder with pulverised coal by using heat supplied by a plasma system. Ore to powder steel making could reduce the energy consumed by 40 percent. Direct steel making could double or triple production rates compared to the blast furnace, and offer a 30% reduction in energy savings. The energy required to produce steel from scrap is less than one-half that required to produce steel from raw materials. However, scrap contains residual elements that have adverse effects on the properties of the steel. The electric arc furnace is a well established technology and because of its increasing market share, improvements such as scrap pre heating, DC arc furnace, induction melting, heat and dust recovery and ladle refining are to be researched.

Fulkerson, W., et al. (1989) research on catalysts, highlights their use in many industries to produce chemical reactions at a lower pressure and temperature, thereby using less efficiency. Better understanding of the basic mechanisms of catalysts may lead to new classes of catalysts. These could be beneficial in the areas of one-step conversion of methane to methanol, photocatalytic reduction of water, combustion enhancement and pollution control.

Williams, R.H. (1978) study gives information on industrial cogeneration. Cogeneration is the simultaneous production of process heat and electric power. Providing moderate or low temperature heat as a by-product of the work from a heat engine is much more efficient than providing heat directly by burning fuel. Most typical cogeneration industry converts only 10 to 15% of the energy into electricity.

Ross, M. (1989) study highlights the need for improving efficiency of electricity use in manufacturing. Intercooled Steam Injected Gas Turbine - a new technology is being developed, incorporating a modern aircraft engine which accommodates variable amounts of steam returned to the turbine combustor, and therefore has a flexible electricity-heat ratio. Steam not returned to the turbine is used as process heat. With a full steam injection, 40% of the energy can be converted into electricity.

Dale, S.J. (1991) assessed energy applications of high temperature superconductivity showed that, variable speed controls for motors are currently available for application on existing and new equipment, to adjust the speed control so that the motor and driven equipment can match the requirement of the process. (Motors account for about 55% of the electric energy consumed in Karnataka state. The potential for conserving energy by applying high temperature superconductors in place of conventional conductors in industrial motors is very large. The advantages include reduction of volume and mass, higher power density, enhanced performance and improved operating efficiency).

Department of Energy (1987) reports that industrial separation processes involving separations of the components in a mixture are highly energy intensive. Advancements of alternate processes, such as, membrane separation, solvent extraction, critical fluid extraction and advanced drying concepts are less energy intensive. This could be beneficial to applications like black liquor concentration in the paper and pulp industry, hot food processing, waste water concentration, dilute soluble food process stream concentration and drying of products such as textiles and paper. Conventional chemical pulping in the paper industry is dominated by a very energy intensive kraft process. The energy required to recycle paper is about one-half that required by the kraft process. Desired improvements in the recycle process concentrate on improving the process to remove color and filler. Improvements in the paper making process focus on improved process control, process physics, and improved materials. These improvements would have a substantial effect on decreasing energy consumption. Bio pulping, chemical pulping with fermentation, and ethanol organosol pulping are the most recent promising advanced processes involving the integration of at least one fermentation process with a conventional pulping process.

9.6.0 Study Area

The primary survey was carried out in industries located in Kumta taluk, in the coastal tract of Uttara Kannada district.

9.7.0 Cashew Industries

Cashew processing industries are selected for this study for the following reasons.

(i) The processing industry, located mainly in the rural areas, depends on local raw materials, tends to rely on local resources for heat energy needed and employs local youth. Eighty five percent of the labourers are women, who also endure the burden of domestic drudgery.

(ii) Fuel wood is used as the main source of heat energy in processing. The quantity of fuel wood required and its impact on environment is enormous. The rising cost and scarcity of fuel wood have forced industrial entrepreneurs to think in terms of better processing devices like boilers, improved driers, as well as, growing fuel wood locally through energy plantation with indigenous species.

(iii) The raw material required for the industry could be grown on barren/already degraded land, unfit for habitation or tillage. The crop does not require any fertiliser or pesticide application or even nursing.

(iv) Cashew plantation helps in restoring the soil condition.

(v) The industry is labour intensive (and hence generates rural employment). The raw materials are processed by hand, which helps in yielding better quality nuts.

The main function of the cashew processing industry is the recovery of kernel from raw nuts by manual or mechanical means. Processing consists of moisture conditioning, roasting, shelling, drying, peeling, grading and packing. The present study covers 142 sample industries from all districts in Karnataka. Detailed investigations of processing were carried out for ten industries in Kumta taluk of Uttara Kannada district.

Cashew processing industry's production flow diagram

```
                Purchase of raw nuts
      (during season) from the surrounding area
                         |
                         !
          Sun drying of raw nut for two days
                         !
      Storage in gunny bags in the factory shed
                         !
                    Conditioning
                         !
              Roasting in steam boilers
                         !
     Shelling: by hand and leg operated machine
                         |
                       !            CNSL
               Shells → expeller ---l
                       |      |
                       |      |------Residue
                       !               !
              Brooma drying -------- Fuel
                         !
                    Hand peeling
                         |
                     ! → testa (for tannery)
                      Grading
                     |of kernels as W 320 (320 nuts/450 gms)
                     |W 240 (240 nuts/450 grams)
                     |W 210
                     |W 180
                     |TH1/2 nut
                     |KH1/JH
                     |LWP small size
                     |SWP very small size
                     |SSP broken
                         !
                 Rehumidification
                         |
                     | Packing in tins: net weight of
                     |   kernels 11 kg 340 gm
                     |   (including tin : 12 kg 340 gm)
                     !
                 Ready for dispatch
```

Some of the production processes are explained as follows.

Conditioning: Conditioning involves sprinkling of water on dried and stored nuts to bring them to an optimum moisture level of 20% - 30%. Roasting is done to make the shell brittle. The mesocarp of the shell consists of honeycomb like cells containing viscous liquid called the cashew-nut shell liquid (CNSL), which provides natural protection to the kernel against insects.

Roasting: Drum roasting is done in most of the industries. The capacity of the drum is 320 kgs. Shells are roasted in steam of 100°C for about 22 minutes. In most of the industries, during the course of the survey, workers were instructed to load nuts to the full installed capacity of the drum. The nuts are kept aside for a day for cooling. Then shelling is done using hand and leg operated shelling machines. For this purpose, female labourers are employed. Normally, a worker shells 17 kgs/day. The wages are paid on pro rata basis, at the rate of Rs. 1.50/kg plus Rs. 1.50 towards transport allowance.

One sample industry surveyed had 12 female workers in this section, most of them in the age group of 15 to 25. The kernel is scooped out by means of a sharp needle. After shelling, the kernels are dried in the drier to reduce the moisture (approximately 10%) and loosen the adhering testa. After drying for 8 - 12 hours, the peeling is done by hand using a knife. Twenty five female workers were employed in this section. Minimum of 8 kg of kernels is given in the morning for peeling to each person, which has to be completed and the same quantity returned, when the worker leaves the work spot. Workers were paid on pro rata basis, at Rs. 1.5 to 2 per kg. Next stage is grading of kernels depending on the specification for exportable grades. The whole nuts are size graded, based on the number of kernels per 450 grams, as W320, W240, W210 & W180 (means 180 kernels/450 grams). Damaged kernels were graded as TH (1/2 size), KH (1/2 TH), LWP (small size), SWP and SSP respectively. This grading is done by experienced female labourers who were paid Rs.300 - 400 per month. Packing is usually done in a tin, which holds 11 kg 340 gms net, which is subsequently evacuated and filled with carbon-di-oxide. A rehumidification process is introduced before packing to overcome possible overdrying. This whole operation has higher yield of 90 to 95% of whole nuts. The raw materials required for these industries can be grown locally. In the next section, cashew plantations are discussed from an ecological perspective in order to assess their viability.

9.7.1 Cashew Plantation: an Ecological Perspective

A cashew plantation has the unique distinction of being cultivated in a wide variety of soils. It can survive on lands of low fertility and can be grown in all soil types, from sandy sea coast to laterite hill slopes, but the best soils are deep, triable, well drained sandy loams without a hard pan. It grows and yields well in shallow and impervious soils as sand hills, where no other economic crop generally grows well. It is tolerant of saline and clayey soils; grows on red and laterite soils. It cannot withstand water stagnation, flooding or bad drainage. In Karnataka, cashew is planted mainly in the coastal belt which receives rainfall ranging from 1300 to 3500 mm.

The area under cashew plantation at the end of 1984-85 in Karnataka was 47,442 hectares. The productivity in Kumta taluk on an average is 13 quintals per hectare. This data regarding productivity is collected by interviewing some farmers randomly selected in the taluk. The total production, as per the estimate of Government Horticulture department, is 22,115 quintals in Kumta taluk during 1992-93 (the area under cashew cultivation being 1595 hectares). Due to favorable weather and rainfall, the scope for increasing the yield is more in the hilly region of the west coast.

Cashew (*Anacardium occidentale* L.) is a small and crooked tree; originally introduced from South America. Every part of cashew is useful to man. The kernels have high nutritive value. They are rich in carbohydrates (22%), unsaturated fats (47%), minerals like calcium (0.55%), phosphorous (0.45%), iron (5.0 mg/100gm), vitamin B (630 mg/100gm) and riboflavin (190 mg/100gm), etc. The kernels supply about 6000 calories of energy per kg as against 3600 by cereals, 1800 by meat and 650 by fresh fruits. The cashew apple juice is rich in vitamin C (261.5 mg/100gm), containing 10.15 to 12.5 % sugar and about 0.35% acid (Malic). The mesocarp of the shell contains black, caustic, oil juice which is rubefacient and vesicant. It contains phenolic compound cardol, anacardic acid and an ether soluble substance. This is a valuable raw material for a number of polymer based industries producing paints, varnishes, resins, industrial and decorative laminates, brake linings and rubber compounding resins. Cashew-nut shell liquid (CNSL) is used extensively in boat manufacturing industries and wood carving centres as it prevents white ants effectively. Industries in Kumta export/sell these shells to Goa and Ankola where boat manufacturers use the CNSL as paint. The delicious and nutritious kernel is commonly eaten roasted, a process which improves flavor. The coating on the kernel yields tannins (24-26%), used in leather industries. The cashew apple and its juice has a medicinal value for scurvy, cough and colds, and is an excellent purgative. The juice is used as an antiscorbutic and diuretic; given in kidney troubles and cholera. The bud and young leaf is used as vegetable and as green manure; leaf contains tannins (23%), used in skin diseases. The alcoholic extract shows hypoglycemic and anti-cancer activity (Nambiar, M.C., et al. 1994).

In addition to these, cashew trees are used for firewood and charcoal. The pulp from the wood is used to fabricate corrugated and hard bound boxes. The bark of the tree yields tannins (9%), possessing anti-hypertensive and hypoglycemic properties; the milky sap, on exposure becomes black and is useful as an indelible ink. Gum from bark, with insecticidal properties, is useful in book binding and pharmaceutical industries.

Cashew, familiarly known as 'money spinner' among plantation crops, due to various uses of the tree, ranging from land reclamation to medicinal purposes, finds a unique place in the regional ecosystem. In Karnataka, 4.04 lakh hectares of land is barren, uncultivable, sandy, slightly alkaline and poor in nutrients. Out of this, 20,068 hectares is in Uttara Kannada district. This could be used for cashew and energy plantations. *Casuarina equisetifolia, Acacia auriculiformis* and *Anacardium occidentale* are best suited for this soil. Cashew, because of its versatility and ability to survive on saline, laterite and red soils, is a very important tree in agro and social forestry schemes. It plays a distinctive role in land reclamation, and is closely planted for effective wide breaks for orchards, and as soil binders.

9.8. DEFINITIONS

9.8.1 Specific Energy Consumption (SEC)

SEC can be identified as the ratio of Energy Consumption (in rupees) to Production (in rupees). Thus

$$SEC = \frac{En}{Pr} \text{ (Rs. / Rs.)}$$

9.8.2 Energy Intensity (EI)

EI is the ratio of Energy Consumption and Production in per unit terms. EI would be helpful in industries where the end products are cash products.

$$EI = \frac{En}{Pr} \text{ (kg / kg) or (kWh / kg)}$$

The intra and inter variation in SEC and EI among the industries producing similar products reveal the extent of inefficient/efficient energy consumption. Higher variation in the value of SEC and EI for the same amount of production among industries producing similar products reveals the extent of disparity in energy consumption.

9.8.3 Percent Production Capacity Utilisation (PPCU)

PPCU is the ratio of actual production to the total installed production capacity. If Pr is actual production and P is total installed production capacity, then

$$PPCU = (Pr/P) * 100.$$

9.9.0 Results and Discussion

9.9.1 The Rural Energy Consumption Model

To study the dynamics of energy utilisation patterns in any industrial plant, SEC and EI would help in predicting minimum energy required to operate an industry efficiently.

9.9.2 Statistical Analyses of Energy Coefficients

Applying standard statistical techniques, relationship between variables En (energy) and Pr (production) can be represented by any of the following mathematical models depending upon the best fit (least error techniques).

The variable En is the energy mix in rural industries comprising mainly fuel wood for heat and electricity for lighting purposes (very minimal). Regression analysis is carried out to look at the role of Energy Consumption and Specific Energy Consumption in this sector.

(i) *Energy Consumption and Production:*

The relationship between the variables Energy Consumption (En) and Production (Pr) is arrived at by least squares method. Results of this analysis are listed in Table 9.8. The best fit is of the form

$$En = A\,(Pr)^{C} \qquad (1)$$

where A is the constant and coefficient c is commonly known as Energy Elasticity.

The data collected from both secondary (for 136 industries) and primary sources (energy auditing in six industries) were compiled and regression analysis was done for this data. The probable relationship based on the least % error of best fit is

$$En = 2.36\,(Pr)^{o.3983}$$

Eqn. (1) could be written as

$$c = \frac{dEn/En}{dPr/Pr} \qquad (2)$$

That is, energy coefficient is expressed in terms of the proportionate change in En to Pr. Lower the value of c, lower is the change in En for corresponding change in product. Differentiating equation (1) with respect to change in production, the dynamic consumption rate obtained is

$$\frac{d\,En}{d\,Pr} = A \times c\,(Pr)^{C-1} \qquad (3)$$

The value of c being less than one, dEn/dPr declines sharply with increase in production in an industry. This might imply better utilisation of installed production capacity of boilers and driers.

(ii) *Specific Energy Consumption and Production:*

Specific Energy Consumption is the ratio of Energy Consumption (in Rs.) to the Production (in Rs.). This ratio would help to predict the minimum Energy required to operate an industrial plant efficiently. The relationship between SEC and Pr for cashew processing sector is tried by least square method. A best fit with least % error of Y est. is chosen and is represented as

$$SEC = 2.369\,(Pr)^{-o.62}$$

The results of the analysis are listed in Table 9.9. The generalised relationship is

$$SEC = A' (Pr)^{-C'} \quad (4)$$

The negative exponent c shows decline in the value of SEC with increase in Pr. This again may be inferred as the improvement of energy efficiency with better efficiency utilisation of installed production capacity.

Table 9.8: Regression Analysis: Energy & Production

Equation	r	% error of Y est	A	B
Linear	0.7286	9.26	19.47	0.0135
Parabola	0.7312	9.10	20.70	0.0108, $c = 0.175 \times 10^{-6}$
Hyperbola	0.2090	30.20	0.1123	0.0001
Exponential	0.4790	12.24	14.88	0.0019
Power law	0.7544	8.02	2.369	0.3983

Table 9.9: Regression Analysis: SEC & Production

Equation	r	% error of Y est	A	B
Linear	0.3947	12.34	0.2923	-0.000108
Parabola	0.4959	11.10	0.3340	-0.000322, $c = 0.596 \times 10^{-7}$
Hyperbola	0.4602	11.89	10.34	-0.0205
Exponential	0.6135	8.46	0.188	-0.000855
Power law	0.7673	4.61	2.369	-0.620

X = independent variable : Pr
Y = dependent variable : SEC

In order to look at the role of installed PPCU on energy utilisation, SEC and hence on energy efficiency of the operation, further analysis is carried out between the variables En and PPCU, and SEC and PPCU.

From equations (3) and (4) it is noticed that energy efficiency improves with increase in production. Normally, with increased production capacity utilisation, production of an industry increases. Regression analysis is done to explore the relationship between EI and PPCU, and SEC and PPCU. By least square method, based on less percentage error of Y estimate, the best fit relationships are,

$$EI = 56.34 (PPCU)^{-0.56} \quad (5)$$
$$SEC = 2.15 (PPCU)^{-0.76} \quad (6)$$

These two relationships just confirm the earlier conclusions, that to improve the energy consumption, that is to improve the energy efficiency of an industry, it is very essential to match the production of the industry with installed production capacity.

The next logical question that arises from these analyses is how much energy can be saved by optimal utilisation of the installed production capacity. Table 9.10 lists SEC and PPCU in the cashew processing sector. Out of 142 industries, it is seen that in 99 industries, the Percent Production Capacity Utilisation (installed) is less than 60%. Examining two industries with PPCU 50 and 80, the SEC for respective values would be (as shown in Eqn.(5))

$$SEC = 2.15\,(50)^{-0.76} = 0.109$$
$$SEC = 2.15\,(80)^{-0.76} = 0.076$$

That is, there is 30.27% variation in SEC, which means 30.27% energy can be saved by proper utilisation of installed production capacity. Figure 9.1 gives the plot of SEC and number of firms with PPCU. The declining trend of SEC with increased PPCU is evident from Figure 9.2. Frequency of number of firms in each PPCU range is a bell shaped curve, which clearly depicts about 69.72% of total samples in PPCU range less than 60%. This gives scope for saving fuel wood with better utilisation of installed production capacity.

Table 9.10: Distribution of Industries in Different PPCU Range

PPCU (%)	Number of Industries	SEC
0-5	1	0.655
5-10	2	0.342
10-20	2	0.264
20-30	6	0.190
30-40	12	0.120
40-50	37	0.100
50-60	39	0.095
60-70	21	0.093
70-80	10	0.092
80-90	8	0.060
>90 - <100	4	0.042
	142	

9.10.0 Energy Auditing: Cashew Processing Industries

Detailed energy audits were carried out in six cashew processing industries located in the taluk by spending one to two days in each industry and measuring actual fuel wood consumption, raw materials, etc. The details of energy consumption, raw materials, and man-power are listed in Table 9.11. The energy analyses were carried out to explore the variation in SEC, EI and PPCU in these industrial units and the results are listed in Table 9.12.

From Table 9.12, it is evident that these industries are labour intensive and mainly employ women. Since the wages are paid based on the amount of work done (that is quantity of nuts dehusked/peeling of kernels, etc.), these labourers seem to be one of the most efficient. The employment opportunity in these industries has created a kind of social awareness among women.

Table 9.11: Energy Auditing in the Cashew Processing Firms

Fuel wood/year for heating	Electricity mainly for lighting	Fuel wood	Raw nut	Output Kernels production	Employees M	F	Wages	Production
Tonnes	kWh	Rs.	kg	kg	No	No	Rs.	Rs.
7.2	180	5400	5914.28	1320	2	8	96000	198000
70	240	49000	31200	8400	4	10	152116	275000
120	400	90000	81200	24010	2	25	195000	3601500
156	25	120000	61200	18000	2	27	210600	2700000
165	600	123750	121200	34200	2	50	267000	5130000
692.5	3000*	519375	577200	53900	4	130	1619700	215460002

* This industry uses electric drier for drying kernels sometimes

M: Male, F: Female

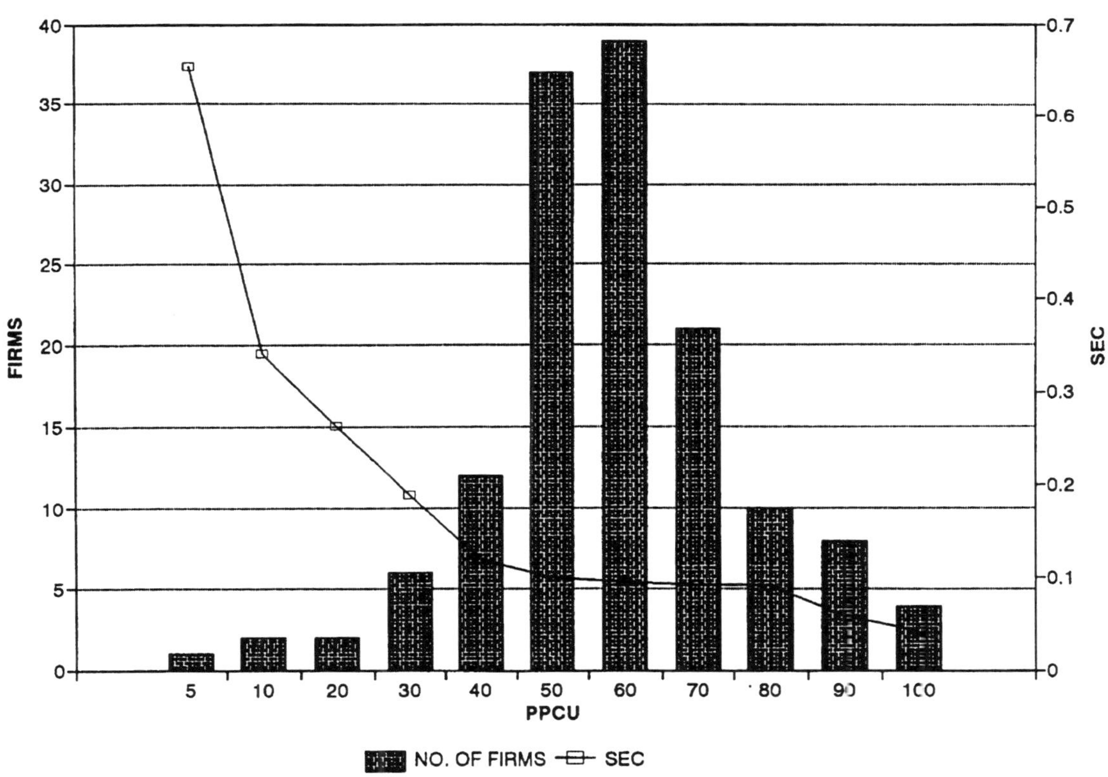

Figure 9.1: SEC and PPCU

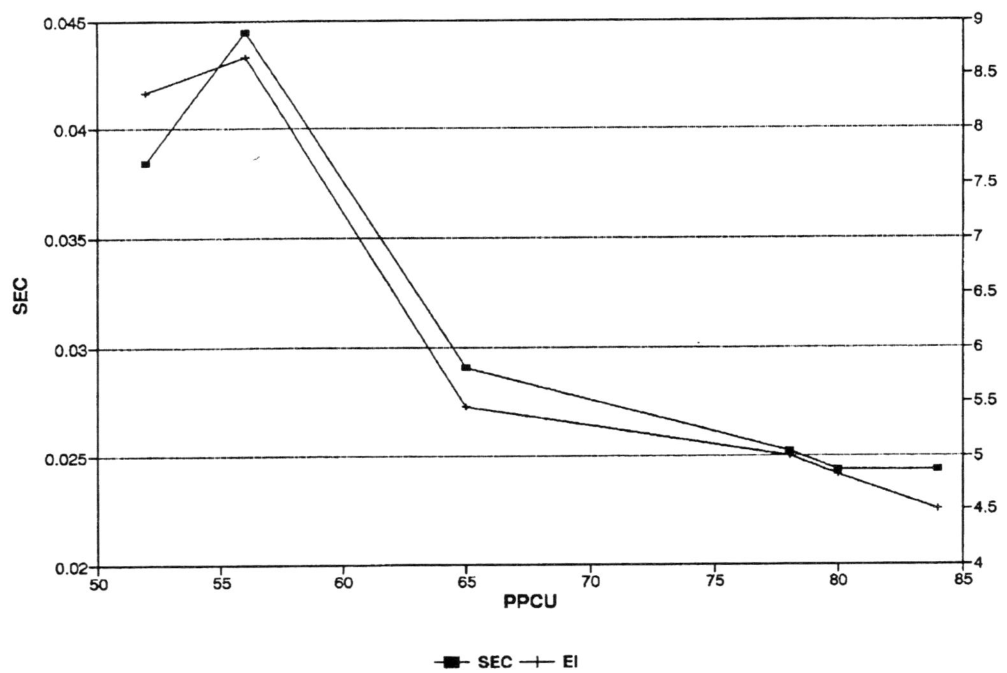

Figure 9.2: SEC and EI Variation for Different PPCU

Since cashew is a cash crop, SEC computed shows very small values and the discussion based on SEC seems to be insignificant. Therefore, EI computed reveals distinctly the disparities in the level of fuel wood consumption in other sectors. It is noticed that the variation ranges from a minimum of 4.49 kg to a maximum of 8.66 kg fuel wood per kg of cashew kernel processed. The PPCU computed reveals that the high level of fuel wood consumption is related to the lower installed production capacity utilisation. Figure 9.2 is the pictorial representation of SEC/EI with PPCU of firms, where detailed survey was carried out.

As listed in Table 9.12, it is evident that an industry with PPCU 52% consumes 8.32 kg of fuel wood while that with PPCU 84% consumes 4.49 kg of fuel wood. This means there is 46% variation in the quantity of fuel wood consumed for processing cashew. This is mainly due to non-utilisation of installed production capacity. Roasting of raw nuts and drying of kernel are the two energy intensive operations in these industries.

Drier is used mainly to remove the moisture (of the order of 10%) in the kernel. High thermal efficiency could be achieved by the combustion of fuel wood in an enclosed chamber over a suitable grate with proper opening for primary and secondary air to generate the highest combustion temperature and thus maximise the heat transferred to the trays. Proper insulation of drier doors and chimney of suitable height and diameter to create draught and disperse smoke are the other essential components in cutting down fuel wood consumption. In two out of six firms, combustion was done in open chamber and drier doors were poorly insulated. Apart from these, drier was not loaded fully in these firms due to the following reasons:

(a) The fuel wood cost is relatively small in the processing of cash crops like cashew compared to the value of the end product. This is one of the reasons for not taking up energy conservation measures by these industries.

(b) Easy accessibility and availability of fuel wood, both on commercial and noncommercial basis, are the main reasons for the rural industries to use fuel wood. However, fast depletion of forests has caused fuel wood scarcity which affects the domestic consumer severely, while these industries continue to get fuel wood at higher/increased cost. Still, cost of fuel wood in the total production cost is very less in this sector. The inefficient usage of fuel wood in the order of 46 to 50% could be conserved by proper maintenance of equipment, better utilisation of installed production capacity and use of renewable sources of energy, like solar energy, for steaming raw nuts.

Table 9.12: Computation of SEC / EI and PPCU

SEC (Rs/Rs)	EI (kg/kg)	PPCU (%)
0.02435	4.824	80
0.02438	4.499	84
0.02521	4.997	78
0.02909	5.454	65
0.04446	8.666	56
0.03843	8.324	52

9.11.0 Energy Utilisation in Energy Intensive Sectors

Brick and tile manufacture and parboiling of paddy are other energy intensive small-scale industries in Uttara Kannada district. The manufacturing process and energy usage are discussed in the following sections.

9.11.1 Brick Manufacture

Brick manufacture is seasonal, labour intensive, and in almost all cases, involves no machine or electrical energy. The output and quality of bricks depend mainly on the skill of labourers. Bricks moulded by hand are dried in the sun and in clamps or kiln using coal or firewood as fuel. The firing process gives strength to the bricks as the formation of complex glass-forms take place at temperature of about 1000°C. Low temperature firing results in under firing while higher temperature results in melting and deformed bricks. Two types of kilns used are either batch type intermittent kiln or continuous kiln with permanent structures. The main stages in brick making are: winning and hauling the clay, crushing, mixing and forming, drying the green bricks, firing the bricks, cooling the fired bricks, storing and delivering the finished products. Batch type intermittent kilns, often wood fired and mostly employed by small production units, have capacities ranging from 50,000 to 100,000 bricks at a time. Clamps are thermally inefficient (about 10%) as substantial amount of heat is lost unutilised in flue gases and cooling of the fired bricks. In a continuous kiln, heat in flue gases is utilised in preheating the bricks. In addition, heat from the cooling zone is utilised in preheating the primary air required for burning of the fuel. The distribution of temperature in general is found uniform throughout the whole cross section. Fire and draught can be easily controlled in continuous kilns. The firing capacity is 25,000 to 30,000 bricks per day. Loss of bricks due to under firing and over firing is higher in the case of clamps. Kilns mainly use coal. Due to scarcity of coal in some parts of India, people have switched over to alternative fuels, such as, fuel wood, agro residues (rice husk, groundnut husk, etc.) and forest residues. In Kumta taluk, clamps are mainly used for manufacturing bricks and fuel wood is used for burning bricks. Survey of four brick manufacturing units shows that fuel wood consumption ranging from 450, 600, 650 to 700 kg for manufacturing 1000 bricks. Study of these four brick works shows that about 25% to 37.41% fuel could be saved.

Use of unburnt compacted dense soil blocks in place of burnt bricks (designed by ASTRA) reduces the fuel wood dependence considerably. The compression is done manually of locally available material such as clay and small quantity of lime. Compaction of soil to make blocks means employment for rural youth. Apart from this, there are improved versions of kilns which can save about 50% of fuel used (present consumption). By using waste heat to dry green ones or for preheating another chamber, about 15-20% fuel can be saved. The thermal efficiency of intermittent kiln is in the range of 10-15%, while of continuous kiln it is 30-35%. Good housekeeping practices like measurement of temperatures, control of dumpers and cleaning of ash pit can save up to 5% of fuel usage, and increase in fire travel distance can reduce fuel consumption. Efficiency of an intermittent kiln is about 13.7%; continuous kiln is about 34.2%. Comparative study of fuel required for brick making in different kilns compiled by UNIDO (UNIDO/ILO 1984) are:

Type of Kiln/Clamp	*Heat Requirement*(MJ/1000 bricks)
Intermittent	
Clamp	7,000
Stove	16,000
Scotch	16,000
Down Draught	15,500
Continuous	
Original Hoffman	2,000
Modern Hoffman	5,000
Bull's Trench	4,500
Tunnel	4,000

9.11.2 Tile Manufacture

Manufacture of roof tiles first started in India at Mangalore (Karnataka) with the help of Basal Mission of Germany to provide employment. The availability of good quality raw materials and favorable conditions helped the establishment of tile industries in coastal places, such as, Mangalore (Dakshina Kannada district), Bhatkal, Kumta and Karwar (Uttara Kannada district). Energy required in the form of fuel wood is about 0.5 tonnes/1000 tiles and kerosene 6 litres/1000 tiles. This is based on the survey conducted in one industry located at Kumta taluk which manufactures about 12,000 tiles per day.

9.11.3 Rice Parboiling

Parboiling on a small-scale involves soaking of paddy overnight in water and then boiling it in a perforated tank on an open pan. The boiled paddy is sun dried and then milled to separate the husk and rice. Larger rice mills use commercial boilers. Parboiling is done in conical shaped vertical tanks, each containing about 13.5 tonnes of paddy. Paddy is fully soaked in hot water at a temperature of 80-85°C, and heated continuously by passing steam through it. It is kept in this conical tank for about 6-8 hours. Later it is transferred to big perforated trays for drying. Hot air at 120-140°C is passed at a rate of 17 cubic meters per minute through the tray for 5-6 hours. The dry parboiled rice is then milled to obtain full grain polished rice, bran and husk. The bran which is a by-product containing 15-20% oil, is purchased by oil mills.

Larger rice mills use rice husk and some quantity of fuel wood and/or other agro residues. Some households use mainly fuel wood. Fuel consumption (in the coastal zone) is in the range of 0.9 (avg)± 1.22 (sd) quintal per quintal of rice. That is, to get a quintal of parboiled rice, fuel wood consumed is in the order of 0.9 quintals. Fuel consumption in interior zone is in the range of 0.7 (avg)± 0.70 (sd) quintal per quintal of rice.

9.11.4 Grain Mills

An earlier study (Ramachandra and Subramanian, 1993), based on a primary survey of grain mills spread over Karnataka state carried out detailed energy analyses (consisted of 8 samples from Uttara Kannada district). Firms using energy resources, such as, fuel wood, diesel, etc. (other than electricity) have higher SEC values in the order of 0.28-0.43, while industries using only electricity have SEC value in the range 0.1 to 0.3. This study shows that there is scope for saving energy of the order of 23.12% in the food grain sector and 18.67% in paddy dehusking sector by structural adjustments.

9.12.0 CONCLUSIONS

These investigations show that majority of the rural industries need heat energy, and rely mainly on wood and agricultural residues. As discussed earlier, in many rural industries, fuel costs and/or shortages are important constraints at existing production levels. It is also noticed that wood is used very inefficiently in most of the rural industries.

1. The wide variations in EI in cashew processing industries, from 4.49 to 8.66 kg of fuel wood/kg of kernel, reveals the scope for energy conservation in the order of 30 to 48%.
2. The relationship between SEC and Pr is power law with an exponent less than one, indicating improvement of SEC with increased production in an industry. Computation of dynamic SEC (dEn/dPr) reveals that the rate of energy consumption is less than the production. The energy consumption rate decreases as the firm's production increases.
3. The use of fuel wood for domestic and rural industrial applications will continue in this region for at least another decade. This study and earlier studies in the domestic sector have revealed that present usage is very inefficient in these sectors. About 40-45% of fuel wood can be saved in domestic cooking and water heating activities. This necessitates a wider use of energy efficient end use devices to conserve energy, which will help in meeting the growing demand of energy in the region.
4. Fuel wood consumption in cashew processing industries could be brought down by using solar water heating devices for roasting the raw nuts and properly designed and insulated driers for drying the kernels.
5. Paddy dehusking and rice milling is an energy intensive sector in Uttara Kannada district. Analysis of sectors 204 and 2042 (National Industry Code: NIC code) revealed that there is scope to save energy of the order 23-38%.
6. Good housekeeping practices would help in saving 25 to 37.41% fuel wood in brick manufacture. Further saving is possible by changing the technology of manufacture, which involves capital investment.
7. Entrepreneurs could be encouraged to grow energy plantations on already degraded and unused lands to meet the fuel wood requirement of the industry in a sustainable way (while setting up energy intensive rural industries), which in turn would help in improving the soil condition by avoiding further degradation and generate employment to rural youth.
8. Decision makers should consider the energy requirement of the industries while emphasising on industrialisation of a region. Decentralised way of meeting energy requirement of industries like solar, wind (depending on availability of source), small hydro and energy plantations would be the most appropriate way of handling energy situation in a region.

9.13.0 ACKNOWLEDGEMENTS

I thank officials of the Directorate of Industries and Commerce at Bangalore, Department of Industries and Commerce at Karwar, Sirsi and Kumta for providing information required for this study.

9.14.0 REFERENCES

1. Subramanian. D.K., 1985. Energy Utilisation in Karnataka, In *State of Environment Report Karnataka*, C.J.Saldanha (Ed), Centre for Taxonomic Studies, Bangalore, India, 1985. pp103-123.
2. Government of Karnataka, 1995. Economic Survey 1994-95, Science and Technology Department, Government of Karnataka, Bangalore, India.
3. Development Commissioner, Ministry of Industry, Government of India, 1990. Second all India Census of SSI units Registered up to 1988 - Vol I & II, Govt. of India, New Delhi.
4. Henderson, P.D., 1975. *India: The Energy Sector*, Oxford University Press, New York.
5. Huang, Jin-ping, 1993. Industry energy use and structural change: A case study of the People's Republic of China. *Energy Economics*, April:131-136.
6. Parikh, J.K., Parikh, K., Gokarn, S., Painuly, J.P., Saha, B. and Shukla, V., 1991. *Consumption Patterns - The Driving Force of the Environmental Stress*. UNCED Report, IGIDR, Bombay, India.
7. Schurr, S.H. and Eliasberg, V.F., 1962. *Energy and Economic Growth in the United States*. Resources for the Future, Washington D.C., USA.
8. Schurr, S.H., 1985. Energy conservation and productivity growth. *Energy Policy*, April:126-132.
9. Janosi, P.E.de and Grayson, L.E., 1972. Patterns of energy consumption and economic growth and structure. *Jour. of Developmental Studies*, 8(Jan):241-249.
10. Sinton, J.E. and Levine, M.D., 1994. Changing energy intensity in Chinese industry: The relative importance of structural shift and intensity change. *Energy Policy*, March:239-255.
11. Ang, B.W., 1991. A statistical analysis of energy coefficients. *Energy Economics*, April:93-110.
12. Hansen, U., 1990. Delinking of energy consumption and economic growth: The German experience. *Energy Policy*, September:631-640.
13. Ramachandra, T.V. and Subramanian, D.K., 1993. Analysis of energy utilisation in the grain mill sector in Karnataka. *Energy Policy*, June:644-655.
14. Ross, M., 1992. Efficient energy use in manufacturing. *Proc. Natl. Acad. Sci, USA*, 89(February):827-831.
15. Siddayao, C.M., Khaled, M., Ranada, J.G. and Saicheua, S., 1987. Estimates of energy and non-energy elasticities in selected Asian manufacturing sectors, Policy implications. *Energy Economics*, April:115-128.
16. Alvarado, S. and Iribarne, J., 1990. Minimum energy requirements in industrial processes: An application of energy analysis. *Energy*, 15(11):1023-1028.
17. Chern, W.S. and James, W.E., 1985. *Energy Demand and Productivity in Selected Asian Countries*. APO Seminar on Energy Conservation in Asian Region, Philippines.
18. Rabadi, N.J., Alsaad, M.A. and Alzu'bi, A., 1991. The potential and strategies for demand side management within the industrial sector in Jordan. *Energy Conversion and Management*, 32(6):577-584.
19. Gardner, D. 1993. Industrial energy use in Ontario from 1962 to 1984. *Energy Economics*, January:25-32.
20. Mahmud, F. and Chishti, S. 1990. The demand for energy in the large-scale manufacturing sector of Pakistan. *Energy Economics*, October:252-253.
21. Wolde-Ghiorgies, W., 1991. Industrial energy utilisation patterns in a developing country: A case study of selected industries in Ethiopia. *Energy Conversion and Management*, 31(3):285-293.
22. McFaddin, S. and Clouse, M., 1993. A new model for sustainable growth in the energy industry. *Energy Economics*, January:3-7.
23. Reitler, W., Rudolph, M. and Schaefer, H., 1987. Analysis of the factors influencing energy consumption in industry: A revised method. *Energy Economics*, July:145-148.
24. Anandalingam, G., 1985. Energy conservation in the industrial sector of developing countries. *Energy Policy*, August:335-339.
25. Parikh, J.K., 1990. Energy conservation in Indian manufacturing industries, *Productivity*, 30(4):379-385.
26. Ramachandra, T.V. and Subramanian D.K., 1991. Analysis of energy utilisation in the small-scale industries in Karnataka. *Energy Management*, January-March:23-47.

27. Levine, M.D., Meyers, S.P. and Wilbanks, T., 1991. Energy efficiency and developing countries. *Environmental Science Technology*, 25(4):585-589.
28. Pasha, H.A., Ghaus, A. and Malik, S., 1989. The economic cost of power outages in the industrial sector of Pakistan. *Energy Economics*, October:301-318.
29. Howarath, R.B., Schipper, L., Duerr, P.A. and Strom, S., 1991. Manufacturing energy use in eight OECD countries: Decomposing the impacts of changes in output, industry structure and energy intensity. *Energy Economics,* April:135-142.
30. Nand, S., 1993. Energy conservation in fertiliser plants. *Energy Management,* July-September:5-10.
31. Ang, B.W., 1987. A cross sectional analysis of energy-output correlation. *Energy Economics*, October:274-286.
32. Ramachandra, T.V. and Subramanian, D.K., 1997. Industrial energy utilisation in Karnataka and potential savings. *Energy Conversion and Management,* 38(6):563-599.
33. Ramachandra, T.V. and Subramanian, D.K. 1992. Energy efficiencies of end use devices in an electro metallurgical industry : A critical study. *Energy Conservation and Management*, 33:899-912.
34. Carlsmith, R.S., Chandler, W.U., McMahon, J.E. and Santini, D.J., 1990. *Energy Efficiency : How Far Can We Go*. ORNL/TM-11441, Martina Marietta Energy Systems Inc., USA.
35. Fulkerson, W., Fairchild, P.D. and Judkins, D.R., 1989. *Energy Technology R & D : What Could Make Difference? Part I - Synthesis Report*. ORNL-6541/V1, Oak Ridge National Laboratory, Oak Ridge.
36. Williams, R.H., 1978. Industrial cogeneration. *Annual Review of Energy*, 3 : 313-356.
37. Ross, M., 1989. Improving the efficiency of electricity use in manufacturing. *Science*, 244:311-317.
38. Dale, S.J., 1991. *Assessment of Energy Applications of High Temperature Superconductivity: Extended Summary Report*. Electricity Power Research Institute, Palo Alto, Calif., USA.
39. DOE, 1987. *The DOE Industrial Energy Conservation Program : Research and Development in Separation Technology*. DOE/NBM-8002773, U.S. Department of Energy, USA.
40. Nambiar, M.C., Rao, E.V.V.B. and Pillai, P.K.T., 1994. Cashew. In *Fruits. Tropical and Sub Tropical*. pp. 386-419. Bose.T.K. and Mitra, S.K. (Eds.), Naya Prokash, Calcutta, India.
41. UNIDO/ILO, 1984. *Small-scale Brickmaking*, Technical Memorandum, No. 5., UNIDO/IS.456, Vienna.
42. op.cit. Ref 13.

Chapter 10

INTEGRATED REGIONAL ENERGY PLANNING

10.0 SUMMARY

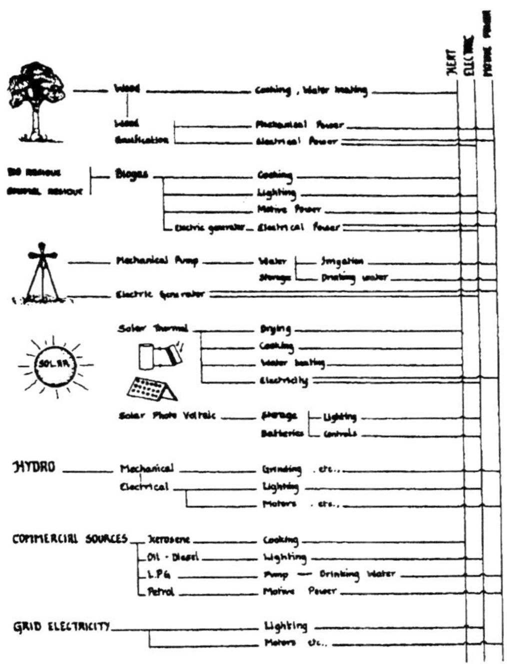

The energy planning endeavors for a particular region involves the finding of a set of sources and conversion devices so as to meet the energy requirement/demand of all the tasks in an optimal manner. This optimality depends on the objective to minimise the total annual cost of energy and the dependence on non-local resources or maximise the overall system efficiency. Factors such as availability of resources in the region and task energy requirements impose constraints on the regional energy planning exercise. Thus, regional energy planning turns out to be a constrained optimisation problem. This chapter describes an optimum energy allocation based on Integrated Energy Planning for Uttara Kannada district and makes a satisfying energy allocation plan for the years 2000, 2005 and 2010. Integrated energy planning gives an optimal mix of new/conventional energy sources and is developed based on Decision Support Systems (DSS) approach.

Regional energy planning exercises need to be flexible (to cope with rapidly changing energy systems) and easy to use. The application of DSS is a new approach to this problem. Towards the goal of implementing analytical methods for integrated planning, computerised decision-system provides useful assistance in the analyses of available information, the projection of future conditions, and the evaluation of alternative scenarios. Some of the features of DSS found particularly useful in regional energy planning are: (i) flexible structure - allows appropriate feasible levels of disaggregation, (ii) integrated nature - promotes a better overall understanding of many processes and concepts involved in planning, allowing planners to concentrate on specific energy subsectors, and (iii) iterative nature and easy scenario testing features - provide guidance in optimising data collection activities.

The application of DSS for Uttara Kananda district energy planning focuses on renewable resources that could be harnessed for energy, land use database, sectorwise energy demand database and optimal allocation of energy resources for various tasks, and then explore the energy use consequences of alternative scenarios, such as, Base case scenarios, high energy intensity and improved end use efficiency options.

Linear programming formulation for optimum allocation based on the cost minimisation objective shows that there is substantial savings of about 19.19% in energy and 36.24% cost reduction in overall energy system. Cost per unit (kWh) of energy with optimal allocation of energy is Rs. 0.31 per kWh (as against Rs. 0.39 per kWh without optimisation). Optimisation carried out with the objective of maximisation of efficiency of 'ijk' combination, for all combinations shows energy saving of 19.98% and cost of energy as Rs. 0.34 per kWh. The scenario analyses reveal that relatively vigorous growth in energy demand in Uttara Kannada district can be accomplished without exceeding available resources.

Keywords: Integrated energy planning, Decentralised energy plans, Decision Support System (DSS), Energy efficiency, Optimisation, Base Case Scenario, Transformation Scenario.

10.1.0 INTRODUCTION

The central theme of energy planning is the preparation of area based decentralised energy plans for meeting energy needs for subsistence and development with least cost to the environment and the economy. Centralised energy planning exercises cannot pay attention to the variations in socio-economic and ecological factors of a region which influence success of any intervention. Decentralised energy planning advocated these days is in the interest of efficient utilisation of resources, ensuring more equitable sharing of benefits from development. The regional planning mechanism takes into account all resources available and demand in a region. This implies that the assessment of the demand and supply, and the intervention in the energy system which may appear desirable due to such exercises, must be at a similar geographic scale. For example, bioresource assessment of supply and demand at the aggregate level is likely to be misleading as scarcity and surplus is always at a localised level. Consequently, the energy interventions in the form of energy supply enhancement, containing demand and/or encouraging alternative fuels may be aimed at the wrong area or target group. Normally, the district is accepted as the appropriate planning level. However, this leaves behind the complex issue of assigning boundaries for energy interventions. This is due to difference in supply and administrative boundary of a district. The geographical boundaries of the supply system in a region, based on collected biofuels, are determined by the biomass resource base and level of demand within the boundaries. In cases where fuel wood is collected from forests, the quality and accessibility of forest resource base are the most influential factors in determining the boundary of the energy system.

Planned interventions to reduce energy scarcity can take various forms, such as:

(a) energy conservation through promotion and use of energy efficient stoves for cooking and water heating, compact fluorescent bulbs in place of ordinary incandescent bulbs,
(b) supply expansions through agroforestry, farm forestry and community forestry, and
(c) alternatives - renewable sources of energy such as micro/mini/small hydropower plants and wind, solar and biomass based systems.

Two basic approaches namely "top-down" and "bottom-up" are practiced in regional energy planning (Rath, et al., 1981). "Top-down" approaches attempt to capture the aggregate behaviour of the energy sector by use of equilibrium or partial equilibrium models which simulate prices, demand, supply and investment interactions to seek long term market equilibrium. Such interdependent relationships are at best difficult to quantify and at worst an inaccurate description of the behaviour of markets. This approach is less suitable for economies of developing nations and smaller nations who have little control over the prices of fossil fuels and other raw materials.

In this connection, a "Bottom-up" or disaggregated approach of energy planning is more suitable for developing countries. In this more pragmatic approach of supply and demand, projections and investigations at disaggregated level using available data, local expertise and experiences of the energy system are possible. This approach allows important (and price independent) effects such as technological innovations, energy transitions, market saturation and other structural shifts to be easily incorporated, which would be virtually impossible using econometric approach. Sometimes, regional macro economic model is used as the basic framework within which demand and supply projections are made.

A range of methodologies is available for modelling energy systems. These include, normative (optimising) models (Meier and Mubayi, 1983), system dynamic models and accounting framework simulation models. Optimising models typically use linear programming techniques to find out a system configuration that maximises and minimises objective functions (such as minimising costs). These have found favor in applications, such as, least cost of electricity planning and regional energy planning studies. System dynamics model makes use of engineering control theory to simulate a system as a series of interconnected stock and flow variables. System dynamics model is a powerful tool for studying the interrelationships of the different parts of a system, but their behavior is dependent on the feedbacks between different variables, and of the relationships between those variables. Small errors in estimates of any of these values will tend to be exaggerated as the model is run for a long planning period. These models are normally not applied as practical tools for examining disaggregated energy systems. Most models of disaggregated energy systems have been based on the accounting framework approach. A set of accounting tools is provided to planners for checking the consequence and consistency of a range of scenarios. These accounting tools are simple, and provide emphasis on data, models and an easy-to-use interface.

The regional energy planning exercise carried out for Uttara Kannada district in this chapter is based on the mathematical approach proposed by Reddy, A.K.N. and Subramanian, D.K. earlier (Reddy and Subramanian, 1980) for the design of rural energy centres. Method involves minimisation of annual cost function to a set of equality and inequality constraints using a linear programming algorithm.

10.2.0 REVIEW OF ENERGY PLANNING STUDIES IN INDIA

Integrated energy planning was recognised as an essential element of development planning in India as early as the sixties. The Government of India constituted the Energy Survey of India Committee (ESIC) in 1963 (GOI, 1965) to study "the present and prospective demands and supplies of energy, both total and in respect of constituents of energy on a national, regional and sectoral basis". The study was expected to provide the Government with the basic material for development planning in the field of energy up to 1981. The committee was specifically required to look into the energy needs of the rural areas. Keeping in view the past trends in energy consumption, the committee estimated the sectoral and regional energy demands for alternate growth scenarios and provided an analysis of the options available for meeting these demands. The committee also made several recommendations on investment planning in energy sector and pricing of different forms of energy. These recommendations provided the agencies concerned with energy planning in the country a greater insight into the long-term problems of the energy sector.

The Fuel Policy Committee (FPC) was appointed by Government of India in 1970 (GOI, 1974) to prepare an outline of the national fuel policy for the next 15 years. Apart from analysing the supply and demand options in the energy sector, the committee was specifically required to look into the technical and organisational aspects of energy planning with special reference to the scope for improving the efficiency of energy use. While the committee was deliberating on the subject in 1973, the world oil market went through major upheavals making it necessary for the committee to consider the implications of those changes for the Indian economy. In the light of this development, the committee made a number of important recommendations on energy policy including several suggestions on substitution of oil by coal and electricity and on energy conservation in general. The committee adopted econometric forecasting techniques and end use analysis in arriving at sectoral energy demand estimates. However, the major thrust of this study, which was finalised in 1974, was more on supply side of energy than on sectoral demand analysis and demand management.

The Working Group on Energy Policy (WGEP) was another expert group (GOI, 1979) constituted by the Government of India in 1977. WGEP was required to outline the national energy policy for the next 5, 10 and 15 years. The report of WGEP was finalised in 1979. WGEP made detailed projections of the demand for both commercial and noncommercial forms of energy up to the end of the century and suggested a number of corrective policy measures to manage the energy demand. The Reference Level Forecast (RLF) and Optimum Level Forecast (OLF) made by the Working Group highlighted the crucial issues of energy planning relevant to energy conservation and inter-fuel substitution. The methodology adopted by WGEP was more or less similar to the one adopted by its preceding expert groups. The sectoral demand estimates in this report were, however, based on a more detailed analysis of the end use requirements of energy. The recommendations of WGEP provided a broad framework for energy sector planning in the Sixth Five Year Plan.

Even though both FPC and WGEP emphasised the need for integrated energy planning, in practice however, no formal institutional mechanism could be evolved on a firm footing for examining the various policy issues on an integrated basis. It was in this context that the Advisory Board on Energy (ABE) was set up in 1983 on the eve of formulation of the Seventh Five Year Plan (GOI, 1984). In addition to several important recommendations on the technical, financial and institutional aspects of energy, the ABE also made detailed projections of energy demand in different regions till 2004 under assumptions of different macro-economic scenarios. These estimates were made based on both end use and regression methods.

The studies undertaken by ABE and the Expert Groups prior to it provided useful policy guidelines for energy sector planning. However, considering the complexity of the investment choices available in energy and energy related sectors, it became necessary to evaluate the various options together with reference to the long-run economic resource costs involved, so as to provide a more precise indication of the optimum energy strategy to be adopted by the Government. It is in this context that long-term energy modelling studies have been undertaken in the Planning Commission to analyse the supply options available in coal, oil, natural gas and electricity with reference to the economic resource costs involved. The sectoral energy demands in this study were estimated on the basis of anticipated growth elasticities. These studies emphasised more on resource cost optimisation on the supply side than on sectoral energy demand analysis. It is, however, becoming increasingly evident that a detailed analysis of the factors that contribute to energy demand in different sectors is essential for evaluating the energy implications of different policy options in the economy. This envisages a much more disaggregated analysis of the sectoral energy demands on the basis of end use requirements of energy in each sector. The present study is an attempt in this direction.

10.3.0 The Decision Support System Approach

A Decision Support System (DSS) may be defined as "a coherent system of computer-based technology (hardware, software and supporting documentation) used by managers as an aid to decision-making in semi-structured tasks". The characteristics of DSS's have been summarised by Parker and Al-Utaibi (1986), and some of these are highlighted below.

1. Flexibility and adaptability to accommodate changes in the environment and the decision-making process of the user.
2. Assist managers in their decision processes in semi-structured tasks - problems for which formal models are useful, but where the planner's judgement is also essential.
3. Support and enhance, rather than replace, managerial judgement.
4. Attempt to combine the use of models of analytical techniques with traditional data access and retrieval functions.
5. Specifically focus on features which make them easy to use by non-computer people in an interactive mode.
6. Organise data and models around decision(s) and are user-initiated and controlled.
7. Present information in a flexible way to support the widely differing requirements and cognitive styles of users. To this end, it should be possible to present data in different contexts and formats, and allow the user to change these contexts.
8. DSS software uses a hierarchical design approach. Rather than attempting to build highly specific software for each study application, the DSS is often designed as a model generator. That is, the model itself is built rather than the model building tools.

From these characteristics, it can be seen that DSS's focus is on providing flexible tools for policy analysis and not on providing models to give answers to structured problems. Indeed, the modelling tools are only one component to DSS, which may be described as comprising three component subsystems.

(i) *The Database Management Subsystem:* This manages an integrated database to drive all the models. Its purpose is to extract and combine information from a variety of sources, display the data structure to the user in a logical way, and handle personal and unofficial data, so that the user can experiment with alternatives based on personal judgement.

(ii) *The Model Management Subsystem:* The DSS offers the user a number of modelling tools. The capabilities of this subsystem include: creating new models, cataloguing and editing of existing models, interrelating of models by links through the database and integrating small model 'building blocks' into larger model systems.

(iii) *The Dialogue Management Subsystem*: Some guidelines for the design of this subsystem (the user-system interface) include;

- the use of a consistent and familiar interface (for example, spreadsheet or word processing programs) throughout the system,
- help functions which are on-line, extensive and context sensitive
- natural language messages, both in normal operation and error conditions,
- data entry forms ready-filled with sensible default values (profiles or templates),
- data entry validation,
- housekeeping functions (data back-ups, copying, deleting files, etc.).

The dialogue style is the main determinant of the usability of the DSS. This design is guided by various methodological considerations, such as:

(a) *Scenario Approach:* Simulates alternative energy and economic futures under a range of different assumptions, such as, replacement of traditional stoves with improved stoves, switching over from incandescent bulbs to compact fluorescent bulbs, improvement in agricultural pumpsets' efficiency, improvement in industrial machinery's efficiency, agroforestry, etc.

(b) *Integrated Energy Environment Planning:* While emphasising a disaggregated approach, design also stresses the importance of integrating the analyses within a comprehensive planning framework. Integration of all types of energy sources, sectorwise energy demands, intermediate conversions of fuels, primary resources, and environmental and economic impacts across separate geographical areas takes place.

(c) *End use, Needs Driven Approach:* Resource requirements and supply side projections are carried out based on energy services required in different economic sectors. This approach places development objectives, such as, the provision of end use goods and services at the foundation of energy analysis. It achieves the goal of integrated energy planning by allowing the development plans of different sectors to guide the evaluation of energy strategies.

(d) *Flexibility and User Friendliness:* Is designed as a set of modules and is flexible, expandable and comprehensive. These provide an expandable data structure which can be adapted to the diverse energy systems with different requirements and developmental views of planners.

The regional energy planning program, designed based on these concepts, consists of mainly four blocks of programs: Energy Scenarios, Computation Models, Aggregation and the Energy Database. This is illustrated in Figure 10.1.1.

The energy scenario module addresses the main components of an integrated energy analysis, namely, sectorwise demand analysis, energy conversion and resource assessment. There are three programs for building scenarios (demand, transformation and resource) and one for comparing and evaluating scenario costs and impacts. The planner uses the scenario building programs to develop current energy balances, projections of supply and demand trends, and scenarios representing the effects of energy policies, plans and actions. The aggregation program assembles area level (Taluk, District, Region, State, Nation) energy projections into multi area results. The detailed structure is given in Figure 10.1.2. Individual components of the model are discussed below:

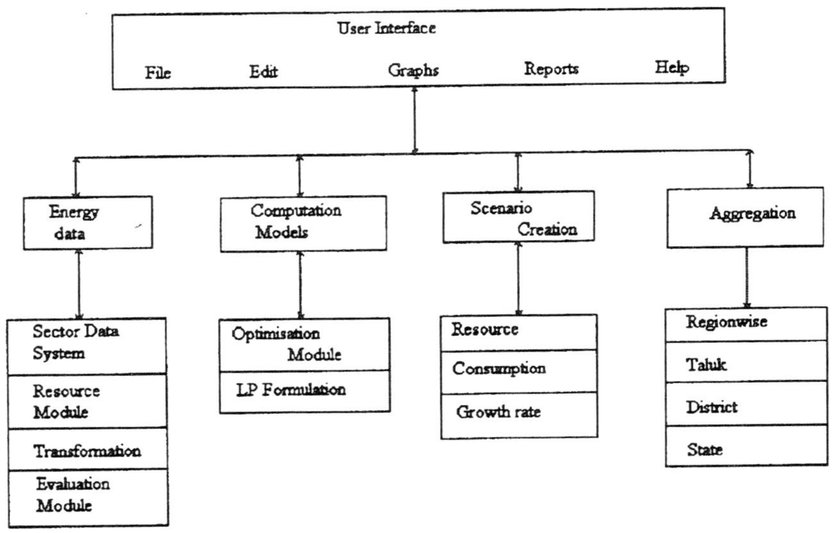

Figure 10.1.1: Regional Energy Planning Model Structure (Overall)

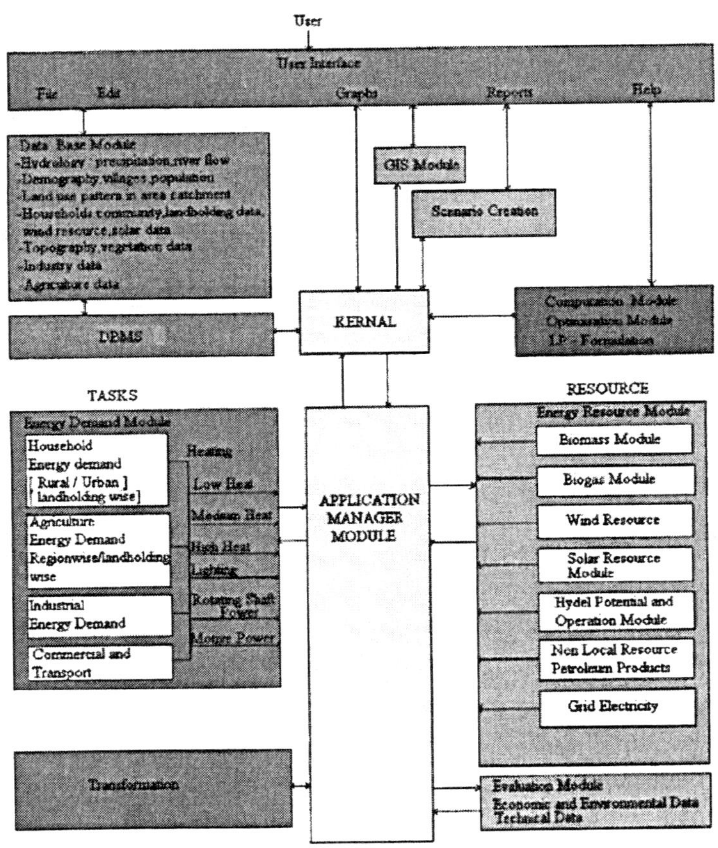

Figure 10.1.2: Detailed Structure of Integrated Regional Energy Planning

10.3.1 The Energy Demand Module

This module provides a disaggregated end use approach to analysis of energy requirements. Data is assembled hierarchically into four levels. These are:

> **Sectors**: household, agriculture, industry, commercial and transport, and each sector is divided into **subsectors** (e.g. high income, low income, large-scale industry, small-scale industry, etc.). Each subsector is divided into **end uses** (such as cooking, water heating, lighting, irrigation in household sector, etc.), and each end use is divided into **devices** (such as traditional stoves, improved cook stoves, kerosene stoves, incandescent lamps, fluorescent bulbs, pumpsets, tractors, etc.). Driving activities to project future demand is computed by growth rates, macro drivers (population), and energy intensity at device level. Demand by devices and sectors are aggregated to get region's demand.

10.3.2 The Transformation Module

This module simulates the energy sector conversion processes that turn primary resources to final power. For example, the biomass requirements (for producer gas, electricity, charcoal, etc.) computed by transformation program are used by the biomass resource program to match biomass resource demands to available wood stock and yields.

10.3.3 The Energy Resource Modules

Various resource modules discussed here are Biomass, Wind Energy, Solar Energy and Hydro Energy available in the region.

(i) *The Biomass Module:*

The biomass module examines the impact of land use changes on the biomass resource base and the impact of bioresources consumption. This, with wood productivity models/details and land management practice details, is designed to assess the present state of bioresources (includes fuel wood, agricultural residues, horticultural residues, dung and other bioenergy resources) in a region to meet the demand. This is outlined in Figure 10.1.3.

Areawise biomass requirements is disaggregated to simulate the requirements for wood and other biofuels in the individual sub areas (taluk) used in the analysis. At the same time, the effects of changing patterns of inter taluk (sub area) such as from hilly to coastal taluks are incorporated. Wood stock and yield data is assigned to each land type to account for spatial variation in availability of resources. Dung resources are projected through an inventory of animals in each area and the quantity of dung produced per animal in each region. The energy demands for firewood, wood used for charcoal and other biomass fuels are taken from the calculations of the demand and transformation programs. Final demand for charcoal in the demand program is met in the transformation program by a mix of traditional and efficient kilns.

Projections of available wood resources, combined with estimates of the energy requirement and scenarios describing the land use changes, allow the biomass module to simulate future wood growth and harvest, and indicate the adequacy of wood and other bioresources. In the final analyses, this module computes supply-demand shortfall if resources are found to be inadequate to meet requirements. This supply-demand shortfall is intended to indicate that responses in the form of energy efficient devices and planting of polyculture plant species are necessary.

(ii) *The Wind Resource Module:*

This module determines the wind potential in a region for (a) water pumping purposes and (b) wind farms - for electricity generation. This assessment is based on wind data - monthly, seasonal and yearly change of wind speeds (based on daily and hourly values) and wind speed at various elevations. Based on available wind resources and extent of land (feasible sites) to set up wind farms, wind resource module projects the wind energy that could be harnessed for electricity generation. Techno-economic analysis decides the viability of wind energy systems in a region.

(iii) Solar Energy Module:

The solar energy module is similar to wind energy module. Based on solar radiation and climatological data, feasible solar energy sites could be determined through this module. Land use details provide the extent of land available for harnessing solar energy through either solar thermal devices or photovoltaic devices. Techno-economic analysis is intended to decide technical, economic, ecological and social aspects, and acceptability of devices in a region.

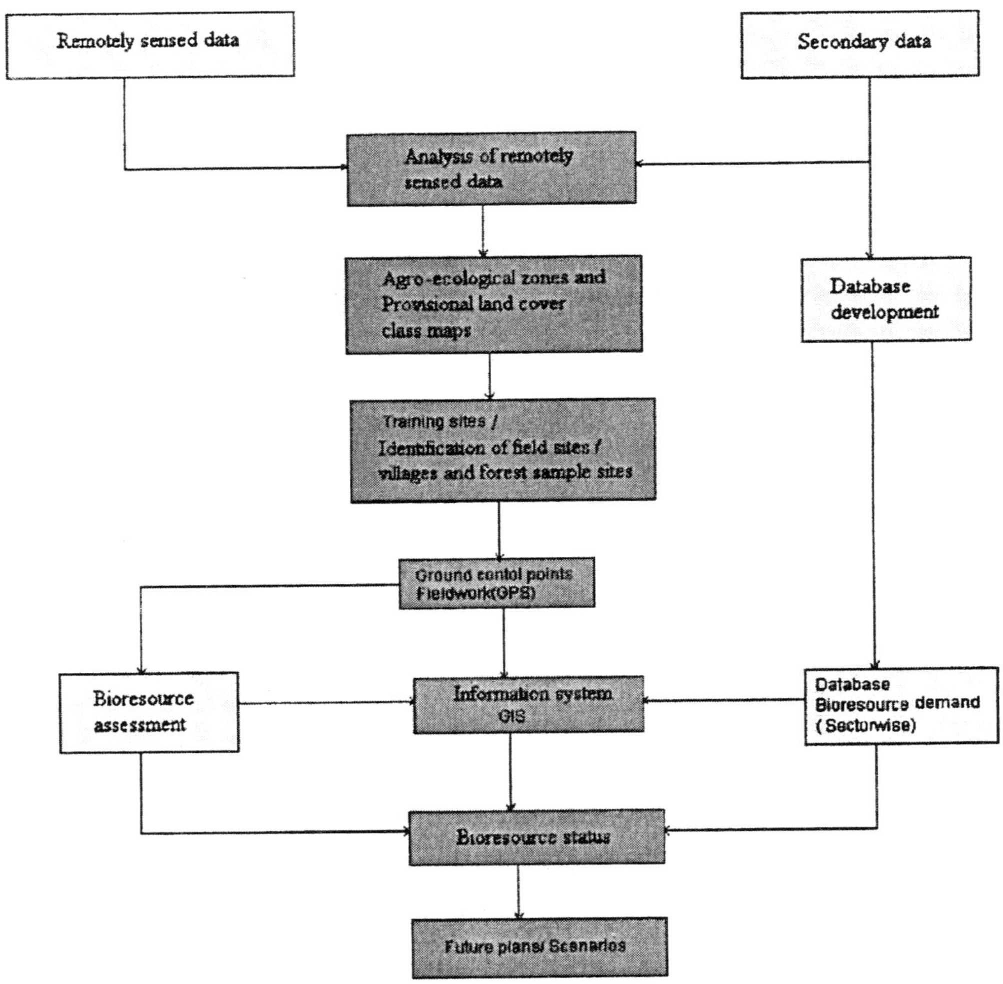

Figure 10.1.3: Methodology for Assessment of Regional Bioresource Status

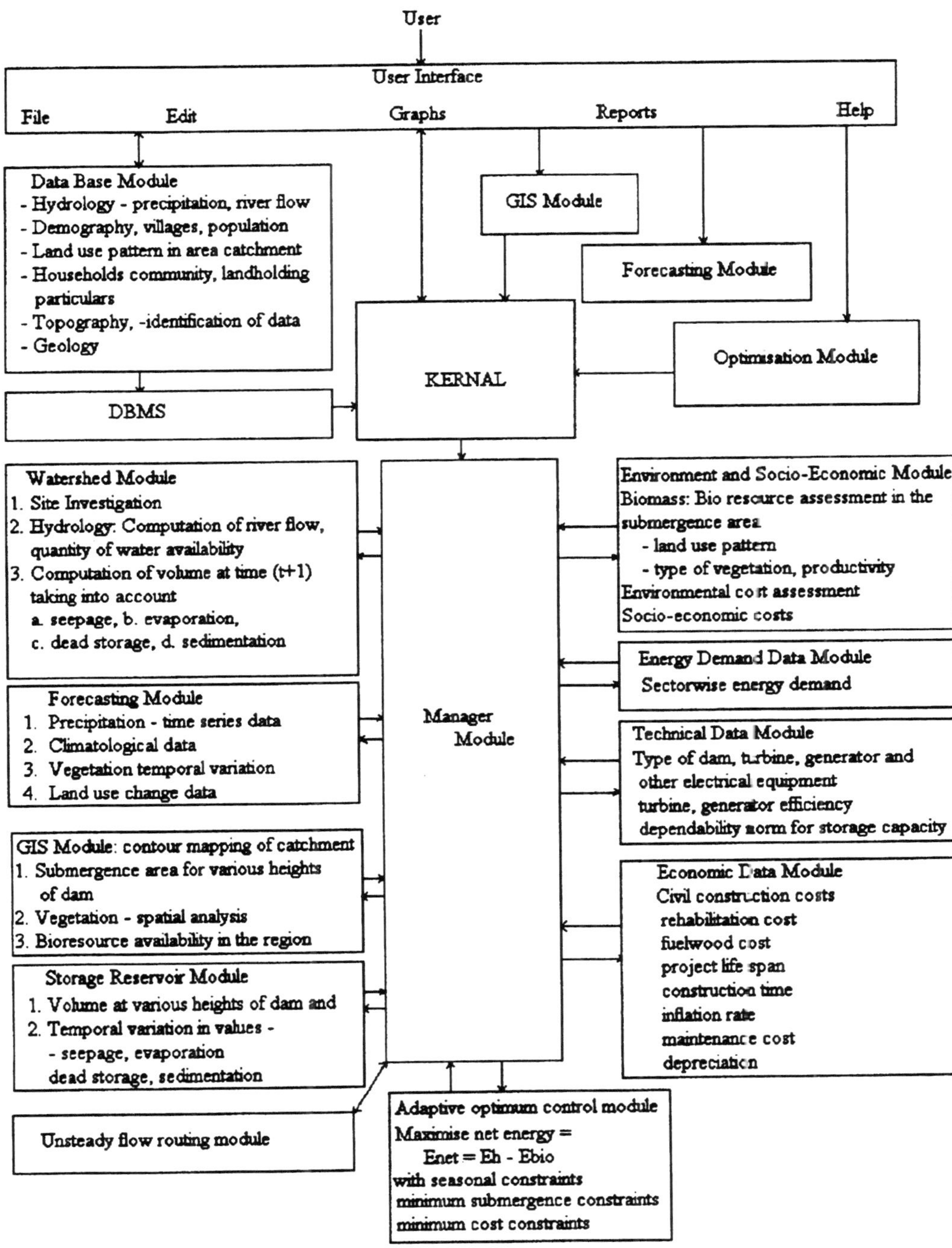

Figure 10.1.4: Design of Hydroelectric Station with Optimal Storage Option – Detailed Operation Module

(iv) Hydroelectric Energy Operation Module:

The main functions of this module involve the answering of basically the following questions: (1) How much energy can be generated using flowing water of a river/stream and what is the reliability of such decisions? (2) How to operate a reservoir on a daily basis, taking seasonal constraints in order to achieve this goal? and (3) How to maximise the net energy available in the region, while reducing submergence and construction costs?

The planning operation module will answer all these questions based on predetermined hydrological flow (database module, environmental and socio-economic module), GIS module, technical data, economic data and a certain user selected reliability level. It also generates adaptive operation policies based on this reliability level. Figure 10.1.4 shows the components of the planning operation module. The sequential operation of this module is as follows:

(1) Start at beginning of the monsoon season,

(2) User determines reliability level,

(3) Manager module querys database record for historical flows of the system, and watershed module to predict inflows in the system,

(4) Manager module starts the statistical module to generate mass curves and statistical parameters,

(5) Manager module starts the environmental and socio-economic module for determining demands and constraints on the system (minimum reservoir level, water demands, downstream flows, etc.), and starts forecasting module to forecast rainfall and/or demands at the defined lead time,

(6) The manager module starts unsteady flow routing module to give initial guess on routings and flow constraints,

(7) GIS module, with contour mapping (elevation) details, provides the submergence area for various dam heights, land use particulars and bioresources availability in the region,

(8) Technical data module provides technical details of electrical machineries, dam, turbine, generator efficiencies, and dependability norm for storage capacity,

(9) Energy demand data module provides energy demand (electricity and fuel wood in the region), which helps in designing operation policies during the lean season,

(10) Economic data module provides civil construction costs for various types and heights of the dam, cost of electrical machineries of various capacity, environmental costs, rehabilitation costs, land use details, and land submergence costs, construction time of each design, inflation rate, operation and maintenance costs, and depreciation costs,

(11) Adaptive optimum control module decides the quantity of water drawn from the reservoir, taking into account seasonal constraints, volume of water stored for particular height of dam, seepage loss, evaporation loss, dead storage capacity of dam, and rate of sedimentation, so as to ensure there is no wastage of water. This also decides on storage capacity of a reservoir during non monsoon season. Net energy available in the region is also presented in this module.

Step 6 is recalled by manager module to update routing and flow constraints. This is repeated till convergence is obtained. Manager module starts GIS module to present changed submergence area for changed water drawn policies (that is for changed height of dam), reservoir levels, thus repeating the process for various reliability levels and dam heights,

(12) Results are tabulated using spreadsheet software, and

(13) Operating rules are presented, and return to main module.

(v) Micro/Mini/Small Hydropower Plants Module:

The potential assessment of feasible sites for seasonal hydropower plants in a region is carried out in this module. Quantity of water, terrain, technical and economic viability of plants decide the implementation of the project.

10.3.4 The Evaluation Module

This consists of Cost Benefit Analysis and Environment Programs - used to compute the overall economic and environmental consequences of alternative energy futures. The cost benefit option allows the incorporation of technology and project costs, inflation and discount rates into a comparison of policy scenarios, as well as, costs of environmental externalities.

10.3.5 Computation Models

The energy planning endeavor for a particular region involves the finding of a set of sources and conversion devices, so as to meet the energy requirements/demand of all the tasks in an optimal manner. This optimality depends on the objective to minimise the total annual cost of energy. Factors such as availability of resources in the region and task energy requirements impose constraints on the regional energy planning exercise. Thus, the regional energy planning turns out to be a constrained optimisation problem. The basic components in the design are,

(i) *Regional Energy System:*

Resources form inputs and consumption by different tasks via varied technologies of conversion form outputs in the regional energy system. The energy sources are both renewable and non-renewable. Renewable resources presently in wide use in Uttara Kannada district are firewood, agro residues and animal wastes. Out of these resources, firewood is fast dwindling because of its uncontrolled and inefficient usage and increasing demand due to increase in population, affecting adversely the quality of life and environment. This is evident from barren hill tops in the densely populated coastal taluks of Karwar, Kumta, Honnavar, Ankola and Bhatkal. There is thus, an urgent need to tap new and technically feasible resources, such as, solar, wind, potential energy of water and biogas, in a decentralised way. The renewable energy resources readily available in the region are:

(1) Bioresources: firewood, agricultural and horticultural residues, animal wastes - either used directly or converted into biogas.

(2) Potential energy of water: used to drive mechanical devices, and converted to electricity

(3) Wind energy: converted to rotary mechanical energy using turbines, and converted to electricity

(4) Solar radiation: can be used directly through thermal device-flat plate collectors or concentration, and converted to electricity using PV (photovoltaic) or thermal devices.

Resources such as tidal wave energies are still in experimental stage and are not estimated/included in the regional energy plan.

(ii) *Tasks:*

The tasks can be categorised based on their thermodynamic characteristics, such as, type and quality of energy needs. These are:

(1) Low grade thermal energy (less than 100°C): needs, such as, domestic water heating, space heating, grain drying, and process heat for small-scale industries.

(2) Medium grade thermal energy (100-300°C):
- cooking in households
- village industries, such as, jaggery making, areca boiling, parboiling, etc.
- small-scale industries, such as, cashew processing, brick kilns, tile industries, etc.

(3) Lighting: domestic, community, industrial, etc.

(4) Mechanical works (rotating mechanical shaft power):
- pumping water for domestic purposes, irrigation, etc.
- small-scale industries like rice mills, etc.

(5) Motive power: agricultural operation and transportation.

(iii) Integration of Sources:

System integration involves mixing of all sources via a number of devices to meet the demand of various tasks. An optimised regional energy system involves proper mix of energy systems, which involves proper mix of energy sources that supply energy required for various tasks through appropriate devices based on the second law of thermodynamics.

The sources may be primary and local in nature (firewood, solar, wind, etc.) or intermediate, which are produced within a region, such as, electricity from solar, wind and bioresources, producer gas from wood, biogas from organic wastes (such as animal dung, green leaves, etc.) or non-local resources transmitted/transported from outside the region, such as, grid electricity, kerosene, diesel, petrol and LPG.

There are various approaches in integration, such as,

(a) *Combining*: In combining sources approach, two or more energy sources act in conjunction to perform a single task. For example, biogas and diesel are both used to run an engine which could be coupled to centrifugal pump to pump water, or to a generator to generate electricity for lighting, pumping, industrial applications, etc.

(b) *Cascading:* Principle of cascading involves usage of heat or waste energy (low grade energy) that is produced while performing a task. This low grade energy could be used for low grade thermal energy tasks. For example, waste heat from an engine may be used to carry out a heating task.

(c) *Time sharing:* Devices make use of same resources for different tasks at different intervals of time.

In an integrated energy system, integration occurs at three levels. Firstly, several energy sources are integrated into a single energy system. Secondly, traditional components are integrated into introduced technologies. Thirdly, energy system is integrated into the fabric of the local society. Such integrated systems change continuously as they are adjusted to new situations of energy availability and utilisation. Hence, the regional energy systems must be dynamic and interactive to be effective. But, Uttara Kannada region is basically rural and economy is sluggish, hence the first two levels of integration suggested are more valid. Integration of various resources and devices to perform various tasks in a region is listed in Figure 10.2.

10.4.0 DESIGN APPROACH

The design of the regional energy system involves finding the quantities of various energy resources required per year and the capacities of devices that minimise the total cost per year.

Let 'M' be the number of resources used/considered to perform 'N' tasks and 'D' be the number of devices.

Let i=1, X_1 = Quantity of Fuel wood/Year
i=2, X_2 = Quantity of Biogas/Year
X_3 = Quantity of Wind
X_4 = Quantity of Solar Energy (Thermal + PV Module)
X_5 = Quantity of Hydroelectric Energy (micro/mini/small and dam option at Unchalli of Aghnashini and Magod of Bedthi river)
X_6 = Electricity from Karnataka Electricity Board
X_7 = Quantity of Kerosene distributed in the district per year
X_8 = Quantity of Petroleum Products

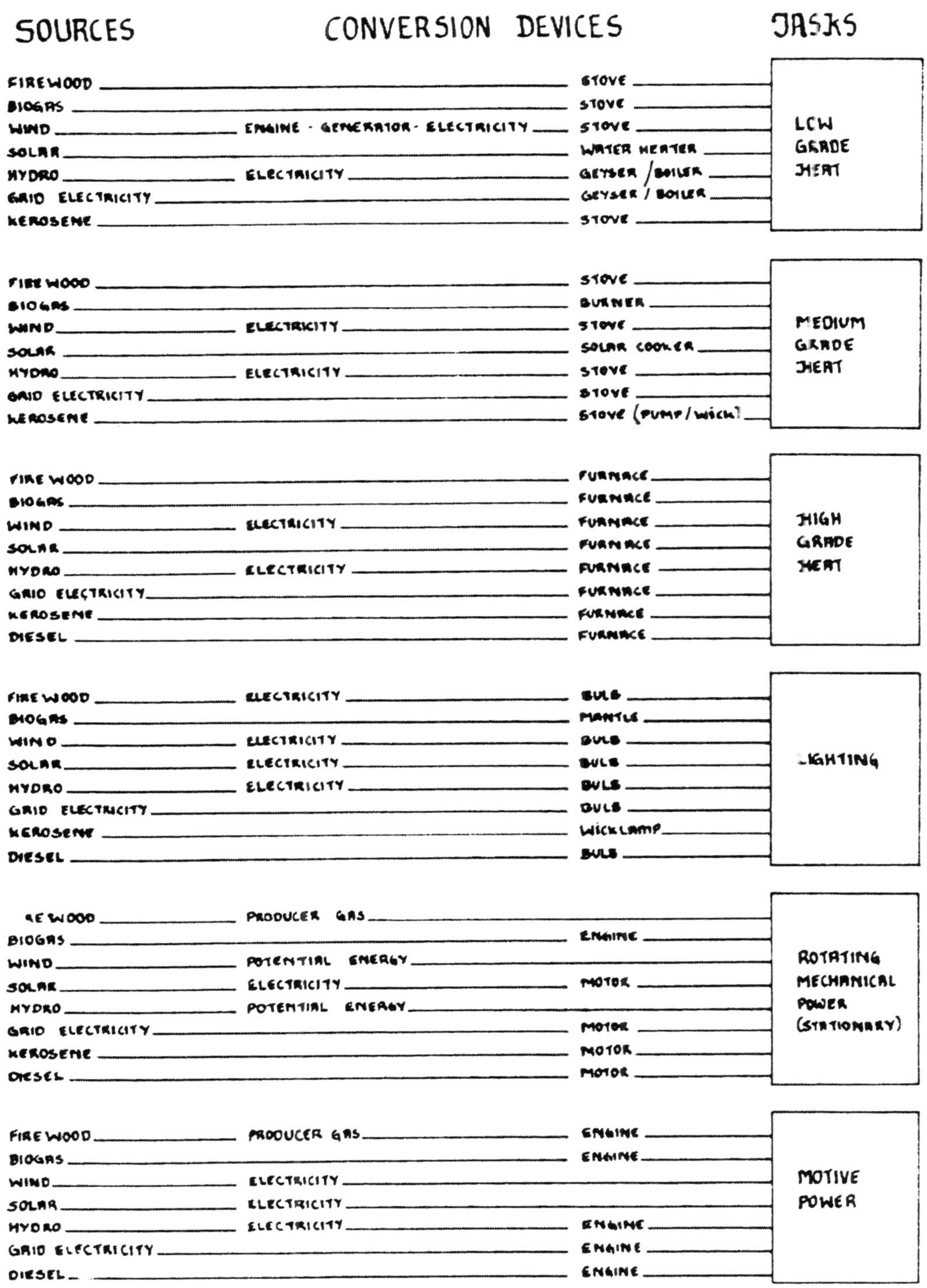

Figure 10.2 Integration of Sources and Devices to Perform Various Tasks

Each of the resources listed above may be used for more than one task, through various devices.

Thus, X_{ijk} is the portion of X_i required for j^{th} task through k^{th} devices.

Therefore, $$X_i = \sum_{j=1}^{N} X_{ijk} \quad (1)$$

Energy Equivalents:
The energy equivalents of the various resources computed are denoted by "R_i"

For e.g. $$E_i = R_i X_i \quad (2)$$

and $$E_{ijk} = R_i X_{ijk} \quad (3)$$

- i=1 R_1 is the energy equivalent of fuel wood per tonne (5112.8 kWh/tonne)
- i=2 R_2 is the energy equivalent of biogas in kWh per cubic meter (5.55 kWh/m^3)
- i=3 R_3 is the annual rotating mechanical energy output of wind turbines in kWh/Sq.m. of swept area
- i=4 R_4 is the annual thermal energy collected by solar thermal collector in kWh/Sq.m.
- i=5 R_5 is the potential energy stored in water in kWh/cubic meter
- i=6 R_6 is the energy equivalent of electricity supplied through grid (=1)
- i=7 R_7 is the energy equivalent of kerosene (8690.717 kWh/kilo ltrs.)
- i=8 R_8 is the energy equivalent of petroleum products (9585.35 kWh/kilo ltrs.)

The regional energy planning design problem involves means to meet the increasing demand for energy of the region from various resources, with the objective of minimisation of total cost and the constraints associated with various resources and loads to be supplied. Thus, this involves design of total annual cost function and the constraint equations.

10.4.1 Problem Formulation

A source may be meeting, either partially or fully, the energy requirements of more than one task at a given time.

The cost C_{ijk} per year thus would be

$$C_{ijk} = (A_{ac})_{ijk} E_{ijk} \quad (4)$$

where $(A_{ac})_{ijk}$ is the annualised cost of ith resources or j^{th} task.

$$(A_{ac})_{ijk} = \left[\frac{r(1+r)^{n_{ijk}}}{(1+r)^{n_{ijk}} - 1} + OC_{ijk} \right] \quad (5)$$

where
r = annual interest
n = time period and
OC = operation and maintenance cost.

For M resources and N tasks, the total cost/year is

$$C = \sum_{i=1}^{M} \sum_{j=1}^{N} \sum_{k=1}^{D} C_{ijk} = \sum_{i=1}^{M} \sum_{j=1}^{N} \sum_{k=1}^{D} (A_{ac})_{ijk} \quad (6)$$

but $E_{ij} = R_i X_{ij}$

$$C = \sum_{i=1}^{M} R_i \sum_{j=1}^{N} \sum_{k=1}^{D} (A_{ac})_{ijk} X_{ijk} \quad (7)$$

The design approach, minimises 'C' total cost subject to a set of constraints.

10.4.2 Constraints

(1) *Task Satisfaction:* It is the first constraint in the regional energy system. The total energy supplied by the sources should be equal to the energy requirements of the tasks.

If U_j is the (useful) energy requirement of task j (based on field data collected from various surveys) and X_{ijk} the quantity of energy obtained from source 'i' via device 'k' for task 'j', then

$$\sum_{i}^{M} \sum_{j}^{N} \sum_{k}^{D} U_{ijk} = \sum_{i=1}^{M} R_i \eta_{ijk} X_{ijk} \geq U_j \quad (8)$$

$j = 1 \ldots\ldots N$, $k = 1 \ldots\ldots D$

X_{ijk} = amount of end use demand supplied for j^{th} task (through k^{th} device by an ith source)

η_{ijk} = efficiency

R_i = energy equivalent of ith source

(2) *Availability of Local Resources:* There is an upper limit on the availability of local resources. Therefore, the total amount of a particular resource consumed for various tasks should be less than or equal to the corresponding maximum value. For i^{th} resource if $E_i max$ is the maximum energy available, then

$$E_i max \geq R_i \sum_{j=1}^{N} \sum_{k=1}^{D} X_{ijk} \quad (9)$$

$j = 1 \ldots\ldots N$, $k = 1 \ldots\ldots D$

Since $E_i max = R_i X_i max$, the above equation can be written as

$$X_i max \geq R_i \sum_{j=1}^{N} \sum_{k=1}^{D} X_{ijk} \quad (10)$$

$i = 1 \ldots\ldots M'$, $j = 1 \ldots\ldots N$ and $k = 1 \ldots\ldots D$

where, M' (< M) be the number of local energy resources and the subset of the set M of all energy resources (that is, $i = 1 \ldots\ldots M'$ and $M' \leq M$). This constraint also takes into account the limited availabilities of some of the local resources.

If M" is the subset of these resources having limited availabilities, R_i is the maximum energy available from source i ($i \varepsilon M''$).

(3) *Non Negativity:* All X_{ijk} values must be non negative,

$$X_{ijk} \geq 0 \quad (11)$$

for $i = 1,2,...M$

$j = 1, 2.....N$

$k = 1,2,.....D$

(4) *Power Requirements:* The fourth constraint is that some tasks have definite power requirement. The constraint imposed by the rate of energy use (power) should also be considered in regional energy planning. If P_{ijk} is the power required for handling the jth task supplied by i^{th} resource via k^{th} device, then the maximum rate of energy use expected of ith resource P_i is

$$P_i = \frac{1}{d_i} \sum_{j=1}^{N} \sum_{k=1}^{D} P_{ijk} \quad (12)$$

where d_i is the diversity factor for the tasks 'j' supplied by the ith resource, and it is ≥ 1

$$d_i = \frac{\text{Sum of individual maximum demand}}{\text{net maximum demand}} \quad (13)$$

If P_j' is the maximum power required for performing j^{th} task , then $X_{ijk} = 0$, if $P_{ijk} < P_j'$. This condition can be used to eliminate the source - device combinations, which cannot yield adequate power for the task. In case, there is a possibility of combining sources in devices, then the condition would be

$$X_{ijk} = 0, \text{ if } \sum_{j=1}^{N} \sum_{k=1}^{D} P_{ijk} < P_j \quad (14)$$

It is also essential that the sum of the power outputs of the ith energy source through all the paths 'ijk' should not exceed maximum power P_i, which the source can develop.

$$P_i \geq \sum_{j=1}^{N} \sum_{k=1}^{D} P_{ijk} \quad (15)$$

$(i = 1,2......... M)$

If the total energy derived per year from the ith resource for all tasks is E_i and if K_i is the effective load factor experienced, then the available power rating is

$$Pa_i = \frac{E_i}{8760\, K_i} \quad (16)$$

$i = 1,2...... M$

The power constraint is $P_i \leq Pa_i$ $(i =1,2....... M)$
But, P_{ijk} is the contribution of power output for j^{th} task from i^{th} resource through k^{th} device.

$$P_{ijk} = \frac{E_{ijk}}{8760\, K_{ijk}} \quad (17)$$

But,

$$E_i max = \sum_{j=1}^{N} \sum_{k=1}^{D} E_{ijk} = R_i \sum_{j=1}^{N} \sum_{k=1}^{D} X_{ijk}$$

$$\text{Therefore } P_{ijk} = \frac{E_{ijk}}{8760\,K_{ijk}} = \frac{R_i\,X_{ijk}}{8760\,K_{ijk}} \tag{18}$$

The power constraint $P_i \le Pa_i$ becomes

$$\frac{1}{k_i}\sum_{j=1}^{N}\sum_{k=1}^{D} X_{ijk} - \frac{1}{d_i}\sum_{j=1}^{N}\sum_{k=1}^{D}\frac{X_{ijk}}{K_{ijk}} \ge 0, \quad i = 1,2......M \tag{19}$$

While, K_{ijk} is the load factor, based on its definition and physical significance, it can be written as

$$K_{ijk} = \frac{E_{ijk}}{8760\,P_{ijk}} = \frac{U_{ijk}}{8760\,P_{ijk}\,\eta_{ijk}} \tag{20}$$

Thus, the regional energy planning exercise becomes an optimisation problem, which is stated as follows:

Minimisation of the annual cost function C,

$$C = \sum_{i=1}^{M} R_i \sum_{j=1}^{N}\sum_{k=1}^{D} (A_{ac})_{ijk} X_{ijk} \tag{21}$$

Subject to the constraints,

(1) $$\sum_{i=1}^{M} R_i\,\eta_{ijk}\,X_{ijk} \ge U_j \tag{22}$$

$j = 1$ ....N, $k = 1$ ....D

(2) $$X_i\max \ge R_i \sum_{j=1}^{N}\sum_{k=1}^{D} X_{ijk} \tag{23}$$

$i = 1$,......M', (where M'< M)

$j = 1$ ....N, $k = 1$ ....D

(3) $$\frac{1}{k_i}\sum_{j=1}^{N}\sum_{k=1}^{D} X_{ijk} - \frac{1}{d_i}\sum_{j=1}^{N}\sum_{k=1}^{D}\frac{X_{ijk}}{K_{ijk}} \ge 0, \quad i = 1,2......M \tag{24}$$

(4) $$X_{ijk} \ge 0 \tag{25}$$

for $i = 1,2...M, j = 1, 2.....N, \quad k = 1,2....D$

10.4.3 Design Implementation

The formulation presented above for regional energy planning is tried for the data collected from extensive field surveys in Uttara Kannada district. Eight resources and six tasks are considered for demonstrating the usefulness of the design approach presented in the previous section. In arriving at the energy requirement for different tasks, aggregation has been done. For example, all the loads requiring rotating shaft power (such as water pumping, some agricultural operations, motor load in small-scale industries) are aggregated into one task (rotating mechanical power). The resources and tasks considered in this design are as follows.

Resources:

i = 1,	fire wood, agricultural residues
= 2,	biogas
= 3,	wind energy
= 4,	solar energy
= 5,	hydropower resources (micro/mini/small, etc.)
= 6,	grid electricity
= 7,	kerosene
= 8,	petroleum products (petrol, diesel, etc.)

Tasks:

j = 1,	low temperature heating (water heating, drying, space heating, etc.)
= 2,	medium temperature heating (primarily cooking, etc.)
= 3,	high temperature heating (furnace, industrial applications, etc.)
= 4,	lighting (domestic, industrial, street lighting, etc.)
= 5,	rotating mechanical power (irrigation, industrial applications, etc.)
= 6,	motive power (transportation, etc.)

10.4.4 Data Used in the Design: Tasks or End-Users

Identification of various tasks and their energy requirements, and also peak power requirements (if any) constitute the first step in regional energy planning.

The regional energy tasks can be categorised as (i) domestic, (ii) agricultural, (iii) industrial, (iv) transport, and (v) commercial and service sectors. They can also be classified according to the quality or grade of energy requirements (thermodynamic characteristics). For example, cooking is a domestic task requiring medium temperature heat, and crop drying an agricultural task requiring low temperature heat.

(i) *Domestic:* Cooking and water heating (for bathing) are the most important domestic activities in Uttara Kannada. Survey carried out in the domestic sector shows that the domestic sector mainly depends on fuel wood. This region is rich in bioresources; the noncommercial sources of energy (fuel wood, agricultural residues and animal waste) consumption constitutes about 97.15% of the district's total consumption. Out of this, firewood's share is about 59.02%, mainly in the domestic sector. Firewood is collected from nearby forests or from *betta* lands twice a year (discussed earlier in detail) in the hilly areas of the district, while in coastal taluks, due to scarcity (depletion of forest resources), villagers make frequent trips to nearby sources to gather fuel wood, mostly in the form of twigs and branches. Apart from these, in the coastal area, the forest department has made arrangement to supply fuel wood at subsidised price, based on ration card, at the rate of 15 kgs/person/month.

Detailed household survey carried out in Kumta taluk (seasonwise) shows that fuel wood consumption ranges from 2.07 ± 0.38 (summer), 2.13 ± 0.37 (winter) to 2.31 ± 0.41 kgs/person/day (monsoon) for cooking.

While for water heating, it ranges from 1.29 ± 0.29 (summer), 1.39 ± 0.36 (winter) to 1.47 ± 0.36 kgs/person/day (monsoon). This shows very minimal seasonal variation in domestic sector's fuel wood consumption.

In the coastal area, due to fuel wood scarcity, households also use kerosene for cooking. It ranges from 0.34 ± 0.79 (coastal) to 0.05 ± 0.19 ltrs/person/month (hilly area).

Due to availability of bioresources in large quantity, households in the domestic sector have switched over to biogas in Sirsi and Siddapur taluks. Biogas consumption ranges from 0.23 (coast) to 0.49 cum/person/day (hilly area).

LPG (liquified petroleum gas) consumption ranges from 0.02 (coastal zone) to 0.01 kg/person/month (hilly zone).

Lighting needs: This also constitutes one of the essential needs in the domestic sector. Most of the villages in the district are electrified except very remote locations, such as, Yana, Sandolli, Medine, Kalave (Kumta taluk), Mogadde and Uddal (Sirsi).

In electrified villages, all households are electrified. Under Bhagyajyothi scheme, the poor and socially backward categories have been provided one bulb in each household (free installation). To meet the lighting requirement in other parts of the house in these households, and due to erratic power supply by the centralised system (KEB), they still use kerosene wick lamps for illumination. These lamps, however, have low efficiencies and poor illumination.

Kerosene consumption for lighting on average ranges from 0.82 (summer) to 0.87 ltrs/person/month (monsoon). Electricity consumption (at present, mainly lighting) is about 2.22 kWh/person/month in this region.

(ii) *Agriculture:* Uttara Kannada district's economy mainly depends on agriculture and horticulture. Land preparation, irrigation, harvesting, threshing and transportation are the main tasks dependent on mechanical form of energy apart from manure input, seed, etc. At present, rainfed paddy is the major crop in agriculture and, hence, less dependent on irrigation. There is scope to grow second crop in this region, which involves irrigation. This can be met either from animate or inanimate sources of energy. The water requirement and associated average daily energy requirements depend on the area irrigated, type of crop, sources of water, total period of irrigation and irrigation efficiency. Uttara Kannada's agricultural sector is active only during monsoon. During lean season, fodder crop is grown in some places (such as Banavasi), which enriches the soil by increasing its nitrogen content.

Apart from this, horticulture work, such as, land preparation and mulch manure application is done during the season when agriculture is not carried out. Most of the horticultural land owners are also agricultural land holders. However, there is scope for second crop in the coastal and plain regions (such as Mundgod).

Cooking, water heating and lighting energy requirements are to be met daily, while the energy requirements of agriculture (such as, land preparation, irrigation, harvesting, etc.) depend on the cropping pattern and season.

(iii) *Industries:* Industries in Uttara Kannada can be grouped as large and medium-scale, small-scale and village industries, such as, brick kilns, jaggery making units, etc. Large and medium scale industries are located mainly in Karwar and Haliyal, while small-scale industries, such as, agro processing units, rice mills, etc., are distributed all over the district. Operations such as jaggery making, etc., are seen only during certain seasons.

Most of the small-scale and village industries depend on fuel wood and agriculture residues to meet their heating needs, since electricity is mainly used for lighting activity. However, large and medium-scale industries such as caustic soda and chemicals (liquid chlorine, hydrochloric acid, etc.) manufacturing sector are dependent mainly on electricity. Alternative sources and technologies to meet the energy needs of the industrial sector are available now. Quantitative analyses of resource availability and techno-economic analyses of technology done in earlier chapters, show that alternatives are available to meet the local energy requirement. The industrial energy requirement is estimated based on the sample survey conducted in Kumta and Sirsi taluks.

10.4.5 Resources Data

With the tasks and corresponding energy requirements identified, the next step in the regional energy planning exercise is to look at the sources, including traditional ones, to perform these tasks.

Firewood is the main source of energy for cooking in Uttara Kannada district. Water heating task is carried out with agriculture residues and firewood. Survey of 1304 households in Kumta taluk reveals that most of them still use traditional stoves, the efficiency of which is in the range of 10-15%. Kerosene and LPG (very small percentage: 0.0025) are among the noncommercial (non-local) resources used in Uttara Kannada. Kerosene is used largely for lighting. Grid electricity is used for lighting and in few households for energising pumpsets to irrigate the plantations during lean season.

Agriculture in Uttara Kannada depends mainly on animate sources of energy, such as, draught animal power, human power, etc. Organic farming is practiced in most sampled households.

Analyses of bioresources (in the third chapter) show that biogas can meet cooking and lighting needs of at least 30% population (in Uttara Kannada district). The biogas potential (based on available animal residue) ranges from 11.42 to 19.45% (in coastal area), and 19.92 to 54.69% (in hilly area).

Producer gas engine can provide necessary motive power required for agricultural pumpsets, tractors and tillers. Wood required for producing producer gas through gasifiers can be raised from energy plantations on degraded land (is necessary). Availability of local resources is limited due to various factors. These are,

(1) Size and number of devices required to harness a source, like the number of wind mills used and the energy they can capture depend on (a) wind regime and the total swept area, and (b) solar radiation size of PV module/collector area (thermal devices).

(2) In the case of animate sources, the available energy depends on their use hours/day and the capacity of sources.

(3) The amount of raw material available for production (e.g., animal wastes for biogas, wood for producer gas, etc.)

(4) In the case of hydropower resources, it depends mainly on precipitation - quantity of water and head of water drop at site.

10.4.6 Assumptions

(i) *Scenario I - Base Case Scenario:*

The Base Case Demand projections assume that energy consumption is basically determined by the growth rate of the population and level of development in the region. The dependency of energy demand was modelled, quite simply as a direct relationship, between demand and population. For this purpose, to project population growth rate, population data provided by directorate of census operation for last 30 years (1961-91) has been used. The energy growth rate used in the Base case and Transformation case, was based on the information provided by The Planning Unit, District Planning Division, Uttara Kannada Zilla Parishad at Karwar. The energy growth estimate used in Scenario I (Base case) and Scenario II is based on the assumption that policies of 1990's will be maintained.

In the Base Case Scenario, changes in demand are assumed to follow changes in key driving parameters. For the domestic sector, future demand was assumed to be proportional to population growth, that is, in the range of 2.07 to 2.34 % per year. Agricultural and commercial changes in demand are linked to growth in overall economy. For the industrial sector, demand in each subsector/group was assumed to increase in proportion to growth in industrial production. Average industrial and overall energy growth levels were assumed to be around 2.8 and 2.45 % per year between 1994 and 2010.

(ii) *Scenario II - High Energy Intensity:*

In Scenario II, for industrial sector, 20% increase in the intensity of useful energy consumption for thermal end use (such as, direct heat, steam generation, etc.) between 1994 and 2000 and 30% increase between 2000 and 2010 is assumed. For the end uses that require electricity (such as, motor, lighting, etc.) a 10% increase from 1994 and 2000 is assumed, with a further 10% increase over base year levels by 2010. The basis of the growth in intensity assumed for the thermal end uses is that Uttara Kannada's regional economy is basically rural and sluggish that characterises this scenario result in inadequate utilisation of industrial capacity.

(iii) *Scenario III - Transformation Case Scenario:*

The Transformation Case Scenario implies substantial changes in attitude about energy consumption within the region. In this aspect, this scenario is granted less apparent probability of occurring than Scenario II. This scenario assumes policies of savings and conservation in almost all sectors, as well as, improvement in the efficiencies of energy consumption and energy transformation equipment. In the industrial sector, the amount of useful energy consumed per unit industrial value added is assumed to decline by 2% between 1994 and 2000, with further 2% decrease by 2010. The result of this assumption is a reduction in the energy intensity for the majority of end uses, and a major proportion of the overall reduction are in end uses that require heat (such as rural industries - cashew processing, etc.). In the domestic sector, 3.5% growth in the electricity consumption per household for end uses, such as, illumination, domestic appliances, etc., is assumed for the period between 1994 and 2010. This increase is based on the growth on income and therefore on the growth in overall economic consumption per capita. For the cooking end use, it is assumed that at least 50% of households using fuel wood, LPG and electricity as fuel by 2010 will have energy efficient devices than those in place of 1994. These efficient devices are assumed to have efficiencies of 35% (wood stoves), 78% (electric) compared with 1994 efficiencies of 7-8% (wood devices) and 60% (electric) respectively. Also, it is assumed that at least 15% of total population would switch over to biogas devices for cooking purposes. (Calculations in an earlier chapter show that biogas from dung resource is sufficient to meet demand of 30% of the population).

(iv) *Scenario IV - State Growth Rate:*
The intent here is to develop a scenario for Uttara Kannada district based on state energy consumption norms. The projection of industrial and total energy consumption norms (industrial, agriculture, etc.) was used. The average growth is significantly higher in this case than in the Base case (5.74% per year for industrial and 4.0% for overall).

The growth rate in coal consumption is about 2.35%, kerosene 3.26%, electricity 8.51%, LPG 19.46%, fuel wood 1.95% and agricultural residues about 3.85% (these growth rates are computed from time series data by regression analyses method). Directorate of Census operation projections were used for total population, persons per household and rural/urban fractions.

(v) *Scenario V; Scenario IV + Efficiency Improvements case:*
This scenario, in addition to policies of scenario IV, assumes the efficiency gains that are quite achievable for a given technology.

In the domestic sector, most households still use incandescent bulbs (about 92% of sampled households in five taluks of Uttara Kannada district). The efficiency of lighting devices is assumed to increase through the introduction of Compact Fluorescent Lamps (CFL). These devices, which produce nearly five times the light per unit energy input compared to incandescent bulbs, and have nearly eight times the life, are assumed to provide 5 percent of lighting by 2000, and 10 percent by 2010. These CFL's have relatively higher initial costs. But taking into consideration the energy saved (due to low wattage requirement), and longer life, calculations indicate saving of about 52% over a period of time. The fuel efficient cooking and water heating devices show saving of about 27% (discussed in the seventh chapter). Solar water heaters in coastal taluks of Uttara Kannada district are found to be viable alternatives. For cooking devices, this scenario assumes that more efficient stoves replace traditional stoves, as in the Transformation scenario. In addition to this, 15% adoption of solar hot water devices by 2005 is assumed. The efficiency of fuel use in services and commercial sector is assumed to improve for all fuels by 10 per cent (2000) and 20 per cent (2010), relative to base year results.

10.5.0 Results and Discussion

Sectorwise energy demand and resource potential compiled in the following section are based on the detailed investigations carried out in the earlier chapters (chapters 2 to 9).

10.5.1 Energy Demand

Compilation of survey data and sectorwise analyses show that noncommercial sources of energy, such as, firewood and agricultural residues are the major components in the districts energy consumption. Table

10.1.1 lists sectorwise energy demand based on detailed investigations carried out in various sectors of Uttara Kannada district. It shows that the domestic sector with 77.96% (8941.56 million units - mkWh) of total energy consumption, is the major energy consuming sector followed by industrial (15.93%) and agricultural (5.13%) sectors. This is illustrated in Figure 10.3.1.

Table 10.1.1: Sectorwise Energy Consumption (mkWh) in Uttara Kannada

	Energy (in mkWh)	% share
Agriculture	588.704	5.133
Commercial	11.168	0.097
Domestic	8941.560	77.959
Transport	100.570	0.877
Industry	1827.598	15.934
	11469.600	100.000

Table 10.1.2 and Table 10.1.3 give sourcewise consumption of energy. Uttara Kannada district mainly depends on noncommercial energy sources (96.23%), such as, firewood (78.09%), dung in the form of biogas (13.12%), and animate sources (5.02%). While, commercial sources, such as, electricity (2.86%), kerosene, diesel, petrol, etc., (0.90%) and LPG (0.0001%) constitute 3.77% of total energy consumption in the district. This is illustrated pictorially in Figure 10.3.2 and Figure 10.3.3 respectively. Sourcewise and sectorwise consumption of energy is listed in Table 10.1.4. Domestic (77.67%), industrial (13.535%) and agricultural (5.024%) sectors mainly depend on noncommercial sources of energy. This aggregated information is depicted in Figure 10.3.4. Energy demand of various sources in the district is listed in Table 10.1.5.

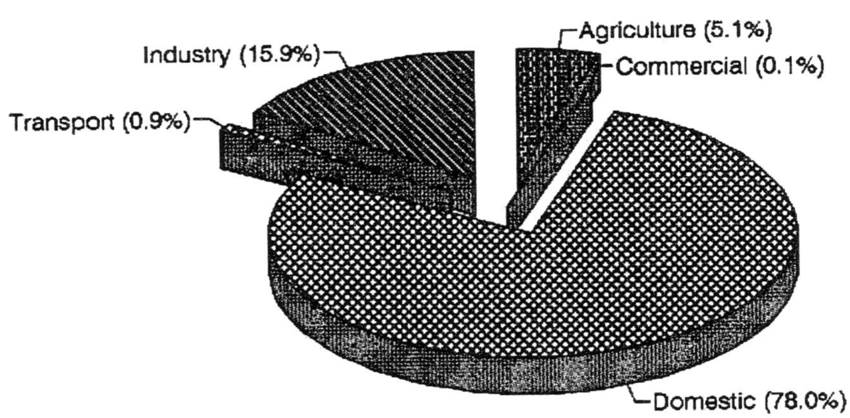

Figure 10.3.1: Sectorwise Energy Consumption - Uttara Kannada District

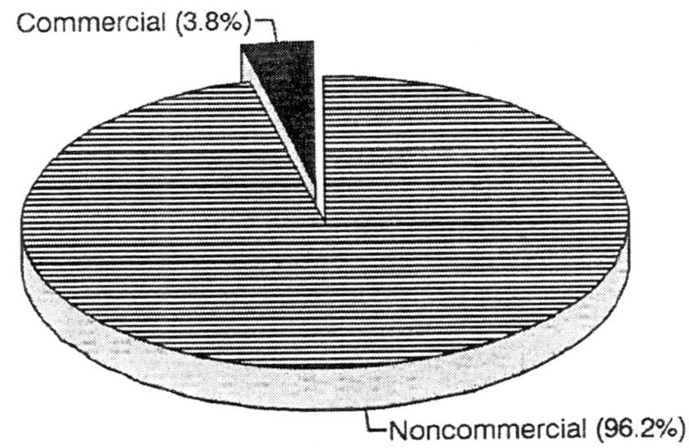

Figure 10.3.2: Energy Consumption Sourcewise – Uttara Kannada District

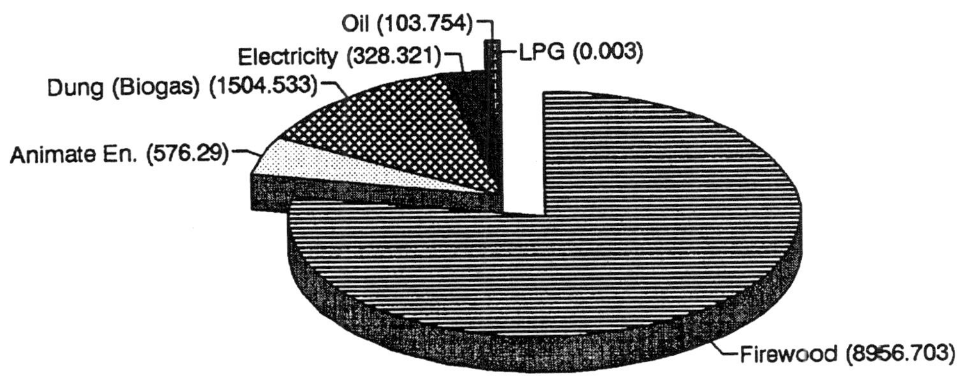

Figure 10.3.3: Sourcewise Consumption of Energy (Source - Types)

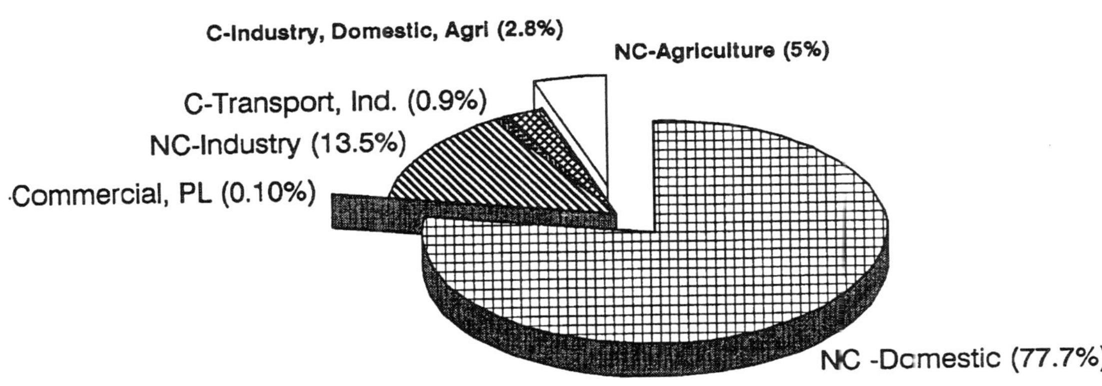

Figure 10.3.4: Sourcewise, Sectorwise (C: Commercial, NC: Non-Commercial)

Table 10.1.2: Sourcewise Composition of Energy (mkWh) in Uttara Kannada

	Energy (in mkWh)	% share
Noncommercial	11037.526	96.233
Commercial	432.074	3.767
	11469.600	

Table 10.1.3: Sourcewise and Typewise Composition of Energy (mkWh) in Uttara Kannada

Noncommercial	Firewood	8956.703	78.091
	Dung(Biogas)	1504.533	13.118
	Animate En.	576.290	5.024
Commercial	Electricity	328.321	2.863
	Oil	103.750	0.904
	LPG	0.003	0.000
		11469.600	100.000

Table 10.1.4: Sectorwise and Sourcewise Consumption of Energy (mkWh) in Uttara Kannada

Sector	Source	Energy (in mkWh)	% share
Domestic	Non Commercial	8908.773	77.673
Industry		1552.459	13.535
Agriculture		576.290	5.024
Commercial+PL*	Commercial	11.168	0.097
Domestic		32.794	0.286
Industry		275.134	2.399
Agriculture		12.410	0.108
Transport		100.572	0.877
	Total	11469.600	100.00

* PL = Public Lighting

Table 10.1.5: Energy Demand (during 1994) Uttara Kannada District (mkWh)

Fuel wood	8908.778
Dung (Biogas)	1552.459
Animate Energy	576.290
Electricity	328.320
Kerosene	0.140
Diesel	69.130
Petrol	34.480
LPG	0.003
	11469.600

10.5.2 Resource Scenario Results

Monthwise availability of solar and wind potential (discussed in detail in earlier chapters) in coastal taluks, and electricity that could be generated by using 5% of barren land are listed in Table 10.2.1. and Table 10.2.2. respectively. While, Table 10.3.1 and Table 10.3.2 give hydropower and bioenergy availability in the region respectively.

Table 10.2.1: Solar Energy Potential in Uttara Kannada District

	Global Radiation (kWh/sq.m.)			Solar Potential
Month	Karwar	Honnavar	Shirali	mkWh
January	5.67	5.47	5.35	8.35
February	6.27	6.10	5.73	8.33
March	6.30	6.24	6.04	9.41
April	6.69	6.39	6.30	9.54
May	6.48	6.23	6.21	9.56
June	5.10	4.93	4.50	7.24
July	4.02	4.17	3.50	5.92
August	4.84	4.86	4.78	7.17
September	5.10	5.27	5.18	7.33
October	5.41	5.35	5.25	7.99
November	5.13	5.03	5.03	7.34
December	5.12	4.97	4.79	7.54
Avg	5.51	5.42	5.22	Total Potential
Sd	0.76	0.66	0.76	95.72 mkWh

Table 10.2.2: Wind Energy Potential in the District and Anticipated Electricity that could be generated by using 5% of waste land

	Wind Potential W/sq.m.				Electricity
Month	Karwar	Honnavar	Shirali	Kumta	Generated
January	6.26	6.23	9.23	6.25	0.16
February	8.27	6.37	9.53	13.79	0.21
March	15.55	6.52	9.95	21.59	0.33
April	30.38	8.06	12.9	28.26	0.47
May	54.84	12.45	16.03	31.99	0.69
June	57.18	13.93	18.88	54.79	0.89
July	106.8	13.78	18.41	66.36	1.25
August	64.63	11.09	16.78	57.77	0.83
September	12.83	3.58	5.35	9.41	0.18
October	4.51	2.68	4.53	8.15	0.12
November	2.85	3.41	8.22	6.65	0.12
December	3.59	6.06	24.11	12.99	0.26
Avg	30.64	7.85	12.83	26.50	Total Potential
Sd	31.72	3.85	5.77	20.82	5.51 mkWh

Table 10.3.1: Monthwise Hydropower Potential in Bedthi and Aghanashini Basins

Hydel Potential in mkWh						
	Aghnashini			Bedthi		
Month	Streams	@Unchalli	Total	Streams	@Magod	Total
January	0.00	4.45	4.45	0.00	5.54	5.54
February	0.00	4.45	4.45	0.00	5.54	5.54
March	0.00	4.45	4.45	0.00	5.54	5.54
April	2.86	4.45	7.31	2.31	5.54	7.85
May	10.19	4.45	14.64	13.35	5.54	18.89
June	114.64	311.24	425.88	104.29	177.28	281.57
July	209.33	311.24	520.57	190.91	177.28	368.19
August	118.07	311.24	429.31	108.61	88.64	197.25
September	33.15	311.24	344.39	40.94	88.64	129.58
October	15.85	4.45	20.30	30.41	5.54	35.95
November	3.21	4.45	7.66	12.17	5.54	17.71
December	0.00	4.45	4.45	0.00	5.54	5.54
Sum	507.30	1280.56	1787.86	502.99	576.16	1079.15

Table 10.3.2: Talukwise Bioresource Demand and Availability taking Biomass Productivity as 3.6 t/ha/yr

					Fuel wood Demand	
Taluk	Agriculture Residue Total	Horticulture Residue Total	Forest Biomass (at 3.6) mkWh	Total Biomass mkWh	at 1.1 t /person/ yr mkWh	at 0.7 t /p/yr mkWh
Bhatkal	17.86	46.86	299.1	363.82	725.60	461.75
Kumta	26.02	96.45	559.0	681.46	754.44	480.10
Karwar	18.24	38.52	753.3	810.06	803.37	511.24
Honnavar	31.78	133.01	819.0	983.83	820.23	521.96
Ankola	23.12	44.32	1076.0	1143.46	513.53	326.79
Mundgod	21.73	7.41	985.0	1014.10	438.34	278.94
Haliyal	66.40	0.82	1124.1	1191.28	530.71	337.72
Siddapur	31.02	156.29	1146.2	1333.54	529.80	337.15
Sirsi	36.22	205.42	1900.0	2141.63	860.12	547.35
Yellapur	22.11	86.63	2087.8	2196.53	457.86	291.36
Supa	23.55	10.25	2956.2	2989.98	559.70	356.17
Total	318.07	825.97	13705.7	14849.70	6993.69	4450.53

Table 10.3.3: Talukwise Bioresource Demand and Availability taking Biomass Productivity as 13.5 t/ha/yr

Taluk	Agriculture Residue Total	Total Horticulture	Forest Biomass (at 13.5) mkWh	Total Biomass	Fuel wood Demand at 1.1 t/person/ yr MkWh	at 0.7 t/p/yr mkWh
Bhatkal	17.86	46.86	532.2	596.97	725.60	461.75
Kumta	26.02	96.45	989.8	1112.23	754.44	480.10
Karwar	18.24	38.52	1333.8	1390.56	803.37	511.24
Honnavar	31.78	133.01	1463.4	1628.19	820.23	521.96
Ankola	23.12	44.32	1905.2	1972.65	513.53	326.79
Mundgod	21.73	7.41	2672.3	2701.48	438.34	278.94
Haliyal	66.40	0.82	2773.9	2841.10	530.71	337.72
Siddapur	31.02	156.29	2069.6	2256.90	529.80	337.15
Sirsi	36.22	205.42	3406.8	3648.44	860.12	547.35
Yellapur	22.11	86.63	3691.3	3800.07	457.86	291.36
Supa	23.55	10.25	5276.0	5309.75	559.70	356.17
Total	318.07	825.97	26114.3	27258.35	6993.69	4450.53

Table 10.3.4: Talukwise Estimated Dung and hence Gas Yield per Year

Taluk	Total (Cattle+buffalo)		Number of persons requirement	Total Population	% of Population that could use biogas
	tonnes	Biogas 000 m^3			
Ankola	61236.78	2204.52	17764.09	91310	19.45
Bhatkal	54289.01	1954.40	15748.62	129017	12.21
Haliyal	143466.90	5164.81	41618.12	94363	44.10
Honnavar	83563.83	3008.30	24240.92	145842	16.62
Karwar	56221.68	2023.98	16309.27	142845	11.42
Kumta	70862.93	2551.07	20556.53	134144	15.32
Mundgod	100492.53	3617.73	29151.74	77939	37.40
Sirsi	279067.32	10046.42	80954.26	152935	52.93
Siddapur	214814.91	7733.34	62315.36	94202	66.15
Supa	68347.05	2460.49	19826.70	99519	19.92
Yellapur	153493.92	5525.78	44526.84	81410	54.69
District Total	1285856.86	46290.85	373012.46	1243526	30.00

Results of solar resource module are listed in Table 10.2.1. It lists monthwise amount of solar energy that could be harnessed through solar conversion devices, by utilising 10% of wasteland presently available in the coastal taluks of Uttara Kannada district. The electrical energy that can be harnessed is of the order of 191.44 million kWh, which constitutes about 65.1% of the present electricity demand from various sectors of Uttara Kannada district. These analyses also show that about 72% of solar energy is available during October to May period.

Table 10.2.2 lists monthwise electrical energy that could be generated by harnessing wind potential along the coastal zone of Uttara Kannada district. The availability of wind resource is quantified and characterised in detail in the second chapter. Due to local necessity of energy in the pre-monsoon period for industrial and irrigation purposes (for agriculture and horticulture), and the availability of wind resources during the same period, the exploitation of wind energy for mechanical and electrical energy purposes becomes feasible and desirable.

Hydropower resource module provides electrical energy potential in Bedthi and Aghnashini river basins of Uttara Kannada district. Results of this module are listed in Table 10.3.1. It is seen that about 1787.61 million units (mkWh) of electrical energy in Aghnashini and 1079.15 million units (mkWh) in Bedthi river basin could be harnessed. Monthwise electrical energy that could be harnessed from streams and major sites at Unchalli (of Aghnashini) and Magod (of Bedthi) are shown in Figure 10.4.1, while Figure 10.4.2 depicts separately the total hydro energy available in the respective river basins. The table also depicts that hydropower resource is available mainly during June to October period. Figure 10.4.3 illustrates the energy available from an area of 95.8 sq.kms by bioresources and hydropower resources through integrated planning approach. Hydroelectric systems for harnessing hydro energy at Magod and Unchalli sites are designed within social and ecological constraints.

The results of biomass module provide talukwise and categorywise biomass availability and demand balances. The biomass availability is computed based on different types of forest patches in the district and respective biomass productivities. Biomass module provides supply scenario for lower and upper ranges of biomass productivity. Similarly, biomass module provides bioresource demand taking into account demand range. Table 10.3.2 and Table 10.3.3 list talukwise biomass availability and demand for higher and lower ranges of productivity. The same is illustrated in Figure 10.5.1 and Figure 10.5.2 respectively. Out of 11 taluks of Uttara Kannada district, coastal taluks Bhatkal, Kumta and Honnavar have bioresource scarcity (Figure 10.5.1), while the remaining taluks have adequate resources to meet the region's demand. Talukwise bioresource availability from livestock population is listed in Table 10.3.4. It shows that if this resource is converted to biogas, it is adequate to meet the requirement of 30% of the total population. If the entire district is considered, bioresource supply appears to be diverse and sufficient in magnitude to comfortably sustain population pressure under the Base case demand scenario investigated. As bioresource (fuel wood) enhancement strategies are pursued through social forestry, etc., complementary policies to promote technically feasible and cost effective improvements in the efficiency of end use equipment are needed. End use efficiency study carried out in randomly selected households in some villages show that there is scope for saving about 27.45% of fuel wood by switching over to energy efficient stoves.

10.5.3 Computation Model Results

An integrated energy system capable of utilising several resources such as wind energy, solar energy, hydro energy and biogas for carrying out various tasks in a region is presented based on the design formulated earlier. The approach is based on minimisation of total annual cost, subject to a set of constraints on the requirements for various tasks.

The efficiency values η_{ijk} used in the formulation for various pre-selected resource-task combinations are listed in Table 10.4.1. The K_{ijk} load factor computed using the relationship (equation.20) is listed in Table 10.4.2. The effective plant load factor K_i and the diversity factor d_i assumed for resources are listed in Table 10.4.3. Capital cost values computed in Rs. per kWh (computed in earlier chapters - techno-economic analyses) are listed in Table 10.4.4. The linear programming problem is solved using Lindo-LP package. The resulting optimal values are listed in Table 10.5.1.

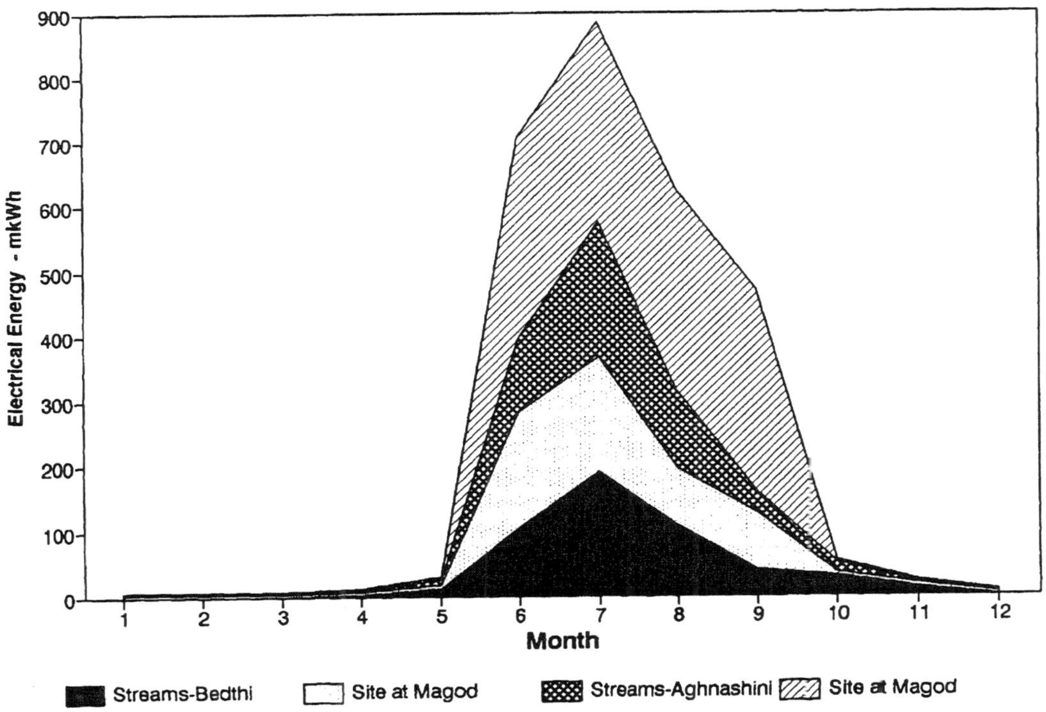

Figure 10.4.1: Hydroelectric Energy (Streams + River) Bedthi and Aghnashini Basins

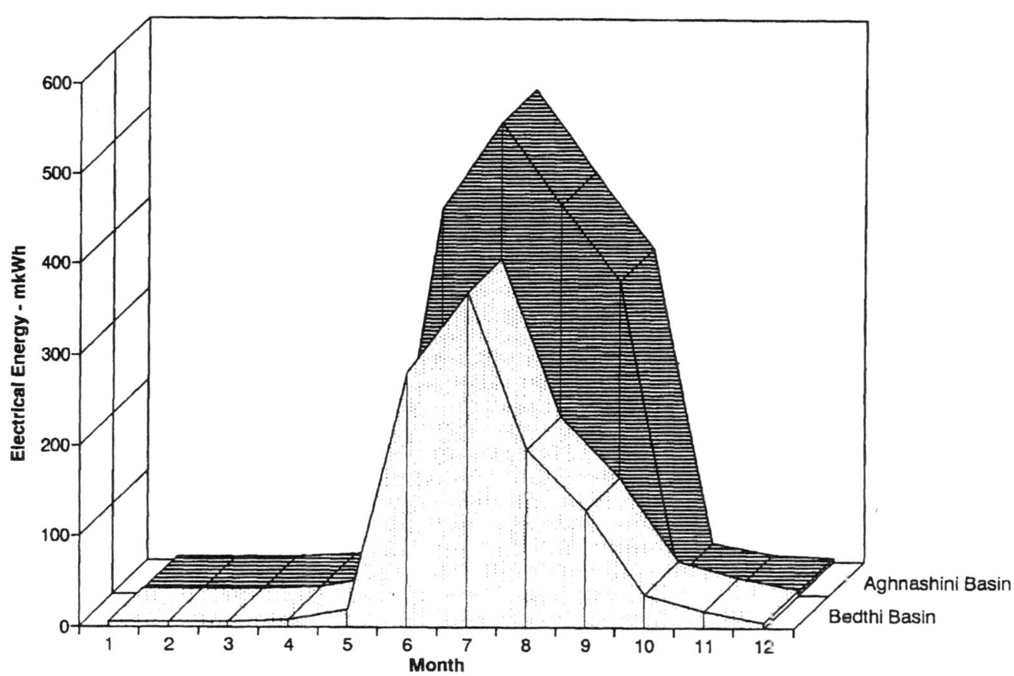

Figure 10.4.2: Total Energy Estimated in Bedthi and Aghnashini River Basins

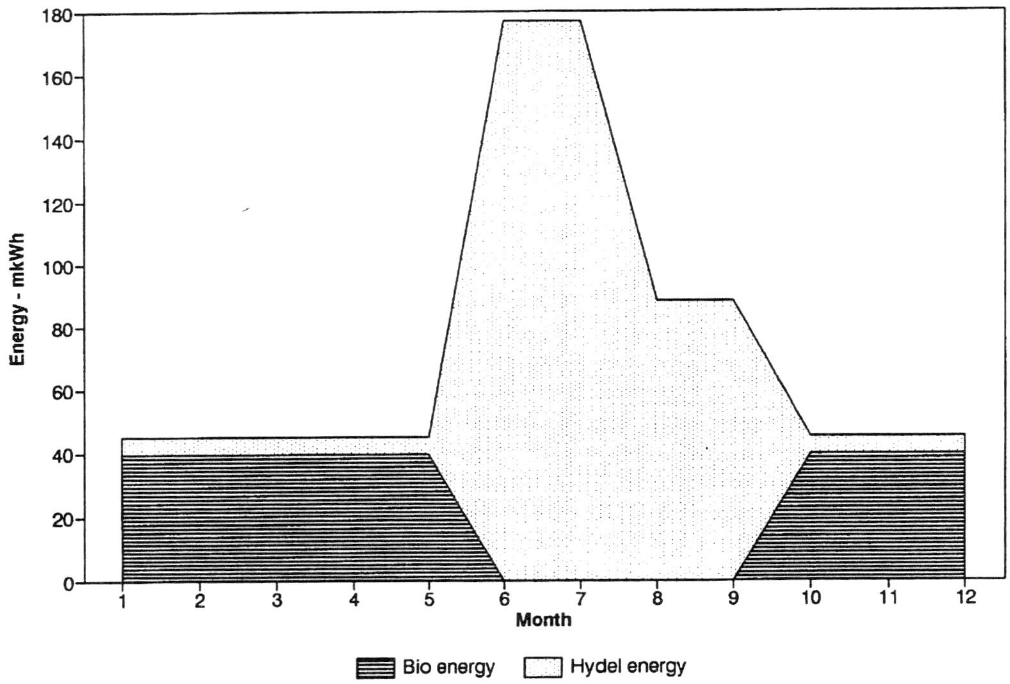

Figure 10.4.3: Net Energy Available in Bedthi Basin - from an Area of 95.8 sq.kms

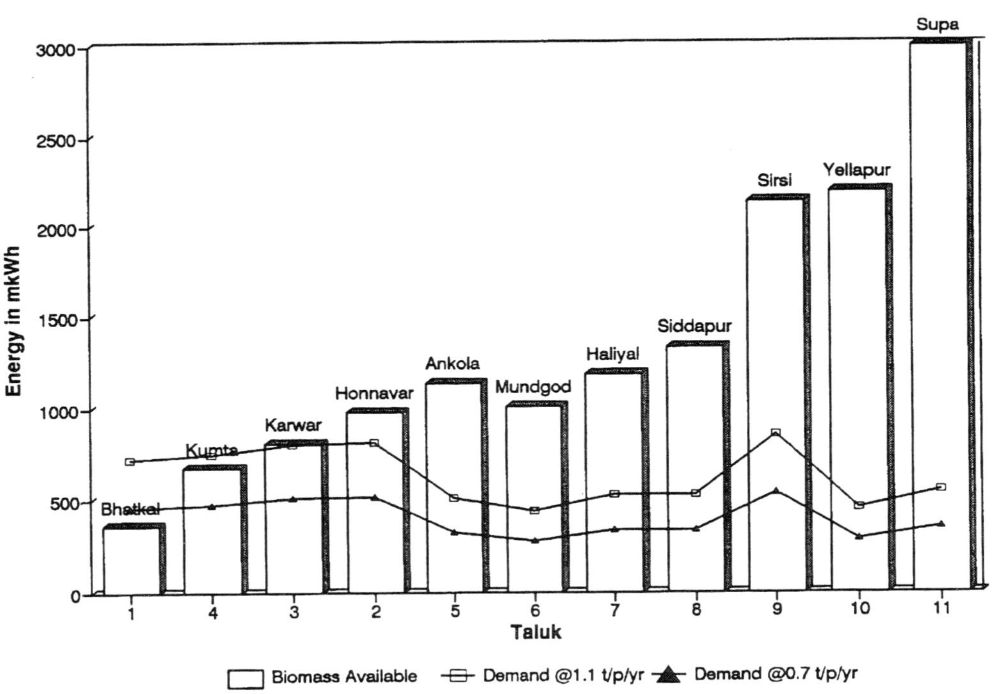

Figure 10.5.1: Biomass Availability and Demand - Talukwise, Uttara Kannada District

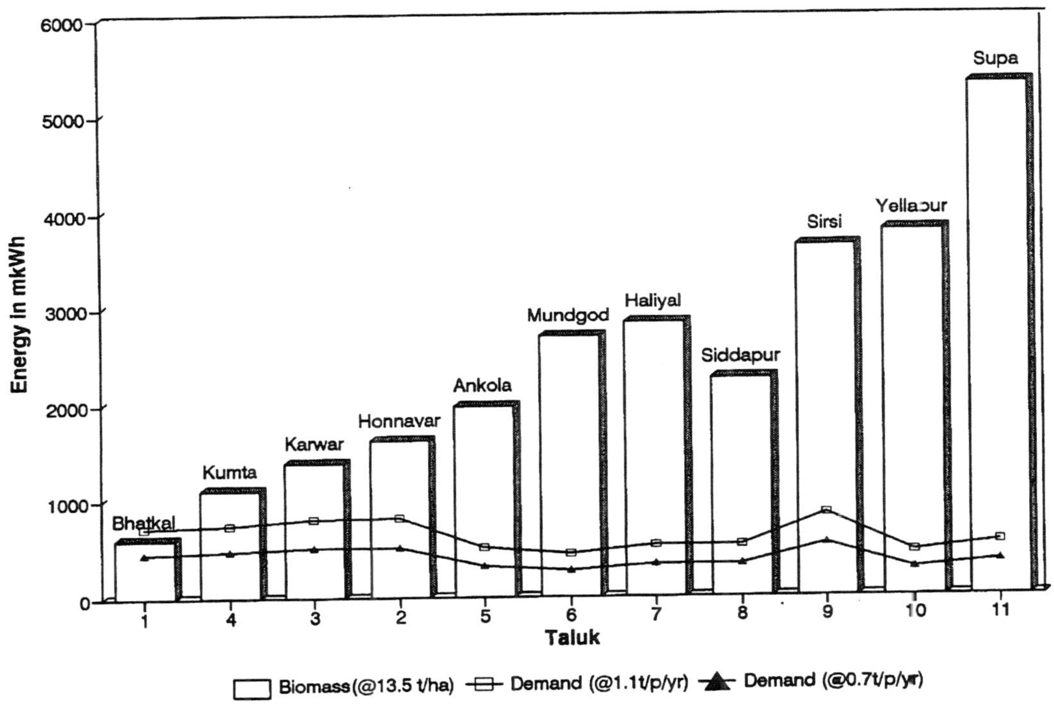

Figure 10.5.2: Biomass Availability and Demand - Talukwise, Uttara Kannada District

Optimum allocation of resources for various tasks is as follows.

E_{11} : 3266.780 million units - mkWh (Wood energy for task 1)
E_{12} : 20795.340 million units (Wood energy for task 2)
E_{13} : 1271.720 million units (Wood energy for task 3)
E_{16} : 1892.580 million units (Wood energy for task 6)
E_{22} : 346.230 million units (Biogas for task 2)
E_{24} : 2588.360 million units (Biogas for task 4)
E_{31} : 33.020 million units (Wind energy for task 1)
E_{35} : 172.530 million units (Wind energy for task 5)
E_{41} : 3742.940 million units (Solar energy for task 1)
E_{46} : 1048.560 million units (Solar energy for task 6)
E_{51} : 1481.580 million units (Hydro energy for task 1)
E_{55} : 1408.030 million units (Hydro energy for task 5)
E_{63} : 280.730.million units (Grid electricity task 3)
E_{64} : 39.260 million units (Grid electricity task 4)
E_{74} : 12.630 million units (Kerosene for task 4)
E_{76} : 2.510 million units (Kerosene for task 6)
E_{85} : 31.380 million units (Diesel for task 5)
E_{86} : 68.610 million units (Diesel for task 6)

Table 10.4.1 Assumed Efficiency Values for Various Resource - Task Combinations

		Task ------------->					
	Resource	j=1	2	3	4	5	6
i=1	Firewood+Residues	0.10	0.10	0.10	0.10	0.25	0.30
2	Biogas	0.30	0.40	0.20	0.10	0.30	0.30
3	Wind	0.10	0.10	0.30	0.10	0.40	0.30
4	Solar	0.60	0.60	0.20	0.10	0.30	0.30
5	Water	0.10	0.10	0.20	0.10	0.60	0.30
6	Grid Electricity	0.10	0.15	0.20	0.10	0.40	0.30
7	Kerosene	0.25	0.15	0.15	0.05	0.25	0.10
8	Petroleum products	0.15	0.15	0.15	0.10	0.35	0.30

The optimal energy allocation is tabulated in Table 10.5.1. Total cost per year is about Rs. 11930.28 million as against Rs. 18711.08 million. Optimisation based on the cost minimisation objective shows that there is saving of about 19.19% in energy and 36.24% cost reduction in overall energy system. Cost per unit (kWh) of energy with optimal allocation of energy is Rs. 0.31 per kWh (as against Rs. 0.39 per kWh without optimisation).

Table 10.4.2: K_{ijk} Values for Various Resource - Task Combinations

		Task ------------->					
	Resource	j=1	2	3	4	5	6
i=1	Firewood+Residues	0.40	0.40	0.35	0.21	0.11	0.11
2	Biogas	0.30	0.29	0.24	0.21	0.11	0.11
3	Wind	0.11	0.11	0.11	0.21	0.21	0.11
4	Solar	0.21	0.24	0.24	0.21	0.21	0.11
5	Water	0.21	0.40	0.24	0.21	0.11	0.11
6	Grid Electricity	0.21	0.21	0.24	0.21	0.21	0.11
7	Kerosene	0.21	0.11	0.11	0.11	0.11	0.15
8	Petroleum products	0.11	0.11	0.11	0.11	0.11	0.24

Table 10.4.3: Assumed Plant Load Factor and Diversity Factor

	Task ------------->							
	j = 1	2	3	4	5	6	7	8
Load factor	0.60	0.30	0.22	0.21	0.21	0.30	0.11	0.21
Diversity factor	1.80	1.40	1.20	1.20	1.20	1.00	1.20	1.20

Table 10.4.4: Cost per Unit of Energy Used in the Design

		Task ------------->					
	Resource	j=1	2	3	4	5	6
i=1	Firewood+Residues	0.15	0.15	0.20	0.88	0.85	0.85
2	Biogas	0.25	0.35	2.75	0.40	2.75	2.75
3	Wind	1.30	1.30	1.30	1.30	0.45	1.30
4	Solar	0.20	0.15	0.45	0.74	1.65	1.65
5	Water	1.09	1.09	1.09	1.09	0.54	1.90
6	Grid Electricity	0.79	0.79	1.30	0.70	1.30	1.90
7	Kerosene	0.65	0.65	0.85	0.35	0.65	1.30
8	Petroleum products	1.85	1.85	1.85	0.82	0.89	0.89

Table 10.5.1: Optimal Allocation of Resources (mkWh) for Various Tasks – Minimisation of Costs - Year 1994

	Tasks --> Resources (mkWh)	1	2	3	4	5	6	Total	Remnant Resources mkWh	%
1	F.W.+Bio residue	3268.78	20795.34	1271.72			1892.58	27228.42	9138.21	25.13
2	Biogas		346.23		2588.36			2934.59	0.00	0.00
3	Wind	33.02				172.53		205.55	0.00	0.00
4	Solar	3742.94					1048.56	4791.50	0.00	0.00
5	Hydro		1481.58			1408.03		2889.61	0.00	0.00
6	Grid Electricity			280.73	39.26			319.99	0.00	0.00
7	Kerosene				12.63		2.51	15.14	0.00	0.00
8	Diesel					31.38	68.61	99.99	0.00	0.00
						Total resources		38484.79		

Table 10.5.2: Optimal Allocation of Resources (mkWh) for Various Tasks – Maximisation of Efficiency Year: 1994 (efficiency of F.W. + Bio residue devices: 10%)

	Tasks --> resources (mkWh)	1	2	3	4	5	6	Total	Remnant Resources mkWh	%
1	F.W.+Bio residue	3039.13	20455.93	1271.74	43.96		2038.94	26849.70	9516.94	26.17
2	Biogas		2256.21		678.37			2934.53	0.00	0.00
3	Wind					172.53	33.02	205.55	0.00	0.00
4	Solar	3742.94					1048.56	4791.50	0.00	0.00
5	Hydro		1481.58			1408.02		2889.60	0.00	0.00
6	Grid Electricity			280.73	39.26			319.99	0.00	0.00
7	Kerosene					12.63	2.51	15.14	0.00	0.00
8	Diesel					31.38	68.61	99.99	0.00	0.00
						Total resources		38106.06		

Table 10.5.3: Optimal Allocation of Resources for Various Tasks – Maximisation of Efficiency and Minimisation of Costs Year: 1994 (increasing efficiency of F.W. + Bio residue devices)

	Tasks --> resources (mkWh)	1	2	3	4	5	6	Total	Remnant Resources mkWh	%
1	F.W.+Bio residue	2966.19	20977.32	524.11			2054.08	26521.69	9850.94	27.08
2	Biogas		346.23		2588.36			2934.59	0.00	0.00
3	Wind					172.53	33.02	205.55	0.00	0.00
4	Solar	3742.94					1048.56	4791.50	0.00	0.00
5	Hydro		531.25	1271.72		1086.63		2889.60	0.00	0.00
6	Grid Electricity			280.74	33.26			314.00	0.00	0.00
7	Kerosene					9.04	6.10	15.14	0.00	0.00
8	Diesel					31.38	68.61	99.99	0.00	0.00
						Total resources		37772.05		

Table 10.5.4: Saving by Optimal Allocation of Energy Sources for Various Tasks

	Energy mkWh/Yr	Cost m.Rs/Yr	Cost/kWh in Rs	% Energy Saving	% Cost Saving	Demand-Resource Ratio
Original Values	47623.00	18711.08	0.39			0.25
Minimisation of Cost	38484.79	11930.28	0.31	19.19	36.24	0.31
Maximisation of efficiency	38106.06	12765.53	0.34	19.98	31.78	0.31

Optimisation is carried out with the objective of maximisation of efficiency of 'ijk' combination for all combinations and results are listed in Table 10.5.2. The efficiency of wood devices for heating is taken as 10%. In this case biogas is first preferred for cooking (about 79%) and the balance is used for lighting. When efficiency of end use devices (for resources-1) is increased (Task 1: efficiency 30%, Task 2: efficiency of 40% and Task 3: efficiency of 30%), then firewood is preferred for cooking and biogas for lighting. The results of this simulation are listed in Table 10.5.3. In the case of maximisation of efficiency criteria, the energy saved is about 19.98% and total cost is about Rs. 12766.53 million. These results are listed in Table 10.5.4.

10.5.4 Demand Scenario Results

The energy demand during the year 2000, 2005 and 2010 is estimated based on the various scenarios discussed earlier. Results of these analyses are:

Scenario I - The Base Case Scenario: The optimal allocations of resources for various tasks, based on maximisation of efficiency criteria for the Base case scenario, are listed in Table 10.6.1 (for the year 2000), Table 10.6.2 (for the year 2005) and Table 10.6.3 (for the year 2010). The remnant renewable resources for each task are computed and listed in these tables.

Table 10.6.4 presents sectorwise projections for all fuels. Overall energy use increases by 26.29 % over 1994 levels, with a nearly 54.66% increase in the use of electricity and a small increase in the annual consumption of fuel wood. Sectorwise, industrial (67.64%), transport (51.48%) and agricultural sectors (48.98%) dominate and show the strongest growth in consumption. While, consumption in domestic sector increases only slightly, as an increase in the information dissemination would result in the substitution of more efficient cook stoves and fuels, such as biogas, for cooking and water heating purposes.

Scenario II: The results of Scenario II are presented in Table 10.7.1 (year 2000), Table 10.7.2 (year 2005) and Table 10.7.3 (year 2010). Sectorwise projection is listed in Table 10.7.4. The increasing energy intensity per unit of industrial output, which is assumed in this case, leads to overall energy demand that is nearly 1852.78 million units (mkWh) higher in 2010 than in the Base case, and industrial demand that is 20 per cent higher than in the Base case. The primary fuels for which demand is higher in this case are electricity, diesel and petrol. This scenario illustrates the effect of not investing in efficient devices as industrial output expands.

Scenario III - Transformation Case: Table 10.8.1, Table 10.8.2 and Table 10.8.3 list the optimal allocation of resources for various tasks in the Transformation scenario. Table 10.8.4 lists sectorwise demand. It indicates that, there is a substantial change in the sectoral shares of energy consumption by 2010, relative to the Base case, and a change in the pattern of fuel use as well. The domestic and transport sectors in the Transformation case use substantially less energy than in the Base case, as a result of significant efficiency improvements. Industrial sector on the other hand uses 9.4% more energy than in the Base case due to higher rates of energy growth and lower rate of efficiency improvement.

Scenario IV: Table 10.9.1 (optimal allocation for the year 2000), Table 10.9.2 (year 2005), Table 10.9.3 (year 2010) and Table 10.9.4 (sectorwise energy demand) show that demand in this scenario is significantly higher than any of the cases discussed earlier. This scenario is based on the energy consumption norms of the state. Overall energy consumption in 2010 is 36.32% higher than in the Base case, and is 54.6% of base year demand. Unlike in the previous scenarios, fuel wood consumption does raise modestly in this scenario as a result of the large number of households. The domestic sector shows 26.05% higher consumption than in the Base case, while the industrial sector also shows similar increase of 29.81%. The commercial and services sector shows growth of 86.74%, the highest growth among all sectors.

Table 10.6.1: Scenario I - Optimal Allocation of Resources for Various Tasks – Maximisation of Efficiency – Year 2000

	Tasks --> resources (mkWh)	1	2	3	4	5	6	Total	Remnant Resources mkWh	%
1	F.W.+Bio residue	2854.66	20740.93	1471.72	74.33		1948.06	27089.70	9276.93	25.51
2	Biogas		2576.21				358.38	2934.59	0.00	0.00
3	Wind					172.53	33.02	205.55	0.00	0.00
4	Solar	3742.94					1048.56	4791.50	0.00	0.00
5	Hydro		1481.58			1408.03		2389.61	0.00	0.00
6	Grid Electricity			280.74	39.26			320.00	0.00	0.00
7	Kerosene					12.63	2.51	15.14	0.00	0.00
8	Diesel					31.38	68.61	99.99	0.00	0.00

Table 10.6.2: Scenario I - Optimal Allocation of Resources for Various Tasks – Maximisation of Efficiency – Year 2005

	Tasks --> resources (mkWh)	1	2	3	4	5	6	Total	Remnant Resources mkWh	%
1	F.W.+Bio residue	3054.65	20634.10	1571.72	108.33		1930.90	27299.70	9066.93	24.93
2	Biogas		2576.21				358.38	2934.59	0.00	0.00
3	Wind					172.53	33.02	205.55	0.00	0.00
4	Solar	3742.94					1048.56	4791.50	0.00	0.00
5	Hydro		1481.58			1408.03		2889 61	0.00	0.00
6	Grid Electricity			280.74	39.26			320 00	0.00	0.00
7	Kerosene					12.63	2.51	15.14	0.00	0.00
8	Diesel					31.38	68.61	99.99	0.00	0.00

Table 10.6.3: Scenario I - Optimal Allocation of Resources for Various Tasks – Maximisation of Efficiency – Year 2010

	Tasks --> resources (mkWh)	1	2	3	4	5	6	Total	Remnant Resources mkWh	%
1	F.W.+Bio residue	3295.66	20871.61	1671.72	143.33		1913.38	27895.70	8470.93	23.29
2	Biogas		2576.21				358.37	2934.58	0.00	0.00
3	Wind					172.53	33.02	205.55	0.00	0.00
4	Solar	3742.94					1048.56	4791.50	0.00	0.00
5	Hydro		1481.58			1408.03		2889.61	0.00	0.00
6	Grid Electricity			280.73	39.26			319.99	0.00	0.00
7	Kerosene					12.63	2.51	15.14	0.00	0.00
8	Diesel					31.38	68.61	99.99	0.00	0.00

Table 10.6.4: Uttara Kannada Scenario 1 – Base Case Demand Results for all Fuels by Sector (million kWh)

	1994	2000	2005	2010
Commercial, Services, PL	11.17	13.10	17.72	25.90
Transport	100.57	115.52	147.02	222.71
Industry	1827.60	2300.56	3392.22	5686.76
Agriculture	588.70	606.09	735.24	1095.39
Domestic	8941.56	9747.63	10725.04	11909.01
Total	11469.60	12782.90	15016.78	18939.76

PL – Public Lighting

Table 10.7.1: Scenario II - Optimal Allocation of Resources for 2000 (million kWh)

	Tasks --> resources (mkWh)	1	2	3	4	5	6	Total	Remnant Resources mkWh	%
1	F.W.+Bio residue	2733.26	20124.76	1879.72	108.16		1925.05	26770.96	9595.68	26.39
2	Biogas		2576.21				358.38	2934.59	0.00	0.00
3	Wind					172.53	33.02	205.55	0.00	0.00
4	Solar	3742.94					1048.56	4791.50	0.00	0.00
5	Hydro		1481.58			1408.03		2889.61	0.00	0.00
6	Grid Electricity			280.73	39.26			319.99	0.00	0.00
7	Kerosene					12.63	2.51	15.14	0.00	0.00
8	Diesel					31.38	68.61	99.99	0.00	0.00

Table 10.7.2: Scenario II - Optimal Allocation of Resources for 2005

	Tasks --> resources (mkWh)	1	2	3	4	5	6	Total	Remnant Resources mkWh	%
1	F.W.+Bio residue	3136.76	20236.35	1879.72	168.16		1893.47	27314.46	9052.17	24.89
2	Biogas		2576.21				358.38	2934.59	0.00	0.00
3	Wind					172.53	33.02	205.55	0.00	0.00
4	Solar	3742.94					1048.56	4791.50	0.00	0.00
5	Hydro		1481.58			1408.03		2889.61	0.00	0.00
6	Grid Electricity			280.73	39.26			319.99	0.00	0.00
7	Kerosene					12.63	2.51	15.14	0.00	0.00
8	Diesel					31.38	68.61	99.99	0.00	0.00

Table 10.7.3: Scenario II - Optimal Allocation of Resources for 2010

	Tasks --> resources (mkWh)	1	2	3	4	5	6	Total	Remnant Resources mkWh	%
1	F.W.+Bio residue	2908.35	20239.65	2479.72	272.16		1824.88	27724.76	8641.87	23.76
2	Biogas		2576.21				358.38	2934.59	0.00	0.00
3	Wind					172.53	33.02	205.55	0.00	0.00
4	Solar	3742.94					1048.56	4791.50	0.00	0.00
5	Hydro		1481.58			1408.03		2889.61	0.00	0.00
6	Grid Electricity			280.73	39.26			319.99	0.00	0.00
7	Kerosene					12.63	2.51	15.14	0.00	0.00
8	Diesel					31.38	68.61	99.99	0.00	0.00

Table 10.7.4: Scenario II - Demand Results for all Fuels by Sector (million kWh)

	1994	2000	2005	2010
Commercial, Services, PL	11.17	13.10	17.27	25.90
Transport	100.57	115.52	147.02	222.71
Industry	1827.60	2459.42	4084.14	8684.90
Agriculture	588.70	650.13	897.23	1449.31
Domestic	8941.56	9415.08	9852.32	11553.69
Total	11469.60	12653.25	14997.97	21936.51

PL – Public Lighting

Table 10.8.1: Scenario III - Optimal Allocation of Resources for 2000 (million kWh)

	Tasks --> resources (mkWh)	1	2	3	4	5	6	Total	Remnant Resources mkWh	%
1	F.W.+Bio residue	2636.41	20797.57	1171.72	78.56		1963.00	26647.26	9719.37	26.73
2	Biogas		2576.21				358.38	2934.59	0.00	0.00
3	Wind					172.53	33.02	205.55	0.00	0.00
4	Solar	3742.94					1048.56	4791.50	0.00	0.00
5	Hydro		1481.58			1408.03		2389.61	0.00	0.00
6	Grid Electricity			280.74	39.26			320.00	0.00	0.00
7	Kerosene					12.63	2.51	15.14	0.00	0.00
8	Diesel					31.38	68.61	99.99	0.00	0.00

Table 10.8.2: Scenario III - Optimal Allocation of Resources for 2005

	Tasks --> resources (mkWh)	1	2	3	4	5	6	Total	Remnant Resources mkWh	%
1	F.W.+Bio residue	3560.44	20474.03	1054.72	145.56		1901.03	27135.79	9230.85	25.38
2	Biogas		2576.21				358.38	2934.59	0.00	0.00
3	Wind					172.53	33.02	205.55	0.00	0.00
4	Solar	3742.94					1048.56	4791.50	0.00	0.00
5	Hydro		1481.58			1408.03		2889.61	0.00	0.00
6	Grid Electricity			280.74	39.26			320.00	0.00	0.00
7	Kerosene					12.63	2.51	15.14	0.00	0.00
8	Diesel					31.38	68.61	99.99	0.00	0.00

Table 10.8.3: Scenario III - Optimal Allocation of Resources for 2010

	Tasks --> resources (mkWh)	1	2	3	4	5	6	Total	Remnant Resources mkWh	%
1	F.W.+Bio residue	3633.26	20705.61	1104.72	190.56		1963.00	27597.16	8769.48	24.11
2	Biogas		2576.21				358.38	2934.59	0.00	0.00
3	Wind					172.53	33.02	205.55	0.00	0.00
4	Solar	3742.94					1048.56	4791.50	0.00	0.00
5	Hydro		1481.58			1408.03		2889.61	0.00	0.00
6	Grid Electricity			280.74	39.26			320.00	0.00	0.00
7	Kerosene					12.63	2.51	15.14	0.00	0.00
8	Diesel					31.38	68.61	99.99	0.00	0.00

Table 10.8.4: Scenario III - Transformation Case: Demand Results for all Fuels by Sector (million kWh)

	1994	2000	2005	2010
Commercial, Services, PL	11.17	13.16	14.79	18.15
Transport	100.57	110.43	117.35	137.81
Industry	1827.60	2319.34	2722.26	3351.92
Agriculture	3254.59	4090.77	5289.21	5745.25
Domestic	9981.52	9449.60	8617.78	8262.17
Total	15175.44	15983.30	16761.39	17515.30

Table 10.9.1: Scenario IV - Optimal Allocation of Resources for 2000 (million kWh)

	Tasks -->	1	2	3	4	5	6	Total	Remnant Resources	
	resources (mkWh)								mkWh	%
1	F.W.+Bio residue	3089.66	21322.74	1843.72	78.17		1926.41	28260.70	8105.93	22.29
2	Biogas		2576.21				358.38	2934.59	0.00	0.00
3	Wind					172.53	33.02	205.55	0.00	0.00
4	Solar	3742.94					1048.56	4791.50	0.00	0.00
5	Hydro		1481.58			1408.03		2889.61	0.00	0.00
6	Grid Electricity			280.74	39.26			320.00	0.00	0.00
7	Kerosene					12.63	2.51	15.14	0.00	0.00
8	Diesel					31.38	68.61	99.99	0.00	0.00

Table 10.9.2: Scenario IV - Optimal Allocation of Resources for 2005 (million kWh)

	Tasks -->	1	2	3	4	5	6	Total	Remnant Resources	
	resources (mkWh)								mkWh	%
1	F.W.+Bio residue	3089.66	21596.39	1943.72	278.17		1852.10	28760.03	7606.60	20.92
2	Biogas		2576.21				358.38	2934.59	0.00	0.00
3	Wind					172.53	33.02	205.55	0.00	0.00
4	Solar	3742.94					1048.56	4791.50	0.00	0.00
5	Hydro		1481.58			1408.03		2889.61	0.00	0.00
6	Grid Electricity			280.74	39.26			320.00	0.00	0.00
7	Kerosene					12.63	2.51	15.14	0.00	0.00
8	Diesel					31.38	68.61	99.99	0.00	0.00

Table 10.9.3: Scenario IV - Optimal Allocation of Resources for 2010 (million kWh)

	Tasks --> resources (mkWh)	1	2	3	4	5	6	Total	Remnant Resources mkWh	%
1	F.W.+Bio residue	3389.66	21704.71	2525.72	478.17		1751.44	29849.70	6516.93	17.92
2	Biogas		2576.21				358.38	2934.59	0.00	0.00
3	Wind					172.53	33.02	205.55	0.00	0.00
4	Solar	3742.94					1048.56	4791.50	0.00	0.00
5	Hydro		1481.58			1408.03		2889.61	0.00	0.00
6	Grid Electricity			280.74	39.26			320.00	0.00	0.00
7	Kerosene					12.63	2.51	15.14	0.00	0.00
8	Diesel					31.38	68.61	99.99	0.00	0.00

Table 10.9.4: Scenario IV - State Energy Norms: Demand Results for all Fuels by Sector (million kWh)

	1994	2000	2005	2010
Commercial, Services, PL	11.17	17.02	34.70	97.17
Transport	100.57	122.48	170.81	274.71
Industry	1827.60	2501.30	4191.24	9121.04
Agriculture	588.70	787.92	1358.97	3293.33
Domestic	8941.56	9790.15	10504.56	12193.61
Total	11469.60	13218.87	16260.27	24979.86

PL – Public Lighting

Table 10.10.1: Scenario V - Optimal Allocation of Resources for 2000 (million kWh)

	Tasks --> resources (mkWh)	1	2	3	4	5	6	Total	Remnant Resources mkWh	%
1	F.W.+Bio residue	2922.13	20431.35	1439.52	123.96		1932.74	26849.70	9516.93	26.17
2	Biogas		2576.21				358.38	2934.59	0.00	0.00
3	Wind					172.53	33.02	205.55	0.00	0.00
4	Solar	3742.94					1048.56	4791.50	0.00	0.00
5	Hydro		1481.58			1408.03		2889.61	0.00	0.00
6	Grid Electricity			280.74	39.26			320.00	0.00	0.00
7	Kerosene					12.63	2.51	15.14	0.00	0.00
8	Diesel					31.38	68.61	99.99	0.00	0.00

Table 10.10.2: Scenario V - Optimal Allocation of Resources for 2005 (million kWh)

	Tasks --> resources (mkWh)	1	2	3	4	5	6	Total	Remnant Resources mkWh	%
1	F.W.+Bio residue	2824.35	20505.51	1716.99	247.96		1874.89	27169.70	9196.93	25.29
2	Biogas		2576.21				358.38	2934.59	0.00	0.00
3	Wind					172.53	33.02	205.55	0.00	0.00
4	Solar	3742.94					1048.56	4791.50	0.00	0.00
5	Hydro		1481.58			1408.03		2889.61	0.00	0.00
6	Grid Electricity			280.74	39.26			320.00	0.00	0.00
7	Kerosene					12.63	2.51	15.14	0.00	0.00
8	Diesel					31.38	68.61	99.99	0.00	0.00

Table 10.10.3: Scenario V - Optimal Allocation of Resources for 2010 (million kWh)

	Tasks --> resources (mkWh)	1	2	3	4	5	6	Total	Remnant Resources mkWh	%
1	F.W.+Bio residue	3013.45	21245.24	1716.99	415.96		1817.05	28208.70	8157.94	22.43
2	Biogas		2576.21				358.38	2934.59	0.00	0.00
3	Wind					172.53	33.02	205.55	0.00	0.00
4	Solar	3742.94					1048.56	4791.50	0.00	0.00
5	Hydro		1481.58			1408.03		2889.61	0.00	0.00
6	Grid Electricity			280.74	39.26			320.00	0.00	0.00
7	Kerosene					12.63	2.51	15.14	0.00	0.00
8	Diesel					31.38	68.61	99.99	0.00	0.00

Table 10.10.4: Scenario V - State Energy Norms+ Efficiency: Base Case Demand Results for all Fuels by Sector (million kWh)

	1994	2000	2005	2010
Commercial, Services, PL	11.17	16.09	29.54	70.34
Transport	100.57	116.43	147.51	203.58
Industry	1827.60	1985.62	2361.25	2993.89
Agriculture	588.70	600.39	632.61	703.61
Domestic	8941.56	8951.28	9355.71	9811.58
Total	11469.60	11669.82	12526.62	13783.01

PL – Public Lighting

Scenario V; Scenario IV + Efficiency improvements scenario: The efficiency improvements inclusion in Scenario IV, by relatively simple efficiency improvement measures proposed, save over 29.91% of energy per year by 2010, when compared with Scenario IV (Case without the improvements). Results of this scenario is listed in Tables 10.10.1 to 10.10.4 respectively.

10.5.5 Sensitivity Analyses

Sensitivity analyses help in identifying the role of a parameter or a set of parameters in optimal allocations. Parametric sensitivity analyses are carried out in order to analyse the effect of approximations. Sensitivity analyses is carried out for tasks by changing efficiencies of end use devices. If a source can supply either partially or fully requirements of more than one task or if the task has access to various resources for its energy needs, the type and number of transitions depend on various critical parameters, such as, unit cost of sources used by a particular task, efficiency of energy conversion of different energy paths, limits on the availability of resources if any, etc. The emerging allocation of resources for various tasks thus depend on the combination of values of critical parameters. Thus, sensitivity analyses give a clear picture as to how a single parameter or a combination of parameters influence optimal scenarios in regional energy planning exercise. This would help in a dynamic situation that may arise while implementing regional energy plan using the optimisation techniques.

When the cost of biogas for cooking is taken as Rs. 0.35/m^3 and wood stove efficiency is around 10%, biogas is preferred for cooking. When efficiency of wood devices is increased in steps of 5%, it is noticed that, at efficiency greater than 30%, wood stoves are preferred for cooking while biogas enters lighting option. This shows that at low level of efficiency, wood stove is not economical, as biogas or electricity generated from hydro resources in a decentralised way is inexpensive. However, this scenario changes with increase in efficiency of wood energy devices beyond 30%. Hence, between 10 and 30% efficiency of wood devices, firewood supplements the available biogas in meeting the domestic needs of a region.

10.6.0 IMPLEMENTATION

The framework of energy planning for a region, discussed so far, can become operational only after overcoming a large number of barriers and constraints, which usually come in the way of integration of energy supply programmes with the energy needs at the local level. These constraints include, among others, lack of mechanism at the local level to carry out assessment with the involvement of the potential beneficiaries, energy planning mechanism, project formulation and implementation. Proper coordination between the energy sector and the overall planning and development machinery is needed in a region for successful implementation of the energy plan. These and related issues are discussed in the following section.

10.6.1 Management Aspects of Energy Planning

The current approach to planning in the energy sector does not offer any significant role to the district or local level institutions. Moreover, the coordination needed between the energy sector and the overall planning and development at district, taluk and village levels is missing. Although forestry planning is carried out by the district forest department, its most significant aspect pertaining to energy is extremely weak and receives very little attention in the planning exercises of the sector. The biomass situation in hilly taluks has, therefore, gradually worsened and has reached a point of crisis. This deteriorating situation obviously demands immediate attention in two directions.

1. Strengthening of local institutions for energy development.
2. Promotion of coordination between institutions concerned with the energy development and overall planning at different hierarchical levels on the one hand and between institutions concerned with research and development on the other.

10.6.2 Strengthening of Local Institutions for Energy Development

As stated in the previous section, local government institutions at the district level (zilla panchayath) and below (village panchayath) can contribute positively in many ways towards energy planning and development. These include: (a) generation of energy databases, (b) promotion of community participation, and (c) extension and training.

10.6.3 Generation of Energy Databases

The elaborate structure of the local government, which goes down to the village panchayath and its electoral wards can be effectively used for the generation of useful data. A simple methodology for obtaining information at both village panchayath and household levels has been demonstrated in this study. It may be pointed out that, besides collection of basic facts and figures, this kind of micro-level survey can also be used to assess problems and solutions perceived by local groups. This was particularly true of fuel wood for which the rapid rate of forest depletion, and not the limited area under forest, was conceived as a major problem or stress. This reckoning obviously is a positive factor and needs to be used effectively by involving representatives of such groups in the energy planning and development processes at the grassroots level. The household survey generated extensive data on energy which provided insights into energy consumption patterns (communitywise and landholding categorywise) use of cooking devices, possession of energy appliances, and the changing trends in energy use in the domestic sector. These surveys also indicated that the existing system of local government can be easily mobilised to build an effective energy information system for planning and development at the district and local levels without incurring any significant additional financial burden.

10.6.4 Promotion of Community Participation

The local government institutions also need strengthening to promote community participation in the planning and development process. These grassroots institutions, besides carrying out local projects with the involvement of communities within their jurisdiction, can also provide a powerful lobby through which communities can negotiate for services from sectoral agencies or departments of the Government.

Unfortunately, the role of local councils in development efforts has been mostly underrated by planners, and is quite often overlooked in the implementation of energy projects, resulting in their failure to achieve the desired objectives. Energy plantation projects provide a case in point. It has been observed that, even when the political will is present and the funds are allocated, implementing a large-scale afforestation (plantation) campaign is an unexpectedly complex and difficult process. Planting millions of trees and successfully nurturing them to maturity is not purely a technical task, like building a dam. Further, tree planting projects almost invariably get enmeshed in the political, cultural, and administrative tangles of a rural locality. The nature of their success, therefore, is largely governed by the intensity of community involvement through local government or other means. Central or State Government stimuli in technical advice and financial assistance in such cases are ineffective (like some afforestation programmes in places like Yana in Kumta taluk) unless community members clearly understand why lands to which they had traditionally free access for grazing and wood-gathering are being demarcated into plantations. Therefore, it is expected that they will view the project with suspicion or even hostility.

Also, the conservation efforts cannot succeed without strong commitment from the community. A major source of energy in the rural hill districts, such as Uttara Kannada, is biomass. Slight improvements in the efficiency of energy use from this source can substantially improve the physical quality of life of the rural people in such districts without any increase, or even decrease, in the supply of primary energy. Adoption of appropriate cooking devices (improved cooking stoves) alone can bring a major change in this direction. Likewise, the replacement of dung fuel with biogas can provide both energy and manure on one hand and improve the quality of the environment on the other.

There is a need to distinguish the difference between subsistence energy requirements and energy required for economic development. As far as the subsistence energy requirement is concerned, the existing energy consumption pattern, with the major chunk being met through bioresources for thermal energy requirements, indicates the importance of intervention in this sector. The low conversion efficiency of energy devices (using bioresources) is a significant aspect in the context of designing energy intervention plans. Bioresource surplus taluks, such as, Sirsi and Siddapur are associated with high levels of consumption. In areas with surplus resources, the implementation of the efficiency measures have to follow an effort to convince and educate local people about the importance of efficient usage of resources. In the biomass scarce region, the significance of the interventions for fuel switching (such as solar energy for water heating, drying, etc.), improvement in end use efficiencies and augmenting supplies (through social forestry and energy plantations on barren land) are to be adopted.

10.6.5 Extension and Training

Usually, energy programmes require a system by which end use devices, such as Improved Cook stoves (IC's) and biogas are disseminated in accordance with the programme objectives. Dissemination mechanisms become extremely important when the end use device is new to the user. Unfamiliar technology obviously raises a number of questions that can be answered only when the technology is successfully implemented and comprehensively evaluated. In order to strengthen the dissemination process, there is a need for rural energy centres. The purpose is to demonstrate the energy techniques that are suitable for local conditions and see that these techniques are practical and reliable from the farmers and locals' point of view. The objectives of the Regional Energy Centres should be to:

(i) study energy supply and consumption patterns to identify village needs,
(ii) demonstrate the use of efficient energy devices and disseminate, monitor and evaluate appropriate technologies,
(iii) augment the fuel supply situation through energy plantations and other decentralised means, and
(iv) analyse the impact of energy-related activities on the socio-economic development of the village and improvement of its environment.

Besides promotion of energy, the local panchayaths by themselves constitute a very strong extension network. Currently, the energy development agencies at the State level have little or no field staff. As a result, their rural energy programmes have suffered. The village panchayat can help fill this gap and can provide valuable assistance through quick dissemination and extension work.

10.6.6 Institutional Coordination and Energy Development Strategy

The implementation of rural energy programmes requires the involvement of a large number of official agencies working under different departments, private and public manufacturers, voluntary agencies, artisans, R and D institutions, village communities, social organisations and private individuals. This requires a considerable degree of coordination at the village, taluk and district levels.

At the district level, this function should be performed by the District Planning Officer who can also assist as a district level Project Implementation Officer. Besides coordination among the different tiers of planning institutions, there is a great need for networking R and D institutions, manufacturers and other concerned groups, e.g., community organisations and NGO's involved in rural energy planning and implementation. Finally, the success or failure of regional energy programmes/projects in a district will also be determined by strategies adopted and logistics provided. After a rural energy technology has been field-tested and found suitable for large-scale implementation in a district, the major task for its implementation will involve formulation of strategies based on the following factors.

(i) Speed or rate of progress of programme implementation,
(ii) Provision of funds and procedures for disbursing loans and subsidies,
(iii) Making raw materials available for installation of energy transmitting devices,

(iv) Creation of infrastructure for manufacturing energy equipment (it has also to be decided whether the public organisations should manufacture it or whether private entrepreneurs should be encouraged by providing incentives),

(v) R & D and maintenance networks to promote the design of energy equipment, training of artisans and mechanics for fabrication, installation as well as maintenance, and

(vi) Monitoring and evaluation to facilitate timely feedback and assess the success of the programme.

10.7.0 CONCLUSIONS

Currently, bioresource is the single most important energy source in Uttara Kannada district, meeting more than 98 per cent of the total energy needs. The over use of forest wood is creating serious ecological problems for the district in the form of deforestation, soil erosion and increased sedimentation. Uttara Kannada district mainly depends on noncommercial energy sources (96.23%), such as, firewood (78.071%), dung in the form of biogas (13.12%) and animate sources of energy (5.024%). While, commercial sources such as electricity (2.863%), kerosene, diesel, petrol, etc., (0.904%) and LPG (0.0001%), constitute 3.767% of total energy consumption in the district.

The availability of wind resource is quantified, and it is found that coastal taluks have a good potential of the order of 30.64(avg)±31.72(Sd) W/sq.m in Karwar and 26.50(avg)±20.82(Sd) W/sq.m in Kumta. Availability of wind resources during pre-monsoon period, and local necessity of energy for industrial and irrigation purposes (for agriculture and horticulture) during this period make the exploitation of wind energy for mechanical and electrical energy purposes feasible and desirable in the coastal tract.

The electrical energy that can be harnessed by solar source is of the order of 191.43 million kWh by utilising 10% of wasteland, presently available in the coastal taluks of Uttara Kannada district. This constitutes about 65.1% of present electricity demand from various sectors of Uttara Kannada district. About 72% of solar energy is available during October to May period, when most of the agro-processing tasks such as drying (areca, cardamom, coconut, etc.) take place.

It is seen that about 1787.61 million units (mkWh) of electrical energy from hydropower sources in Aghnashini and 1079.17 million units (mkWh) in Bedthi river basin can be harnessed in an ecologically sound way.

Bioresources available in the region is of the order of 13705.7 million units (mkWh) from forests, 318.07 million units (mkWh) in the form of agricultural residues and 825.97 million units (mkWh) as bio residues from areca, coconut and cashew plantations.

Bioresource availability from livestock population is about 1.28 million tonnes, which is equivalent to 46.29 million cubic meter of biogas. This is adequate to meet the requirement of 30% of the total population.

End use efficiency study carried out in randomly selected households in some villages shows that there is scope for saving about 27.45% of fuel wood by switching over to energy efficient stoves.

If the entire district is looked at, bioresource supply appears to be diverse and sufficient in magnitude to comfortably sustain population pressure under the Base case demand scenario investigated. As bioresource (fuel wood) enhancement strategies are pursued through social forestry, etc., complementary policies to promote technically feasible and cost effective improvements in the efficiency of end use equipment are needed.

The Decision Support System (DSS) approach is adopted for designing integrated regional energy planning for Uttara Kannada district. Integration is emphasised across all types of resources in the system and across all parts of the energy system (such as sectorwise energy demand, seasonal constraints in availability, environmental and economic impacts). Thus, the proposed integrated energy system is capable of meeting all energy needs from energy sources, taking into consideration the seasonal constraints in the availability.

There could be substantial saving in energy and overall cost, if efficient matching of resources and tasks is implemented. Optimisation based on the objective of cost minimisation, shows cost saving of 36.24%, and energy saving that could be of the order of 19.19%.

Optimisation carried out with the objective of maximisation of efficiency of 'ijk' combination for all combinations, shows the energy saved to be about 19.98% and the saving in total cost as about 31.78%.

The scenario analyses show that potential renewable resources in Uttara Kannada district can meet the energy requirements of the district even under the assumption of high growth rates. That is, the relatively vigorous growth in energy demand in Uttara Kannada district can be accomplished without exceeding available resources in the region.

The process of energy planning at present, however, is a highly centralised activity, and district and local level institutions are not playing any significant role in the process. As a result, the energy crisis in rural areas, and particularly in hilly districts like Uttara Kannada, is not adequately reflected in national level planning.

In addition, energy development and conservation programmes are not being effectively implemented. This applies to a wide spectrum of programmes, ranging from the enhancement of social forestry to the introduction of energy-saving devices, e.g., improved cooking stoves. Hence, for energy plan to be optimal, there is a need to look at all sources available in the region and scrutinise the demand for energy services intensively.

This study has shown that the objective of effective implementation of energy planning cannot be achieved without decentralisation and community involvement. India is fortunate that it has a wide network of local government institutions at the district and lower levels. However, the system can be effective only if these institutions are assigned their due role.

The existing energy planning capability of the district is very weak. It is extremely important to amend this state of affairs by strengthening the energy planning and management capabilities from district to institutional level. This study has suggested some measures that can be adopted by existing institutions to generate an adequate database. This information can be used for effective energy planning. The study also showed that community participation, generated through union council (panchayath) involvement, can produce feasible projects within the broad goals set by district planning.

10.8.0 Acknowledgements

My sincere thanks to Prof. B.G. Raghavendra, Mr. Mathirajan and Mr. Balachandra for providing LP package and for useful discussions. I am grateful to Prof. I.G. Sharma and Prof. Vedula for useful tips on simulations and OR techniques.

10.9.0 References

1. Rath, N.S. and Voss, A., 1981. Energy models for planning and policy assessment. *European Journal of Operational Research*, 8:99-114.
2. Meier, P. and Mubayi, V., 1983. Modelling energy-economic interactions in developing countries: a linear programming approach. *European Journal of Operational Research*, 13:41-59.
3. Reddy, A.K.N. and Subramanian, D.K., 1980. The design of rural energy centres, In *Rural Technology*. Indian Academy of Sciences, Bangalore, India, pp. 109-130.
4. GOI, 1965. *Report of the Energy Survey of India Committee*. Planning Commission, Government of India, New Delhi.
5. GOI, 1974. *Report of Fuel Policy Committee*. Planning Commission, Government of India, New Delhi.
6. GOI, 1979. *Report of Working Group on Energy Policy*. Planning Commission, Government of India, New Delhi.
7. GOI, 1984. *A Perspective on Demand for Energy in India Up to 2004-05*. Advisory Board on Energy, Government of India, New Delhi.
8. Parker, B.J. and Al-Utaibi, G.A., 1986. Decision support systems: the reality that seems hard to accept. *International Journal of Management Science*, 14(2):135-143.

SUMMARY

The central theme of the integrated energy plan would be to prepare an area based decentralised energy plan to meet the energy needs for subsistence and development of alternate energy sources at least cost to the economy and environment. Centralised energy planning exercises cannot pay attention to the variations in socio-economic and ecological factors of a region which influence success of any intervention. Decentralised energy planning is in the interest of efficient utilisation of resources. The regional planning mechanism takes into account various resources available and demand in a region. This implies that the assessment of the demand and supply and the intervention in the energy system which may appear desirable due to such exercises must be at a similar geographic scale. In this regard, the district is accepted as the appropriate planning level. Uttara Kannada district, Karnataka State, is chosen for carrying out the present study. Planned interventions to reduce energy scarcity can take various forms such as (a) energy conservation through promotion and use of energy efficient stoves for cooking and water heating, compact fluorescent bulbs in place of ordinary incandescent bulbs, (b) supply expansions through energy plantations and (c) alternatives - renewable sources of energy such as micro/mini/small hydropower plants and wind, solar and biomass based systems. Ecologically sound development of the region is possible when energy needs are integrated with the environmental concerns at the local and global levels. For this purpose an integrated planning framework is necessary.

The main objective of this research is to develop an "***Ecologically Sound Integrated Energy Plan***" at district level to arrive at an optimal mix of energy sources that can meet the energy demand in a region by keeping in view its development priorities. The energy plan consists of the following components:

(i) ***Assessment of Supply Situation***: This involves analysis of present energy supply system; assessment of woody and other bioresources; potential assessment of renewable sources such as solar, wind, hydro, woody and other bioresources; study of supply system of commercial energy.

(ii) ***Sectorwise Energy Demand***: This involves survey of the present energy consumption in different sectors for various end uses covering the type of fuel, magnitude and composition of fuel, trends, seasonal constraints and preferences in consumption and estimating energy demand based on the survey data.

(iii) ***Development of Energy Plan***: Based on the estimated supply and demand, an energy plan to meet the energy demand is worked out in accordance with the development priorities. Techno-economic analyses of various energy technologies would be carried out to find out technical and economic viability of the system. Energy plan at district level is proposed based on the Decision Support System approach.

(iv) ***Suggestions for Implementation and Management***: With the knowledge of administrative structure at district level and agencies implementing energy development programmes, a suitable institutional structure would be suggested for implementing and managing the energy plan. Analyses of the importance of community participation in energy conservation, planning and identification of measures to enhance the level of participation are suggested.

This book presents a conceptual design for an energy system which would meet the demand of all sectors in the region. The proposed design would in principle, supply enough energy for sustainable development of a region. The energy planning endeavor involves finding a set of sources and conversion devices to meet the energy requirements/demand of all the tasks in an optimal manner. This optimality depends on the objective; to minimise the total annual costs of energy or minimisation of non-local resources or maximisation of overall system efficiency. Factors such as availability of resources in the region and task energy requirements impose constraints on the regional energy planning exercise. Thus, regional energy planning turns out to be a constrained optimisation problem. This exercise describes an optimum energy allocation based on Integrated Energy Planning approach for the district and makes a satisfying energy allocation plan for the year 2000, 2005 and 2010. Integrated energy planning developed based on Decision Support Systems (DSS) approach is flexible, adaptable, ecologically sound and gives an optimal mix of new renewable/conventional energy sources.

Uttara Kannada district is located in the mid-western part of Karnataka state, India. It lies between 74° 9' to 75° 10' east longitude and 13° 55' to 15° 31' north latitude and extends over an area of 10,291 sq. kms, which is 5.37 per cent of the total area of the state, and ranks 10th in the state. Most of the district is hilly and thickly forested. The population was 12,43,526 in the last census. For administrative purposes the district has been divided into 11 taluks. There are 13 towns and 1,283 inhabited villages and 55 uninhabited villages in the district.

Assessment of Supply Situation: The energy sources presently in wide use in the district are firewood, agro residues and animal wastes. Out of these resources, firewood is fast dwindling because of its uncontrolled and inefficient usage and increasing demand due to increase in human population. There is thus an urgent need to tap in new technically feasible resources such as solar, wind, potential energy of water and biogas in a decentralised way. The renewable energy resources readily available in the region are:

***Wind Energy*:** The monthly average hourly wind speed with monthly average power density (W/sq.m.) is computed for Kumta and Sirsi based on hourly wind speed data of 24 months. The monthly average hourly wind speed is maximum in the month of July for Kumta (13 kmph) and Sirsi (18 kmph). The mean annual wind speed at Kumta is 9 kmph and corresponding power density is 26 W/sq.m. At Sirsi, annual mean wind speed is 9 kmph and corresponding power density is 45 W/sq.m. The mean annual wind speed at Karwar is 8 kmph and power density, 30 W/sq.m. With this background, an attempt is made to assess the potential along the coastal tract of the district with the assumption that 5 to 10% of wastelands could be made available for harnessing wind energy. It is found that about 0.75 to 2 MW could be generated in Karwar and Kumta taluks during seasons in this tract.

***Bioresources*:** The resource base in the district under each sector such as forests, agriculture, horticulture and animal residues are assessed talukwise. Total biomass availability in the district is about 14800 mkWh. Based on detailed investigation of biomass resource availability and demand, the district can be categorised into two zones: (a) Biomass surplus zone, consisting of hilly taluks such as Sirsi, Siddapur, Yellapur and Supa and (b) Biomass deficit zone, consisting of thickly populated coastal taluks such as Bhatkal, Kumta, Ankola, Honnavar and Karwar. Fuel wood is mainly used for cooking, while horticulture residues from coconut and arecanut trees are used for water heating purposes. Animal residues available is about 1285 thousand tonnes, which can be used for biogas generation (46 million cubic meter) in Sirsi, Siddapur and Yellapur taluks. This is sufficient to meet the demand of 30% of the total population and promises a viable alternative for cooking and lighting fuel, and also as useful fertiliser.

***Hydroenergy Potential*:** Hydro potential at five feasible sites in the streams of Bedthi and Aghnashini river catchments in the district of Western Ghats is assessed based on stream gauging carried out for a period of 18 months. Relationship between stream run off and precipitation in the region is determined by regression analyses. This relationship is used to find out the hydroenergy potential of ungauged streams in the catchments. Computations of discharge on empirical basis/rational method based on precipitation history of last 90 years and therefore power and energy values computed are in conformity with the power calculations based on stream gauging. This method is used for calculations of power in ungauged streams in Bedthi and Aghnashini catchment areas. It is estimated that if all streams are harnessed for energy, electricity generated annually would be in the order of 720 mkWh in Bedthi basin and 510 mkWh in Aghnashini catchment area. This exercise provides insight to meeting the regional energy requirement through integrated approaches like harnessing hydro power in decentralised way during seasons and meeting lean season requirement through solar, thermal or other options. Most of these streams are seasonal, and demands for alternatives to meet the region's requirement during lean season. In view of this, an ecologically friendly design of hydroelectric plant is proposed at Magod (of Bedthi river) which generates maximum electricity during seasons and has options to store water so as to generate electricity sufficient to meet the region's requirement in lean season.

***Ecologically Friendly Optimal Design of Hydroelectric Projects*:** The planning of water resources depends upon types and size of projects, the ecological factor involved, etc. Emphasis is placed on presenting an overview of water resources in the district by providing meteorological, hydrological, ecological and economic data. The economic data includes all costs and benefits, specifically including hitherto under estimated environmental and social costs and benefits. This alternative design is proposed for sites at Magod (Bedthi river) and Unchalli (Aghnashini river). A model is proposed to minimise submergence and maximise net energy in a region with the seasonal power generation, reservoir storage capacity and installed generation capacity as the decision variables.

Parametric optimisation techniques are used to compute energy from hydro source at Magod site of Bedthi river. If no seasonal variation is allowed between P_{tm} (peak hydropower generation during season) and P_{td} (peak hydropower generation during dry season), height of the dam required is 107 m and corresponding submergence area 95.03 sq.kms. The energy generated annually is 746 mkWh. More hydroelectric energy can be harnessed by drawing the water during the monsoon on run-of-river basis and storing small quantity to meet the region's requirement during the lean season. This saved area with bioresources would also help in meeting the thermal energy requirement of the region like firewood, etc., leaf manure for agriculture and horticulture lands. This can be achieved by allowing seasonal variation between P_{tm} and P_{td} (that is P_{tm} to P_{td} ratio as 3). More of inflow during wet season can be used directly without storing, reducing the storage capacity requirements. It is noticed that the submergence area saved is about 69.97%. That is, height of the dam is 87 m and corresponding submergence area 28.59 sq.kms. by allowing $P_{tm} = 3*P_{td}$. Reduction in the height of the dam from 107 m to 77 m, changes the submergence area from 95.05 sq.kms. to 16.27 sq.kms. Further reduction by 10 meters (dam height of 67 meters) brings down the submergence area to 5.70 sq.kms. It is noticed that by allowing seasonal variation there is an increase in electric energy generated. That is, when $P_{tm} = P_{td}$ energy generated in a year is 746 mkWh, and when $P_{tm} = 3*P_{td}$ energy generated is 810 mkWh.

Net energy analyses carried out by incorporating biomass energy lost in the land submergence show that land saving could be achieved and maximisation in net energy at a site made possible if the hydroelectric station generation capacity is adjusted according to seasonal variations in the river's discharge/runoff. This analysis obtains the upper bound on the height of dam and therefore on reservoir area for the project to yield a positive net energy. Analyses carried out for different heights of the dam for biomass productivity of 6.5 t/ha/yr, show that net energy increases from 190 to 550 mkWh with reduction in 40 meter in dam height. The net energy gain is 65% for a reduction of 37% in dam height. That is, increase in net energy from 520 to 896 mkWh by hydro thermal combined option in a land area of 95.03 sq.kms.

The cost per unit for various designs of the dam and for biomass productivity of 6.5 t/ha/yr shows 40.5% reduction in cost (Rs. 1.53 per unit to Rs. 0.92 per unit) for a dam height reduction of 37%. If costs of bioresources are excluded and electricity from hydro resource alone considered, then cost reduction is from Rs. 1.53 per unit to Rs. 1.09 per unit. Benefit to Cost ratio computed considering forest land value as Rs. 1,11,200 per hectare is less than 1.5 for dam height of 107 m, while for heights 87, 77 and 67 meters, it is greater than 1.5. Benefit to Cost ratio increases from 0.82 to 1.68 by reducing the submergence area from 95.03 sq.kms. (107 meter height dam) to 5.70 sq.kms. (67 meter height dam).

Solar Energy: The empirical relationships between global radiation (GR) and climatological parameters such as sunshine, mean daily temperature, relative humidity, specific humidity, minimum and maximum temperature and rainfall (based on data at Goa and Mangalore) are used to compute global radiation (GR-kWh/sq.m.) for Karwar (based on 37 years of climatological data), Honnavar (50 years of climatological data) and Bhatkal (Shirali: 15 years of climatological data). Karwar has GR range of 5.5-6.5 during January-May and 4-5 during monsoon months, July-September. While at Honnavar, GR range during January-May is 5.47-6.5 and a minimum of 4-5 during monsoon months.

Amount of solar energy that could be harnessed by utilising 4.77 sq.kms. (5% of present wasteland in coastal taluks) as solar collector area, is found to be of the order of 95.72 mkWh annually from coastal belt of the district alone. That means solar energy can meet at least 32.5% of present electricity demand of the district. Cost of electricity generated from Solar-Thermal technology ranges from Rs. 6.85 (at 30% efficiency, 0.45 load factor), Rs. 4.74 (at 30% efficiency, load factor 0.65) to Rs. 3.62 (at 30% efficiency and load factor 0.85).

The abundance of the solar resource can be illustrated by comparing the land requirements of solar (thermal or PV) with those of hydro or energy plantation projects. Except for run-of-river projects and for high head sites in deep gorges, the land requirements for hydroelectric projects at today's conversion efficiencies, average around 25 to 45 times. This means that solar energy is capable of meeting 5 to 10 times the electricity demand of Karnataka state while occupying land area less than that currently used by hydroelectric projects.

Sectorwise Energy Demand: Compilation of survey results from various sectors shows that noncommercial sources of energy such as firewood, agricultural and animal residues are the major components in the district's energy scenario. Sectorwise energy demand based on detailed investigations carried out in various sectors of the district shows that the domestic sector with 77.959% (8941.56 mkWh) of total energy consumption is the major energy consuming sector followed by industries sector (15.934%) and agriculture sector (5.133%). The district mainly depends on noncommercial energy sources (96.233%) such as firewood (78.091%) and dung (13.118%). While, commercial sources such as electricity (2.863%), kerosene, diesel, petrol, etc., (0.904%) and LPG (0.0001%) constitute 3.66% of the total energy consumption. Sourcewise and sectorwise consumption of energy shows that domestic (77.673%), agriculture (5.02422%) and industry (13.535%) sectors mainly depend on non commercial sources of energy.

Domestic Sector: Fuel wood is the main source of energy in the domestic sector. Comparative analyses of village level energy consumption pattern in the domestic sector across coastal, interior, hilly and plain zones of the district are carried out. The sample of 1304 households from 90 villages in Kumta taluk shows that most households [97 (for cooking) to 98% (for water heating) of the sample] still use traditional stoves for cooking and water heating.

End use efficiency experiments carried out in some households of Masur village (based on 4 days experiment) of Kumta taluk, show variation in fuel wood consumption from 0.835(avg)±0.225(Sd) [in improved energy efficient stoves] to 1.15(avg)±0.65(Sd) kgs/person/day [in traditional stoves]. There is scope for saving energy of the order of 27% to 42% by switching over to improved devices designed with objectives to maximise combustion rate, radiant heat transfer, convection, conduction and user satisfaction.

Fuel wood consumption analysis illustrates regional variation in fuel wood consumption. Average consumption ranges from 2.01±1.49 (coast) to 2.32±2.09 (hilly) kgs/person/day. Seasonwise fuel wood requirement for the coastal zone ranges from 1.98±1.40 in summer to 2.11±1.73 in monsoon. Similarly for the hilly region the range is 2.22±1.56 in summer to 2.51±2.72 kgs/person/day in monsoon. Data also reveals variation in fuel wood requirement for water heating (for bath and washing) purposes. The variation in consumption ranges from 1.17±0.02 (coast) to 1.63±0.05 (hilly). Seasonal fuel wood requirement ranges from 1.12±0.50 (summer) to 1.22±0.54 (monsoon) for the coastal zone. While for the hilly zone the range is from 1.53±0.63 (summer) to 1.73 ±0.75(monsoon). The overall picture of other sources of energy for domestic purposes shows that in the coastal zone kerosene consumption per person per day for cooking ranges from 0.05±0.19 (hilly) to 0.34±0.79 (coast) and for lighting from 0.74±0.46 (coast) to 0.98±0.53 ltrs/person/month (hilly). The hilly zone is more dependent on kerosene for lighting due to non-electrification.

The sample of 1304 households consists of landless category (2.68%), and those owning land from 0-0.4 hectare (24.76%), 0.4-0.8 hectare (32.13%), 0.8-1.2 hectare (17.48%), 1.2-1.6 hectare (6.53%), 1.6-2.0 hectare (5.23%) and >2 hectare (11.19%). Fuel wood consumption for cooking varies from 1.83±0.96 (landless) to 2.31±2.69 (for >2 hectare) and for water heating from 1.23±0.69 (landless) to 1.60±0.70 kgs/person/day (for >2 hectare).

Electricity consumption seems to have increased with increase in the landholdings. The range is from 1.99±3.15 (landless), 2.14±2.83 (0-0.4 hectare) to 3.59±3.82 (>2 hectare category) kWh/person/day. Kerosene usage for lighting has decreased with an increase in landholding. It varies from 0.87±0.74 (landless), 0.86±0.47 (0-0.4 hectare) to 0.75±0.47 (> 2 hectare category) ltrs/person/month.

Based on the energy consumption norms (regionwise, seasonwise, end usewise) obtained from detailed surveys, it is estimated that in the domestic sector of the district (a) Fuel wood required (for cooking, water heating, space heating, jaggery making, parboiling) works out to be 1.65 million tonnes/year, (b) Electricity demand (excluding irrigation) about 32.65 mkWh/year and (c) Kerosene demand for cooking and water heating about 15,855 Kl ('000 litres) per year. The coastal area's climate is favorable to set up solar water heating devices to meet the requirement of hot water for bathing purposes. If 30% of population switch over to solar water heating devices, fuel wood saved would be of the order of 66.77 kilo tonnes per year from five coastal taluks of the district.

Agriculture Sector: Land preparation, irrigation, harvesting, threshing and transportation are the main tasks dependant on mechanical form of energy apart from manure input, seed, etc. At present, rainfed paddy is the major crop in agriculture and hence less dependent on irrigation. Energy input in the agriculture sector is about 5500 mkWh, the major share is from manure in the form of organic matter. The energetics in rainfed paddy cultivation in three zones (coast, interior and hilly) and for various landholding categories are analysed. Farm Yard Manure (FYM) constitute a major component input energy in paddy cultivation. However, the level of input changes from zone to zone and across the various categories of landholding. Detailed analyses of energy input in various categories of farmers show that the marginal farmers get higher yield compared to any other categories of landholding and energy input in the form of Farm Yard Manure is almost twice compared to quantity/acre applied by the large farmers (>2 hectare). The greater quantity of FYM usage by marginal farmers may be attributed to larger quantity of dung availability (in marginal farmers category livestock per acre is almost twice that of medium farmers or four times that of large farmers).

Industries Sector: Energy is a vital component in the rural industries such as agro processing industries, brick and tiles industries, etc. Rural industries like cashew processing, etc., are major consumers of fuel wood in rural coastal areas in the district. Energy auditing carried out in sample rural industries have revealed wide disparity in energy consumption to produce same quantity of similar products. About 60% of sample units have Percentage Production Capacity Utilisation (PPCU) less than 50% and higher Specific Energy Consumption (SEC) and Energy Intensity (EI). The wide variations in EI in cashew industries from 4.49 to 8.66 kg of fuel wood/kg of kernel reveals the scope for energy conservation in the order of 30 to 48%. It is noticed that optimal utilisation of installed capacity (better PPCU) would help in energy conservation. The energy analyses carried out reveal the relationship between SEC and Pr being power law with an exponent less than one. This indicates improvement of SEC with an increase in the production which in turn implies the utilisation of installed production capacity optimally. Since these industries are seasonal, operating at peak load during pre-monsoon period (that is from February to June) every year, it is possible to switch over to solar energy for conditioning, mild roasting and Broma drying operations (solar energy is also available in plenty during this period) in cashew industries. Fuel wood consumption in cashew processing industries could be brought down by: (a) use of solar water heating devices for roasting the raw nuts, and (b) use of insulated, properly designed drier for drying the kernels.

Paddy dehusking and rice milling is one of the energy intensive sectors in the district. Analysis of the sector revealed that there is scope to save energy of 23-38%. About 25 to 37% could be saved in brick manufacture by good housekeeping practices. Further saving is possible by changing the technology of manufacture, which involves capital investment.

Integrated Energy Planning: The energy planning endeavours for a region involve finding a set of sources and conversion devices so as to meet the energy requirements/demand of all the tasks in an optimal manner. This optimality depends on the objectives such as minimising the total annual cost of energy or minimisation of non-local resources or maximisation of system overall efficiency. Factors such as availability of resources in the region and task energy requirements impose constraints on the regional energy planning exercise. Thus, the regional energy planning turns out to be a constrained optimisation problem. Integrated energy planning gives an optimal mix of new/conventional energy sources and is developed based on Decision Support Systems (DSS) approach. This exercise provides an optimum energy allocation for the district and make a satisfying energy allocation plan for the year 2000, 2005 and 2010.

The application of DSS for the district energy planning takes into account the renewable resources that could be harnessed for energy, land use and sectorwise energy demand databases, optimal allocation of energy resources for various tasks and then explore the energy use consequences of alternative scenario's such as base case scenario, high energy intensity and improved end use efficiency options. Linear programming formulation for optimum allocation based on cost minimisation objective shows that there is substantial savings of about 19.19% in energy and cost reduction in overall energy system is about 36.24%. Cost per unit (kWh) of energy with optimal allocation of energy is Rs. 0.31 per kWh (as against Rs. 0.39 per kWh without optimisation). Optimisation carried out with the objective of maximisation of efficiency of 'source-task-device' (ijk) combination for all combinations shows energy saving of 34.08% and cost of energy as about Rs. 0.335 per kWh. The scenario analyses reveal that potential renewable energy resources can meet the energy requirements of the district even under the assumption of high growth rates.

INDEX

A

Advisory Board on Energy, 244, 284, 370, 411
Agenda 21, xvi
agricultural economy, 1
agricultural residues, xvi, 3, 55, 56, 81, 86, 89, 91, 231, 238, 240-243, 245, 246, 279, 287, 332, 347, 364, 373, 384, 387, 410
agriculture sector, 416
agro-climatic zones, 238, 246, 247, 283
agrowastes, 3, 247
air radio-navigation systems, 9
alternative engines, 9
amortised total annual cost (ATAC), 51
anemometers, 27, 29
Angstrom's equation, 187
arable land, xv
ASTRA stoves, 273, 283, 287, 294

B

bagasse, 56, 77
Base Case Scenario, 368, 386, 400
bioenergy, vii, xiii, xv, 24, 55, 56, 66, 68, 77, 80, 85, 87, 91, 127, 128, 132, 143, 151, 155, 157, 257, 373, 391
biofuels, 55, 286, 368, 373
biogas, ix, xi, xiii, 3, 4, 12, 55, 56, 59, 80, 81, 85-87, 91, 93, 245-247, 257-259, 264, 265, 268, 269, 271, 275-279, 283, 287, 297, 299, 301-303, 307, 333, 377, 378, 380, 384, 386-388, 390, 393, 394, 398-410, 414
biomass, x, xii, xv, 3, 4, 5, 11, 55, 56, 60, 64-66, 77, 80, 85, 87-92, 101, 127, 128, 133-135, 137, 151, 152, 154, 156-158, 166, 168, 172, 174-176, 244, 246, 247, 283, 294, 332, 347, 368, 373, 394, 407-409, 413-415
brick making, 55, 362, 363

C

capacitor banks, 51
cash crops, 9, 77, 361
cashew processing, 55, 56, 232, 348, 352, 356, 358, 364, 377, 387, 417
catchment area, ix, xiv, 99, 103, 108-112, 115-117, 119, 120, 124, 125, 133, 140, 184
catchment characteristics 99
catchment shape, 108
catchment slope, ix, 100, 109-112
catchment yield, 97
cellulose, 56, 80
Centre for Renewable Energy Development Agency (CREDA), 14, 53
channel slope, 100, 108
charcoal, 55, 79, 89, 332, 345, 354, 373
chemical energy, 2, 4
coal, xv, xvi, 1, 2, 3, 4, 11, 87, 128, 240, 242, 243, 247, 332, 348, 351, 352, 369, 370, 387
coal extraction, 4
coastal zone, 239, 256, 283, 322, 363, 384, 393, 416
Coconut husk, 79
coefficient of variation (COV), 51
collector arrays, 8
Communitywise Energy Consumption, 259
conversion devices, 367, 377, 393, 413, 417
cookstoves, xv, 85, 259, 278, 290
cost estimates, 51, 133
Cost-Benefit Analyses, 127
cotton, 10, 77, 82
crop preservation, 9
crop residues, xv, 82, 241, 242, 246, 247, 283, 309
cumulative distribution function, 27
cup counter anemometers, 29

D

dam height, 143, 146
dam site, xiv, 143
decentralised energy plans, 368

Decision Support System (DSS) , xix, 12, 14, 24, 367, 368, 370, 371, 410, 413, 414, 417
deforestation, xv, 12, 226, 238, 286, 410
depletable resources, 3, 11
depletion of fertility, 12
desalination, 5
desertification, xv
diffuse radiation, 7, 8, 188, 198, 206
Diffuse Solar Radiation, xi, 188, 193, 198, 202, 206, 207
Draught Animal Power (DAP), 10
Draught Animal Transport, 10
draught power, 10, 309

E

ecologically sound design, 127
economic development, xv, 1, 19, 91, 348, 408, 409
economic production, xv, 56
efficiency improvements, 348, 400, 406
Einstein, A., 3
electrical energy, 2
energetics, vii, xiv, 304, 309, 319, 320
energy conservation, xvi, 13, 23, 233, 331, 332, 348, 349, 350, 361, 364, 368, 369, 413, 417
energy consumption, xvi, 11, 12, 25, 56, 66, 157, 237-243, 245-248, 256, 257, 269, 273, 283, 286, 310-312, 317, 318, 331, 348-352, 355, 357, 358, 364, 365, 369, 386-388, 400, 401, 408, 410, 413, 416, 417
energy consumption pattern, xii, 239, 275, 317, 319, 323, 324
energy conversion, 5, 25, 51, 137, 151, 311, 371, 407
energy database, 371
energy demand, xvi, xvii, 11-13, 24, 28, 66, 98, 101, 158, 176, 240-242, 244, 245, 256, 286, 312, 331, 332, 349, 367-369, 370, 376, 386-388, 400, 401, 410, 413, 415, 417
energy efficiency, xvii, 3, 238, 331, 366, 368
energy efficient devices, 55, 273
energy gatherers, 2
energy inputs, xii, 309, 318
Energy Intensity (EI), 331, 348, 355, 417
energy pattern factor, ix, 25, 28, 41
energy planning, xvi, 237, 238
energy plantation, 55
energy production, 4, 238, 348
energy supply, xvi, 5, 11, 12, 13, 55, 91, 246, 286, 331, 368, 407, 409, 413
entropy, 3
environmental costs, xiii, xvii, 132, 134, 173, 376
environmental protection, 51
ESCI Study, 240
ethanol, 55, 56, 87, 352
evaporation loss, 143, 376

F

farm operations, 10, 80, 311
farm yard manure, 22, 309, 316, 317, 319, 324, 329, 416
fit test, 107
flat plate collectors, 8
fluorescent bulbs, 368, 371, 373, 413
fodder, 52, 55, 56, 60, 61, 66, 73-77, 81, 82, 85, 91, 158, 256, 287, 297-299, 314, 385
food energy, 1, 311
forest submersion, 166
fossil fuel, xv, xvi, 1, 2, 3, 7, 11, 55, 226, 311, 350, 368
fuel consumption, 14, 55, 85, 91, 237, 238, 239, 247, 257, 259, 272-275, 278, 279, 283, 287, 289, 291, 292, 294, 350, 362
fuel consumption pattern, 247, 289
fuel efficient stoves, 12, 55, 85, 91, 93, 239, 279, 287, 288, 289
Fuel Policy Committee (FPC), 241, 369
fuel wood, xv, 1, 11, 13, 23, 24, 52, 56, 66, 79, 80, 85, 86, 87, 90, 127, 128, 135, 137, 157, 158, 160, 162, 164, 166, 172, 175, 176, 231, 232, 237-247, 256, 257, 259, 269, 270, 272-280, 283, 286-292, 294, 296, 297, 331, 332, 347, 352, 356, 358, 361-364, 368, 373, 376, 380, 384, 385, 387, 394, 400, 401, 408, 410, 416, 417
fuel wood consumption, xi, xii, xiv, 55, 232, 260, 261, 262, 269-274, 277, 278, 280, 281, 291, 292

G

geothermal reservoirs, 3
global energy, xvii
global radiation (GR), 187-189, 198, 204, 205, 209, 211, 213, 219, 220, 233, 415
global solar radiation, x, xi, xiii, xiv, 188, 190, 197, 198, 199, 205, 206, 208, 211, 214-217
grain drying, 377
grain mills, 363
Gravitational energy, 2
greenhouse effect, 9
gross domestic product, 11
ground water table, 12, 104, 133
groundnut, 22

H

heat energy, 2
heights of the dam, 133-135, 146, 151, 158, 168, 172, 176, 376, 415
Hertzian relays, 9
hilly zone, 237, 314, 317, 318, 322, 328, 384, 416
Honnavar, ix, xi, xiii, xiv, 13, 15, 16, 20, 21, 23, 25, 29, 31, 34-37, 45-50, 52, 55, 57, 60, 62, 63, 68-84,

91, 92, 98, 104-107, 187-189, 214, 219, 220, 223, 225, 226, 228, 229, 232, 233, 281, 282, 310, 314, 315, 377, 391-394, 414, 415
horticulture residues, 55, 91, 257, 414
hours of sunshine, 198, 219
household energy, 66, 238, 241, 242, 245, 286
human density, 15, 314, 327
hydel plants, 23, 24, 97
hydel resources, 51
hydraulic ram, 7
hydro energy, vii, xiii, 3, 5, 6, 14, 97, 98, 121, 127, 131, 136, 155, 156, 373, 394
hydro plants, 6, 7, 23, 98, 128, 131
hydro technology, 7
hydroelectric dams, 11
hydroelectric development, 128
hydroelectric energy operation module, 376
hydroelectric projects, vii, 127, 131, 134, 177, 414
hydrology, 121, 133, 177
hydropower generation, 6
hydropower station, 98

I

improved stoves, 85, 238, 247, 278, 283, 286, 287, 290, 291, 294, 371
incandescent lamps, 373
India Meteorological Department (IMD), 25, 29, 103, 233
industries sector, 417
inorganic fertilisers, 310, 312, 322
integrated energy Planning, xvii, 55, 367, 368, 369, 413, 414, 417
irrigation, 1, 4, 5, 7, 10, 16, 22, 52, 61, 90, 101, 129, 136, 139, 230, 237, 245, 246, 257, 271, 279, 283, 284, 305, 309-311, 314-317, 328, 333, 373, 377, 384, 385, 394, 410, 416

K

Kaplan turbines, 7
Karnataka Forest Act, 61
Karwar, ix, xi, xiii, xiv, 13-16, 19-21, 23, 25, 26, 29, 31, 32, 34-38, 45-52, 55, 57, 60, 62, 63, 66, 68-76, 78, 79, 81-84, 91, 92, 98, 104-107, 133, 177, 187-189, 214, 219, 220, 223, 225, 226, 228, 229, 232, 233, 239, 281, 282, 284, 310, 311, 314, 315, 328, 363, 364, 377, 385, 386, 391-393, 410, 414, 415
kerosene stoves, 259, 275, 373
kinetic energy, 2, 5, 25, 26
Kumta, ix, xi, xii, xiii, xiv, 12-16, 20, 21, 23, 25-30, 35-38, 40-42, 44-53, 55-57, 62, 63, 65, 66, 68-79, 81-84, 86, 87, 90, 91, 98, 104-107, 172, 176, 214, 226, 228, 229, 231-233, 237-239, 247, 248, 250-258, 274-276, 278, 279, 281-288, 294, 310, 311, 314, 315, 319, 320, 328 331, 333, 348, 352, 354, 362-364, 377, 384, 385, 391-394, 408, 410, 414, 416

L

land acquisition, 51
Land use Analyses, 127
Land use Pattern, 256
landholding, 13, 14, 238, 239, 246, 248, 256, 269, 283, 284, 287, 288, 310, 311, 318, 332, 408, 416
Light energy, 2
livestock density, 15, 256, 322
livestock holding, 238
livestock population, 15, 55, 80, 91, 239, 394, 410
living standards, xvi

M

manures, 21, 22
mapping of contours, 133
mean daily temperature, 187, 415
mean temperature, 187, 219, 235
mechanical energy, 2, 3, 4, 5, 6, 7, 131, 309, 377, 380
mechanical engine, 2
methane, 55, 77, 86, 89, 351
methanol, 56, 87, 89, 351
microchips, xv
MTCE, (million tonnes of coal equivalent), 3
MTOE (million tonnes of oil equivalent), 3
muscular energy, xv

N

National Commission on Agriculture (NCA), 242
natural gas, xvi, 1, 2, 3, 89, 370
net energy analyses, 127, 133, 148, 176, 415
net primary productivity (NPP), 66
Newton, Isaac, 2
nongovernmental organisations (NGO's), 12
non-local resources, 367, 378, 413, 417
nuclear energy, 2
nuclear fusion, 3

O

oil, xii, xv, xvi, 1, 2, 3, 4, 7, 11, 56, 68, 77, 80, 91, 166, 227, 230, 240, 247, 332, 333, 350, 354, 363, 369, 370
oil refinement, 4
optimal design, vii, 127, 414
optimisation, 130, 177, 368, 398, 400, 410, 417
Optimum Level Forecast (OLF), 243, 369
overgrazing, xv, 12, 82

P

paddy cultivation, 22, 73, 288, 309, 310, 312, 314, 315, 317, 328, 329, 416
parametric analyses, 133
parametric optimisation, 127, 136
parboiling, 274, 282, 363
passive systems, 7
passive thermal systems, 7
peat, 3
Pelton turbine, 7
per capita fuel consumption, xii, 55, 247, 279, 287
Percentage Production Capacity Utilisation (PPCU) , xii, xiv, 331, 348, 355, 357-362, 417
perennial streams, 100
photocells, 9
photosynthesis, xv, 1, 56
photovoltaic devices, 188, 230
photovoltaic generators, 8
plant efficiency, 102
potential energy, 2, 6, 377
precipitation, ix, 97, 104, 107, 115
primary energy, 3, 137, 245, 408
probability density function (pdf), 27
productivity, ix, x, 65, 66, 67, 79, 92, 159-161, 163, 169, 170, 309, 365, 392, 393
pumpsets, 90, 310, 333, 371, 373, 385, 386

R

radiation climatology, 188, 189
radiation energy, 7, 234
rainfall, 20, 22, 23, 51, 57, 103, 104, 106-108, 127, 128, 132, 133, 135, 137, 140, 143, 176, 187, 188, 219, 233, 235, 312, 314, 322, 354, 376, 415
rain-gauge stations, 104, 140
Reference Level Forecast (RLF), 242, 243, 369
regional energy planning exercise, 237, 367, 369, 377, 383, 385, 407, 413, 417
regional energy system, 377
rehabilitation, x, 127, 167, 168
rehabilitation costs, 127
relative humidity, 187, 188, 219, 220, 233, 235, 415
renewable resources, xvi, xvii, 3, 11, 23, 238, 286, 367, 400, 410, 417
rice husk, 73
river discharge, 140
run off, 97, 103, 121
rural energy, 132, 232, 238, 240, 245, 246, 332, 347, 369, 409, 411
rural industries, 331

S

Sahyadrian Tract, 20
salination, 12
secondary energy, 3, 4, 5
seepage loss, 143, 376
Shirali, ix, xi, xiii, xiv, 25, 29, 35-37, 45-50, 52, 104, 187-189, 219, 220, 223, 225, 233, 261, 263, 265, 391, 415
siltation, 12, 87, 132, 134, 135, 143, 167, 331
silting capacity, 143
Sirsi, ix, xiii, 12-16, 20-23, 25-33, 36, 38, 39, 41-43, 45, 53, 55, 57, 62, 64, 65, 66-69, 70-72, 74-87, 90-92, 98, 104-107, 113, 121, 134, 139, 167, 168, 176, 177, 214, 237-239, 278, 279, 281-288, 310, 314, 328, 364, 384, 385, 392, 393, 409, 414
small-scale industries, xii, 335-338
snow ablation, 188
soil erosion, xv, 12, 26, 80, 167, 314, 410
soil fertility, 80, 238, 332
solar collectors, 7
solar cookers, 9, 245
solar dryers, 9
solar electricity, 9
solar energy, vii, xi, xii, xix, 1, 3, 5, 7-10, 24, 25, 56, 187, 188, 204, 213, 219, 220, 226-229, 231-233, 361, 378, 374, 384, 391, 393, 394, 397, 409, 410, 415, 417
solar energy module, 374
solar radiation, 8, 9, 104, 187-189, 198, 204, 206, 208, 211, 213, 214, 219, 220, 233-235, 247, 374, 386
solar thermal devices, 188, 230
solar water heating, 188
space heating, xv, 23, 237, 244, 247, 256, 274, 277, 279, 283, 290, 377, 384, 416
Specific Energy Consumption (SEC), 331, 348, 355, 417
specific humidity, 187, 220, 233, 235, 415
steam power, 56, 87
steam turbines, 9
storage reservoir, 97, 127, 135, 158
storage systems, 135
stream discharge data analyses, 133
stream flow, 104, 107, 121, 128, 130, 131
stream flow measurement, 108
stream slope, ix, 100, 108-112
streams, ix, x, xiv, 109-112, 115-120, 122, 124, 125, 149, 150, 178, 179, 184, 392, 395
submergence area, x, xiii, 143, 144, 145, 146, 152, 153
submersion costs, 127
sugar-cane, 10, 56, 77, 274, 279
sunshine duration, 187, 204, 219, 233, 234
sustainable development, xvi, 24, 238, 331, 348, 413

T

techno-economic analyses, 55, 85, 188, 227
thermodynamic conversion, 9
thermodynamics, 2, 3, 378

time sharing, 378
tobacco, 4, 10, 346
topographical features, 26, 100
tracking collectors, 8
traditional stoves, 55, 85, 237, 238, 259, 273, 274, 278, 289-291, 371, 373, 385, 387, 416
transformation scenario, 368
transpiration, 143, 188
tributaries, xiv, 178, 179, 184
turbine efficiency, 51
turbine size, 51

U

UN Bruntland Committee, xvi
useful energy, 3, 4, 5, 85, 244, 332, 348, 386, 387
Uttara Kannada district, 12, 20, 22, 24, 28, 45, 52, 55, 56, 60, 64-66, 73, 77, 79, 81, 86, 91, 97, 104, 107, 113, 121, 127, 132, 134, 139, 168, 176, 187, 188, 205, 237, 238, 278-280, 283, 294, 310, 331, 333, 348, 352, 355, 362-364, 367-369, 377, 383, 385-388, 393, 394, 410, 413, 414

V

vegetation, 12, 13, 20, 22, 26, 57, 60, 62-66, 81, 82, 83, 91, 103, 104, 107, 132, 143, 158, 166, 176, 188, 238, 286
Villagewise Consumption of Energy, 259
volcanoes, xv

W

water discharge, ix, x, 115-120, 141, 142
water heating stoves, 289
water pump, 5
water pumping, 5, 98, 373, 383
water resources, xix, 135, 177
watershed, 12, 20, 99, 104, 107, 376
waterwheel, 2
Weibull Distribution, 25, 27, 38
weirs, 6, 7, 134
wind data, 14, 27, 29, 36, 41, 51, 52, 373
wind energy, vii, xii, 5, 6, 25, 26, 53, 54, 373, 391, 414
wind farms, 25
wind frequency distribution, 27
wind generators, 6
wind power technology, 51
wind resource module, 373
wind speed, 5, 25, 26-29, 31, 36, 37, 38, 45, 51, 188, 235, 373, 414
wind turbine, xv, 6, 26, 27, 51, 246, 380
wind velocity, 27, 28
windfarms, 51
windmills, 2, 6
windpumps, 5
wood gasification, 55, 89
wood gasifier, 87, 90, 91, 172
World Meteorological Organization, 27